Inhaltsverzeichnis.
Contents. — Table des matières.

FORTSCHRITTE DER CHEMIE ORGANISCHER NATURSTOFFE

PROGRESS IN THE CHEMISTRY OF ORGANIC NATURAL PRODUCTS

PROGRÈS DANS LA CHIMIE DES SUBSTANCES ORGANIQUES NATURELLES

HERAUSGEGEBEN VON   EDITED BY   RÉDIGÉ PAR

L. ZECHMEISTER
CALIFORNIA INSTITUTE OF TECHNOLOGY, PASADENA

ELFTER BAND   ELEVENTH VOLUME   ONZIÈME VOLUME

VERFASSER   AUTHORS   AUTEURS

A. ALBERT · K. BRÜCKNER · R. B. COREY · K. FREUDENBERG
H. H. INHOFFEN · R. LEMBERG · L. PAULING · S. PEAT
H. SCHMID · W. A. SCHROEDER

MIT 67 ABBILDUNGEN   WITH 67 ILLUSTRATIONS   AVEC 67 ILLUSTRATIONS

WIEN · SPRINGER-VERLAG · 1954

ISBN-13: 978-3-7091-8016-7   e-ISBN-13: 978-3-7091-8014-3
DOI: 10.1007/978-3-7091-8014-3

Starch: Its Constitution, Enzymic Synthesis and Degradation.

By STANLEY PEAT, Bangor, U.K.

With 4 Figures.

Contents.

I. Introduction.

It is interesting to speculate on the unique position occupied by *D*-glucose in the life processes of both animals and plants. We cannot yet begin to understand why the constitution and configuration of this sugar should fit it to play the predominating role it does. The facts remain that glucose is the only sugar found in the blood of animals and that enzymic mechanisms exist in animal tissues for the conversion of almost every other monosaccharide into glucose. In the plant the story is similar, but here glucose is associated with the closely related ketose, *D*-fructose and the disaccharide, sucrose. It would appear that glucose and fructose are interchangeable in many of the sequences of plant metabolism. Indeed, interconversion of glucose and fructose in plants is an established fact and the enzyme system responsible has been closely studied. In the majority of plants the energy derived from photosynthesis is stored as carbohydrate and the most abundantly distributed representative of this class of reserve carbohydrate is starch, a polyglucose. Polymerised fructose (fructosans) is by no means an unusual mode of storage and in some plants, e. g. sugar cane or beet, the favoured form of storage is sucrose.

Glucose is more than a vehicle of energy, it is a structural unit also, inasmuch as the basic constituent of the cell walls of all higher plants is cellulose which, like starch, is built entirely of glucose units. For the carbohydrate chemist interested in the metabolism of plants, glucose is the focal point but the raw material for chemical investigation would obviously be the polysaccharides, starch and cellulose, since these are synthesised every year in enormous quantities by vegetation all over the globe.

From the metabolic point of view, starch is of greater significance than cellulose. The latter is a skeletal material, giving strength and rigidity to plants, and once it has been laid down in the cell walls very little happens to it until the plant dies. Starch, on the other hand, comes and goes. The resting cereal seed or potato is gorged with it but when the seed germinates or when the potato sprouts, the starch is rapidly converted into forms which can be directly assimilated by the growing and respiring plant. It is the purpose of this article to review the attempts that have been, and are being, made to elucidate the interrelated problems of the constitution of starch and of the metabolic changes of synthesis and degradation which it undergoes in the plant.

The granular form in which starch is deposited makes easy its isolation in a pure state while the colour reaction with iodine simplifies its detection and distinction from other tissue constituents. Its basic structure, as a chain polymer in which the glucose units are mutually combined by 1 : 4-glucosidic links, was established by the classical methods of carbohydrate chemistry.

The value of n in the molecular formula $(C_6H_{10}O_5)_n$ was a matter of controversy for many years. On the one hand, chemical end-group assay indicated a value for n of 25–30 while, on the other, the physical properties of starch patently required a molecular weight many times this value. These observations were not irreconcilable if the assumption were made that starch possesses a branched structure, and ultimately this assumption proved to be correct. It was generally held at this time that starch was molecularly homogeneous but in the early 1940's it was proved incontrovertibly that most starches consist of at least two distinct molecular species to which MAQUENNE's names, amylose and amylopectin were given. Amylopectin which constitutes 75–80% of the starch has the branched structure formerly ascribed to the whole starch while the minor component, amylose, was currently held to be constituted of unbranched chains only. The proportion of non-reducing end-group in amylopectin is 1 : 20; in amylose it varies between wide limits, e. g. between 1 : 200 to 1 : 1000.

Purely chemical or physical methods could not carry the investigation of a complex high polymer like starch much beyond the establishment of the basic constitution just described. The elucidation of the finer details of structure required the more intimate probing of biological agents. The study of the starch-metabolising enzymes advanced side by side with that of starch itself, the two lines of investigation being inextricably interwoven. The tools, in this case the enzymes, had to be studied and improved at the same time as they were being used. This interdependence necessitates the greatest caution in the drawing of conclusions either with respect to the structure of starch or to the mode of action of the enzymes on it. For instance, it appeared to be an established fact that amylose was converted completely into maltose by β-amylase and indeed this fact constituted the main evidence for the linearity of amylose chains. Nevertheless, it has been recently shown that the action of crystalline β-amylase (from sweet potato) on amylose ceases at a conversion limit of 70% (as maltose) unless there is also present a supplementary enzyme, Z-enzyme, when the normal 90–100% conversion is observed.

It is not the primary purpose of this review to examine the patterns of action of these enzymes as an independent subject. It will be appreciated however that enzymological studies have played and are playing an important and necessary part in elucidating the constitution of starch and as such find a place here.

II. Chemical Investigations of the Basic Structure of Starch.

Starch is essentially a high polymer, the building unit of which is D-glucose. It is described therefore as a polyglucose or a glucosan or a glucan according to fancy—an example of the chaotic state of the present-day nomenclature of polysaccharides. Starch grains contain traces of other constituents such as phosphate (to which reference will be made later), fatty acids, silica, protein, but none of these plays a significant part in the pattern of starch structure.

The mode of mutual combination of the glucose units in starch natur-ally attracted the attention of the early investigators and a strong hint was given by the well-known fact that the disaccharide, maltose, is formed by the action of the amylolytic enzymes. In favourable circum-stances, as much as 80% of starch is converted into maltose, an indication that this sugar exists preformed in starch. Karrer and Nägeli (*100*) were able to convert starch largely into hepta-acetyl maltose by acetolysis with acetyl bromide and concluded that the main inter-glucose linkage in the polysaccharide was the maltose link. When Haworth established finally that maltose was 4-(α-D-glucopyranosyl)-D-glucopyranose (*79*) and that the disaccharide was not formed by reversion synthesis during the hydrolysis of starch (*82*), it became evident that the linkages between the glucopyranose units of starch were mainly of the α-1 : 4-glucosidic type. In view of the high yield of maltose and since no disaccharide other than maltose was formed in comparable amount, it followed that the linkages joining maltose residues are also of the α-1 : 4-type. The conception of starch structure reached in this way was that of continuous chains constituted of α-D-glucopyranose members linked by 1 : 4-glucosidic bonds.

1. End-Group Assay.

The methylation technique used so successfully by Haworth in elucidating the constitution of the disaccharides was then applied to the study of the polysaccharides. Trimethyl starch was prepared and hydro-lysed and the main product was 2 : 3 : 6-trimethyl glucose, isolated in 85% yield (*77*). The constitution of starch as a chain polymer was confirmed. The next developmental step was the proof that starch, like cellulose, was composed of terminated chains and not of endless loops. It was shown (*82*) that the hydrolysate of methylated starch contained, in addition to the main product, a small percentage of tetramethyl glucopyranose which could have come only from the end of a chain of glucose residues. In other words, the proportion of tetramethyl glucose in the hydrolysate was a measure of the proportion of glucose units constituting non-reducing chain ends. A small amount of dimethyl glucose was also found in the hydrolysate of methylated starch but its formation could be ascribed to the incompleteness of the methylation. The picture of starch which emerged from these experiments was that of a congerie of chains of α-1 : 4-linked glucose units, the average length of the chains, *i. e.* the degree of polymerisation (*D.P.*), being *ca.* 25 glucose units (M. W. 4000 .

End-group assay had previously revealed (*78*) that cellulose was also constituted of terminated chains of glucose units joined by 1 : 4-links but that the average length of the chains was very much greater than in starch. The main structural difference between the two polysaccharides lay however in the fact that the basic unit in cellulose is β-glucopyranose whereas that of starch is α-glucopyranose.

That the configuration of the chain-forming link in starch is α-glucosidic was deduced from a number of considerations, namely, the high optical rotation of starch and of the maltose produced from it; the anomeric relationship of maltose and cellobiose; the fact that the former is hydrolysed by α-, the latter by β-glucosidases, and in particular from the kinetic studies of the hydrolysis of starch and cellulose made by FREUDENBERG (*59, 56*).

2. The Molecular Size of Starch.

The end-group assay method seemed to indicate a lower molecular weight (4000) for starch than might be expected from its physical properties. Estimations of the molecular size of starch and its derivatives by osmotic pressure methods (*43*), viscosimetry (*167*) and ultracentrifugal studies (*107*) all suggested a value many hundred times that given by the assay of non-reducing end groups. Moreover, an attempt to estimate the proportion of *reducing* end groups by measuring the very small reducing power of starch yielded results corresponding to average chain lengths of 460–1470 (*160*). The obvious conclusion was drawn that starch, unlike cellulose, consisted of short chains (*ca.* 25 glucose members) associated together by physical or chemical forces to form the macromolecules which were the entities involved in the determination of molecular weight by the physical methods described. The hypothesis of physical aggregation of short basic chains proved untenable and it became evident that the reducing end of one chain was linked chemically to one of the glucose members of a second chain and that starch therefore possessed a branched structure. Three methods of formulation of such a branched structure competed for favour at that time and will be examined in a later Section (p. 25) of this article. The simplest is the so-called "laminated" formula of Haworth, Hirst and Isherwood (75) which assumes single branching, *i. e.* one branch per chain. Until very recently this simple formula has expressed our knowledge of starch as adequately as the more complicated formulae of Staudinger (*168*) and K. H. Meyer (*118*).

It follows, if starch has a branched structure, that at least two different types of glucosidic linkage must be present, namely, the chain-forming link (the α-1 : 4-link) and the branch-forming link. The precise nature of the branch links has been the subject of study in many laboratories but consideration of this problem will be postponed until the question of the homogeneity of starch has been discussed.

III. Amylose and Amylopectin.

It was commonly held during the pre-war period when the investigations described in the previous Section were in progress, that starch (and cellulose) consisted of one molecular species only, not in the usual sense of all the molecules being alike but in the sense that each sample of starch is a mixture of molecules built to the same chemical pattern but of different sizes—a polymeric-homologous series in fact [Staudinger (*168*)]. Nevertheless, hints were not lacking that starch consisted of at least two polyglucoses. From the botanists' point of view, starch was certainly composed of two different substances, one of which formed the envelope of the grain and the other the substance contained by the envelope. A. Meyer (*116*) named these substances α- and β-amylose respectively, while Maquenne (*111*) described the membrane as "amylopectin" and the contents as "amylose". The issue was obscured by the peculiar physical properties of starch solutions, *e. g.* the formation of gels, the spontaneous precipitation (retrogradation) which occurs when a solution is kept, and Maquenne certainly believed that amylose and amylopectin were chemically the same substance differing only in their degrees of physical aggregation. Nevertheless it was Maquenne who showed that it was the "amylose" which was responsible for the blue colour given by starch with iodine. The chemists who followed Maquenne in this field accepted the aggregation theory without question and sought to determine the basic unit from which these aggregates were built. It thus happened that for some thirty years the "unitarian" theory (if we may call it such) of starch structure held sway.

1. Separation.

A large number of methods which purported to effect the separation of the components of starch have been described in the literature but most of these processes have no present-day significance inasmuch as their designers were not seeking to separate two chemically distinct types but rather to fractionate physical aggregates of the same chemical unit. A retrospective glance at the older methods shows that only three achieve a separation of two chemically distinct entities from starch. Two of these methods originated with TANRET (*172*) and one with SAMEC (*163*). One of TANRET's methods, involving the leaching out of solubles from a partly gelatinized starch in which the particles are swollen but intact, was developed by BALDWIN (*9*) and applied by K. H. MEYER (*120*) to whom belongs the credit of being the first to demonstrate, by the methylation procedure, that amylose has a linear, and amylopectin a branched, structure. The second of TANRET's methods, which took advantage of the selective adsorption of amylose by cotton, was improved by PACSU (*136*) who used charcoal, fuller's earth or alumina as adsorbents alternative to cotton. Cold water elutes the amylopectin fraction while the amylose is recovered by hot water extraction of the cotton. The method of electrophoretic separation of SAMEC (*163*) depends on the presence of polar groups in one or other of the components. This effects a satisfactory separation of potato starch into amylose (*syn.* amyloamylose) and amylopectin (*syn.* erythroamylose) because, in potato starch, the latter component is esterified with small amounts of phosphoric acid whereas the amylose is not. The method has been successfully employed by HOPKINS *et al.* (*97*) who showed that the amyloamylose of potato starch was almost completely converted into maltose by barley amylase.

These methods are open to objections on practical grounds; yields are poor and separation of the components is far from complete. SCHOCH (*164*) ascribed the lack of agreement on procedure for isolating the components of starch and also with respect to their physical and chemical characteristics and relative proportion in whole starch to (i) incomplete dispersion of the starch; (ii) slow separation methods, such as prolonged leaching or electrophoresis, which allow the starch to retrograde; (iii) hydrolysis during separation; (iv) interference by non-carbohydrate constituents, *e. g.* the fatty acid in maize starch. A method devised by this author was based on the selective precipitation of fraction A (*i. e.* amylose) by butanol and was free from the disadvantages specified above. Numerous other methods which effect true separation of amylose and amylopectin have been developed [see the review by SCHOCH (*165*)] but the "butanol" method remains one of the cleanest and simplest methods of fractionation and as such it is widely used. The writer prefers to use thymol as precipitating agent since, in his view, it is even simpler in application than is butanol. The discovery that thymol preferentially precipitates the linear component (amylose) arose from the observation that thymol, added as a preservative to starch dispersions, causes the starch to flocculate *(81)*. From this beginning, the "thymol" method was developed on a quantitative basis *(34)*. It is convenient in certain circumstances to be able to reverse the order of separation and to precipitate the amylopectin fraction, leaving the amylose in solution. This is accomplished by the coprecipitation of amylopectin and aluminium hydroxide *(35)*. From the supernatant liquid an amylose of high purity is precipitated by alcohol. The technique currently employed in the writer's laboratory for the isolation of high-grade amylose is a combination of the "thymol" and "aluminium hydroxide" methods *(90)*. Amylose-free amylopectin is obtained by a further fractionation with methanol of the crude amylopectin prepared by the thymol method.

2. Comparison of Properties.

Maquenne's names amylose and amylopectin, have for want of better come into general usage to designate respectively the linear and branched components of starch. This nomenclature is clumsy and misleading; the name "amylopectin" is a particularly unfortunate choice since "pectin" connotes an entirely unrelated group of polysaccharides. Nothing short of an international agreement on polysaccharide nomenclature however could at this late stage rid us of the incubus of these terms.

a) Solubility.

A high-grade amylose is only very sparingly soluble in hot water and separates again in an insoluble form when the solution is kept. On the other hand, amylose disperses readily in warm dilute alkali and the solution is stable, although neutralisation of the alkali leads in a short time to precipitation. Amylose solutions are usually prepared in this way, namely by solution in alkali followed by neutralisation. The method must be used with circumspection because of this rapid spontaneous precipitation (retrogradation) from a neutral solution and because amylose is degraded, even under mild conditions, by alkali *(33)*. According to K. H. Meyer *(117)*, the phenomenon of retrogradation is to be ascribed to the structure of amylose as a long-chain linear polymer with a high content of hydrophilic groups. Associative forces (hydrogen-bonding?) cause these unbranched chains to aggregate parallelwise until colloidal dimensions are exceeded and precipitation ensues. Schoch explains the selective precipitation of amylose by butanol in similar terms; the amylose forms an alcoholate which presents a hydrophobic surface to the aqueous solvent *(165)*.

Amylopectin, unlike amylose, dissolves readily in water and the solution shows very little tendency to gel or to undergo retrogradation. Solutions of native starches gel readily but they retrograde only very slowly despite their content of amylose (20–25%). It would appear that a starch sol is to be regarded as a solution of colloidally unstable amylose "peptized" by the presence of the stable amylopectin.

b) Degree of Polymerisation.

The measurement of molecular weight of starch and its components by physical methods presents serious difficulties, both experimental and theoretical, so it is not surprising that quoted values vary over a very wide range (182). Thus, figures given for the molecular weights of the components of a variety of starches are: amylose, 10000–200000; amylopectin 200000–6000000. It is more convenient to express molecular size in terms of degree of polymerisation (D. P.), i. e. of the number of glucose residues constituting one molecule of polysaccharide. In these terms, the D. P. of amylose varies from 60 to 1200, and of amylopectin, from 1200 to 36000. Broadly speaking, the macromolecule of amylopectin is 20–30 times the size of that of amylose.

c) Percentage of Non-reducing End-Groups.

In striking contrast with the relative degrees of polymerisation of amylose and amylopectin are the proportions of terminal non-reducing glucose units which they contain. In the case of amylose, the proportion of non-reducing end-group is one in 200–1000 glucose units, whereas in amylopectin it is about one in 20–25. The high degree of branching in the amylopectin molecule is obvious from these figures.

d) Reaction with Iodine.

Amylose in low concentration reacts with iodine to give an intense blue solution; in more concentrated solution a blue-black precipitate is formed. Amylopectin also stains with iodine but the colour is not nearly so intense and is reddish-purple. It appears that adsorption complexes are formed between iodine and unbranched chains of glucose units linked by α-1:4-bonds provided these chains are above a minimum length. Chains of length 8–12 units give a red colour (absorption peak at ca. $500\,m\mu$) and as the chain-length increases, the colour of the iodine complex passes through a series of transitional colours until at chain-length 30–35, it becomes blue with light absorption peak at 600 $m\mu$. Synthetic amylose (D. P. > 450) has an absorption peak at 645 $m\mu$ and a native amylose at 650 $m\mu$ (170, 7). The "outer" chains of amylopectin, i. e. the portions of chains between non-reducing ends and the first branch links, have an average length of 10–12 glucose units which is just above the minimum length for iodine-complex formation and it may be these outer chains which are partly responsible for the red colour given by amylopectin with iodine (maximum absorption at ca. 550 $m\mu$).

Since about 200 mg. of iodine are bound per gram of amylose, and since the adsorption of iodine by amylopectin is negligible, a method for the quantitative estimation of amylose in starch is provided by potentiometric titration with dilute iodine (*21*). Many spectrophotometric methods have also been devised for the same purpose (*105, 114, 10, 36*). The method of Hassid and McCready (*114*) has been extensively used in the writer's laboratory for many years and the need was felt for an expression of colour intensity in terms of definable units. The term *A. V.* (absorption value) is used to describe the optical density ($\times$ 4) of a solution of a polysaccharide-complex, the wavelength of the light used being stated whenever an absorption value is quoted, *e. g. A. V.* (680 mμ). *A. V.* is not a property characterising an individual polysaccharide, since no stipulation is made with regard to concentrations of either polysaccharide or iodine. For characterisation, the blue value (*B. V.*) is measured; *B. V.* is defined, as the *A. V.* (680 mμ) of a solution containing unit concentrations of polysaccharide and iodine (*36, 6*).

Most pure amyloses show *B. V.* 1.40, although higher values have been recorded, whereas the *B. V.* of an amylopectin is of the order of 0.16 (cf. *90*).

e) Crystallinity.

Amylose has been obtained in crystalline form by repeated precipitation as the butanol complex (*104, 32, 4*). It can be obtained in a number of micro-crystalline forms (*102*), presumably due to sub-fractionation (*44*). X-Ray examination confirms the crystallinity of amylose and also that of the amylose-iodine complex (*162*). The typical "fibre" diagram given by amylose distinguishes it from amylopectin which exhibits no crystallinity. These observations support the view that amylose is essentially unbranched whereas amylopectin is highly branched. Nevertheless, X-ray data indicate that when amylopectin is precipitated its branched molecules associate in an orderly manner (*99*). It is interesting to find that amylose triacetate is fibrous and cannot be distinguished by appearance from cellulose acetate. Stretched films of amylose acetate show high crystallinity, giving an X-ray diffraction pattern of the "fibre" type (*181*). On the other hand, amylopectin acetate is an amorphous free-flowing powder, which does not form films and cannot be spun into fibres.

3. The Amylose-Amylopectin Ratio.

The quoted compositions of starches with respect to amylose and amylopectin content had little significance prior to the recognition that these components were structurally different and until adequate methods for their estimation had been devised. The results of these more recent analyses indicate that the amylose/amylopectin ratio is surprisingly constant over a wide range of starch varieties. Nearly all starches consist of amylose and amylopectin approximately in the proportion of one part by weight of amylose to four parts of amylopectin (*182*). Some exceptions are known, notably the starches from the waxy varieties of maize, rice, sorghum, barley and millet which, when genetically pure, contain none of the linear amylose component (*123, 39*). Consequently waxy maize starch is now in general use as a ready-made source of an amylose-free amylopectin.

Another starch of unusual interest is that of the pea. The starch of the smooth species of pea is not much different in composition from the cereal starches; it contains 30-35% amylose (*140, 156*). In starch from a number of varieties of wrinkled peas, the amylose is however the main component and may constitute 60–70% of the starch (*88*). In one variety of wrinkled pea ("Steadfast"), PEAT, BOURNE and NICHOLLS (*140*) found a starch apparently constituted of 98% amylose. Other workers (*156*) have been unable to confirm this high value and, having regard to the difficulties experienced in isolating amylose from this starch, the writer is prepared to agree that "Steadfast" pea starch may contain some amylopectin. The high amylose content is reflected in the physical properties of wrinkled pea starch. For example, the starch does not form pastes even after being heated with water at 120°; suspensions are formed rather than gels. Moreover, "Steadfast" pea starch cannot be separated completely from protein, with which it is intimately associated in the starch granules.

IV. Structure of Amylose.

Evidence has already been presented which shows that amylose is a chain polymer constituted of glucose members joined by α-1 : 4-links. Investigation of the fine structure of amylose has been concerned mainly with questions of chain-length and of branching.

1. Molecular Size.

To K. HESS (*85*) belongs the credit of being the first to furnish evidence that starch is chemically heterogeneous and that the proportion of end-group in amylose is only about one-tenth of that in amylopectin. The method of methylation and end-group assay was applied to each of the components of potato starch, separated by the electrophoretic method of SAMEC (*163*)—a fortunate choice since it is one of the few methods which really effects a separation. HESS' analysis showed that whereas the proportion of end-group in amylopectin was 1 : 33, the proportion in amylose was 1 : 283. In the same year (1940), MEYER *et al.* (*127*) using the "leaching out" method of separating the components of maize starch, showed by end-group assay (methylation) that the percentage of end-group in amylopectin was 3.5 and only 0.3 in amylose. Furthermore, these authors concluded that the correspondence of *D. P.* as determined by end-group assay and by osmotic pressure measurement indicated that amylose possessed an unbranched structure, like that of cellulose. This view was disputed by Hess and Steurer (*86*) who, on the basis of a similar comparison decided that amylose was not entirely linear, but Meyer (*128*) ascribed the discrepancy to the presence of a little of the amylopectin fraction in the amylose separated by Samec's method, and this is perhaps the correct explanation.

The "synthetic soluble starch" of Hanes (*72*) (p. 26) was shown by Haworth, Heath and Peat (*74*) not to be a whole starch but rather the amylose component only and there were indications that this synthetic amylose also was linear in structure. The same conclusion was reached by Hassid and McCready (*114*, *73*) with respect to potato amylose isolated by the methods of both Meyer (*120*) and Samec (*163*).

At about this time the first practical method of end-group assay alternative to the Haworth methylation procedure was developed by Hirst and his associates (*41*, *68*). The method makes use of the well-known Malaprade reaction which involves the specific oxidation of α-glycol groups with periodic acid. The non-reducing end-group of an amylaceous polysaccharide (*i. e.* the glucose unit which gives rise to tetramethyl glucose in Haworth's method) is distinguished from the other units of the chain in possessing three adjacent unsubstituted secondary alcohol groups and in consequence yields one mole of formic acid when it is oxidised with periodate. The reducing end-group similarly yields two moles of formic acid, so that the quantitative determination of the formic acid liberated gives a measure of the degree of polymerisation (*D. P.*). If amylose is unbranched then a total of three moles of formic acid should be produced from one mole of amylose. The advantages of this method of end-group assay are so obvious that it has been widely used in preference to the cumbersome methylation method. Nevertheless the periodate method must be used with caution, mainly because of the tendency to over-oxidation beyond the theoretical limit (*67*, *124*). It should also be pointed out that this method gives more accurate values for amylopectin than it does for amylose for the reason that in a highly branched structure the formic acid contributed by the reducing end-group is so very small in comparison with that from the non-reducing ends that it may be neglected. In an unbranched chain, as in amylose, the ratio of reducing to non-reducing ends is however 1 : 1, and two-thirds of the formic acid measured comes from the end which is much the more prone to over-oxidation.

POTTER and HASSID (*152, 153*) instituted a comparison between molecular weight determined by osmotic pressure measurement and that estimated by the periodate method for amyloses prepared from the starches of six different plants. The conclusion was reached that the amyloses of potato and Easter lily bulb consist of unbranched chains whereas in other amyloses (tapioca, wheat, maize, sago) two or three chains may be linked to form one amylose molecule, indicating a slight degree of branching. The chain-lengths of these amyloses, determined by periodate oxidation, ranged from 420 to 980 glucose residues. MEYER and RATHGEB (*124, 125*) from similar experiments also concluded that at least one sub-fraction of the amylose of potato starch is unbranched. SCHOCH et al. (*108*) devised a technique for the sub-fractionation of the amylose component of a number of starches by successive partial precipitation with octyl alcohol and a study of these sub-fractions indicated that they consist of a continuous series of homologous linear polymers rather than a limited number of discrete components. Attempts were also made to correlate reducing value with chain-length for these linear polymers but the important conclusion was reached that none of the oxidising agents—hypoiodite, bromine, alkaline copper, ferricyanide or alkaline dinitrosalicylate—commonly employed for reducing end-group assay is specific for terminal aldehyde groups. This valuable paper also relates retrogradation tendency to chain-length. It appears that "the ease of retrogradation of a linear starch substance is inversely related to its chain-length" and that "linear sub-fractions of equal intrinsic viscosity have the same retrogradation tendency, irrespective of their source." POTTER and HASSID (*154*) also examined these sub-fractions of amylose with respect to their degrees of polymerisation determined by (i) osmotic pressure measurements and (ii) periodate oxidation. The close agreement between the *D. P.* determined by the two methods for the subfractions of potato amylose indicated the substantially unbranched character of this species of amylose. The *D. P.* of maize amylose sub-fractions determined osmotically was higher however than the *D. P.* estimated by end-group assay—a suggestion that some degree of branching existed in maize amylose.

2. Evidence from Amylolysis.

While this is not the appropriate place for a discussion of enzyme mechanisms it is nevertheless necessary to refer to the use of the starch-splitting enzymes as diagnostic tools; in particular, valuable information concerning the linearity of the amylose chain structure is forthcoming from a study of the action of β-amylase upon it.

There is abundant evidence in favour of the conception that β-amylase as normally prepared from cereal grains, malt or soya beans operates by an endwise attack on the non-reducing ends of chains of α-1 : 4-linked glucose units, initiating the hydrolytic scission of the penultimate linkages and the liberation of maltose. The enzyme action progresses along the chains and if these are strictly linear, *i. e.* formed by α-1 : 4-linkage only, a more or less complete conversion into maltose is to be expected. It is in fact found that β-amylase, as normally prepared, effects an almost quantitative ($>$90%) conversion of amylose into maltose (*95*). In contrast, β-amylolysis of amylopectin ceases at a conversion limit of 50–60% (as maltose). It would appear that the branch linkages of amylopectin act as "barriers" to the hydrolytic progress of β-amylase along the chains and therefore protect the "inner" chain linkages from attack by this enzyme.

3. Z-Enzyme.

There is now reason to believe that the β-amylase preparations used in the above experiments consisted not of one enzyme but of two. Soya beans provide a convenient source of a β-amylase preparation which is practically free from α-amylase and as such has been used in the writer's laboratory for many years under the designation "stock soya β-amylase" (*142*). In 1946, Balls *et al.* announced the isolation in crystalline form of the β-amylase of sweet potato (*11, 12*). The writer was privileged to receive a sample of this highly pure β-amylase from Dr. Balls* and to institute a comparison with the stock soya bean preparation. It was at once apparent that the crystalline β-amylase is different from the amorphous soya bean specimen inasmuch as it converts potato amylose to the extent of only 70% into maltose (*147, 141, 142*). Under the same conditions, *i. e.* at p_H 4.8 the soya bean enzyme converted 90% of the amylose into maltose. This striking difference of behaviour is confined to amylose as substrate; amylopectin is hydrolysed to the same limit (52–53%) by each of the β-amylase preparations. The possibility that retrogradation of the amylose was responsible for the difference was eliminated by a series of control experiments (*141*) and the discrepancy was traced to the presence of a second carbohydrase (named Z-enzyme) in the stock soya preparation. At p_H 3.6, the crystalline enzyme and the stock soya bean β-amylase behave identically inasmuch as hydrolysis of amylose ceases abruptly at 70% conversion in each case, an indication that Z-enzyme is an enzyme which supplements the hydrolytic action of β-amylase on amylose at p_H 4.8 but which is inhibited at p_H 3.6. Methods were developed whereby soya bean β-amylase could be prepared free from Z-enzyme (*142*) and Z-enzyme itself isolated in a pure state, *i. e.* uncontaminated with β-amylase (*143*).

* I cannot let this opportunity pass of expressing the warmest thanks to Dr. Balls who, in a strikingly generous fashion, has presented specimens of crystalline β-amylase to very many carbohydrate laboratories in Europe and America.

These observations point unmistakably to the conclusion that "anomalies" exist in the amylose structure which hinder true β-amylolysis and that these anomalous structures are removable by the action of the Z-enzyme usually found in β-amylase preparations. The nature of these anomalies is an intriguing problem. They do not consist of branch linkages of the type present in amylopectin since the limit of β-amylolysis of amylopectin is unaffected by Z-enzyme addition. Ester-phosphate groups are present in potato starch and these are known to obstruct β-amylolysis (*151*). If this were the explanation of the observations described then Z-enzyme would be a phosphatase, but the pure Z-enzyme isolated by PEAT, THOMAS and WHELAN (*143*) was devoid of phosphatase activity. Furthermore it was shown that the fortuitous resemblance between Z-enzyme and a weak α-amylase had no basis in fact; Z-enzyme has no α-amylolytic activity.

A clue to the nature of the activity of Z-enzyme arose from the observation of DILLON and O'COLLA (*51*) that wheat β-amylase as normally prepared contains a second enzyme (laminarinase) which is capable of hydrolysing laminarin with liberation of glucose. It seemed likely that the occurrence of Z-enzyme was not limited to the soya bean and that laminarinase and Z-enzyme were identical. Strong support for this view came from the observation that stock soya bean β-amylase hydrolysed laminarin whereas the purified enzyme (*i. e.* freed from Z-enzyme) did not. Laminarin is a ramified polyglucose in which all the links have the β-configuration, the main chain-forming links being β-1 : 3 glucosidic bonds (*20a, 146*), and when it was shown that pure Z-enzyme would hydrolyse not only laminarin but also cellobiose, gentiobiose and sophorose (all β-linked disaccharides) it became clear that Z-enzyme is a β-glucosidase. When it was further observed that the β-glucosidase of almonds (emulsin) could replace Z-enzyme as the factor required to supplement the hydrolysis of amylose by pure β-amylase the evidence became conclusive (*143*).

If Z-enzyme is a β-glucosidase, then it follows that the Z-labile linkages in potato amylose are β-links. The location of these anomalous links in the amylose chains is still a matter for speculation. It does appear however that Z-enzyme operates by an endwise attack on terminal β-glucosidic links since its action on laminarin is distinguished by the liberation of glucose as the only saccharide of low molecular weight (*52, 143*). The simplest picture of the constitution of amylose suggested by these facts is that of essentially linear chains of α-1 : 4-linked glucose units bearing a limited number of branch points which prevent the complete degradation of the chains by β-amylase. These branch links have the β-configuration and unlike the α-branch links of amylopectin, seem to join *single* glucose residues to the main chain.

There is a further point. Z-Enzyme (or emulsin) acts independently of the presence of β-amylase on amylose; in one experiment, a Z-treated amylose was subsequently converted into maltose to the extent of 98% by crystalline β-amylase, whereas before the treatment with Z-enzyme the β-amylolysis limit was 68%. It should be noted that the proportion of branch β-links necessary to explain the observed facts will depend on the location of the branch points in the chain and may be exceedingly small. It seemed probable that the amount of glucose liberated by Z-enzyme would not be detectable. A paper-chromatographic method has however been developed which, it is claimed, is sufficiently sensitive to detect this proportion of glucose, if it were liberated by Z-enzyme, but in fact none was detected (96). This failure to detect the liberation of glucose does not mean that Z-labile linkages are not present in amylose or that Z-enzyme is an artefact but simply implies that the suggestion (143) limiting the β-linked branches to single glucose "stubs" may be incorrect.

4. The Maltosaccharides.

It is proposed to use the term *"maltosaccharide"* (or *"maltodextrin"*) as a generic description of oligo- and polysaccharides constituted entirely of glucopyranose residues joined by α-1 : 4-glucosidic links ("maltose links"). The name "amylosaccharide" is used with the same connotation by some authors but it is the writer's opinion that nomenclature will be simplified if "maltosaccharide" has the restricted meaning given above and if "amylosaccharide" (or "amylodextrin") is used in a more general sense to describe either branched or unbranched polyglucoses derived from starch.

Further evidence of the essential linearity of amylose chains comes from a study of the products of their partial hydrolysis by acid. The hydrolysis proceeds by a random attack on the chain-forming links so that the amylose is broken into fragments of varying length. These chain-fragments can be separated with absolute precision, by an application of the charcoal chromatographic technique and isolated as chemically homogeneous maltose, malto-triose, -tetraose, -pentaose, -hexaose and -heptaose. Glucose and higher maltosaccharides than those mentioned are of course also formed but the latter have not yet been separated. An examination by Whelan *et al.* (178) of the malto-saccharides so obtained from amylose, with respect to reducing power, molecular rotation, rate of flow on papergrams and action of β-amylase proved that they constitute a polymer-homologous series of chain molecules formed of glucose residues united by one type of bond only, *viz.*, by α–1 : 4 bonds. In other words, the short chain molecules derived from amylose are strictly linear. It is important for practical purposes to note that the first member of this series of maltodextrins is maltose and not glucose. The monomer has often in the past been regarded as the first member but such an assumption is unjustified and can lead to errors of interpretation. When the molecular rotations of the malto-saccharides are plotted against $D. P.$ using FREUDENBERG's relationship (57) a straight line graph is obtained and it is noteworthy that the amylose point (as well as the glucose point) is not located on this straight line (178). One explanation would be the presence in native amylose of structural peculiarities which are absent from the maltodextrin fragments.

The action of crystalline β-amylase on the maltosaccharides is of interest (178). Maltotriose resists the hydrolytic activity of low concentrations of β-amylase whereas the higher saccharides are rapidly hydrolysed: maltotetraose is converted into maltose (2 moles); maltopentaose into maltose (1 mole) and maltotriose (1 mole); maltohexaose into 3 moles of maltose. β-Amylase is without action on maltose but in high concentrations it slowly converts maltotriose into maltose (1 mole) and glucose (1 mole). The completed action of β-amylase on maltopentaose therefore yields maltose (2 moles) and glucose (1 mole). It follows, incidentally, that maltose will not be the sole product of the completed β-amylolysis of amylose. The latter consists of chains of odd and even numbers of glucose members, presumably in a 1 : 1 ratio, and the reducing terminal unit of each "odd" chain will appear as glucose in the hydrolysate. This glucose (which of course is unconnected with the hypothetical "stubs" removed by Z-enzyme) will represent such a small fraction of the long amylose chain as not to be detectable. An average amylose ($D. P.$ 400) would yield maltose and glucose in the molar ratio, 400 : 1.

The question of branching in amylose is a vexed one on which there is at present no unanimity of opinion. For example, KERR and CLEVELAND (103), assuming that maize crystalline amylose is linear, concluded from the rates at which maltose is produced by β-amylase that in potato amylose there is an average of one or two branches per molecule and 2–3 branches per molecule in tapioca amylose. On the other hand, as already stated, POTTER and HASSID (152, 153, 154) using a different line of attack, concluded that potato amylose was unbranched whereas maize amylose was slightly branched.

In concluding this survey of the constitution of amylose a brief reference to work concerned with the conformation of the molecule may be helpful. It can be shown by inspection of models, or by calculation, that a chain of glucose units linked by α-1 : 4-links will tend to assume a spiral form. X-ray and other studies of crystalline amylose (162) and butanol-precipitated amylose (161, 166) support the helical conception and indicate that the amylose chains are coiled in tight spirals, each loop of which contains six glucose members. It is supposed that the formation of insoluble complexes with iodine, butanol or fatty acids tightens and stabilises the helix. KATZBECK and KERR (101) are of the opinion that complex-formation is not an adsorption phenomenon, but that the long amylose chains form helices around the complexing agents. HUSEMANN and BARTL (*98*) regard precipitated amylose as possessing an extended chain structure and that when it is dissolved the chains undergo contraction. Acetylation causes extension of the chains, presumably because of the size of the acetyl groups introduced. The minimum chain length required for the formation of insoluble complexes of this kind was investigated by WHISTLER *et al.* (*53*) who found that it varied according to the nature of the complexing agent. With butanol the minimum length of maltodextrin chain which would form a complex was 20 glucose units.

V. Structure of Amylopectin.

It will have been gathered from the foregoing that the early chemical investigations of the constitution of starch were in fact studies of its major component, amylopectin. The presence of 20% of long-chain amylose did not materially affect the results with respect to molecular size and degree of branching of the whole starch and, when later the amylopectin was separated from the amylose and could be examined independently, end-group assay, determinations of molecular weight and β-amylolysis limits yielded values for amylopectin which were of the same order as those already found for whole starch.

1. Molecular Size and Degree of Branching.

Amylopectin is virtually without reducing power but assay of the non-reducing ends by the methylation or periodate techniques showed the presence of 4–4.5% of end-group, *i. e.* 1 glucose unit in 20–25 is a terminal unit (*82, 160, 85, 127, 73, 152, 153, 125, 42*). The only explanation of these two facts is that the amylopectin molecule has a highly branched structure. This may be pictured, in its simplest form, as being developed by the mutual glucosidic union of basic repeating chains which are composed, on the average, of 20–25 glucose units (the HAWORTH "laminated" formulation, *Fig. 4a*, p. 25). A branch link, *i. e.* the glucosidic bond between two repeating chains, interferes with the progress of β-amylolysis in the sense that the action of the enzyme in splitting off successive units of maltose from the free ends or the chains ceases when a branch point is reached (*70, 118, 36*). The β-amylase can neither hydrolyse nor by-pass a branch link and in consequence only about 55% of amylopectin is convertible into maltose. The resistent remainder is a high-molecular dextrin, named limit β-dextrin (*syn.* erythrogranulose, x-amylodextrin, dextrin-A) which retains many of the physical characteristics, including colour with iodine, of the original amylopectin. The ratio of terminal to non-terminal glucose units is 1 : 12 (approx.) as might be expected since half of the amylopectin molecule is converted into maltose (*76*).

Some attention has been given to the question of the homogeneity of amylopectins from different sources with respect to degree of branching. MEYER and GIBBONS (*121*) fractionated amylopectin by electrodialysis and found the fractions not to differ significantly in their chemical properties. Similarly, POTTER and HASSID (*155*) sub-fractionated (by methanol precipitation) the amylopectins isolated from maize and tapioca starches and showed, by periodate oxidation, that the sub-fractions possessed roughly the same degree of branching as the parent amylopectin (tapioca, 26.5 glucose residues per terminal group; maize, 25.5). On the other hand, MEYER and SETTELE (*126*) were able to sub-fractionate waxy maize amylopectin and to show by extent of β-amylolysis and by periodate oxidation that the "outer" chain-length increases (from 9 to 17 glucose units) with the *D. P.* of the sub-fraction, whereas the length of the "inner" chains is approximately constant (7.5 ± 1.5). It has been shown (*3*) that the β-amylolysis limit of the amylopectin from the "assimilation" starch of sweet potato leaves is lower than that of the amylopectin from the "reserve" starch in the tubers. Further evidence that amylopectins from different sources may differ in degree of branching has recently been provided by potentiometric studies of the iodine-uptake of a number of branched α-1 : 4-glucosans (*2*). This method segregates amylopectins and glycogens into two clearly marked groups and moreover within the first group, amylopectins of different origin, *e. g.* barley and oat, are readily distinguished.

2. Nature of the Branch Link.

In order to define the branch-forming bond it is necessary to know (i) whether it is a covalent glucosidic bond or whether weaker forces, *e. g.* hydrogen bonding are responsible, (ii) which hydroxyl groups and which glucose members of the basal chains are involved in the linkages and (iii) the configuration (α- or β-) of such glucosidic linkages.

Conclusive evidence that the branch link is in fact a glucosidic link formed between the reducing group ($C_{(1)}$) of one glucose unit and $C_{(6)}$ of a second and that it has the α-configuration has been forthcoming only within the last year or two. Before presenting this evidence however, a mention of earlier work directed to this end may have historical interest. Whatever may be the conformation of the branched molecule, it is obvious that the number of non-reducing chain ends will be the same as the number of branch points and that therefore a completely methylated amylopectin will yield equimolecular proportions of tetramethyl glucopyranose (end-group) and a dimethyl glucose (involved in the branch link). Further, the location of the methyl groups in the dimethyl glucose would determine the position of the branch linkage, a position which can only involve $C_{(2)}$, $C_{(3)}$ or $C_{(6)}$, since $C_{(1)}$ and $C_{(4)}$ are occupied by the chain-forming bonds and $C_{(5)}$ is part of the pyranose ring. The very serious drawback to this line of reasoning is that it is impossible in practice to methylate all of the hydroxyl groups not involved in chain- or branch-formation. Consequently it is usually found that the amount of dimethyl glucose in the hydrolysate of a methylated starch (or amylopectin) is greater than the amount of tetramethyl glucose. Nevertheless, it seemed likely that an examination of this dimethyl fraction would afford some indication of the kind of branch linkage, and evidence was presented almost simultaneously by HIRST (23, 13), by FREUDENBERG (58) and by MYRBÄCK (131) that the branch linkage probably involved $C_{(6)}$.

HIRST et al. partly degraded a methylated rice starch by mild acid treatment and overcame the difficulty mentioned by further methylation of the fragments (23). Hydrolysis of this fully methylated degraded starch yielded a dimethyl glucose (3%) which was shown to consist almost entirely of 2 : 3-dimethyl glucose (13). The branch linkage was therefore 1 : 6-glucosidic. FREUDENBERG and BOPPEL (58), claiming that methylation in liquid ammonia produced a fully methylated starch (later shown to be degraded), found that the dimethyl glucose obtained from it by acid hydrolysis contained both 2 : 3- and 2 : 6-dimethyl glucose. Control experiments showed that the latter was produced by de-methylation of 2 : 3 : 6-trimethyl glucose and the reasonable assumption was made that only the 2 : 3-dimethyl glucose came directly from the methylated starch. MYRBÄCK and AHLBORG (131) isolated from the products of α-amylolysis of maize starch, a trisaccharide which on methylation and hydrolysis yielded tetramethyl glucose (1 mole), 2 : 3 : 6-trimethyl glucose (1 mole) and 2 : 3 : 4-trimethyl glucose. The chemical evidence of the constitution of the last-named is not strong but it is sufficient to indicate that the glucosidic linkages in the trisaccharide were 1 : 4- and 1 : 6-links. Furthermore, the high specific rotation of the trisaccharide suggested that both links had the α-configuration, a suggestion which was supported by observations of MEYER (118) on the action of yeast maltase on limit β-dextrin.

In 1947, HIRST and his co-workers developed a novel method of approach to the problem, which avoided the difficulties of methylation, by the submission of amylopectin to direct oxidation with periodate (69, 89). It will be appreciated from *Fig. 1*, that if the side-chain is attached at $C_{(6)}$, as shown, the glucose unit bearing it will contain one α-glycol group at $C_{(2)}$, $C_{(3)}$ and that the bond between these carbon atoms will be disrupted by periodate oxidation. Subsequent hydrolysis of the oxidised starch will yield no glucose in these circumstances. On the other hand, if the side-chain is attached at either $C_{(2)}$ or $C_{(3)}$ (the only alternative positions),

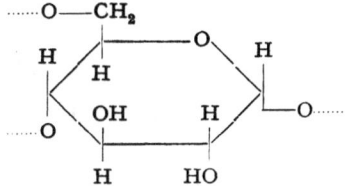

Fig. 1. The glucose unit at a branch point.

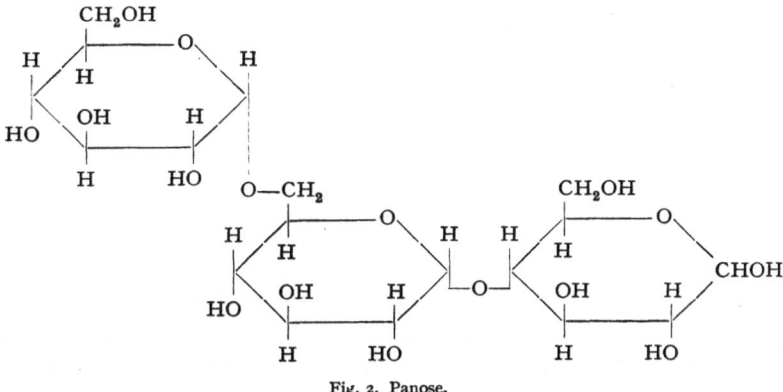

Fig. 2. Panose.

the glucose residue will not be attacked by periodate and will be recoverable as glucose after hydrolysis. This procedure was applied to a number of starches and to the limit β-dextrin from waxy maize starch. In every case only a small amount ($<1\%$) of glucose was detected by a paper chromatographic method (54), a convincing proof that the greater proportion ($>75\%$) of the branch-linkages are of the 1:6-type. This method was applied with slight modification to the amylopectins of potato and maize by GIBBONS and BOISSONNAS (61) who concluded that more than 97–98% of the branches originated at position $C_{(6)}$.

The most conclusive proof of the character of the branch link would be afforded by the isolation from amylopectin of a recognisable low molecular saccharide actually containing the branch link. Apart from MYRBÄCK's evidence (131, 1) proof of this kind was first provided by MONTGOMERY et al. (129) who were able to isolate a crystalline disaccharide, which was not maltose, from the products of hydrolysis of waxy maize starch by an amylase preparation (impure) from Aspergillus oryzae. This disaccharide, which was also isolated in the same way from tapioca and maize starches, was proved to be 6-[α-D-glucopyranosyl]-D-glucose (isomaltose) (130), identical with the disaccharide isolated by WOLFROM et al. (60, 183) from the products of partial hydrolysis of dextran and of waxy maize starch. The name "isomaltose" is an unfortunate choice

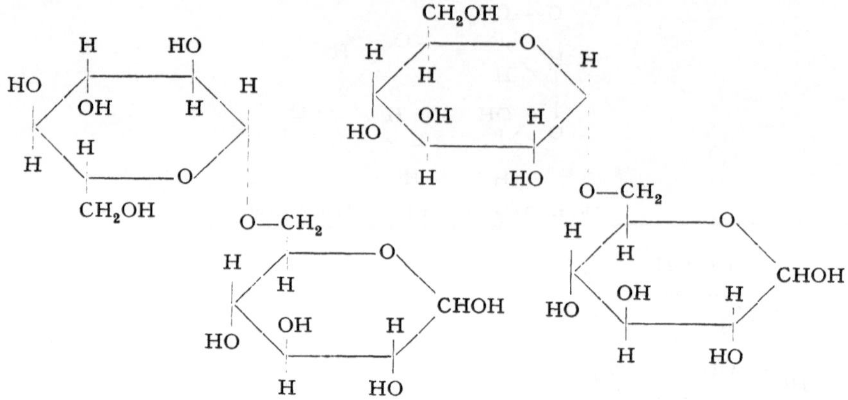

(a) Gentiobiose.   (b) Isomaltose.

Fig. 3.

for this disaccharide and the alternative name "brachiose" has been proposed (*130*); but traditional names die hard and "isomaltose" was E. FISCHER's name for the disaccharide obtained in his attempts to synthesise maltose by acid treatment of glucose. It was later shown that FISCHER's "isomaltose" consists mainly of gentiobiose (*27*), of which isomaltose proper is the α-linked analogue (*Fig. 3*). Gentiobiose has been converted into isomaltose by chemical means (*110*).

The doubt is always present that these alleged products of hydrolysis are not preformed in the starch but are produced by reversion synthesis from glucose. This dubiety is stronger, of course, when acid is the hydrolysing agent but even enzymic methods of hydrolysis are not above suspicion; for example, an application of chromatographic techniques has led to the isolation of five pure disaccharides (β-links) from the products of the action of emulsin on glucose (*145*). Nevertheless it has been shown that the action of malt α- and β-amylases on amylopectin does not yield isomaltose (*183, 184*) and the weight of evidence favours the view that the isomaltose link (α-1 : 6) pre-exists in the amylopectin molecule. The partial hydrolysis of amylopectin (waxy maize starch) by dilute acid leads to the liberation of glucose, maltose and isomaltose which were isolated by WOLFROM *et al.* as the crystalline β-acetates (*188*). At the same time it was shown that amylose (from maize starch) yielded no isomaltose under the same conditions of hydrolysis. Had any α-1 : 6-links been present in the amylose, isomaltose should have been detected because the isomaltose link (α-1 : 6) is four times as resistant to hydrolysis as is the maltose (α-1 : 4) link (*185*) and isomaltose would tend to accumulate.

The evidence that the branch linkage in amylopectin is α-1 : 6 is reinforced by the isolation of a trisaccharide in addition to isomaltose (*174*) from a partial acid hydrolysate. This trisaccharide was shown to be identical with that synthesised from maltose by *Aspergillus niger* and isolated by PAN *et al.* (*137, 138*), after whom it has been named "panose". Panose is 4-α-isomaltosyl-*D*-glucose (*Fig. 2*, p. 21), *i. e.*, it is a trisaccharide containing an α-1 : 4- and an α-1 : 6-link (*187*). WOLFROM has shown that some slight degree of acid reversion to isomaltose occurs under the usual conditions of hydrolysis of amylopectin but that the amount of crystalline β-isomaltose octa-acetate isolable from amylopectin is 200 times as great as the small amount formed by acid reversion under like conditions (*175*). A statistical analysis of the hydrolysis of amylopectin based upon the relative rates of hydrolysis (4 : 1) of the α-1 : 4- and α-1 : 6-linkages indicates that appreciable amounts of isomaltose would be liberated only in the final stages of hydrolysis. About 8% of the glucosidic links in amylopectin are branch links and calculation shows that at 91% hydrolysis, 42% of these links should appear as isomaltose ($\equiv$3.4% of the amylopectin). The actual amount of pure isomaltose acetate isolated corresponded to 1% of the amylopectin, a value in fair agreement with the theoretical (*189*).

3. R-Enzyme: a Debranching Enzyme.

There was one curious feature revealed by the earlier investigations of the starch-degrading enzymes: none of them was capable of splitting the branch linkages of amylopectin. These links are not attacked by β-amylase and, as we have seen, they actually impede normal β-amylolysis. α-Amylase, which operates by random scission of chain (α-1 : 4)-linkages (*180*) is equally incapable of hydrolysing the branch links although it can "by-pass" them and attack the α-1 : 4-links of the "inner" chains (*179*). The branch-link is also resistant to phosphorolysis (p. 26). Indeed it appeared at one time that Nature had omitted to provide for the complete conversion into sugar of the principal "reserve" polysaccharide. This proved however not to be the case. In 1950 came the demonstration of the presence of a "debranching" enzyme in plants [R-Enzyme; HOBSON, WHELAN and PEAT (*91*)], in animal tissues [amylo-1 : 6-glucosidase; CORI and LARNER (*50*)], and in a micro-organism [FRENCH and KNAPP (*55*)]. Earlier hints of a debranching action came from the work of MEYER (*119*) and others (*106, 150, 113*).

R-Enzyme, which is isolated from the potato or the broad bean, is a purely hydrolytic enzyme, the sole function of which appears to be the hydrolysis of α-1 : 6-glucosidic linkages in certain chain polymers (*94*). It is evident that R-enzyme is without action on α-1 : 4-links from the observations that its action on amylose of high blue value is negligible and that its activity with respect to whole starch is proportional to the amylopectin content of the latter. The action of R-enzyme on amylopectin or on the limit β-dextrin derived from it is accompanied by a pronounced fall of specific viscosity in aqueous solution and by a *rise* in blue value. This rise in *B. V.*, which is not due to the synthesis of longer chains, provides a convenient means of estimating debranching activity (*94*). It has been shown that R-enzyme completely debranches waxy maize starch inasmuch as the combined action of R-enzyme and β-amylase effects the quantitative conversion of this amylopectin into a mixture of maltose and maltotriose (*173*). The maltose/maltotriose ratio can be accurately determined and provides an alternative method of calculating the proportion of terminal to non-terminal glucose units in the amylopectin molecule. Potato amylopectin, unlike waxy maize, is largely, but not completely, degraded to maltose and maltotriose by R-enzyme + β-amylase (*94, 173*). The incompleteness of the breakdown in this case is probably due not so much to the failure of R-enzyme to split some of the α-1 : 6-linkages but rather to the presence in potato amylopectin of obstacles other than branch points (*e. g.* phosphoric ester groups) to β-amylolysis. This difference between potato and waxy maize amylopectins needs, however, further investigation, especially as it is known (*173*) that R-enzyme has no hydrolytic action on glycogen, despite the abundance of α-1 : 6-links in the latter. If the glycogen is first fragmented by a short treatment with α-amylase, the branch links then become susceptible to scission by R-enzyme (*173*). It is worthy of note that the debranching enzyme of animal tissues has no action on native amylopectin (*50*).

4. Conformation of the Amylopectin Molecule.

It is time now to return to the question of the order and arrangement of chains and branches in the amylopectin structure. Three formulations of the branched structure have been proposed, *viz.* the "laminated" formula of Haworth (*75*) (*Fig. 4a*), the "herring-bone" formula of Staudinger (*168*) (Fig. 4b) and Meyer's "arborescent" formula (*118*) (Fig. 4c).

The Haworth formula represents a series of short linear chains mutually joined by α-1 : 6-linkages in which the average position of the branches is at the centres of the repeating chains. This implies an ordered system of single branching whereas the Meyer method of formulation envisages multiple and random branching. In the Staudinger formula all the branch links are carried by a single main chain. Little evidence has hitherto been available which would distinguish between these formulae and although the Meyer formulation is teleologically more satisfactory, some workers have preferred to represent amylopectin by the laminated formula because of its simplicity and because it adequately accounted for most of the chemical properties of amylopectin.

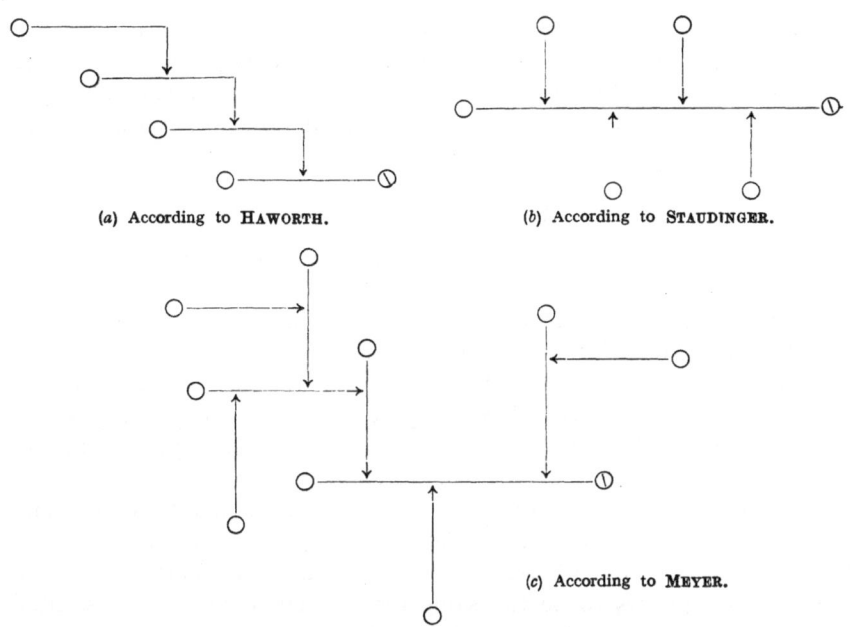

(a) According to HAWORTH.

(b) According to STAUDINGER.

(c) According to MEYER.

Fig. 4. Formulation of Amylopectin.

(○ = Non-reducing chain end; ◐ = Reducing chain end; ↓ = α – 1:6-link.)

Hints have not been lacking however that the HAWORTH formulation is an over-simplification. For example, it was pointed out by MEYER et al. (*122*) that if amylopectin had the singly branched structure proposed by HAWORTH, then the limit β-dextrin derived from it should consist if extended of long chains bearing "stubs" of one or two glucose units. This amylose-like structure is not at all in keeping with the physical properties of the β-dextrin. It is perhaps unnecessary to report other arguments in this controversy (*e. g.* see reference *50*) because the writer believes that convincing evidence has now been provided which shows that the randomly-branched tree-like formula of MEYER represents the true structure of amylopectin. This experimental evidence is reported in a paper by PEAT, WHELAN and THOMAS (*149*) and as the argument is unavoidably complex, the reader is referred to the original paper for details. In essence, the distinction between the three formulae is made on the basis of an examination of the maltosaccharides liberated when the limit β-dextrin, prepared by the action of β-amylase on waxy maize starch, is submitted to the debranching action of R-enzyme.

VI. Enzymic Synthesis of Starch.

It would be supererogatory to give a detailed treatment of the subject of the enzymic synthesis of starch in this article inasmuch as it was reviewed in 1948 by Hassid and Doudoroff (72a), in 1951 by Hehre (83), by the writer (139) and by Bernfeld (28), and in 1953 by Barker and Bourne (16). It is appropriate however that consideration be given to the relationship between studies on degradation and synthesis with respect to the constitution of starch.

1 Synthesis of Amylose.

It was Bodnar (31) who gave the first hint that starch is degraded by the agency of a phosphorylase as well as by the amylases. The end product of phosphorolytic breakdown is α-glucose-1-phosphate which was isolated as the dipotassium salt by Cori (46), using the phosphorylase of muscle tissues. It soon became evident that the phosphorolytic action was reversible and in 1940, Hanes reported (71) the successful synthesis of "granular starch" by the agency of a plant phosphorylase. This so-called granular starch proved to be amylose (74).

Phosphorylase is wide-spread in plant (and animal) tissues and although there are differences of detail in the composition and activity of phosphorylases from different sources, nevertheless the basic mechanism of synthesis appears to be the same for all. There are three essential participants in the synthetic reaction: (i) the enzyme, (ii) a donor substrate from which the enzyme "transfers" a glucosyl ($C_6H_{11}O_5$) residue (45) to (iii) an acceptor.

a) The Donor.

No phosphorylase has yet been found which can utilise for the synthesis of amylose a donor substrate other than the α-form of D-gluco-pyranose-1-phosphate (the divalent ion, $C_6H_{11}O_5 \cdot O \cdot PO_3^=$) (186, 157, 115). The reaction catalysed by phosphorylase is reversible and an equilibrium is established between the divalent anions of glucose-1-phosphate (G-1-P) and phosphoric acid. With potato phosphorylase, the ratio $HPO_4^=/C_6H_{11}O_5 \cdot O \cdot PO_3^=$ is $2 \cdot 2$ at equilibrium and is independent of p_H (72, 176).

b) The Acceptor.

From the time of the earliest observations of Hanes (72) it has been recognised that phosphorylase is without action on pure glucose-1-phos-phate and that synthesis occurs only if a third substance (a "primer", "starter", "co-substrate") is also present. Even with crude preparations of phosphorylase an induction period in the synthesis occurred and this was more prolonged the higher the purity of the enzyme preparation. The initial lag was abolished by the addition of a little starch (72) or glycogen (49) or by the products of the partial hydrolysis of these poly-saccharides although the ultimate products of such hydrolyses, i. e. maltose and glucose had no priming activity (66).

It appeared that before synthesis of amylose by the agency of phosphorylase could begin a pattern of the chains ultimately synthesised must pre-exist in the digest—a suggestion made as early as 1941 by CORI and CORI (*47*). The development of this theme is described in the review mentioned above (*139*) and it is now generally recognised that synthesis of chains of α-1 : 4-linked glucose units by phosphorylase takes the form of a lengthening of priming chains of identical constitution by the apposition of single glucose residues (derived exclusively from glucose-1-phosphate) to the non-reducing ends, thus,

Priming chain		Glucose-1-phosphate			Mineral phosphate	
G_n	+	G-1-P	$\rightleftarrows$	G_{n+1} +	P	
G_{n+1}	+	G-1-P	$\rightleftarrows$	G_{n+2} +	P	etc.

It is often more expeditious to express these reactions in diagrammatic form, as follows,

$$\bigcirc\overset{\alpha\ 4}{\text{—}}\bigcirc \ldots\ldots \bigcirc\overset{\alpha\ 4}{\text{—}}\oslash \ + \ \text{G-1-P} \ \rightleftarrows \ \bigcirc\overset{\alpha\ 4}{\text{—}}\bigcirc\overset{\alpha\ 4}{\text{—}}\bigcirc \ldots\ldots \bigcirc\overset{\alpha\ 4}{\text{—}}\oslash \ + \ \text{P}$$

in which a circle represents a glucose residue and a crossed circle the *reducing* terminal glucose unit.

That the apposition takes place at the non-reducing and not at the reducing end of a priming chain has been established by a number of converging lines of evidence; in particular by HESTRIN's demonstration (*87*) that oxidation of the aldehydic group of a maltosaccharide acceptor has little or no effect on its priming activity.

The specific requirement for the synthesis catalysed by plant phosphorylases is that the acceptor shall consist of a chain of glucose members linked by α-1 : 4-glucosidic bonds, *i. e.* possess the maltosaccharide structure. Only amylose and its derivative maltosaccharides or the "outer" chains of amylopectin and glycogen can function as acceptors in the synthetic reaction. The limit β-dextrin from amylopectin, for example, is practically without priming activity because the outer chains have been removed by β-amylolysis and the "stubs" remaining (one or two glucose units) are presumably too short to act as acceptors (*40, 171*). This brings us to the question of the length of the maltosaccharide chain requisite for priming activity. The first member of the maltosaccharide series, namely maltose, cannot act as an acceptor. A chain of three glucose members as in maltotriose can however function in this way, albeit very slowly (*177, 8*). The preparation of a series of pure maltosaccharides made possible an accurate determination of the minimum length of chain required in an acceptor and it has been shown (*8*) that although maltotriose can initiate synthesis, a chain of three units is of decidedly sub-optimal length. Maltotetraose and the higher maltosaccharides are efficient acceptors which show *inter alia* very little

difference in degree of efficiency in priming the synthetic action of phosphorylase.

It will be appreciated that, because there is a higher proportion of
non-reducing end groups in amylopectin than in amylose, the former
has, weight for weight, a much higher priming power than has the latter,
and that random hydrolysis, by acid or α-amylase, of both components
of starch will be attended by enhanced priming power per unit weight,
the proportion of chain-ends being increased by such hydrolysis. Otherwise
expressed, the branched amylopectin structure carries a greater number
of "accepting centres" than does linear amylose, these acceptors being
the non-reducing termini of chains of more than three α-1 : 4-linked
glucose members. No other structure can function as an acceptor in the
phosphorylase reaction. Thus neither glucose, fructose, sucrose nor
maltose has priming activity (66), nor has dextran (40, 87, 177). Potato
amylopectin contains a small amount of organically bound phosphate
and POSTERNAK (151) was able to isolate from potato starch by α-amylo-
lysis, a "phosphohexaose" and a "phosphotetraose". He showed that
these were respectively maltohexaose and maltotetraose phosphorylated
at $C_{(6)}$ since (i) 1 : 6-links were present in neither phosphate and (ii) glucose-
6-phosphate was formed from the tetraose phosphate by acid hydrolysis.
POSTERNAK further concluded that the phosphate groups are not in the
vicinity of the reducing end groups or of the branch points. In this
context, the most significant observation of. POSTERNAK's is that the
priming power of the phospho-hexaose (or -tetraose) is only one-fifth
of the power of the malto-hexaose (or -tetraose) obtained from it by
dephosphorylation with kidney phosphatase. Clearly the phosphoric
ester group militates against the performance of the linear chain as an
acceptor, which indicates that the acceptor, as well as the donor, is
involved in complex-formation with the enzyme.

It must not be forgotten that phosphorylase catalyses a reversible
reaction. Indeed, it is now recognised that there exist in plants two
distinct enzyme systems each capable of degrading starch to simple
sugars; the one operates by a process of phosphorolysis, the other, the
amylolytic system, by hydrolysis. The reason for this apparent lavishness
on the part of Nature has yet to be found, but it is a source of much
interesting speculation. The action-patterns of phosphorylase and
β-amylase are very similar. Each functions by attack at the non-reducing
chain-ends, the former enzyme cleaving the terminal α-1 : 4-linkage
(giving glucose-1-phosphate), the latter hydrolysing the penultimate link
(liberating maltose). The progress of each enzyme along a linear chain
is halted by a branch point, although it follows from the observation
that the limit-dextrin produced from amylopectin by phosphorylase
is degraded to a further extent by β-amylase, that phosphorolysis ceases

at a point more remote from the branch than does β-amylolysis (*40, 171*).
Again, R-enzyme, which "opens up" the structure of amylopectin
β-dextrin by hydrolysing the branch links and so making it more susceptible to degradation by β-amylase, also enables phosphorolysis of the
dextrin to proceed in the normal way (*173*). The similarity is still further
stressed by the fact that the anomalous linkages in amylose which
impede hydrolysis by pure β-amylase and which are removed by Z-enzyme,
also act as barriers to the complete phosphorolysis of amylose by phosphorylase (*143*).

With the advent of the pure maltosaccharides as primers in the
synthetic action of phosphorylase, it has become possible to prepare
synthetic amyloses of any desired chain-length since the length to which
the chain grows is determined by (i) the relative concentrations of glucose-
1-phosphate and primer and (ii) the chain-length of the primer. Chains as
long as 1000 glucose units have been synthesised (*5*) and there is evidence
that these synthetic products each have a fairly compact distribution
of molecular weights around the average (*7*).

c) The Enzyme.

It has already been mentioned that the phosphorylases are not
necessarily identical proteins and indeed there are marked differences
of composition and mode of action between potato and muscle phosphorylase. An often quoted difference is relative ease of crystallisability. The
muscle enzyme was obtained in crystalline form soon after its discovery (*65*)
whereas plant phosphorylase defied many efforts to bring it to crystallisation (*66, 30, 19, 169, 112, 132, 64, 158*). BAUM and GILBERT (*22*), con-
sequent upon the discovery that potato phosphorylase is strongly and
selectively precipitated by alcohol in the presence of amylose, have now
achieved the crystallisation of this plant phosphorylase. In the same
paper the crystallisation of the branching-enzyme, Q-enzyme, is also
described.

2. The Synthesis of Amylopectin.

The discovery of the reversibility of phosphorolysis solved the problem
of the biological synthesis of the minor component of starch but did not
indicate how the branched component originated. It was obvious that
another enzyme system was involved when HAWORTH, PEAT and
BOURNE (*80*) showed that an enzyme fraction from potato juice, which
was practically free from phosphorylase, had very little action on glucose-
1-phosphate until phosphorylase was added to the digest. The resulting
polysaccharide stained red-purple with iodine and was later shown to
have the characteristics of a natural amylopectin (*38*). The enzyme
responsible—the first "branching" enzyme to be isolated—was named
Q-enzyme.

a) The Branching Enzyme.

Crude preparations of Q-enzyme proved to be very unstable and suffered spontaneous loss of branching activity, so that the enzyme needed to be freshly prepared from potatoes for each experiment. Methods were however developed (*17, 19*) which yield potato Q-enzyme as a dry, enzymically-stable powder suitable for storage and almost free from phosphorylase and the amylases. These preparations do however contain R-enzyme in small proportion (*91*) and also D-enzyme (*148*). Gilbert and Patrick have now succeeded in crystallising Q-enzyme (*62, 63*) and it remains to be seen whether this crystalline specimen is enzymically pure. It is very probable the Q-enzyme occurs in all starch-forming organisms; as well as from the potato, it has been isolated from pea and broad bean (*92*), from green gram (*158*), from *Polytomella coeca* (*24, 25*) and its presence was demonstrated in *Neisseria perflava* (*84*). The corresponding enzyme of animal tissues is the branching factor shown to be present in liver and muscle (*48, 109*), which brings about the synthesis of glycogen.

b) The Mechanism of Branching.

It seemed at first that Q-enzyme played two distinct and independent roles; the one concerned with the synthesis of branch links, the other with the scission of α-1 : 4-linkages. In appearance, amylopectin could be synthesised by alternative routes: (i) by the simultaneous action of phosphorylase and Q-enzyme on glucose-1-phosphate in the presence of a primer (*38*) and (ii) by the direct action of Q-enzyme alone on amylose (*37*). These two aspects of Q-enzyme activity are seen however not to be independent when it is recognised that Q-enzyme operates by the simultaneous scission of α-1 : 4-links in an appropriate substrate (linear chains of the amylose type) and synthesis of α-1 : 6 links. Q-enzyme *per se* is without action on glucose-1-phosphate and the synthesis of amylopectin from the Cori ester results from two consecutive reactions, the first being the synthesis of linear chains, catalysed by phosphorylase, and the second, the action of Q-enzyme on these linear chains whereby appropriately situated α-1 : 4-chain links are converted into α-1 - 6-branch links and the ramified structure of amylopectin results (*18, 20*). Q-enzyme is in no sense a phosphorylase and phosphate-transfer plays no part in the synthesis of branch links (*20*). The alternative view that amylopectin was the product of the combined action of two phosphorylases, one synthesising α-1 : 4-links and one, α-1 : 6-links from a common substrate, glucose-1-phosphate [Bernfeld and Meutémédian (*29*)] has been proved to be incorrect by Bailey and Whelan (*6*).

Beckmann and Roger (*26*) came to the surprising conclusion that Q-enzyme is an artifact for the reason that some of the properties of amylopectin can be simulated by an amylose-fatty acid complex. The Q-enzyme preparation is supposed to contain fatty acid and the conversion of amylose into amylopectin is pictured by these authors as being the formation of a complex of amylose with fatty acid. The work of the writer and his collaborators, discussed in this Section, leaves no doubt however but that the conversion of amylose into amylopectin is a true enzymic reaction. Moreover, BOURNE *et al.* (*24, 25*) have used BECKMANN's method of distinguishing between amylopectin and amylose-fatty acid complexes to show that the similarity between synthetic and native amylopectin is actually very close, while NUSSENBAUM and HASSID (*134*) have confirmed the enzymic character of the amylose-amylopectin conversion and have shown that the amylopectin synthesised by Q-enzyme contains no fatty acid.

c) Substrate for Q-Enzyme.

The only substrate upon which Q-enzyme can act is a chain of glucopyranose units joined by α-1 : 4-glucosidic links. The minimum length of such a substrate chain is of critical importance. Thus although amylose itself is rapidly converted into amylopectin, no branching is observed when Q-enzyme is allowed to act on maltosaccharides containing 2–9 units or on α-dextrins (from amylose) containing 25 or fewer glucose units per chain (*93*). Similarly it has been found (*135*) that potato Q-enzyme is without action on linear dextrins of average chain-lengths 23, 30 and 42 units but converts a 116-unit dextrin into amylopectin. Finally, the availability of pure maltosaccharides as primers in phosphorylase synthesis (*8, 178*) and the recognition that synthesis of amylose proceeds by the apposition of glucose residues simultaneously to all the primer molecules (*7*) made possible the enzymic synthesis of linear chains of accurately known length which could serve as substrates for the study of the branching action of Q-enzyme. It was thus shown (*144*) that the minimum length for *rapid* branching by Q-enzyme is about 40 glucose units. A slow branching action is observed however with substrates of shorter chain-length; for example, a synthetic 28-unit linear dextrin was branched by Q-enzyme but at a much slower rate than was a 58-unit dextrin. Even with the synthetic maltodextrins there is a certain small "spread" of molecular weight (*7*) and it may be that the Q-enzyme is branching only the longer chains in the 28-unit dextrin.

It is a dictum that all enzyme reactions are theoretically reversible but much research has established that the branching action of Q-enzyme is irreversible in the practical sense (*20*), *i. e.* that Q-enzyme cannot break the α-1 : 6-branch link, although the earlier work was hampered by the failure to recognise that most Q-enzyme preparations contain also a true debranching enzyme, R-enzyme (*91, 94*).

LARNER (*109*) has demonstrated by a radioactive labelling method that Q-enzyme and the branching factor of animal origin are able to utilise the outer chains of branched amylaceous polysaccharides as substrate and thus effect a further degree of branching. It is claimed that Q-enzyme will cause branching of the outer chains if the *average* length of such chains is 14 glucose units or more, which is near the average outer chain-lengths of amylopectins.

d) Q-Enzyme: a Transglucosylase.

The branching enzyme operates by splitting a maltosaccharide chain (*D. P.* at least 40 glucose units) into two fragments, one of which (that carrying the newly exposed reducing end) it "transfers" to an appropriate acceptor chain, establishing an α-$1:6$-linked branch. Q-enzyme was therefore classed as a transglucosidase (*20*). These carbohydrases (including phosphorylase) are however more accurately described as "transglucosylases"(*83*) since it appears to be the glucosyl residue $(R \cdot O \cdot C_6H_{10}O_4 \cdot)$ and not the glucosyloxy residue $(R \cdot O \cdot C_6H_{10}O_4 \cdot O \cdot)$ which is transferred in the formation of the new link (*45*).

The transglucosylation effected by Q-enzyme may be represented diagrammatically as follows (*Table 1*).

Table 1. Transglucosylation Effected by Q-Enzyme.

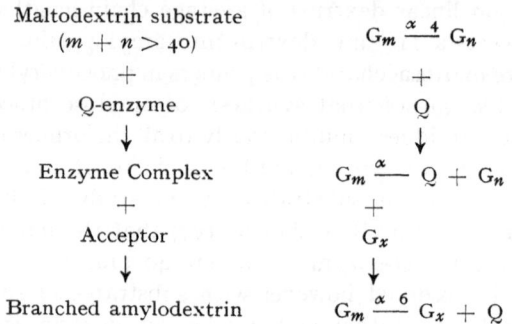

e) The Acceptor.

It has been observed by BOURNE and his associates (*14*) that when the Q-enzyme of *Polytomella coeca* is incubated with amylose, the rate of branching is inversely related to the blue value of the amylose. Indeed, with amyloses of the highest purity and therefore highest B. V., there is a time-lag in the initiation of the reaction which thereafter proceeds autocatalytically. The induction period is abolished by the addition of amylopectin. Similar conclusions have been drawn with respect to the action of potato Q-enzyme (*159*). The inference is clear. The conversion of amylose into amylopectin can be "primed", just as the synthesis of amylose by phosphorylase needs to be primed. This of course is to be expected since the transglucosylation effected by Q-enzyme implies the presence of an acceptor. The induction period shows amylose itself to be a poor acceptor but the products of Q-enzyme action upon it function as acceptors and consequently the branching proceeds at an ever increasing rate. BOURNE et al. (14) found that the rate of conversion of amylose into amylopectin was increased by the addition of amylopectin, glycogen, maltodextrins and commercial maltose but was not affected by the addition of glucose, cellobiose or dextran, nor by galactose, lactose, fructose, sucrose, inulin or xylan. It has further been shown (15) that maltose is a true acceptor in the reaction catalysed by the protozoal Q-enzyme, by the device of adding ^{14}C-labelled maltose to the digest. The resulting branched polysaccharide was radioactive.

It is inferred from these experiments that an acceptor in the transglucosylation reaction catalysed by Q-enzyme must be constituted of at least two glucopyranose residues joined by an α-1 : 4-link (maltose). Furthermore, since amylopectin is, weight for weight, a better acceptor than is amylose, it would appear that free chain ends are involved. This view derives strong support from the observation that whereas the linear maltosaccharides act as primers, the cyclic maltosaccharides (the SCHARDINGER dextrins) do not (16). Here is yet another indication that in transfer-reactions the acceptor, as well as the donor, is associated in complex-formation with the enzyme.

f) D-Enzyme.

While the conception of Q-enzyme as a transglucosylase is no longer in doubt, the recent discovery in potato juice of another type of carbohydrase may necessitate revision of current views of the mechanism of starch synthesis. The new factor, named provisionally D-enzyme, manifested itself when short-chain maltosaccharides (178) were digested with a Q-enzyme preparation. It was observed (148) that iodine-staining products were synthesised from achroic maltosaccharides, and paper chromatographic analysis revealed that the linear substrates had been disproportionated into products of higher and lower molecular weight, the latter including glucose. Glucose is not a substrate for D-enzyme and maltose also appears not to be so, but the enzyme brings about rapid disproportionation of malto-triose, -tetraose, -pentaose, -hexaose and -heptaose. It was further established, by α-amylolytic examination (see reference 180) that none of the products of the disproportionation is branched. Of the known disproportionating enzymes, D-enzyme most closely resembles the SCHARDINGER dextrinase of NORBERG and FRENCH (133).

The isolation of D-enzyme in an enzymically pure form has not yet been achieved so that it would be premature to speculate about its place in the biological scheme of starch metabolism. It is nevertheless obvious that caution must be exercised in interpreting the observations made on the activity of the branching enzyme, when probably every specimen of Q-enzyme used in the past has contained this disproportionating enzyme. Although much has yet to be learnt about D-enzyme, it is clear that it catalyses redistribution reactions among the linear maltosaccharides which are acceptors in the synthetic reactions initiated by both phosphorylase and Q-enzyme.

Acknowledgement. *I wish to express my thanks to my colleague, Dr.* W. J. WHELAN, *for reading the manuscript and for many helpful discussions on this subject.*

References.

1. AHLBORG, K. und K. MYRBÄCK: Über Grenzdextrine und Stärke, 12. Mitt. Darstellung und Konstitutionsbestimmung eines schwer hydrolysierbaren Disaccharids („Isomaltose") aus Stärke. Biochem. Z. **308**, 187 (1941).
2. ANDERSON, D. M. W. and C. T. GREENWOOD: The Characterization of Branched α: 1-4-Glucosans. Chem. and Ind. **1953**, 642.
3. BABA, A. and Y. SHIMABAYASHI: Enzymic Degradation of Assimilation Starch in Green Leaves. 2. On the Amylopectin from the Green Leaves of Sweet Potato Plant. J. Agric. Chem. Soc. Japan **26**, 502 (1952).
4. BABA, A., Y. SHIMABAYASHI and S. OKUBO: Preparation of Amylose Crystals from Sweet Potato Leaves. Sci. Report Shiga Agr. Coll. No. 1, 42 (1952).
5. BAILEY, J. M.: Starch Metabolising Enzymes. Ph. D. Dissertation. Univ. of Wales, 1952.
6. BAILEY, J. M. and W. J. WHELAN: The Enzymic Synthesis and Degradation of Starch. XI. Isophosphorylase. J. Chem. Soc. (London) **1950**, 3573.
7. — — Action Pattern of Potato Phosphorylase. Biochemic. J. **51**, xxxiii (1952).
8. BAILEY, J. M., W. J. WHELAN and S. PEAT: Carbohydrate Primers in the Synthesis of Starch. J. Chem. Soc. (London) **1950**, 3692.
9. BALDWIN, M. E.: Separation and Properties of the Two Main Components of Potato Starch. J. Amer. Chem. Soc. **52**, 2907 (1930).
10. BALDWIN, R. R., R. S. BEAR and R. E. RUNDLE: The Relation of Starch-Iodine Absorption Spectra to the Structure of Starch and Starch Components. J. Amer. Chem. Soc. **66**, 111 (1944).
11. BALLS, A. K., R. R. THOMPSON and M. K. WALDEN: A Crystalline Protein with β-Amylase Activity Prepared from Sweet Potatoes. J. Biol. Chem. **163**, 571 (1946).
12. BALLS, A. K., M. K. WALDEN and R. R. THOMPSON: A Crystalline β-Amylase from Sweet Potatoes. J. Biol. Chem. **173**, 9 (1948).
13. BARKER, C. C., E. L. HIRST and G. T. YOUNG: Linkage between the Repeating Units in the Starch Molecule. Nature (London) **147**, 296 (1940).
14. BARKER, S. A., A. BEBBINGTON and E. J. BOURNE: Carbohydrate Primers for Q-Enzyme. Nature (London) **168**, 834 (1951).
15. — — — The Mode of Action of the Q-Enzyme of *Polytomella coeca*. J. Chem. Soc. (London) (in press).

16. BARKER, S. A. and E. J. BOURNE: Enzymic Synthesis of Polysaccharides. Quart. Rev. Chem. Soc. (London) 7, No. 1, 56 (1953).

17. BARKER, S. A., E. J. BOURNE and S. PEAT: The Enzymic Synthesis and Degradation of Starch. IV. The Purification and Storage of the Q-Enzyme of the Potato. J. Chem. Soc. (London) 1949, 1705.

18. BARKER, S. A., E. J. BOURNE, I. A. WILKINSON and S. PEAT: The Enzymic Synthesis and Degradation of Starch. V. The Action of Q-Enzyme on Starch and its Components. J. Chem. Soc. (London) 1949, 1712.

19. — — — — The Enzymic Synthesis and Degradation of Starch. VI. The Properties of Purified P- and Q-Enzymes. J. Chem. Soc. (London) 1950, 84.

20. — — — — The Enzymic Synthesis and Degradation of Starch. VII. The Mechanism of Q-Enzyme Action. J. Chem. Soc. (London) 1950, 93.

20a. BARRY, V. C.: Constitution of Laminarin. Isolation of 2 : 4 : 6-Trimethyl Glucopyranose. Sci. Proc. Roy. Dublin Soc. 22, 59 (1939).

21. BATES, F. L., D. FRENCH and R. E. RUNDLE: Amylose and Amylopectin Content of Starches Determined by their Iodine Complex Formation. J. Amer. Chem. Soc. 65, 142 (1943).

22. BAUM, H. and G. A. GILBERT: A Simple Method for the Preparation of Crystalline Potato Phosphorylase and Q-Enzyme. Nature (London) 171, 983 (1953).

23. BAWN, C. E. H., E. L. HIRST and G. T. YOUNG: Nature of the Bonds in Starch. Trans. Faraday Soc. 36, 880 (1940).

24. BEBBINGTON, A., E. J. BOURNE, M. STACEY and I. A. WILKINSON: The Q-Enzyme of Polytomella coeca. J. Chem. Soc. (London) 1952, 240.

25. BEBBINGTON, A., E. J. BOURNE and I. A. WILKINSON: The Conversion of Amylose into Amylopectin by the Q-Enzyme of Polytomella coeca. J. Chem. Soc. (London) 1952, 246.

26. BECKMANN, C. O. and M. ROGER: The Question of the Branching Enzyme in Potatoes. J. Biol. Chem. 190, 467 (1951).

27. BERLIN, H.: The Identity of Isomaltose with Gentiobiose. J. Amer. Chem. Soc. 48, 1107 (1926).

28. BERNFELD, P.: Enzymes of Starch Degradation and Synthesis. Adv. Enzymology 12, 379 (1951).

29. BERNFELD, P. et A. MEUTÉMÉDIAN: Sur les enzymes amylolytiques. VII. L'isophosphorylase et la formation de polysaccharides ramifiés. Helv. Chim. Acta 31, 1735 (1948).

30. BLISS, L. and N. M. NAYLOR: The Phosphorylase of Waxy Maize. Cereal Chem. 23, 177 (1946).

31. BODNÁR, J.: Biochemie des Phosphorsäurestoffwechsels der höheren Pflanzen. I. Über die enzymatische Überführung der anorganischen Phosphorsäure in organische Form. Biochem. Z. 165, 1 (1925).

32. BOIS, E. and G. VALLIÈRES: The Crystallization of Amylose. Canad. J. Res. 23, 214 (1945).

33. BOTTLE, R. T., G. A. GILBERT, C. T. GREENWOOD and K. N. SAAD: Degradation of Potato Amylose in Neutral and Alkaline Solution. Chem. and Ind. 1953, 541.

34. BOURNE, E. J., G. H. DONNISON, W. N. HAWORTH and S. PEAT: Thymol and Cyclohexanol as Fractionating Agents for Starch. J. Chem. Soc. (London) 1948, 1687.

35. BOURNE, E. J., G. H. DONNISON, S. PEAT and W. J. WHELAN: The Fractionation of Potato Starch by Means of Aluminium Hydroxide. J. Chem. Soc. (London) 1949, 1.

36. Bourne, E. J., W. N. Haworth, A. Macey and S. Peat: The Amylolytic Degradation of Starch. A Revision of the Hypothesis of Sensitisation. J. Chem. Soc. (London) **1948**, 924.

37. Bourne, E. J., A. Macey and S. Peat: The Enzymic Synthesis and Degradation of Starch. II. The Amylolytic Function of the Q-Enzyme of the Potato. J. Chem. Soc. (London) **1945**, 882.

38. Bourne, E. J. and S. Peat: The Enzymic Synthesis and Degradation of Starch. I. The Synthesis of Amylopectin. J. Chem. Soc. (London) **1945**, 877.

39. — — The Amylose Component of Waxy Maize Starch. J. Chem. Soc. (London) **1949**, 5.

40. Bourne, E. J., D. A. Sitch and S. Peat: The Enzymic Synthesis and Degradation of Starch. III. The Role of Carbohydrate Activators. J. Chem. Soc. (London) **1949**, 1448.

41. Brown, F., S. Dunstan, T. G. Halsall, E. L. Hirst and J. K. N. Jones: Application of New Methods of End-Group Determination to Structural Problems in the Polysaccharides. Nature (London) **156**, 785 (1945).

42. Brown, F., T. G. Halsall, E. L. Hirst and J. K. N. Jones: The Structure of Starch. The Ratio of Non-Terminal to Terminal Groups. J. Chem. Soc. (London) **1948**, 27.

43. Carter, S. R. and B. R. Record: The Osmotic Pressure of Solutions of Polysaccharide Derivatives. II. The Osmotic Pressure of Derivatives of Lichenin, Inulin, Glycogen, Starch, and Starch Dextrin. J. Chem. Soc. (London) **1939**, 644.

44. Cleveland, F. C. and R. W. Kerr: Osmotic Pressure Studies on Corn Amylose. J. Amer. Chem. Soc. **71**, 16 (1949).

45. Cohn, M.: Mechanisms of Cleavage of Glucose-1-phosphate. J. Biol. Chem. **180**, 771 (1949).

46. Cori, C. F. and G. T. Cori: Mechanism of Hexose Monophosphate Formation in Muscle and Isolation of a New Phosphate Ester. Proc. Soc. Exp. Biol. **34**, 702 (1936).

47. — — Carbohydrate Metabolism. Annu. Rev. Biochem. **10**, 151 (1941).

48. Cori, G. T. and C. F. Cori: Crystalline Muscle Phosphorylase. IV. Formation of Glycogen. J. Biol. Chem. **151**, 57 (1943).

49. — — The Activating Effect of Glycogen on the Enzymatic Synthesis of Glycogen from Glucose-1-phosphate. J. Biol. Chem. **131**, 397 (1939).

50. Cori, G. T. and J. Larner: Action of Amylo-1,6-Glucosidase and Phosphorylase on Glycogen and Amylopectin. J. Biol. Chem. **188**, 17 (1951).

51. Dillon, T. and P. O'Colla: Hydrolysis of Laminarin by Wheat β-Amylase. Nature (London) **166**, 67 (1950).

52. — — The Enzymic Hydrolysis of 1 : 3-Linked Polyglucosans. Chem. and Ind. **1951**, 111.

53. Dvonch, W., H. J. Yearian and R. L. Whistler: Behavior of Low Molecular Weight Amylose with Complexing Agents. J. Amer. Chem. Soc. **72**, 1748 (1950).

54. Flood, A. E., E. L. Hirst and J. K. N. Jones: Quantitative Estimation of Mixtures of Sugars by the Paper Chromatogram Method. Nature (London) **160**, 86 (1947).

55. French, D. and D. W. Knapp: The Maltase of *Clostridium acetobutylicum*. Its Specificity Range and Mode of Action. J. Biol. Chem. **187**, 463 (1950).

56. Freudenberg, K.: The Kinetics of Long Chain Disintegration Applied to Cellulose and Starch. Trans. Faraday Soc. **32**, 74 (1936).

57. Freudenberg, K. und G. Blomqvist: Die Hydrolyse der Cellulose und ihrer Oligosaccharide. Ber. dtsch. chem. Ges. **68**, 2070 (1935).

58. FREUDENBERG, K. und H. BOPPEL: Die Stelle der Verzweigung der Stärkeketten. Naturwiss. **28**, 264 (1940).

59. FREUDENBERG, K., W. KUHN, W. DÜRR, F. BOLZ und G. STEINBRUNN: Die Hydrolyse der Polysaccharide. 14. Mitt. über Lignin und Cellulose. Ber. dtsch. chem. Ges. **63**, 1510 (1930).

60. GEORGES, L. W., I. L. MILLER and M. L. WOLFROM: The Crystalline Octa-acetate of 6-α-D-Glucopyranosido-β-D-glucose. J. Amer. Chem. Soc. **69**, 473 (1947).

61. GIBBONS, G. C. et R. A. BOISSONNAS: Recherches sur l'amidon. 48. Nature de la liaison d'embranchement du glycogène et de l'amylopectine. Helv. Chim. Acta **33**, 1477 (1950).

62. GILBERT, G. A. and A. D. PATRICK: Purification and Crystallisation of Q-Enzyme. Nature (London) **165**, 573 (1950).

63. — — Enzymes of the Potato Concerned in the Synthesis of Starch. 1. The Separation and Crystallization of Q-Enzyme. Biochemic. J. **51**, 181 (1952).

64. — — Enzymes of the Potato Concerned in the Synthesis of Starch. 2. The Separation of Phosphorylase. Biochemic. J. **51**, 186 (1952).

65. GREEN, A. A., G. T. CORI and C. F. CORI: Crystalline Muscle Phosphorylase. J. Biol. Chem. **142**, 447 (1942).

66. GREEN, D. E. and P. K. STUMPF: Starch Phosphorylase of Potato. J. Biol. Chem. **142**, 355 (1942).

67. HALSALL, T. G., E. L. HIRST and J. K. N. JONES: The Structure of Glycogen. Ratio of Non-Terminal to Terminal Glucose Residues. J. Chem. Soc. (London) **1947**, 1399.

68. — — — Oxidation of Carbohydrates by the Periodate Ion. J. Chem. Soc. (London) **1947**, 1427.

69. HALSALL, T. G., E. L. HIRST, J. K. N. JONES and A. J. ROUDIER: Structure of Starch: Mode of Attachment of the Side-Chains in Amylopectin. Nature (London) **160**, 899 (1947).

70. HANES, C. S.: The Action of Amylases in Relation to the Structure of Starch and its Metabolism in the Plant. New Phytologist **36**, 101, 189 (1937).

71. — Breakdown and Synthesis of Starch by an Enzyme System from Pea Seeds. Proc. Roy. Soc. (London) B **128**, 421 (1940).

72. — Reversible Formation of Starch from Glucose-1-phosphate Catalysed by Potato Phosphorylase. Proc. Roy. Soc. (London) B **129**, 174 (1940).

72 a. HASSID, W. Z. and M. DOUDOROFF: Enzymatically Synthesized Polysaccharides and Disaccharides. Fortschr. Chem. organ. Naturstoffe **5**, 101 (1948).

73. HASSID, W. Z. and R. M. McCREADY: The Molecular Constitution of Amylose and Amylopectin of Potato Starch. J. Amer. Chem. Soc. **65**, 1157 (1943).

74. HAWORTH, W. N., R. L. HEATH and S. PEAT: Constitution of the Starch Synthesised *in vitro* by the Agency of Potato Phosphorylase. J. Chem. Soc. (London) **1942**, 55.

75. HAWORTH, W. N., E. L. HIRST and F. A. ISHERWOOD: Polysaccharides. XXIII. Determination of the Chain Length of Glycogen. J. Chem. Soc. (London) **1937**, 577.

76. HAWORTH, W. N., E. L. HIRST, H. KITCHEN and S. PEAT: Polysaccharides. XXV. α-Amylodextrin. J. Chem. Soc. (London) **1937**, 791.

77. HAWORTH, W. N., E. L. HIRST and J. T. WEBB: Polysaccharides. II. The Acetylation and Methylation of Starch. J. Chem. Soc. (London) **1928**, 2681.

78. HAWORTH, W. N. and H. MACHEMER: Polysaccharides. X. Molecular Structure of Cellulose. J. Chem. Soc. (London) **1932**, 2270.

79. HAWORTH, W. N. and S. PEAT: The Constitution of the Disaccharides. XI. Maltose. J. Chem. Soc. (London) **1926**, 3094.

80. HAWORTH, W. N., S. PEAT and E. J. BOURNE: Synthesis of Amylopectin. Nature (London) **154**, 236 (1944).

81. HAWORTH, W. N., S. PEAT and P. E. SAGROTT: A New Method for the Separation of the Amylose and Amylopectin Components of Starch. Nature (London) **157**, 19 (1946).

82. HAWORTH, W. N. and E. G. V. PERCIVAL: Evidence of Continuous Chains of α-Glucopyranose Units in Starch and Glycogen. J. Chem. Soc. (London) **1931**, 1342.

83. HEHRE, E. J.: Enzymic Synthesis of Polysaccharides: A Biological Type of Polymerization. Adv. Enzymology **11**, 297 (1951).

84. HEHRE, E. J., D. M. HAMILTON and A. S. CARLSON: Synthesis of a Polysaccharide of the Starch-Glycogen Class from Sucrose by a Cell-Free Bacterial Enzyme System (Amylosucrase). J. Biol. Chem. **177**, 267 (1949).

85. HESS, K. und B. KRAJNC: Die Endgruppenbestimmung bei den Stärkekomponenten. X. Mitt. über Stärke. Ber. dtsch. chem. Ges. **73**, 976 (1940).

86. HESS, K. und E. STEURER: Vergleich von Endgruppengehalt, Viscosität und osmotischem Druck bei Stärke und ihren Komponenten. XII. Mitt. über Stärke. Ber. dtsch. chem. Ges. **73**, 1076 (1940).

87. HESTRIN, S.: Action Pattern of Crystalline Muscle Phosphorylase. J. Biol. Chem. **179**, 943 (1949).

88. HILBERT, G. E. and M. M. MacMASTERS: Pea Starch, a Starch of High Amylose Content. J. Biol. Chem. **162**, 229 (1946).

89. HIRST, E. L., J. K. N. JONES and A. J. ROUDIER: Structure of Acorn Starch. J. Chem. Soc. (London) **1948**, 1779.

90. HOBSON, P. N., S. J. PIRT, W. J. WHELAN and S. PEAT: The Enzymic Synthesis and Degradation of Starch. XIII. Improved Methods for the Fractionation of Potato Starch. J. Chem. Soc. (London) **1951**, 801.

91. HOBSON, P. N., W. J. WHELAN and S. PEAT: A 'Debranching' Enzyme in Bean and Potato. Biochemic. J. **47**, xxxix (1950).

92. — — — The Enzymic Synthesis and Degradation of Starch. X. The Phosphorylase and Q-Enzyme of Broad Bean. The Q-Enzyme of Wrinkled Pea. J. Chem. Soc. (London) **1950**, 3566.

93. — — — The Enzymic Synthesis and Degradation of Starch. XII. The Mechanism of Synthesis of Amylopectin. J. Chem. Soc. (London) **1951**, 596.

94. — — — The Enzymic Synthesis and Degradation of Starch. XIV. R-Enzyme. J. Chem. Soc. (London) **1951**, 1451.

95. HOPKINS, R. H.: The Actions of the Amylases. Adv. Enzymology **4**, 389 (1946).

96. HOPKINS, R. H. and R. BIRD; PEAT, S. and W. J. WHELAN: The Z-Enzyme in Amylolysis. Nature (London) **172**, 492 (1953).

97. HOPKINS, R. H., E. G. STOPHER and D. E. DOLBY: The Fractionation of Potato Starch by Electrophoresis. J. Inst. Brewing **46**, 426 (1940).

98. HUSEMANN, E. und H. BARTL: Über die Größe und Gestalt der Amylosemoleküle. Makromolek. Chem. **10**, 183 (1953).

99. JEANES, A., N. C. SCHIELTZ and C. A. WILHAM: Molecular Association in Dextran and in Branched Amylaceous Carbohydrates. J. Biol. Chem. **176**, 617 (1948).

100. KARRER, P. und C. NÄGELI: Polysaccharide. VI. Die Konstitution der Stärke und des Glykogens. Helv. Chim. Acta **4**, 263 (1921).

101. KATZBECK, W. J. and R. W. KERR: Amylose Complexes. J. Amer. Chem. Soc. **72**, 3208 (1950).

102. KERR, R. W.: Some Crystalline Forms of Amylose. J. Amer. Chem. Soc. **67**, 2268 (1945).

103. KERR, R. W. and F. C. CLEVELAND: The Structure of Amyloses. J. Amer. Chem. Soc. **74**, 4036 (1952).

104. KERR, R. W. and G. M. SEVERSON: On the Multiple Amylose Concept on Starch. III. The Isolation of an Amylose in Crystalline Form. J. Amer. Chem. Soc. **65**, 193 (1943).

105. KERR, R. W. and O. R. TRUBELL: Spectrophotometric Analyses of Starches. Paper Trade J. **117**, No. 15, 25 (1943).

106. KNEEN, E. and J. M. SPOERL: Limit Dextrinase Activity of Barley Malt. Proc. Amer. Soc. Brewing Chemists **1948**, 20.

107. LAMM, O.: Über die Charakterisierung von Stärke durch Dispersoidanalyse. Naturwiss. **24**, 508 (1936).

108. LANSKY, S., M. KOOI and T. J. SCHOCH: Properties of the Fractions and Linear Sub-Fractions from Various Starches. J. Amer. Chem. Soc. **71**, 4066 (1949).

109. LARNER, J.: The Action of Branching Enzymes on Outer Chains of Glycogen. J. Biol. Chem. **202**, 491 (1953).

110. LINDBERG, B.: Synthesis of β-Isomaltose Octa-acetate. Nature (London) **164**, 706 (1949).

111. MAQUENNE, L. et E. ROUX: Sur la constitution, la saccharification et la rétrogradation des empois de fécule. C. R. hebd. Séances Acad. Sci. **140**, 1303 (1905).

112. MARUO, B.: The Enzymic Formation and Degradation of Starch. III. Dry Preparation of Potato Phosphorylase. J. Agric. Chem. Soc. Japan **23**, 271 (1950).

113. MARUO, B. and T. KOBAYASHI: The Enzymic Formation and Degradation of Starch. I. Examination of Amylosynthease. J. Agric. Chem. Soc. Japan **23**, 115 (1949).

114. McCREADY, R. M. and W. Z. HASSID: The Separation and Quantitative Estimation of Amylose and Amylopectin in Potato Starch. J. Amer. Chem. Soc. **65**, 1154 (1943).

115. MEAGHER, W. R. and W. Z. HASSID: Synthesis of Maltose-1-Phosphate and D-Xylose-1-Phosphate. J. Amer. Chem. Soc. **68**, 2135 (1946).

116. MEYER, A.: Untersuchungen über Stärkekörner. Jena. 1895.

117. MEYER, K. H.: Recent Developments in Starch Chemistry. Adv. Colloid Sci. **1**, 143 (1942).

118. MEYER, K. H. and P. BERNFELD: Recherches sur l'amidon. V. L'amylopectine. Helv. Chim. Acta **23**, 875 (1940).

119. — — Recherches sur l'amidon. XXI. Sur les enzymes amylolytiques de la levure. Helv. Chim. Acta **25**, 399 (1942).

120. MEYER, K. H., W. BRENTANO et P. BERNFELD: Recherches sur l'amidon. II. Sur la nonhomogénéité de l'amidon. Helv. Chim. Acta **23**, 845 (1940).

121. MEYER, K. H. et G. C. GIBBONS: Fractionnement de l'amylopectine. Recherches sur l'amidon. 47. Helv. Chim. Acta **33**, 213 (1950).

122. MEYER, K. H., P. GÜRTLER and P. BERNFELD: Structure of Amylopectin. Nature (London) **160**, 900 (1947).

123. MEYER, K. H. et P. HEINRICH: Recherches sur l'amidon. XXIV. La composition de quelques espèces d'amidon. Helv. Chim. Acta **25**, 1639 (1942).

124. MEYER, K. H. and P. RATHGEB: Recherches sur l'amidon. 42. Dosages des acides formés lors de l'oxydation des polyols par le periodate. Helv. Chim. Acta **31**, 1540 (1948).

125. — — Recherches sur l'amidon. 43. Le dosage des groupes terminaux de l'amidon et du glycogène. Helv. Chim. Acta **31**, 1545 (1948).

126. MEYER, K. H. et W. SETTELE: Recherches sur l'amidon. 52. Sur l'inhomogénéité des amylopectines de tapioca et de waxy maize. Helv. Chim. Acta **36**, 197 (1953).

127. Meyer, K. H., M. Wertheim et P. Bernfeld: Recherches sur l'amidon. IV. Méthylation et détermination des groupes terminaux d'amylose et d'amylopectine de maïs. Helv. Chim. Acta **23**, 865 (1940).

128. — — — Recherches sur l'amidon. XIII. Contribution à l'étude de l'amidon de pommes de terre. Helv. Chim. Acta **24**, 378 (1941).

129. Montgomery, E. M., F. B. Weakley and G. E. Hilbert: Crystalline Derivatives of 6-α-D-Glucopyranosido-β-D-glucose from Starch. J. Amer. Chem. Soc. **69**, 2249 (1947).

130. — — — Isolation of 6-[α-D-Glucopyranosyl]-D-glucose (Isomaltose) from Enzymic Hydrolyzates of Starch. J. Amer. Chem. Soc. **71**, 1682 (1949).

131. Myrbäck, K. and K. Ahlborg: Über Grenzdextrine und Stärke. VIII. Die Konstitution eines Stärkedextrins. Nachweis α-glykosidischer 1,6-Bindungen in Dextrin und Stärke. Biochem. Z. **307**, 69 (1940/41).

132. Nakamura, M.: The Enzymic Formation and Degradation of Starch. XIII. Phosphorylase of Sweet Potatoes. J. Agric. Chem. Soc. Japan. **25**, 413 (1952).

133. Norberg, E. and D. French: Studies on the Schardinger Dextrins. III. Redistribution Reactions of *Macerans* Amylase. J. Amer. Chem. Soc. **72**, 1202 (1950).

134. Nussenbaum, S. and W. Z. Hassid: Enzymic Synthesis of Amylopectin. J. Biol. Chem. **190**, 673 (1951).

135. — — Mechanism of Amylopectin Formation by the Action of Q-Enzyme. J. Biol. Chem. **196**, 785 (1952).

136. Pacsu, E. and J. W. Mullen: Separation of Starch into its Two Constituents. J. Amer. Chem. Soc. **63**, 1168 (1941).

137. Pan, S. C., A. A. Andreasen and P. Kolachov: Enzymic Conversion of Maltose into Unfermentable Carbohydrate. Science (New York) **112**, 115 (1950).

138. Pan, S. C., L. W. Nicholson and P. Kolachov: Isolation of a Crystalline Trisaccharide from the Unfermentable Carbohydrate Produced Enzymically from Maltose. J. Amer. Chem. Soc. **73**, 2547 (1951).

139. Peat, S.: The Biological Transformations of Starch. Adv. Enzymology **11**, 339 (1951).

140. Peat, S., E. J. Bourne and M. J. Nicholls: Starches of the Wrinkled and the Smooth Pea. Nature (London) **161**, 206 (1948).

141. Peat, S., S. J. Pirt and W. J. Whelan: The Enzymic Synthesis and Degradation of Starch. XV. β-Amylase and the Constitution of Amylose. J. Chem. Soc. (London) **1952**, 705.

142. — — — The Enzymic Synthesis and Degradation of Starch. XVI. The Purification and Properties of the β-Amylase of Soya Bean. J. Chem. Soc. (London) **1952**, 714.

143. Peat, S., G. J. Thomas and W. J. Whelan: The Enzymic Synthesis and Degradation of Starch. XVII. Z-Enzyme. J. Chem. Soc. (London) **1952**, 22.

144. Peat, S., W. J. Whelan and J. M. Bailey: The Enzymic Synthesis and Degradation of Starch. XVIII. The Minimum Chainlength for Q-Enzyme Action. J. Chem. Soc. (London) **1953**, 1422.

145. Peat, S., W. J. Whelan and K. A. Hinson: Synthetic Action of Almond Emulsin. Nature (London) **170**, 1056 (1952).

146. Peat, S., W. J. Whelan and H. G. Lawley: The Structure of 'Insoluble' Laminarin. Biochemic. J. **54**, xxxiii (1953).

147. Peat, S., W. J. Whelan and S. J. Pirt: The Amylolytic Enzymes of Soya Bean. Nature (London) **164**, 499 (1949).

148. Peat, S., W. J. Whelan and W. R. Rees: D-Enzyme: a Disproportionating Enzyme in Potato Juice. Nature (London) **172**, 158 (1953).

149. PEAT, S., W. J. WHELAN and G. J. THOMAS: Evidence of Multiple Branching in Waxy Maize Starch. J. Chem. Soc. (London) 1952, 4546.

150. PETROVA, A. N.: The Enzymic Hydrolysis and Synthesis of Muscle Glycogen. Biochimiya (U. S. S. R.) 13, 244 (1948).

151. POSTERNAK, T.: On the Phosphorus of Potato Starch. J. Biol. Chem. 188, 317 (1951).

152. POTTER, A. L. and W. Z. HASSID: Starch. I. End-Group Determination of Amylose and Amylopectin by Periodate Oxidation. J. Amer. Chem. Soc. 70, 3488 (1948).

153. — — Starch. II. Molecular Weights of Amyloses and Amylopectins from Starches of Various Plant Origins. J. Amer. Chem. Soc. 70, 3774 (1948).

154. — — Starch. IV. The Molecular Constitution of Amylose Subfractions. J. Amer. Chem. Soc. 73, 593 (1951).

155. — — Starch. V. The Uniformity of the Degree of Branching in Amylopectin. J. Amer. Chem. Soc. 73, 997 (1951).

156. POTTER, A. L., V. SILVEIRA, R. M. McCREADY and H. S. OWENS: Fractionation of Starches from Smooth and Wrinkled Seeded Peas. Molecular Weights, End-Group Assays and Iodine Affinities of the Fractions. J. Amer. Chem. Soc. 75, 1335 (1953).

157. POTTER, A. L., J. C. SOWDEN, W. Z. HASSID and M. DOUDOROFF: α-L-Glucose-1-phosphate. J. Amer. Chem. Soc. 70, 1751 (1948).

158. RAM, J. S. and K. V. GIRI: Starch-Synthesizing Enzymes of Green Gram *(Phaseolus Radiatus)*. Arch. Biochemistry 38, 231 (1952).

159. REES, W. R.: Personal communication.

160. RICHARDSON, W. A., R. S. HIGGINBOTHAM and F. D. FARROW: Reducing Power and Average Molecular Chainlength of Starch and its Hydrolysis Products, and the Constitution of their Aqueous Pastes. J. Textile Inst. 27, T 131 (1936).

161. RUNDLE, R. E. and F. C. EDWARDS: The Configuration of Starch in the Starch-Iodine Complex. IV. An X-Ray Diffraction Investigation of Butanol-Precipitated Amylose. J. Amer. Chem. Soc. 65, 2200 (1943).

162. RUNDLE, R. E. and D. FRENCH: The Configuration of Starch and the Starch-Iodine Complex. II. Optical Properties of Crystalline Starch Fractions. J. Amer. Chem. Soc. 65, 558 (1943).

163. SAMEC, M. und A. MAYER: Studien über Pflanzenkolloide. XI. Elektrodesintegration von Stärkelösungen. Kolloidchem. Beihefte 13, 272 (1921).

164. SCHOCH, T. J.: Fractionation of Starch by Selective Precipitation with Butanol. J. Amer. Chem. Soc. 64, 2957 (1942).

165. — Fractionation of Starch. Adv. Carbohydrate Chem. 1, 247 (1945).

166. SPARK, L. C.: Some Notes on the Structure of B-Starch. Biochim. Biophys. Acta 8, 101 (1952).

167. STAUDINGER, H. und H. EILERS: Über hochpolymere Verbindungen. 136. Mitt. Über den Bau der Stärke. Ber. dtsch. chem. Ges. 69, 819 (1936).

168. STAUDINGER, H. und E. HUSEMANN: Über hochpolymere Verbindungen. 150. Mitt. Über die Konstitution der Stärke. Liebigs Ann. Chem. 527, 195 (1937).

169. SUMNER, J. B., T. C. CHOU and A. T. BEVER: Phosphorylase of the Jack-Bean: its Purification, Estimation, and Properties. Arch. Biochemistry 26, 1 (1950).

170. SWANSON, M. A.: Studies on the Structure of Polysaccharides. IV. Relation of the Iodine Color to the Structure. J. Biol. Chem. 172, 825 (1948).

171. SWANSON, M. A. and C. F. CORI: Studies on the Structure of Polysaccharides. III. Relation of Structure to Activation of Phosphorylases. J. Biol. Chem. 172, 815 (1948).

172. TANRET, CH.: Sur la pluralité des amidons. C. R. hebd. Séances Acad. Sci. 158, 1353 (1914).

173. Thomas, G. J.: The Branch Linkages of Starch. Ph. D. Dissertation, Univ. of Wales, 1952.

174. Thompson, A. and M. L. Wolfrom: Degradation of Amylopectin to Panose. J. Amer. Chem. Soc. **73**, 5849 (1951).

175. Thompson, A., M. L. Wolfrom and E. J. Quinn: Acid Reversion in Relation to Isomaltose as a Starch Hydrolytic Product. J. Amer. Chem. Soc. **75**, 3003 (1953).

176. Trevelyan, W. E., P. F. E. Mann and J. S. Harrison: The Phosphorylase Reaction. I. Equilibrium Constant: Principles and Preliminary Survey. Arch. Biochemistry **39**, 419 (1952).

177. Weibull, C. and A. Tiselius: The Starch Phosphorylase of Potato. Ark. Kemi, Mineral. Geol. A **19**, No. 19 (1945).

178. Whelan, W. J., J. M. Bailey and P. J. P. Roberts: The Mechanism of Carbohydrase Action. I. The Preparation and Properties of Maltodextrin Substrates. J. Chem. Soc. (London) **1953**, 1293.

179. Whelan, W. J. and P. J. P. Roberts: Action of Salivary α-Amylase on Amylopectin and Glycogen. Nature (London) **170**, 748 (1952).

180. — — The Mechanism of Carbohydrase Action. II. α-Amylolysis of Linear Substrates. J. Chem. Soc. (London) **1953**, 1298.

181. Whistler, R. L. and N. C. Schieltz: Orientation in Stretched Films of Amylose Triacetate. J. Amer. Chem. Soc. **65**, 1436 (1943).

182. Whistler, R. L. and C. L. Smart: Polysaccharide Chemistry. New York: Academic Press. 1953, p. 251.

183. Wolfrom, M. L., L. W. Georges and I. L. Miller: Crystalline Derivatives of Isomaltose. J. Amer. Chem. Soc. **71**, 125 (1949).

184. Wolfrom, M. L., L. W. Georges, A. Thompson and I. L. Miller: Enzymic Hydrolysis of Amylopectin. Isolation of a Crystalline Trisaccharide Hendecaacetate. J. Amer. Chem. Soc. **71**, 2873 (1949).

185. Wolfrom, M. L., E. N. Lassettre and A. N. O'Neill: Degradation of Glycogen to Isomaltose. J. Amer. Chem. Soc. **73**, 595 (1951).

186. Wolfrom, M. L., C. S. Smith, D. E. Pletcher and A. E. Brown: The β-Form of the Cori Ester (*d*-Glucopyranose 1-Phosphate). J. Amer. Chem. Soc. **64**, 23 (1942).

187. Wolfrom, M. L., A. Thompson and T. T. Galkowski: 4-α-Isomaltopyranosyl-*D*-glucose. J. Amer. Chem. Soc. **73**, 4093 (1951).

188. Wolfrom, M. L., J. T. Tyrree, T. T. Galkowski and A. N. O'Neill: Acid Degradation of Amylopectin to Isomaltose. J. Amer. Chem. Soc. **72**, 1427 (1950).

189. — — — — Acid Degradation of Amylopectin to Isomaltose and Maltotriose. J. Amer. Chem. Soc. **73**, 4927 (1951).

(Received, November 20, 1953.)

Neuere Ergebnisse auf dem Gebiete des Lignins und der Verholzung.

Von K. FREUDENBERG, Heidelberg.

Mit 3 Abbildungen.

I. Einleitung.

Im folgenden wird die Frage nach der Konstitution und Entstehung des Lignins behandelt. Dieser Naturstoff, der ein Fünftel bis ein Viertel und mehr des Holzes ausmacht und damit als einzelne organische Substanz mengenmäßig unmittelbar hinter der Cellulose steht, stellt der Forschung eine pflanzen- und strukturchemische Aufgabe von großem Umfang. Eigenartig und neu ist das Prinzip der Polymerisation, nach dem diese Substanz aufgebaut ist. Auch ist die Ligninforschung merkwürdig durch die Wege und Umwege, die sie hat gehen müssen, um zum heutigen Stand der Kenntnis zu gelangen.

Technische Fragen können hier nicht behandelt werden. Man wird sie in dem Maße beantworten können, wie die Konstitutionsforschung fortschreitet.

In seinem Buch „The Chemistry of Lignin" hat Brauns (19) mit großer Gründlichkeit das experimentelle Material gesammelt, das bis 1952 erarbeitet war. Ein Jahr vorher ist das umfassende Buch von Hägglund, „Chemistry of Wood" (77), erschienen. Auch das Werk von Wise und Jahn, "Wood Chemistry" (135a) ist hier zu nennen. Vorkommen, Kennzeichnung, Isolierung und Analyse des Lignins sind von Freudenberg (1954) in „Moderne Methoden der Pflanzenanalyse" beschrieben (43a). Diese Bücher entheben mich der Aufgabe, die ältere Literatur anzuführen. Auf diese wird zurückgegriffen, wo es der Zusammenhang mit der neuesten Entwicklung nötig macht. Auch aus der neuesten Literatur wird im folgenden nur das ausgewählt, was für das Konstitutionsproblem von unmittelbarer Bedeutung ist.

In diesen *Fortschritten* wurden Übersichtsreferate über das Lignin von Freuden-berg (39 a) (1939) und von Brauns (18 a) (1948) veröffentlicht.

II. Der Stoff.

Obwohl das Laubholzlignin leichter von den Polysacchariden abzutrennen ist, beschränkt sich die Untersuchung vorwiegend auf das Lignin der Coniferen, und zwar hauptsächlich der Fichte *(Picea excelsa)*. Der Grund dafür ist der einfachere Bau des Coniferenlignins, dessen aromatischer Anteil fast ausschließlich aus einer Brenzcatechin-komponente besteht, während das Laubholzlignin außerdem in erheblicher Menge eine Pyrogallol-komponente enthält (58).

Ein kleiner Teil des Fichtenlignins, etwa 2% des Gesamtlignins, kann nach Brauns (20, 23) dem Holze mit organischen Lösungsmitteln bei gewöhnlicher Temperatur als „natives Lignin" entzogen werden. Er wird im folgenden „*lösliches Lignin*" genannt. Wir benutzen Aceton von 20° als Extraktionsmittel. Das lösliche Lignin besitzt ein Molekulargewicht von 1000—3000 und darf als eine niedermolekulare und deshalb lösliche Fraktion des Gesamtlignins angesehen werden. Konstitutionelle Unterschiede vom Hauptbestandteil des Lignins, dem „*unlöslichen Lignin*", haben sich bisher noch nicht zu erkennen gegeben (s. dagegen 65a). Möglicherweise enthält dieses weniger Phenolhydroxyl und Konstitutionswasser. Nach Nord (128 a—d) lassen sich größere Mengen Lignin durch kalte organische Lösungsmittel extrahieren, wenn das Holz der Wirkung holz-zersetzender Pilze ausgesetzt war. Tschudakoff (133) nimmt dabei eine Veränderung des Lignins an.

Etwa 15% des unlöslichen Lignins werden in löslicher Form gewonnen, wenn nach Entfernung des löslichen Lignins das Holzmehl 20 Tage mit

einer Mischung geschüttelt wird, die aus gereinigtem Dioxan [BRAUNS (22)],
2—2,5% Chlorwasserstoff und 3% Wasser besteht, das in dem Holz
vorhandene Wasser eingerechnet [STUMPF und FREUDENBERG (130)].
Das entstandene Produkt (44) wird im folgenden ,,*Dioxan-Salzsäure-
Lignin*'' genannt. Es unterscheidet sich von dem löslichen Lignin durch
einen etwas höheren Methoxylgehalt sowie den Gehalt an 1,5—2% Chlor.
Wir erblicken in ihm ein Spaltstück des höhermolekularen unlöslichen
Lignins des Holzes. Die Ausbeute von 15% des Gesamtlignins läßt sich
erhöhen, wenn etwas stärkere Salzsäure angewendet und länger geschüttelt
wird. Die Präparate sind aber nicht mehr so hell wie die erste Fraktion
von 15%. Versuche von STUMPF, WEYGAND und GROSSKINSKY (131)
mit radioaktivem Dioxan haben ergeben, daß bei diesen Extraktionen
kein Dioxan in das gelöste Lignin eintritt. Etwa ein Viertel des im
Fichtenholz vorhandenen Lignins läßt sich mit kalter Ameisensäure
extrahieren, die auf 100 ccm 5 ccm einer bei 0° gesättigten wäßrigen
Salzsäure enthält. Das erhaltene Präparat ist teilweise mit Ameisensäure
verestert, die durch Verseifung leicht entfernt werden kann. Wenn zur
Extraktion radioaktive Ameisensäure verwendet wird, so bleibt nach der
Verseifung inaktives Lignin zurück. Die Ameisensäure bildet demnach
mit dem Lignin kein unverseifbares Kondensationsprodukt (44).

Folgende in der Wärme hergestellten Präparate haben für die Untersuchung des Lignins nur beschränkten Wert.

Eine heiße Mischung von Eisessig mit wenig konzentrierter wäßriger
Magnesiumchloridlösung entzieht dem Fichtenholz nach SCHÜTZ und
KNACKSTEDT (129) das meiste Lignin in teilweise acetylierter Form (62).
Dieses Präparat ist nach der Abspaltung der Acetylgruppen ärmer an
Wasser als die bei gewöhnlicher Temperatur gewonnenen Ligninpräparate.
Unverseifbare Essigsäure tritt nicht in das Lignin ein, wie der Versuch
mit radioaktiver Essigsäure ergeben hat (44). Dagegen enthält es noch
Reste von Kohlehydraten. In Übereinstimmung mit den vorigen Präparaten, aber in Gegensatz zu den folgenden, ist dieses ,,*Essigsäurelignin*''
in Aceton, Dioxan und einigen anderen organischen Lösungsmitteln,
insbesondere in der Gegenwart von geringen Mengen von Wasser, in
der Kälte größtenteils löslich. Mit einer heißen 30proz. wäßrigen Lösung
von *m*-xylolsulfonsaurem Natrium wird aus Fichtenholzmehl nur eine
geringe Menge eines Ligninpräparats extrahiert, während Buchenholz
reichliche Mengen abgibt. Aber auch dieses Buchenlignin ist wasserärmer
als Präparate, die auf anderem Wege bei gewöhnlicher Temperatur hergestellt sind.

Im Gegensatz zu den beiden folgenden sind die bisher geschilderten
Präparate ebenso wie das Lignin im Holze unter Wasser thermoplastisch.
Sie verlieren diese Eigenschaft beim Kochen mit verdünnter oder beim
Stehen mit konzentrierter Mineralsäure.

Lange Zeit diente zur Untersuchung des Lignins das „*Cuproxam-*" oder „*Cuoxam-lignin*" (*58, 53, 48*), das aus dem Holze durch abwechselnde Behandlung mit sehr verdünnter heißer Schwefelsäure und Cuoxamlösung gewonnen wird. Nur durch mehrfach wiederholte Behandlung mit diesen Reagentien kann der Kohlehydratanteil entfernt werden. Das Präparat hat durch die heiße Säure zweifellos Kondensation erlitten, ebenso wie das „*Salzsäurelignin*", das nach WILLSTÄTTER und ZECHMEISTER (*135*) durch hochkonzentrierte Salzsäure bei 0° gewonnen wird. HÄGGLUND und JOHNSON (*79*) und andere haben gezeigt, daß die von WILLSTÄTTER und ZECHMEISTER angegebene Extraktionsdauer abgekürzt werden kann. Cuoxam- und Salzsäure-Lignin haben zu ihrer Zeit, als es galt, die Grundzüge der Ligninchemie festzulegen, gute Dienste geleistet. Für die neuere Ligninchemie jedoch, die zu spezielleren Fragestellungen vorgedrungen ist, müssen die zuerst geschilderten Präparate herangezogen werden.

Es gibt eine Anzahl Umwandlungsprodukte, die für die Erforschung des Lignins eine Rolle spielen, sich jedoch von verändertem Lignin ableiten. Dies gilt für jenen Teil des Lignins, der mit Alkoholen in Gegenwart von Mineralsäure und bei höherer Temperatur dem Holze entzogen wird, ferner für die Ligninsulfonsäure und alle übrigen schwefelhaltigen Ligninderivate. Hierzu zählt auch ein *Methyllignin*, das aus methyliertem Holz durch Abbau mit Ameisensäure, die geringe Mengen Chlorwasserstoff enthält, gewonnen wird. Dieses bei gewöhnlicher Temperatur bereitete Präparat hat den Vorzug, in Benzol und Chloroform löslich zu sein (*59*).

III. Analyse und Reaktionen.

Von der älteren und seither bewährten Auffassung ausgehend, daß das Lignin ein Phenylpropanderivat ist, hat HOLMBERG die Gepflogenheit eingeführt, die Zusammensetzung der Ligninpräparate auf die Phenylpropan-Einheit zu beziehen. Will man die C_9-Formel eines Ligninpräparats mit der des Coniferylalkohols vergleichen, so addiert man zur Formel des Coniferylalkohols soviel Wasser, daß der Sauerstoffgehalt des wasserhaltigen Coniferylalkohols derselbe wird wie der des Lignins.

Coniferylalkohol .... C_9H_9 O_2 $(OCH_3)_1$
Lösliches Fichtenlignin .... $C_9H_{8,1}O_{2,4}$ $(OCH_3)_{0,9}$
Coniferylalkohol + 0,4 H_2O .... $C_9H_{9,8}O_{2,4}$ $(OCH_3)_1$
Defizit im Lignin............... $H_{1,7}$ $(OCH_3)_{0,1}$

Der Unterschied im Wasserstoff zeigt an, daß das lösliche Lignin um zwei oder nahezu um zwei Wasserstoffatome ärmer ist als der Coniferylalkohol. Das gilt auch für andere Ligninpräparate und ihre Derivate, z. B. Thioglykolsäure-lignin (*84—89*).

Die Aufgliederung der Sauerstoff-funktionen des Lignins ist eine der vordringlichen Aufgaben der Ligninchemie. Von den 2,4 Atomen Sauerstoff des löslichen Lignins gehören 1,4 acetylierbarem Hydroxyl an[*]. Ein erheblicher Teil dieser Hydroxylgruppen ist phenolischer Natur. Zu ihrer Bestimmung wurden früher die Toluolsulfoester des Lignins herangezogen. Bei der Verkochung aromatischer Toluolsulfoester mit Hydrazin entsteht Toluolsulfohydrazid und aus diesem durch das heiße Hydrazin Toluolsulfinsäure, die durch ihre Addition an Dibenzalaceton nachgewiesen werden kann. Dieses Verfahren, das für einfache Phenole brauchbar ist (70), versagt bei diesen Modellversuchen, wenn in der Kochflüssigkeit Lignin anwesend ist. Es ist daher auch auf das Lignin selbst nicht anwendbar.

Bessere Ergebnisse liefert das Reagens von SANGER (126), das Dinitrofluorbenzol [vgl. ZAHN und WÜRZ (136)]. In aceton-wäßriger Lösung setzt es sich in Gegenwart von Alkalibicarbonat vorzugsweise mit Phenolhydroxylen um unter Bildung von Dinitrophenyläthern. Mit Benzolcarbonsäuren und Zimtsäuren reagiert es langsamer, noch langsamer mit Zimtalkoholen, kaum mit anderen Alkoholen (44). Aus dem Umsetzungsprodukt ist zu schließen, daß das lösliche Lignin pro C_9-Einheit etwa 0,5 phenolisches Hydroxyl enthält (65). Der gleiche Betrag wurde für die Sulfonsäure aus löslichem Lignin gefunden (44). Auch bei der Umsetzung mit Chlorkohlensäureester wurden ähnliche Werte erhalten (44). In recht guter Übereinstimmung hiermit findet AULIN-ERDTMAN (9—14) mit Hilfe eines weiter unten geschilderten optischen Verfahrens für lösliches Lignin (*Picea mariana*) und die daraus hergestellte Sulfonsäure pro C_9 etwa 0,5 phenolisches Hydroxyl. Im löslichem Lignin von *Tsuga heterophylla* findet sie 0,65. Bei der Ligninsulfosäure des Holzes tritt dagegen ein Widerspruch auf, der unten besprochen wird.

Um eine rechnerische Grundlage zu haben, bleiben wir zunächst bei dem präparativ gefundenen Wert von etwa 0,5 Phenolhydroxyl im löslichen Lignin. Die übrigen Hydroxylgruppen, 0,9 pro C_9, sind aliphatischer Natur. Der größte Teil davon ist primär. Der Nachweis wird erbracht durch Umsetzung des Toluolsulfoesters mit Natriumjodid in siedendem Aceton. Hierbei tauschen vorzugsweise primäre Alkohole ihr ursprüngliches Hydroxyl gegen Jod aus, aber auch sekundäre Carbinole können in geringem Umfang diese Umwandlung erleiden, so daß ihre Anwesenheit in geringer Menge nicht ausgeschlossen ist. Von den 2,4 Atomen Sauerstoff in der C_9-Einheit des löslichen Lignins sind somit 1,4 Atome aufgeschlüsselt. Ein geringer Teil Carbonylsauerstoff ist nachweisbar durch Umsetzung mit Hydroxylaminsulfat und Titration der durch Oximbildung frei gesetzten Schwefelsäure. Der Betrag liegt zwischen 0,1 und 0,2 Atom Sauerstoff pro C_9-Einheit. Der Rest des Sauerstoffs (0,8—0,9 Atom) ist ätherartig gebunden. Hier ist zwischen Diaryläther, Arylalkyläther und Dialkyläther zu unterscheiden. Für

---

[*] Bestimmt mit einem Gemisch von Acetanhydrid und Pyridin (1 : 1 Mol), 40 Stdn., 50°, potentiometrisch titriert (44).

Diaryläther-gruppierungen liegt bisher kein Anzeichen vor. Dennoch wird man die Möglichkeit ihres Vorkommens im Auge behalten müssen, nachdem Asahina (8, 8a) diese Gruppierung in den Depsidonen der Flechten und Mayer (108) in der Dehydro-digallussäure des Kastaniengerbstoffes angetroffen haben. Sicher sind dagegen gemischte aliphatisch-aromatische Äther und Dialkyläther vorhanden. Ihre mengenmäßige Verteilung läßt sich folgendermaßen abschätzen: pro C_9-Einheit steht ein para-ständiges Phenolsauerstoffatom zur Verfügung, das teils frei, teils ätherartig gebunden vorliegt. Wir haben oben gesehen, daß pro C_9-Einheit 0,5 Phenolhydroxyl vorhanden sind. Demnach sind 0,5 Phenolsauerstoff für Ätherbindungen zur Verfügung. Der Rest, das sind 0,3 bis 0,4 Atome Äthersauerstoff pro C_9-Einheit, muß beiderseits mit aliphatischen Gruppen verbunden sein.

Demnach läßt sich die Zusammensetzung des löslichen Fichtenlignins folgendermaßen aufschlüsseln*: $C_9H_{6,7}$ph$(OH)_{0,5}$al.$(OH)_{0,9}(OCH_3)_{0,9}$Carbonyl-$O_{0,15}$Arylalkyl-Äther-$O_{0,5}$Dialkyläther-$O_{0,35}$.

Mit Diazomethan läßt sich das Lignin partiell methylieren. Für die Unterscheidung von aromatischem und aliphatischem Hydroxyl ist jedoch die Umsetzung mit Diazomethan bekanntlich nicht geeignet. Der größte Teil des acetylierbaren Hydroxyls ist tosylierbar. Von dem Rest ist ungewiß, ob er Hydroxylen angehört, die zur Tosylierung unfähig sind, wie tertiäres Hydroxyl oder Phenylcarbinol, oder ob die Tosylierung aus topochemischen Gründen unvollständig ist. Die im Lignin vorhandenen Hydroxylgruppen sind bis auf einen kleinen Rest mit Dimethylsulfat und Alkali bei gewöhnlicher Temperatur methylierbar.

Lösliches Lignin, Cuproxam- und Salzsäure-lignin weisen gegenüber dem Coniferylalkohol ein Defizit von 0,08 Methoxyl auf. Entweder sind Fremdstoffe einkondensiert, die nicht zur Coniferylgruppe gehören, oder es ist durch Oxydation auf dem Wege über o-Chinonbildung Entmethylierung eingetreten.

Für die Farbreaktion des Lignins ist nach Adler (3, 4), Kratzl (97, 97a), Pew (111) eine Zimtaldehydgruppe verantwortlich, die jedoch mengenmäßig stark zurücktritt (auf 40—60 Einheiten eine solche Gruppe).

Wenn Lignin der Oxydation mit Chromsäure-Schwefelsäure nach Kuhn und Roth zum Zwecke der C-Methylbestimmung unterworfen wird, so werden kleine, aber schwankende Beträge an Essigsäure erhalten. Es besteht der Verdacht, daß die C-Methylgruppen ganz oder zum Teil erst während der Operation durch Umlagerungsreaktionen entstehen.

Von den Derivaten des Lignins ist mit Vorrang am wichtigsten die *Ligninsulfonsäure*. Über den Sulfitaufschluß und die Bindung der Sulfongruppen sind aus den Laboratorien von Hägglund und Erdtman in den letzten Jahren wertvolle Arbeiten hervorgegangen, in denen ins einzelne

---

* ph = phenolisches und al. = aliphatisches OH.

gehende Aussagen gemacht werden konnten [Übersichten bei HÄGGLUND (78), BRAUNS (21) und LINDGREN (102)]. Durch Vergleich mit Modellsubstanzen (HOLMBERG u. a.) wurde ermittelt, daß im Sulfitaufschluß das im Lignin vorhandene Hydroxyl einer Phenylcarbinolgruppe durch die Sulfogruppe ersetzt wird und daß aliphatische Phenylcarbinoläther unter Bildung von Sulfogruppen aufgespalten werden. Eine dritte durch Vergleich mit Modellen ermittelte Reaktion beruht wahrscheinlich auf der Wechselwirkung eines aliphatisch-aromatischen Äthers mit Bisulfit.

Der Gehalt der Ligninsulfonsäure an Phenolgruppen ist zur Zeit umstritten. Wir haben mit Hilfe von Dinitrofluorbenzol und Chlorkohlensäureester etwa 0,6 Phenolhydroxyl pro C_9 gefunden (65, 44), während AULIN-ERDTMAN (11) ihren weiter unten beschriebenen optischen Beobachtungen entimmt, daß nur 0,2—0,3 Phenolgruppen pro ursprüngliche Einheit vorhanden sind. RICHTZENHAIN und ALFREDSSON (125) finden bei Versuchen über die Wechselwirkung von Ligninsulfosäure mit Hypochlorit höhere Phenolwerte als AULIN-ERDTMAN, während ADLER (1) aus einer Reaktion mit Perjodsäure die Richtigkeit der Werte von AULIN-ERDTMAN folgert. Soviel scheint festzustehen, daß der Phenolgruppengehalt des Lignins bei der Bildung der Ligninsulfonsäure nicht oder nur in untergeordnetem Maße zunimmt. In welchem Umfange bei der Bisulfitreaktion die Molekülgröße des Lignins verändert wird oder andere Umbildungen eintreten, kann nicht gesagt werden.

Beim Studium der Wechselwirkung des Fichtenlignins mit Schwefelwasserstoff in neutralem oder alkalischem Medium bei erhöhter Temperatur begegnet man denselben reagierenden Gruppen, jedoch sind sie schwerer zu unterscheiden. Mit besonderer Leichtigkeit setzt sich, wie HOLMBERG (84—89) sowie HEDÉN und HOLMBERG (81) und HOLMBERG und GRALÉN (90) gefunden haben, Thioglykolsäure (Mercaptoessigsäure) mit dem Lignin des Holzes in Gegenwart von wäßriger Salzsäure bei erhöhter Temperatur um. Die entstandene Säure ist offensichtlich zunächst mit anderen Holzbestandteilen verestert, denn sie läßt sich dem Holz nicht mit organischen Lösungsmitteln, sondern nur durch warmes Alkali entziehen. Danach ist sie in organischen Lösungsmitteln löslich. Sie enthält bis zu eine Thioglykolgruppe pro C_9. Die Ligninarten verschiedener Pflanzen und Bäume weisen erhebliche Unterschiede auf (86).

IV. Abbau.

Aus dem nach den verschiedenen Verfahren isolierten Fichtenlignin, der Ligninsulfosäure und dem im Holz befindlichen Lignin werden durch Oxydation mit Nitrobenzol und heißem Alkali 20—25% Vanillin gewonnen (40, 41, 60 a). Laubholzlignin liefert ein Gemisch von Vanillin und Syringaaldehyd. Häufig erhält man aus Laubholzlignin mehr Syringaaldehyd als Vanillin [CREIGHTON u. a. (27, 28)]. In einigen Eukalyptusarten überwiegt die Syringa-komponente bedeutend [Bland, Ho und Cohen (*18*)]. Zu diesen Aldehyden tritt bei gewissen Ligninarten, insbesondere den Gramineen, der *p*-Oxybenzaldehyd hinzu. Von Kratzl (*97, 97a*) wurde in der Ligninsulfosäure als Spaltstück neben Vanillin Acetaldehyd gefunden. Wenn Fichtenlignin oder Methyllignin mit starkem Alkali in der Hitze behandelt und dann methyliert wird, so liefert das Produkt bei der Oxydation neben Veratrumsäure einige Prozente Isohemipin-säure (I) (*58*). Metahemipinsäure (II) tritt in geringen Mengen auf, aber nur, wenn das Ausgangsmaterial vorher mit Mineralsäure in Berührung war [Richtzenhain (*122—124*)]. Ein Teil der Isohemipinsäure wird nach Richtzenhain (*123, 124*) auch aus nicht mit heißem Alkali behandeltem methyliertem Lignin unmittelbar durch Oxydation gewonnen. Diese letztere Fraktion der Isohemipinsäure entstammt einer in geringem Umfang vorhandenen Gruppierung (III), während bei dem anderen Teil der Isohemipinsäure das zu methylierende Phenolhydroxyl erst durch das heiße Alkali freigelegt wird.

Zur Erklärung dieser Reaktion wird im Lignin ein Phenylcumaran-system (IV) angenommen. Ein Chromansystem (V) kann aus Gründen, die unten angeführt werden, ausgeschlossen werden. Neuerdings wurde gefunden (*44*), daß manche Phenylcumaransysteme aufgespalten werden,

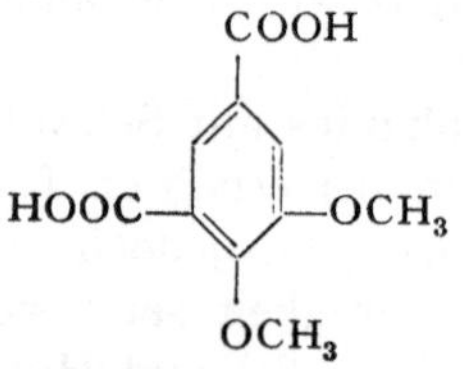

(I.) Isohemipinsäure.

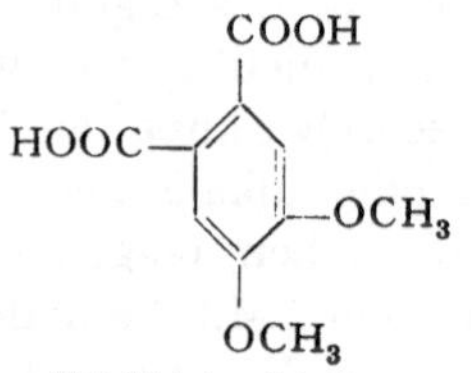

(II.) Metahemipinsäure.

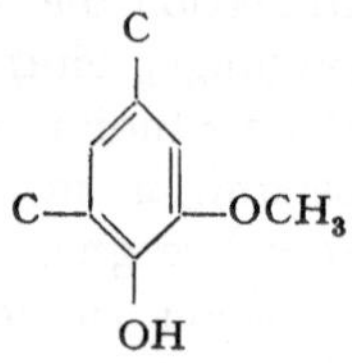

(III.)

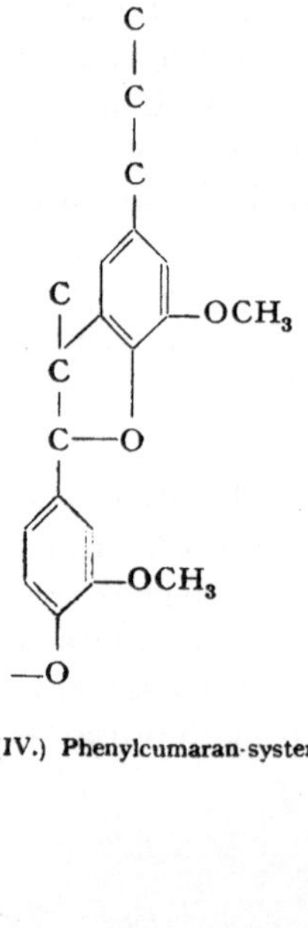

(IV.) Phenylcumaran-system.

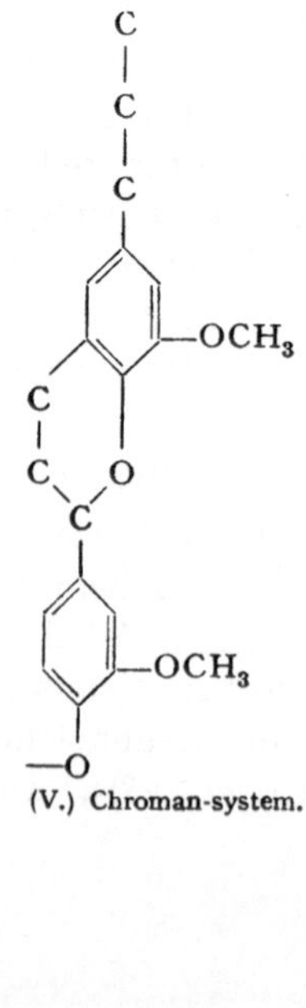

(V.) Chroman-system.

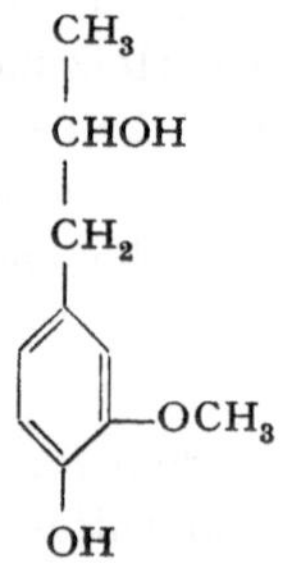

(VI.) 1-Guajacyl propanol-2.

wenn man sie bei gewöhnlicher Temperatur mit einer Mischung von Acetanhydrid und Eisessig behandelt, der eine Spur Überchlorsäure beigegeben ist. Bei der Aufspaltung entstehende Phenolgruppen werden hierbei acetyliert. Zugleich entsteht eine Doppelbindung, an die sich unter Umständen Essigsäure anlagert (vgl. hierzu 65 a).

In den genannten Oxydationsprodukten des Lignins ist die Seitenkette abgebaut. In anderen Fällen hat jedoch der Abbau bei C_9 haltgemacht. So hat SCHORYGINA (127, 128) bis zu 13% 1-Guajacylpropanol-2 (VI) erhalten, als sie Fichtenholz in flüssigem Ammoniak mit Natrium behandelte. Andere C_9-Körper hat HIBBERT (19, 82) bei der Behandlung des Lignins oder Holzes mit warmen Alkoholen erhalten, denen geringe Mengen an Mineralsäure zugesetzt waren. Dabei bleibt ein großer Teil des Lignins im Holz. Ein anderer Teil wird in lösliche Formen übergeführt und enthält zusätzliche Alkylgruppen, die durch Verätherung von Phenylcarbinolen oder Umätherung von Phenylcarbinoläthern entstanden sind. Hier begegnen uns dieselben Gruppen wie bei der Bisulfitreaktion. Gleichzeitig treten einige Prozente monomerer Bausteine auf. Es handelt sich um α-Diketone und Ketoläther, Substanzen, die· bei der Isolierung zweifellos Umlagerungen erlitten haben und keine unmittelbaren Schlüsse auf die Art des Einbaus dieser Gruppen zulassen. Die Versuche von SCHORYGINA und von HIBBERT sagen jedoch aus, daß ein kleiner Teil von C_9-Einheiten nicht durch C-C-Kondensation, sondern über Ätherbindungen in den Hauptteil des Lignins eingebaut ist. Bei den hydroxylierten Cyclohexyl-propanen, die ADKINS und andere Bearbeiter der Ligninhydrierung erhalten haben, kann die Auftrennung sowohl durch Lösung von Ätherbindungen wie durch Krackung von C-C-Bindungen erfolgt sein.

Lignin spaltet bei der Behandlung mit heißen, starken Mineralsäuren 1,5—2% *Formaldehyd* ab. Der Betrag erhöht sich bis auf mehr als 3%, wenn das Lignin vorher an einzelnen empfindlichen Stellen oxydiert wird, wie das beim Cuproxamlignin der Fall ist (71). Untersuchungen von Modellsubstanzen (71) und von radioaktivem künstlichem Lignin (47) haben ergeben, daß der Formaldehyd dem primären Carbinol des ursprünglichen Coniferylalkohols entstammt.

Aussichtsreich für die Chemie des Lignins ist die von ZIEGLER und SNATZKE gefundene Spaltung mit Diazoniumverbindungen (138). Eine andere Reaktion von ZIEGLER und GARTLER bedient sich des Chinonimidchlorids (137), das neuerdings von GIERER und ADLER auf das Lignin angewendet wurde (74). In beiden Fällen werden Guajacylgruppen abgetrennt.

V. Modellsubstanzen.

Seit HOLMBERG die schon erwähnte Reaktionsfähigkeit der Phenylcarbinole und ihrer aliphatischen Äther gegenüber Bisulfit festgestellt und mit der Bildung der Ligninsulfonsäure verglichen hat, sind immer ein-

4*

gehendere Versuche mit Modellen für diese und andere Reaktionen des Lignins angestellt worden [KRATZL (96), LINDGREN und SAEDÉN (103)] [Übersicht bei LINDGREN (102)]. Auf diesem Wege konnten die oben geschilderten Strukturelemente ermittelt werden, die Voraussetzung für das Zustandekommen der Sulfitreaktion sind. Darüber hinaus ist durch Vergleich mit Modellen im Lignin ein Äther zwischen dem mittelständigen Hydroxyl des Guajacylglycerins und der Phenolgruppe des Coniferyl-alkohols vermutet worden [ADLER und LINDGREN (5, 6)]. Für die Abspaltung des Formaldehyds wurde im 2,3-Di-phenylpropandiol-(1,3) das günstigste Modell gefunden (71). Eine Anordnung dieser Art ist tatsächlich, wie weiter unten dargetan wird, im Lignin vorhanden.

Eine besondere Rolle unter den Modellen für die Chemie des Lignins spielt das von COUSIN und HÉRISSEY (26) aus Isoeugenol (VII) gewonnene *Dehydro-diisoeugenol*. Es entsteht aus Isoeugenol durch Dehydrierung mit alkoholischem Eisenchlorid oder mit Pilzredoxase. Später hat ERDTMAN (31, 32) dieses Produkt als ein Phenylcumaran erkannt. Die endgültige Formel ist (VIII) [FREUDENBERG und RICHTZENHAIN (67); vgl. ERDTMAN (30)]. Da bereits vorher Cumaransysteme als Bestandteile des Lignins erörtert wurden [KLASON (94, 95, 95a); FREUDENBERG (39)] und später durch die Auffindung der Isohemipinsäure eine starke Stütze fanden, gewannen das Dehydro-diisoeugenol und seine Abkömmlinge eine steigende Bedeutung für die Chemie des Lignins. Insbesondere interessiert, daß der Cumaranring dieser Verbindung unter Bedingungen des Sulfitaufschlusses geöffnet wird und eine Sulfonsäure bildet. Auch Isohemipinsäure wird mit geringer Ausbeute aus dieser Sulfonsäure, wie auch aus Ligninsulfonsäure gewonnen. Für die Genese des Lignins ist die Substanz, wie weiter unten dargetan wird, sehr aufschlußreich. Neuerdings wurde diese Substanz und verschiedene ihrer Derivate für das Studium der Ringaufspaltung mit Hilfe von Acetanhydrid/Essigsäure/Perchlorsäure herangezogen (44).

Für die Beurteilung der Spektra des Lignins und seiner Derivate im Infrarot und Ultraviolett ist eine unübersehbare Zahl von Modellsubstanzen benutzt worden.

VI. Optisches Verhalten.

Es ist eine auffallende und immer wieder bestätigte Tatsache, daß Lignin der verschiedensten Art und alle seine Derivate optisch inaktiv sind, obwohl zahlreiche asymmetrische Kohlenstoffatome nachweisbar sind. Diese Feststellung bedarf einer besonderen Erklärung, die weiter unten gegeben wird.

In den Infrarotspektren des Lignins und seiner Derivate und ebenso des künstlichen Lignins und seiner Derivate (69, 51) kommen die Hydroxylgruppen, die zum Teil auf die Methoxylgruppen zurückzuführenden CH-Frequenzen sowie die in geringem Umfang vorhandenen Carbonylgruppen zum Ausdruck. Wichtiger ist der Vergleich der gesamten Spektren. Im Infrarotspektrum des Dehydrierungs-Polymerisats („DHP") des Coniferylalkohols finden sich alle Banden des Spektrums des löslichen Lignins wieder, wenn auch zum Teil mit verschiedener Intensität. Das IR-Spektrum der Ligninsulfonsäure und der Sulfonsäure des DHP stimmen nach Lage und Intensität durchaus überein. Dasselbe gilt für das nach dem oben geschilderten Ameisensäureverfahren hergestellte Methyllignin und das ebenso bereitete methylierte DHP.

Lösliches Lignin hat im Ultraviolett ein Maximum bei etwa 230 mμ, ein Minimum bei 260, ein Maximum bei 280 und Ausbuchtungen zwischen 300 und 350 mμ, die konjugierten Systemen angehören, zu denen bei 340 mμ die Komponente eines im Kern durch Sauerstoff substituierten Zimtaldehyds hinzukommt. In der Ligninsulfonsäure sind diese Ausbuchtungen eingeebnet, im übrigen aber stimmen die Spektra überein.

Nach eigenen und in der Literatur vorliegenden Messungen lassen sich die UV-Spektra des künstlichen und natürlichen Lignins und ihrer Sulfonsäuren etwa folgendermaßen ordnen.

In der Coniferylreihe setzen sich die Spektra im wesentlichen aus Propylbrenzcatechin-derivaten zusammen, die in der Seitenkette a) gesättigt oder b) mit einer dem Kern benachbarten Doppelbindung (Äthylenoder Carbonylgruppe) oder c) mit konjugierten Doppelbindungen zunächst dem Kern versehen sind.

Modifikationen treten auf, wenn außerdem Propylpyrogallol-komponenten vorhanden und vor allem, wenn die Bausteine zu größeren Ver-

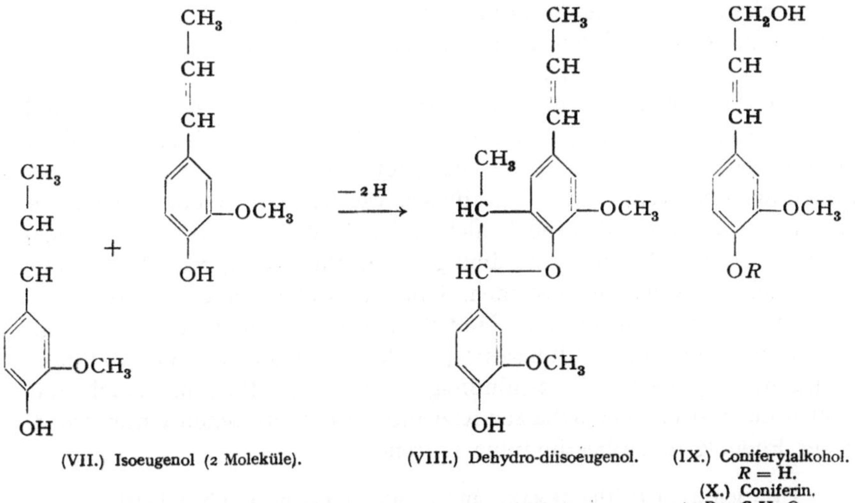

(VII.) Isoeugenol (2 Moleküle).   (VIII.) Dehydro-diisoeugenol.   (IX.) Coniferylalkohol. $R =$ H.
(X.) Coniferin. $R =$ C$_6$H$_{11}$O$_5$.

bänden vereinigt sind. In diesem Falle beobachtet man eine Verflachung der Minima, die mit einer Verbreiterung der Banden zusammenhängt.

FLEXSER, HAMMETT und DINGWALL (36) sowie MORTON und STUBBS (109) und COGGESHALL und GLESSNER (25) haben die Ultraviolettspektren von Phenolen in Gegenwart von Mineralsäure und von Alkali untersucht und eine Abhängigkeit vom pH gefunden. Diese Erscheinung wurde von AULIN-ERDTMAN (9—13), ENKVIST und ALFREDSSON (29) sowie GOLDSCHMID (75) auch am Lignin festgestellt. Wenn bei Phenolen die Differenz zwischen der Absorption in alkalischem und neutralem Gebiet gegen die Wellenlänge aufgetragen wird, so werden charakteristische Kurven gewonnen, in denen eine Erhöhung der Absorption im alkalischen Gebiet und eine mehr oder weniger ausgeprägte Verschiebung der Maxima nach dem längeren Wellengebiet zum Ausdruck kommt. Liegen Gemische von verschiedenen Phenolen vor, so überdecken sich die Maxima und Minima teilweise. Auch Lignin und Ligninsulfonsäure zeigen diese Erscheinung.

Vergleicht man den Effekt, der beim löslichen Lignin, seiner Sulfonsäure und der Ligninsulfonsäure aus Holz gefunden ist, mit dem entsprechenden Effekt solcher Phenole und Phenolsulfonsäuren, deren Gruppierungen im Lignin vermutet werden, so kann auf diesem Wege der Phenolgruppengehalt des Lignins und der Ligninsulfonsäure abgeschätzt werden. Aus ihren zahlreichen und sehr sorgfältigen Beobachtungen folgert AULIN-ERDTMAN, daß im löslichen Lignin *(Picea mariana)* und der zugehörigen Sulfonsäure pro C_9-Einheit 0,5 Phenolgruppe, und in der Ligninsulfonsäure des Fichtenholzes 0,2—0,3 Phenolgruppe vorhanden seien. Der Widerspruch zu den oben geschilderten präparativen Ergebnissen, die durch Anwendung von Fluordinitrobenzol und Chlorkohlensäureester erhalten wurden und bei der Ligninsulfosäure 0,6 Phenolhydroxyl ergeben haben (44, 65), ist noch nicht geklärt*. Das präparative Verfahren setzt voraus, daß unter den gewählten Bedingungen außer den Phenolgruppen andere Hydroxyle nur in geringem Umfange mit der Fluorverbindung oder dem Chlorkohlensäure-ester reagieren. Diese Frage muß durch weitere Modellversuche geklärt werden. Auch das optische Verfahren ist mit gewissen Unsicherheiten behaftet. Man kennt zwar, wie oben ausgeführt und weiter unten spezifiziert werden wird, mehrere Gruppierungen des Lignins, aber noch nicht alle. Man kann bisher nur solche Modelle wählen, die den Anordnungen der Phenolgruppen des Lignins, soweit man sie kennt, entsprechen. Obendrein mag der Umstand, daß in einem Falle niedermolekulare Modelle herangezogen werden, im Falle des Lignins und seiner Sulfonsäure jedoch hochmolekulare Verbindungen vorliegen, gleichfalls die Schlußfolgerungen beeinflussen. Auch hier können nur weitere Versuche zur Klärung der aufgetretenen Unterschiede in der Phenolgruppenbestimmung führen.

---

* Titration nach H. BROCKMANN und E. MEYER ergibt 0,5 Phenol-OH (44).

VII. Biosynthese des Lignins.

Vor 80 Jahren hat TIEMANN die Konstitution des im Cambialsaft der Coniferen vorkommenden Coniferins (X) aufgeklärt und die Eigenschaften des Coniferylalkohols (IX) beschrieben, dessen Fähigkeit, mit Säuren amorphe Produkte zu bilden, auffiel. Diese Kondensate oder Polymerisate des Coniferylalkohols haben mancherlei Ähnlichkeit mit dem Lignin. TIEMANN und MENDELSOHN (*132*) haben eine Beziehung zwischen Lignin und Coniferylalkohol vermutet. Zu Ende des vergangenen Jahrhunderts hat KLASON gefolgert, daß das Lignin ein Abkömmling des Coniferylalkohols sei (*89a*). Später hat KLASON den Gedanken erörtert, daß ein Teil des Fichtenlignins ein Kondensationsprodukt von zwei oder drei Molekülen Coniferylaldehyd sein könnte (*94, 89a*). In Anlehnung an die Catechine erwog er einen sauerstoffhaltigen Ring und schlug ein Phenylcumaransystem vor. FREUDENBERG, der in der Folgezeit die Bedeutung der Ätherbindungen im Lignin betonte, gelangte zu ähnlichen Vorstellungen (*39*), allerdings mit dem Unterschied, daß er das von E. FISCHER an den Peptiden und Polysacchariden entdeckte Prinzip der bifunktionellen Verknüpfung auf das Lignin übertrug und entsprechend seiner Vorstellung von der Cellulose (*38;* 1921) an Stelle kleiner, abgeschlossener Moleküle nach der Konstitution der Kettenglieder des Lignins suchte. Gleichfalls zu der Oxydationsstufe des Coniferylaldehyds gehörte das bald darauf von FREUDENBERG, ZOCHER und DÜRR (*72*) als Baueinheit des Lignins erörterte Guajacylglycerin. Allen diesen Vorstellungen liegt die Erkenntnis zugrunde, daß das Bauelement des Lignins einer höheren Oxydationsstufe angehört als der Coniferylalkohol und daß im Lignin cyclische oder offene Ätherbindungen vorkommen.

Später hat KLASON (*95, 95a*) nochmals den Zusammenhang des Lignins mit dem Coniferylalkohol erörtert und gefolgert, ,,daß Coniferylalkohol ... sich selbst bei gewöhnlicher Temperatur zu einer trimeren Form des Coniferylaldehyds oxydiert, die identisch ist mit dem Lignin des Holzes". ,,Nun ist ... der Coniferylalkohol autoxydabel zu Coniferylaldehyd, der sich dann zu einem lignin-ähnlichen Körper polymerisiert."

Als einige Jahre später ERDTMAN (*31, 32*) die bereits besprochene Beobachtung machte, daß das chemisch oder enzymatisch entstehende Dehydrierungsprodukt des Isoeugenols, das Dehydro-diisoeugenol von COUSIN und HÉRISSEY, ein Phenylcumaran ist, brachte er diesen Vorgang mit der Ligninchemie in Verbindung durch die Folgerung, daß das Lignin durch enzymatische Dehydrierung eines in der Seitenkette oxydierten Phenylpropanderivates entstehe. Es ist bemerkenswert, daß damit erneut eine Parallele zu den Catechinen und Gerbstoffen und ihren durch oxydierende Fermente entstehenden Kondensationsprodukten, den Gerbstoffen und Gerbstoffroten [GORIS (*76*), FREUDENBERG (*37*)] hergestellt war.

Durch die Auffindung der Isohemipinsäure wurde einige Zeit später das Vorkommen von sauerstoffhaltigen Ringen im Lignin und damit der Zusammenhang mit dem Dehydro-diisoeugenol gestützt. Zunächst wurde in einer mit Richtzenhain begonnenen (68) und von ihm weitergeführten (120, 121, 121a) Untersuchungsreihe die enzymatische Dehydrierung zahlreicher den Ligninbausteinen nahestehender Phenolderivate untersucht. Als Ferment diente die von Cousin und Hérissey (26) am Isoeugenol erprobte Pilzredoxase, die auf die verdünnte wäßrige Lösung des Substrats von 20° anfänglich bei pH 8, später bei pH 7 und darunter zur Anwendung gebracht wurde. In zahlreichen Fällen konnte teils an amorphen, teils an kristallinischen Dehydrierungsprodukten die Bildung von cyclischen und offenen Ätherbindungen und damit eine Analogie zum Lignin festgestellt werden. Danach wurden die Versuche auf den Coniferylalkohol (IX) und später auf andere Oxyzimtalkohole, insbesondere den Sinapinalkohol (XI) ausgedehnt (42, 43, 45—47, 49—52, 54—57, 59—61, 65, 66, 69, 71). Hierbei hat sich folgende Versuchsanordnung herausgebildet (56, 54, 57).

$$H_2COH$$
$$|$$
$$CH$$
$$|$$
$$CH$$

$$H_3CO——OCH_3$$

$$OR$$

(XI.) Sinapinalkohol, $R = H$.
(XII.) Syringin, $R = C_6H_{11}O_5$.

Die sehr verdünnte wäßrige Lösung des betreffenden Oxyzimtalkohols (0,2—0,5%) wird mit der wäßrigen Lösung des gereinigten Ferments versetzt und unter Durchleiten von Luft bei 20° gehalten; pH 5,5—6,5 ist am günstigsten. Die geeignete Wasserstoffionenkonzentration kann durch Puffer eingestellt werden, im allgemeinen ist jedoch Puffer unnötig, weil die Wasserstoffionenkonzentration von selbst innerhalb des gewünschten Bereichs erhalten bleibt. Sterilisationsmaßnahmen haben sich nicht als nötig erwiesen, da niemals das Wachstum von Mikroorganismen beobachtet worden ist, auch wenn die Versuche über Wochen ausgedehnt wurden.

Je nach der Stärke des Ferments beginnt sich die Lösung des Coniferylalkohols nach wenigen oder mehreren Stunden zu trüben, worauf sich ein sehr schwach bräunlichgelb gefärbter Niederschlag absetzt. Nach wenigen Tagen ist fast der ganze Coniferylalkohol in Gestalt dieses amorphen Pulvers ausgefallen, das abfiltriert und bei gewöhnlicher Temperatur getrocknet wird. Es enthält geringe Mengen von Stickstoff und Mineralbestandteilen. Zur Entfernung dieser aus der Fermentlösung herstammenden Beimengungen wird in Aceton gelöst, das wenige Prozente Wasser enthält. In die filtrierte Lösung wird unter Rühren Benzol eingetropft und die erste Fällung entfernt. Aus dem Filtrat wird durch Zusatz von weiterem Benzol das Hauptprodukt niedergeschlagen. Es wird gesammelt, bei gewöhnlicher Temperatur von Aceton und Benzol befreit, mit Wasser von gewöhnlicher Temperatur durchgerieben, filtriert und bei gewöhnlicher Temperatur getrocknet. Falls der Stickstoffgehalt o,3% übersteigt, wird die Reinigungsoperation wiederholt.

Die Ausbeute beträgt 60—80% des eingesetzten Coniferylalkohols.

Dieses Verfahren zielt auf die Gewinnung des acetonlöslichen Anteils der Reaktionsprodukte hin. Es besteht jedoch kein Zweifel, daß auch die entfernten unlöslichen Anteile von derselben Art sind wie die löslichen und sich von ihnen vermutlich nur durch einen höheren Kondensationsgrad unterscheiden. Der p-Cumaralkohol (p-Oxyzimtalkohol) (54) verhält sich ebenso. Auch der Kaffeealkohol (3,4-Dioxy-zimtalkohol) (55) liefert ein entsprechendes Produkt, das jedoch von dunklerer Farbe ist und in etwas geringerer Ausbeute entsteht. Der m-Cumaralkohol (m-Oxyzimtalkohol) reagiert unter diesen Bedingungen überhaupt nicht, während der o-Oxyzimtalkohol zwar ein helles Produkt liefert, aber nur äußerst langsam reagiert. Völlig anders verhält sich der Sinapinalkohol (3,5-Dimethoxy-4-oxyzimtalkohol)(XI) (60, 57). Die Lösung färbt sich dunkelbraun und wird schließlich schwarz. Ein Niederschlag tritt, wenn überhaupt, so nur in geringem Umfange auf. Als Reaktionsprodukt konnte in geringer Menge Dimethoxy-p-chinon isoliert werden. Beachtenswert ist, daß eine hälftige Mischung von Coniferyl- und Sinapinalkohol in guter Ausbeute ein helles Produkt liefert, das aus beiden Komponenten zusammengesetzt ist. Die überstehende Lösung färbt sich hierbei nicht dunkel. Wird die Mischung von Coniferyl- und Kaffeealkohol (m,p-Dioxyzimtalkohol) in der geschilderten Weise dehydriert, so entsteht ein Reaktionsprodukt, das beide Komponenten enthält (55).

Am besten ist bisher das *Produkt aus Coniferylalkohol* untersucht worden. Es enthält noch sämtliche Kohlenstoffatome des Coniferylalkohols und entspricht, auf die C_9-Einheit umgerechnet, einem Coniferylalkohol, der nahezu zwei Wasserstoffatome verloren hat (57). Die Einheit ist ein zweifach dehydrierter Coniferylalkohol. Man ist daher berechtigt, von einem Polymerisat des dehydrierten Coniferylalkohols zu sprechen. Wir haben es *Dehydrierungspolymerisat* („DHP") genannt. Bei den meisten Polymerisationen, wie der des Styrols oder Isoprens, ist die der Polymerisation unterliegende Einheit als Substanz faßbar. Dies ist hier nicht der Fall, sondern die Einheit oder besser gesagt die Einheiten werden erst durch die Dehydrierungsreaktion gebildet und sind so instabil, daß sie nicht gefaßt werden können und sich sofort zusammenlagern. Der Vorgang ist mit der Bildung der Stärke aus Glucose-1-phosphorsäureester vergleichbar. Hier entsteht durch Abspaltung von Phosphorsäure ein nicht faßbares Zwischenprodukt, das man, wenn man will, als Diradikal formulieren kann, und das sich nunmehr zum Polysaccharid polymerisiert.

Im folgenden wird der Ausdruck Dehydrierungspolymerisation beibehalten. Man hat auch die Bezeichnung Polykondensation erwogen, aber sie ist irreführend, denn das künstliche Lignin ist kein Kondensat des Coniferylalkohols, das etwa durch Wasserverlust entstünde. Daß diese Polymerisation aus einer ganzen Reihe mesomerer Formen des dehydrierten Coniferylalkohols heraus entsteht und die Einheiten daher sehr verschieden sind, bedeutet, wie noch gezeigt werden wird, eine erhebliche Komplikation, ändert jedoch nichts am Gesamtbild des Vorgangs. Näheres hierüber wird mitgeteilt, wenn die Zwischenstufen der Dehydrierungspolymerisation besprochen werden (S. 61, 62).

Das Dehydrierungspolymerisat des Coniferylalkohols (Coniferylalkohol-DHP) ist dem löslichen Fichtenlignin außerordentlich ähnlich (42, 43, 56, 59, 69, 51). Das gilt für die Elementarzusammensetzung, die Farbenreaktionen, Löslichkeit, Menge des abspaltbaren Formaldehyds, Bildung von Isohemipinsäure, Menge und Art der Hydroxylgruppen, Bildung und Zusammensetzung der beiderseitigen Sulfonsäuren, Thermoplastizität (die in beiden Fällen durch Kochen mit verdünnter Mineralsäure verlorengeht), Infrarotspektrum der freien Substanzen, ihrer Methyläther und Sulfonsäuren usw. Die Übereinstimmung ist so groß, daß nur die Unterschiede interessieren.

Eine kleine Differenz besteht im Methoxylgehalt. Während das lösliche Lignin wie alle Coniferenlignine pro C_9-Einheit ein Defizit von etwa 0,08 Methoxyl aufweist (0,92 Methoxyl statt 1), besitzt das künstliche Lignin den vollen Methoxylgehalt. Es ist möglich, daß das natürliche Lignin diesen kleinen Teil seines Methoxyls durch nachträgliche Oxydation verloren hat. Wichtiger ist der Unterschied im Gehalt an Doppelbindungen oder genauer gesagt, an hydrierbaren Gruppen. Er ist beim künstlichen Lignin größer als beim nativen. Hierbei kann es sich um Äthylen- oder um Carbonylbindungen handeln. Aulin-Erdtman (11) findet im DHP 1 Zimtaldehydgruppe auf 20 Einheiten, während nach Adler im löslichen Lignin eine auf 40—60 Einheiten vorkommt (3, 4).

Mit den ungesättigten Gruppen hängt wohl auch ein Unterschied in der Reaktion mit Diazobenzolsulfonsäure (in Carbonatlösung) zusammen. Fichtenholz und lösliches Lignin werden leuchtend orange gefärbt, Coniferylalkohol-DHP gibt ein stumpfes Rot.

Der verschiedene Gehalt an Doppelbindungen gibt sich auch im Ultraviolettspektrum zu erkennen (55, 44). Bei einem bei pH 7 hergestellten Coniferylalkohol-DHP ist die Bande von 280 mμ wahrnehmbar, aber sie ist überlagert durch eine bei 270 mμ liegende Bande, deren Extinktion etwa doppelt so stark ist wie die von 280. Wird das DHP bei pH 5,5—6 hergestellt, so ist der Unterschied zum löslichen Lignin bedeutend geringer. Das Maximum der Absorption liegt jetzt bei 275—278 mμ, ist also dem des Lignins (280 mμ) nähergerückt. Außerdem ist die Extinktion nur noch die 1,3-fache der des Lignins. Bei einzelnen DHP-Präparaten, die längere Zeit an feuchter Atmosphäre gelagert haben, ist ein weiteres

Absinken der Extinktion und zugleich eine Annäherung an das Maximum von 280 mμ beobachtet worden. Daraus wurde geschlossen, daß sich die Spektra durch Alterung einander nähern können. Das Ultraviolettspektrum des Methyllignins, das nach dem oben geschilderten Verfahren mit Ameisensäure gewonnen war, steht in Übereinstimmung mit dem UV-Spektrum des methylierten künstlichen Lignins. Auch die beiderseitigen Sulfonsäuren stimmen wie im Infrarot auch im Ultraviolett überein.

Coniferylalkohol, der am $C_{(1)}$-Kohlenstoffatom, das der Carbinolgruppe angehört, radioaktiv markiert ist, liefert ein radioaktives DHP (47). Der aus diesem abgespaltene Formaldehyd enthält die dem Ausgangsmaterial entsprechende Radioaktivität. Dies ist ein weiterer Beweis dafür, daß der Formaldehyd der endständigen Carbinolgruppe entstammt. Die aus dem radioaktiven DHP gewonnene Isohemipinsäure ist radioinaktiv. Demnach ist an ihrer Bildung das endständige Kohlenstoffatom der Phenylpropankette nicht beteiligt. Die Isohemipinsäure entstammt, wie schon oben ausgeführt worden ist, einem Cumaransystem; die dem Methoxyl benachbarte Carboxylgruppe der Isohemipinsäure wird vom mittelständigen Kohlenstoffatom der Seitenkette geliefert. Diese Feststellung führt zu dem Schluß, daß im künstlichen Lignin keine Chromanringe vorkommen (47). Bei der über Erwarten großen Übereinstimmung des künstlichen Lignins mit dem natürlichen Lignin darf hier wie in anderen Fällen angenommen werden, daß chemische Feststellungen, die an einem von beiden gemacht werden, auch für das andere gelten. Des weiteren kann aus diesen Versuchen gefolgert werden, daß tatsächlich das *Lignin ein Dehydrierungspolymerisat des Coniferylalkohols ist.*

Wenn somit festgestellt ist, daß das lösliche natürliche und das künstliche Lignin ganz oder fast ganz übereinstimmen, so erheben sich am künstlichen Lignin dieselben Konstitutionsfragen wie am natürlichen, und es wäre hierdurch für das Konstitutionsproblem nicht viel gewonnen, denn die Schwierigkeiten der Konstitutionsermittlung sind bei beiden Präparaten dieselben. Aber der Einblick in die Genese des Lignins ermöglicht nunmehr, die Entstehung des Lignins von den ersten Stufen an zu studieren und hierdurch der Forschung eine erfolgversprechende Wendung zu geben. Ein Vergleich mag diese Erweiterung der Methodik erläutern.

Man stelle sich vor, daß die Herstellungsverfahren und die Kenntnis der Phenolharze verlorengegangen seien. Ein Chemiker, der einen Block dieses Materials findet, stellt sich die Aufgabe, die Konstitution dieser Substanz zu ermitteln. Sein Arbeitsplan wird zunächst ungemein dürftig sein. Er steht wie der Ligninchemiker zunächst einem fensterlosen Monolithen gegenüber. Nur langsam wird es ihm gelingen, eine ungefähre Zuordnung seiner Substanz in das System der organischen Verbindungen zu finden. Schließlich aber dringt er so weit in die Materie ein, daß er sich Vorstellungen über ihre Entstehung machen kann. Sobald er den Gedanken erfaßt, daß sie ein Kondensat von Phenolen mit Aldehyden sein könnte, gewinnt er freie Fahrt. Jetzt kann er mit definierten Ausgangs-materialien die ersten und später die weiteren Stufen der Synthese unter-suchen und auf diesem Umwege endgültige Vorstellungen über die Kon-stitution seines Materials gewinnen.

VIII. Vorstufen der Ligninbildung. Die sekundären Bausteine.

Im vorigen Kapitel wurde geschildert, wie das Dehydrierungspoly-merisat (DHP) des Coniferylalkohols entsteht. Zwischen der Zusammen-gabe der Reaktionsteilnehmer und dem Beginn der Ausscheidung ver-gehen meistens einige Stunden. Dieses Stadium der Reaktion wurde papierchromatographisch untersucht. Es zeigte sich, daß der Coniferyl-alkohol verschwindet, wenn ein Teil ausgeflockt ist. Er wird erkannt am R_F-Wert und der violetten Farbe, die auftritt, wenn das Chromato-gramm mit verdünnter Diazobenzolsulfonsäurelösung besprüht wird. Nach seinem Verschwinden enthält die filtrierte Lösung im wesentlichen vier Fraktionen A—D, deren R_F-Werte in Wasser, das mit Benzol gesättigt ist, unten zusammengestellt sind (*Tabelle 1*).

Schüttelt man in diesem Stadium die zentrifugierte Reaktionslösung wiederholt mit kleineren Mengen Methylenchlorid aus, so wird haupt-sächlich Fraktion A und B mit der mengenmäßig geringen Fraktion D extrahiert, während der Hauptanteil der Fraktion C erst durch wieder-holtes Ausschütteln mit größeren Mengen Methylenchlorid völlig der Lösung entzogen wird. Aus den ersten Fraktionen läßt sich nach Ein-engen und Zusatz von Petroläther die Hauptmenge der Fraktion A in

Tabelle 1. Vorstufen der künstlichen Ligninbildung.

Substanz	Formel	R_F	Farbe (Diazoreaktion)	Geschätzte Menge (%)
Fraktion A, Dehydro-diconiferylalkohol	(XIII)	0,57	stumpfes Rot	45
Fraktion B, DL-Pinoresinol..........	(XIV)	0,51	stumpfes Rot	15
Fraktion C......................		0,67	rötlichgelb	20
Fraktion D		0,45	keine; fluoresziert	10
Olivil........................	(XV)	0,67	rot	
Coniferylalkohol	(IX)	0,57	violett	
β-[2-Methoxy-4-(γ-oxypropenyl)-phen-oxy]-coniferylalkohol	(XVI)	0,5	rotbraun	

kristallisierter Form abscheiden. Sie besteht aus Dehydro-diconiferyl-alkohol (XIII) (57). Die Mutterlauge wird in Aceton übergeführt und in Gegenwart von Bicarbonat mit Dinitrofluorbenzol umgesetzt. Dabei scheidet sich der Dinitrophenyläther des DL-Pinoresinols (XIV) kristallinisch ab (65); aus ihm kann das DL-Pinoresinol gewonnen werden.

Die Substanz C ist noch nicht kristallinisch isoliert worden. Ihr R_F-Wert ist dem des Olivils (XV) gleich. Sie könnte daher wie dieses ein Dimeres mit vier Hydroxylgruppen sein. Vom Olivil unterscheidet sie sich jedoch durch die Farbreaktion mit Diazobenzolsulfonsäure. An Modellsubstanzen wurde festgestellt, daß der gelbe Ton stets auftritt, wenn dem Guajacylrest eine Carbinolgruppe benachbart ist. Für die Substanz C, die empfindlicher ist als die übrigen, kommen unter anderem folgende Formulierungen in Betracht: ein Pinoresinol, bei dem ein dem Guajacylrest benachbartes Wasserstoffatom durch OH ersetzt ist [Halbacetal; vgl. AULIN-ERDTMAN, BJÖRKMAN, ERDTMAN und HÄGGLUND (14)]; ein durch Hydrolyse halbseitig geöffnetes Pinoresinol; ein Guajacylglycerin, in dem das mittelständige Hydroxyl der Seitenkette mit dem Phenolhydroxyl des Coniferylalkohols veräthert ist (Hydrat von XVI; vgl. ADLER (2, 6, 7)]. Die Fraktion D kann monomolekular sein. Sie tritt der Menge nach hinter den anderen Fraktionen zurück und enthält unter anderem eine aldehydische Komponente.

Wenn wir den (dehydrierten) Coniferylalkohol als den primären Baustein des Lignins erkennen, so können die (dehydrierten) Substanzen A—D als *sekundäre Bausteine* bezeichnet werden.

Die Substanzen A—D sind optisch inaktiv. Die Konstitution des Dehydro-diconiferylalkohols wurde durch Umwandlung in das hydrierte

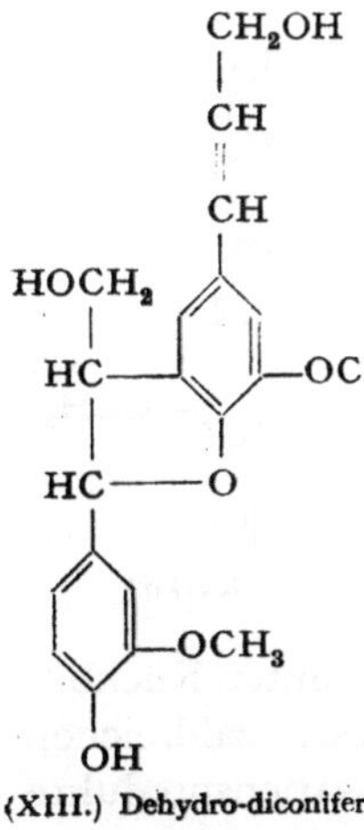

(XIII.) Dehydro-diconiferylalkohol. 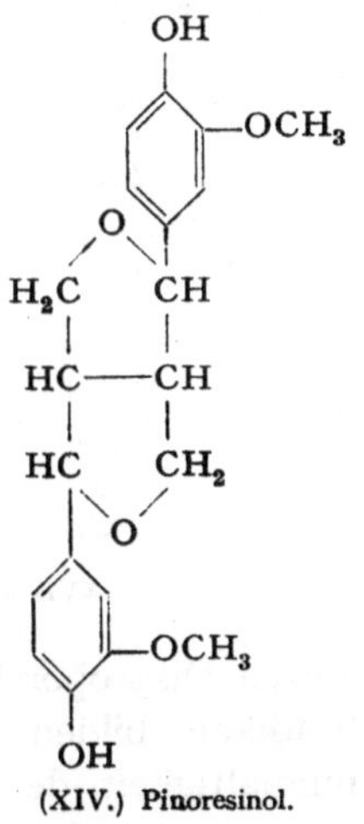   (XIV.) Pinoresinol. 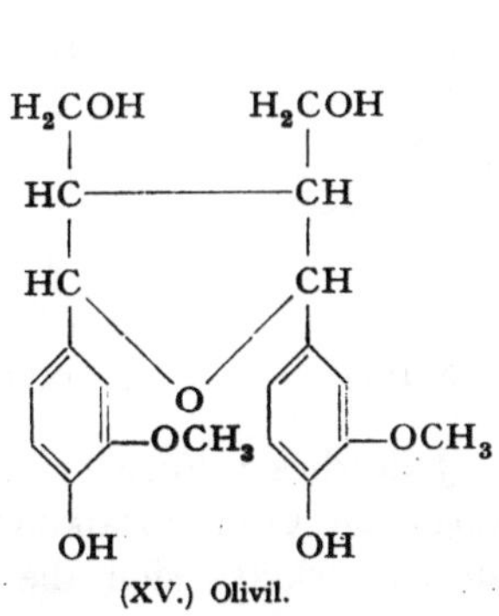   (XV.) Olivil.

und methylierte Dehydro-diisoeugenol (VIII) gesichert (*57*). Der Dimethyläther des *DL*-Pinoresinols ist identisch mit dem hälftigen Gemisch des
natürlichen Eudesmins (*L*-Pinoresinol-dimethyläther) mit dem Dimethyläther des *D*-Pinoresinols (*65*). Außerdem wurde die Konstitution des
DL-Pinoresinols durch Synthese bekräftigt (*50*).

Das in der Tabelle 1 (S. 60) angeführte Olivil ist natürlicher Herkunft, während
die Substanz (XVI) ein synthetisches Produkt ist (*61*).

Das von uns verwendete Pilzferment ist eine *Phenolredoxase*, deren
Wirkung auf den Coniferylalkohol zunächst in der Wegnahme des Phenol-
wasserstoffatoms besteht. Das entstandene Radikal kann in vier mesomeren
Formen (Resonanzformen) ausgedrückt werden [(XVII a—d); vgl. ERDT-
MAN (*33, 34*)].

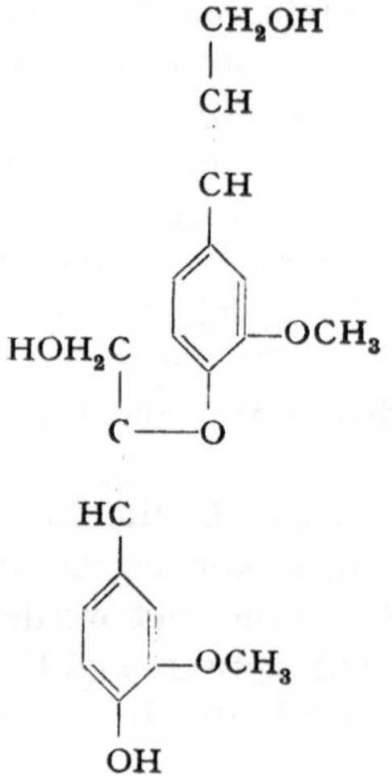

$$CH_2OH$$
$$|$$
$$CH$$

$$CH$$

HOH$_2$C ——OCH$_3$
|
C —— O

HC

——OCH$_3$

OH

(XVI.) *β*-[2-Methoxy-4-(*γ*-oxypropenyl)-phenoxy]-coniferylalkohol.

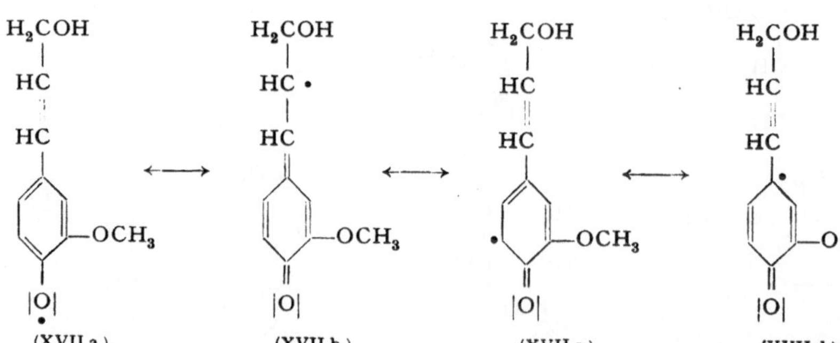

H_2COH	H_2COH	H_2COH	H_2COH
HC	HC •	HC	HC
HC	HC	HC	HC
——OCH$_3$	——OCH$_3$	——OCH$_3$	——OCH$_3$
|O|	|O|	|O|	|O|
(XVII a.)	(XVII b.)	(XVII c.)	(XVII d.)

Jede dieser Formen kann durch Disproportionierung unter Rückbil-
dung von Coniferylalkohol Diradikale bilden. Aus diesen zahlreichen
Formen erklärt sich die Mannigfaltigkeit der Kondensationsprodukte

des Coniferylalkohols und außerdem ihre optische Inaktivität; denn die Wirkung des Ferments ist mit der ersten Stufe, der Wegnahme des Phenolwasserstoffs, beendet. Die weiteren Reaktionen verlaufen automatisch, d. h. ohne Mitwirkung des Ferments. Die Bildung des Dehydrodiconiferylalkohols ist zu erklären aus der Zusammenlagerung von *b* und *c* oder des Coniferylalkohols mit einem aus *c* entstandenen Diradikal. *DL*-Pinoresinol entsteht durch Zusammenlagerung von zwei Radikalen *b*.

Der Sinapinalkohol erleidet, wie oben schon ausgeführt ist, durch die Pilzredoxase eine tiefgreifende Veränderung. Aber er wird durch eine andere Dehydrase mit größter Leichtigkeit in ein dimeres Dehydrierungsprodukt übergeführt, das dem Pinoresinol entspricht und Syringaresinol (XVIII) genannt wurde. Diese Redoxase ist im rohen Emulsin enthalten. Als versucht wurde, das Glucosid des Sinapinalkohols, das Syringin*, mit rohem Emulsin zu spalten, entstand statt des Sinapinalkohols das Syringaresinol. Auch Sinapinalkohol selbst wird durch dieses Fermentgemisch in Syringaresinol verwandelt. Diese Mandelredoxase ist keine Phenolredoxase. Das *DL*-Syringaresinol, das inzwischen auch synthetisch zugänglich geworden ist, wird durch Pilzredoxase ebenso wie der Sinapinalkohol in dunkelbraune wasserlösliche Produkte verwandelt. Wenn es jedoch zusammen mit Coniferylalkohol (1 : 1 Gew.-Teil) der Pilzredoxase ausgesetzt wird, bildet sich ohne Dunkelfärbung ein schönes Dehydrierungspolymerisat, das beide Komponenten enthält. Es verhält sich demnach gegenüber Pilzredoxase wie der Sinapinalkohol.

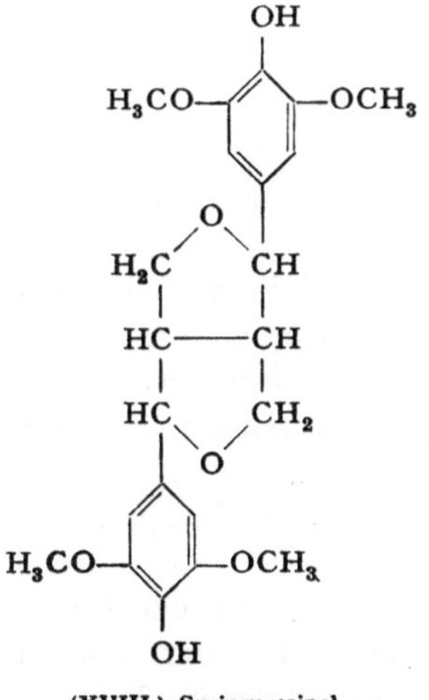

(XVIII.) Syringaresinol. 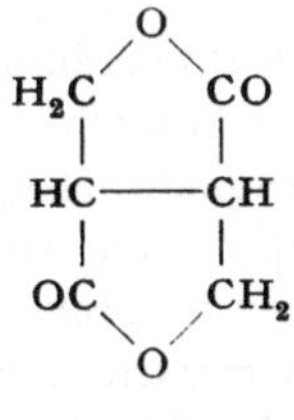   (XIX.) Dimethylol-bernsteinsäure-dilacton.

* Syringin findet sich nach REINITZER (*119*) in der Rinde von *Lonicera, Robinia pseudacacia* und der Oleaceen *Syringa, Ligustrum, Fraxinus, Olea Forsythia, Phillyrea* und *Jasminum*.

IX. Verknüpfung der sekundären Bausteine zum Lignin.

Wenn man Dehydro-diconiferylalkohol (XIII) für sich allein (57) und ebenso DL-Pinoresinol (XIV) für sich allein (49, 65, 50) der Wirkung der Pilzredoxase aussetzt, so entstehen in beiden Fällen ligninartige amorphe Dehydrierungspolymerisate in guter Ausbeute. Sie sind wasserstoffärmer als ihre Ausgangsmaterialien. Somit ergibt sich folgendes Bild von der Entstehung des Fichtenlignins.

Der Coniferylalkohol, den wir als den primären Baustein bezeichnen, wird unter Dehydrierung zunächst in die sekundären dimeren Bausteine A, B und C verwandelt, zu denen die vielleicht monomere Fraktion D hinzukommt. Sie sind ihrerseits der Dehydrierung fähig und bilden dabei ein höhermolekulares gemischtes DHP, das Lignin. Bisher liegen nur unbestimmte, unten erwähnte Anhaltspunkte dafür vor, welche Wasserstoffatome bei dieser zweiten Reaktion, der Dehydrierung der sekundären Bausteine, austreten, oder, was dasselbe ist, welcher Art die Bindungen sind, durch welche die sekundären Bausteine miteinander verknüpft werden; soviel aber dürfte feststehen, daß die speziellen Gruppierungen der sekundären Bausteine auch im endgültigen Kondensat im wesentlichen erhalten bleiben.

Es ist nicht ausgeschlossen, daß neben der Dehydrierungspolymerisation in geringem Umfang eine von den Radikalen eingeleitete Polymerisation des Coniferylalkohols selbst einhergeht, und daß solche vermutlich nach Art der Polystyrole gebauten Stücke in das DHP eingebaut sind. Die Wasserstoffbilanz verlangt jedoch, daß die Dehydrierung bei weitem überwiegt.

Erinnern wir uns daran, welche Gruppierungen aus der analytischen Untersuchung des Lignins, der Ligninsulfosäure und anderer Derivate erschlossen oder wahrscheinlich gemacht worden sind. Da ist das Phenylcumaran-system zu nennen: es findet sich im Dehydro-diconiferylalkohol, der außerdem für sich allein Formaldehyd abspaltet und als eingebautes Bauglied des Lignins zum mindesten für einen Teil des abspaltbaren Formaldehyds verantwortlich ist. Des weiteren werden aus der analytischen Ligninchemie aliphatische Äther von Phenylcarbinolen gefordert: sie sind in der Pinoresinol-komponente vorhanden. Es ist bekannt, daß das Pinoresinol ähnlich wie die entsprechende Gruppierung des Lignins mit Bisulfit reagiert. Möglicherweise ist in der Komponente C gleichfalls eine Äthergruppierung enthalten. Ohne Zweifel aber findet sich in ihr jenes freie Phenylcarbinol, das durch die analytische Untersuchung des Lignins gefordert wird. Auch für die wenigen Prozente einer aldehydischen Komponente des Lignins findet sich der erwartete Baustein, und zwar in der Fraktion D. Wenn man die Annahme machen will, daß bei der unter Dehydrierung verlaufenden Zusammenschweißung der sekundären Bausteine die Zahl und der Charakter der Hydroxyl- und Äthergruppen nicht wesentlich verändert wird, so kann man aus der Kenntnis des Lignins oder DHP einerseits und den Komponenten *A* und *B* anderseits die ungefähre Zusammensetzung der Komponenten *C* und *D* berechnen. Dabei ergibt sich, daß sie hydroxyl-reicher sein müssen als *A* und *B* und daß darin eine Substanz vorkommt, die etwa dem entspricht, was oben (S. 61) für die Substanz *C* vermutungsweise geäußert worden ist. Das monomere Spaltprodukt von SCHORYGINA könnte der Fraktion *C* oder *D* entstammen. SCHORYGINA gibt für das 1-Guajacyl-propanol-2 13% des Lignins an (*127*, *128*); der genauen Bestimmung dieser Zahl kommt einige Bedeutung zu.

Viele Mühe wurde darauf verwendet, die Pinoresinol-komponente im Fichtenlignin und die Syringaresinol-komponente im Buchenlignin unmittelbar festzustellen. Eine Handhabe hierfür bietet sich durch eine von ERDTMAN und GRIPENBERG (*35*) gefundene Abbaureaktion des Pinoresinols, die auch auf das Syringaresinol anwendbar ist. Durch geeignete Oxydation des methylierten und bromierten Pinoresinols wurde dessen inneres Ringsystem in Gestalt des Dimethylolbernstein-säure-dilactons (XIX) herausgearbeitet. Das Syringaresinol verhält sich entsprechend (*49*). Die Versuche, dieses Dilacton aus dem Fichten- oder Buchenlignin zu gewinnen, sind jedoch mißlungen. Die Erklärung hierfür hat die enzymatische Dehydrierung des Pinoresinols ergeben: das Pinoresinol-DHP läßt sich gleichfalls nicht zum Dilacton abbauen. Offenbar wird bei der Dehydrierung des Pinoresinols das innere Gerüst an irgendeiner Stelle angegriffen und hierdurch die besagte Reaktion unmöglich gemacht. Diese Feststellung gibt einen ersten Anhalt für die Art, wie ein sekundärer Baustein, das Pinoresinol, durch erneute Dehydrierung in das Lignin eingebaut sein kann. Ein weiterer Hinweis für die Verknüpfung von sekundären Bausteinen untereinander ist der durch Vergleich mit Modellen auf optischem Wege gewonnene Befund von AULIN-ERDTMAN, daß im Lignin etwa 5% der Coniferylbausteine durch Diphenylbindung in Nachbarschaft zum Phenolhydroxyl verknüpft sein könnten.

Für die Verknüpfung der sekundären Bausteine ist oben die vereinfachende Annahme gemacht worden, daß hierbei die Hydroxyl- und Ätherfunktionen dieser Bausteine im wesentlichen unverändert bleiben. Dies kann nur zutreffen, wenn die Zahl der Phenolgruppen und alkoholischen Hydroxylgruppen ungefähr den Werten entspricht, die mit den geschilderten präparativen Methoden ermittelt worden sind. Wenn dagegen der geringere, auf optischem Wege ermittelte Gehalt an Phenolgruppen zutrifft, so müssen die Vorstellungen über den Bau des Lignins abgeändert werden.

Die hier entwickelte Vorstellung läßt eine kettenförmige Kombination zu, die verzweigt sein kann. Ein verzweigtes System ist wahrscheinlicher als ein unverzweigtes. Man kann sich Gedanken darüber machen, ob ein solches System, das in Wasser unlöslich sein muß, von den Fermenten durch weitere Dehydrierung höher kondensiert werden kann. Vielleicht wird es durch andere Zellbestandteile zunächst in kolloider Lösung gehalten und dann durch Fermente weiter dehydriert. Es hat derart viele empfindliche Stellen, daß es sich auch beim Lagern durch Wasserentzug oder durch Luftoxydation weiter kondensieren kann.

X. Beziehung der natürlichen Ligninarten untereinander und zum künstlichen Lignin.

Die wichtige Frage, ob die Bildung des Lignins in der Pflanze auf dieselbe Weise wie die des künstlichen Lignins verläuft, und zwar über dieselben sekundären Dehydrierungsprodukte, kann für das Fichtenlignin bejaht werden. Erstens hat sich gezeigt, daß die Redoxasen der Cambialzone der Fichte dasselbe leisten wie die Pilzredoxase. Darüber wird weiter unten berichtet. Zum anderen haben sich im Methylenchloridauszug des frischen Cambialsaftes der Fichte außer freiem Coniferylalkohol sämtliche vier Fraktionen A bis D feststellen lassen (44)*. Sie kommen nur in geringer Menge vor, weil sie vermutlich sofort weiterverarbeitet werden. Aber sie konnten ausreichend durch die R_F-Werte und Farbenreaktionen identifiziert werden. Es kann daher ausgesprochen werden, daß *nicht nur die Bildung des künstlichen, sondern auch des natürlichen löslichen Lignins über die geschilderten Zwischenstufen verläuft.*

In das Verhältnis des löslichen Lignins zum unlöslichen gewinnt man einen Einblick durch das „Dioxan-Salzsäure-Lignin", das offenbar ein Spaltstück des unlöslichen Lignins ist. Der Chlorgehalt läßt auf eine Molekülgröße von 1000—3000 schließen. Das ist dieselbe Größenordnung wie die des löslichen Lignins, dem es auch in den Löslichkeitseigenschaften entspricht. Die Untersuchung der Hydroxylgruppen ist noch nicht genügend durchgeführt. Das Gesamtlignin ist in der Ligninsulfosäure vertreten. Ihre Zusammensetzung läßt den Schluß zu, daß sich das unlösliche Lignin vom löslichen im wesentlichen nur durch die Molekulargröße unterscheidet (s. dagegen 65a).

Die verschiedenen Ligninpräparate stehen somit wahrscheinlich in folgendem Verhältnis zueinander.

Das lösliche Lignin, das nur wenige Prozent des Gesamtlignins bildet, unterscheidet sich von dem Hauptteil, dem unlöslichen Lignin, im wesentlichen nur durch das geringere Molekulargewicht. Das künstliche Lignin ist zur Hauptsache in organischen Lösungsmitteln löslich und steht dem natürlichen löslichen Lignin außerordentlich nahe. Der geringe unlösliche Anteil des künstlichen Lignins ist wegen seiner Ver-

---

* MANSKAJA (106) gibt an, im Cambialsaft der Coniferen Coniferylalkohol und Vanillin durch Farbreaktionen festgestellt zu haben. Vanillin ist uns nicht begegnet.

mengung mit Anteilen des Ferments noch nicht untersucht worden. Das Dioxan-Salzsäure-Lignin ist ein Spaltstück des unlöslichen Lignins und hat große Ähnlichkeit mit dem löslichen natürlichen Lignin. Das methylierte Lignin, das aus Methylholz durch Ameisensäure-Chlorwasserstoff abgetrennt wird, entspricht dem löslichen Lignin und dem Dioxan-Salzsäure-Lignin.

XI. Die Bindung des Lignins im Holze.

Die soeben gegebene Übersicht geht von der Vorstellung aus, daß der Hauptanteil des Lignins an sich in organischen Lösungsmitteln unlöslich ist und durch gelinde chemische Mittel unter Zerlegung teilweise in lösliche Substanzen umgewandelt werden kann. Wenn diese Vorstellung zutrifft, so braucht man zur Erklärung für das Verhalten des Lignins im Holze keine Bindung zwischen Lignin und Kohlehydraten anzunehmen. Zugunsten einer solchen Bindung lassen sich eine Anzahl Gründe anführen, die aber bei näherer Betrachtung zum Teil an Beweiskraft verlieren.

Es ist längst bekannt, daß aus Fichtenholz, mag es noch so fein verteilt sein, durch cellulose-lösende Mittel, wie Cellulase, ammoniakalische Kupferlösung, Xanthogenatmischung oder heiße 80proz. Calciumrhodanid-lösung, nur geringe Anteile in Lösung gebracht werden können. Das gilt auch für heißes Wasser, wenn es nicht bei solchen Temperaturen angewendet wird, bei denen einzelne Holzbestandteile zersetzt werden. Auch methyliertes Holz läßt sich nicht durch Lösungsmittel aufteilen. Die mehr oder weniger großen Anteile, die bei allen diesen Versuchen in Lösung gebracht werden, enthalten meistens die Lignin- und Polysaccharidkomponenten in ungefähr demselben Verhältnis, wie sie im Holze vorliegen. Mit Cellulase läßt sich bestenfalls eine teilweise Anreicherung des Ligninanteils im Rückstand erreichen (*63, 64, 112—118*). Man kann diese Beobachtungen mit einer Bindung zwischen Lignin und Polysacchariden erklären. Man kann aber auch annehmen, daß die Polysaccharide nicht aus dem Holzgefüge herausdiffundieren können, und daß die geringen gelösten Anteile Partikelchen sind, in denen das Lignin durch die anderen Holzbestandteile in Lösung gehalten wird. In vielen Fällen enthalten diese gelösten Anteile nicht mehr die Holzbestandteile in unveränderter Form. Am löslichen Lignin lassen sich Peptisierungen in Wasser leicht beobachten. Man braucht nur eine Acetonlösung mit Wasser zu versetzen und das Aceton bei gewöhnlicher Temperatur zu entfernen, um wahrzunehmen, daß ein erheblicher Anteil in der wäßrigen Lösung bleibt. Wenn demnach aus dem löslichen Lignin in so einfacher Weise kolloide wäßrige Lösungen erhalten werden können, so ist es vorstellbar, daß auch unlösliches Lignin durch Schutzkolloide in Lösung gehalten werden kann.

Holzzerstörende Pilze vermögen in monatelanger Einwirkung das Holz so zu verändern, daß wesentliche Anteile des Lignins in organischen Lösungsmitteln löslich werden [NORD (*28a*; *128a—d*; *99a—c*; *129a—d*); TSCHUDAKOFF (*133*)]. Hier kann es sich um die Freilegung an sich löslicher Ligninanteile handeln, die erst durch den Abbau der übrigen Holzbestandteile für das Lösungsmittel zugänglich werden. Es ist auch möglich, daß mit der Tätigkeit des Pilzes ein oxydativer Abbau des unlöslichen Ligninanteils verbunden ist, durch den Ligninanteile löslich werden, obwohl der Abbau unter der Grenze der analytischen Wahrnehmbarkeit liegen kann. Eine andere Erklärung wäre, daß durch die Wirkung der holzabbauenden Mikroorganismen etwa vorhandene Lignin-Polysaccharidbindungen gesprengt werden.

Das Lignin ist in der Gegend der Mittellamelle* sehr stark angereichert. Dies hat BAILEY (*15, 16*) festgestellt. Durch Mikromanipulation hat er aus dem Bereich der Mittellamelle Anteile isoliert, die bis zu 71% Lignin enthielten. Optische Versuche von LANGE (*100, 101*) haben gleichfalls ergeben, daß in diesem Teil des Gewebes das Lignin sehr stark angereichert ist. Hier stößt die Annahme einer Verbindung des Lignins mit Polysacchariden bereits auf Schwierigkeiten durch die mengenmäßige Verteilung der beiden Komponenten.

Auch wenn man die Cellulose betrachtet, entstehen Schwierigkeiten, vor allem in dem Teil, in dem sie zu festen Kristalliten gepackt ist, in denen kein Platz für einen Fremdkörper ist. Ein Blick auf elektronenmikroskopische Aufnahmen der delignifizierten und der intakten Holzwand [z. B. von HODGE und WARDROP (*83*)] lehrt, daß der größte Teil des Lignins auf und zwischen die Fibrillen gepackt und eine Verbindung beider Anteile schon aus topochemischen Gründen schwer vorstellbar ist [vgl. BAILEY (*16*)]. Dazu kommt folgende chemische Überlegung. Man kennt bisher keine glucosidischen Verbindungen von Polysacchariden mit irgendwelchen nicht zuckerartigen Aglykonen. Man wird wohl annehmen dürfen, daß vor der Verholzung das Gewebe der Cellulose bereits fertig gebildet ist. Eine nachträgliche glucosidische Verbindung zwischen Cellulose und dem zwischen die Fibrillen oder auch in die Oberflächen der Fibrillen hineingepackten hochmolekularen Lignin ist schlechterdings nicht vorstellbar.

Man kann einwenden, daß statt der Cellulose auch die Hemicellulosen mit dem Lignin verbunden sein könnten [HARRIS (*80*)]. Obwohl diese Vorstellung eher annehmbar ist, vermag man sich auch hier eine Glucosidierung des Polysaccharids mit dem hochmolekularen Lignin kaum vorstellen. Es ist oben erwähnt worden, daß Cuproxamlignin oder das nach dem Ameisensäureverfahren hergestellte Methyllignin mit erheblicher Zähigkeit Kohlehydratanteile festhalten. Methyliertes Cuproxamlignin, das noch

---

* Schicht zwischen den Holzzellen.

einige Prozent Kohlehydratanteil enthielt, wurde hydrolysiert. Der methylierte Kohlehydratanteil bestand aus 2,3,6-Trimethylglucose, rührt also von der Cellulose her (40). Es besteht der Verdacht, daß Celluloseanteile oder -bruchstücke während der Isolierung dieser Präparate, bei der Säuren verwendet werden müssen, in das Lignin einkondensiert werden.

Um sich eine chemische Bindung zwischen Lignin und Polysacchariden vorzustellen, muß man sehr spezielle Erklärungsversuche anstellen. Sie wurden oben bereits angedeutet. Es könnte sein, daß chinonmethidartige Zwischenstufen des Lignins mit Hydroxylen von Kohlehydraten zu Ätherbindungen reagieren, an denen Phenylcarbinolgruppen des Lignins beteiligt wären. Es wäre auch denkbar, daß das halbacetalartige Hydroxyl einer möglichen Oxypinoresinolgruppe, geeignete Quellungszustände vorausgesetzt, mit dem Hydroxyl einer benachbarten Polysaccharidkette unter Wasseraustritt reagiert. Eine Verbindung zwischen Lignin und Polysacchariden wird man demnach nicht in glucosidartigen Bindungen suchen dürfen, sondern in Reaktionen der geschilderten Art, an denen alkoholische Hydroxyle der Polysaccharide beteiligt sein könnten. Versuche, lösliche Polysaccharide während der Bildung des künstlichen Lignins einzukondensieren, haben bisher zu keinem Ergebnis geführt. Vielleicht ist es falsch, die Alternative: Bindung des Lignins an die Polysaccharide oder keine Bindung, zu stellen. Wenn solche Bindungen vorliegen, so dürften sie wahrscheinlich nur einen Teil des Lignins erfassen, so daß möglicherweise beide Auffassungen zutreffen.

Wertvolle Beiträge zu der Struktur der verholzten Faser finden sich bei FREY-WYSSLING und MÜHLETHALER (72 a) sowie BUCHER (24 a) und BUCHER und WIDERKEHR-SCHERB (24 b).

XII. Der Vorgang der Verholzung (66, 47, 44).

Das Mittelstück in der Abb. 1 zeigt einen Schnitt durch den Bereich des Cambiums einer Conifere. Links ist die Rinde, rechts das Holz. Unmittelbar innerhalb des mit Zellkernen versehenen Cambiums, also im Bilde rechts vom Cambium, finden sich die jungen, noch nicht verholzten Zellen. Dann folgt ein Gebiet, in dem die Zellen kräftiger werden, rechts im Bilde sind sie verholzt und haben ihre endgültige Form erreicht. Im unteren Abschnitt des Bildes ist das Gewebe mit Phloroglucin-Salzsäure betupft. Lignifizierte Zellen werden mit dunkelroter Farbe sichtbar. Ihnen folgen (von rechts nach links) einige Zellreihen schwacher Lignifizierung, dann hört gegen das Cambium hin das Lignin völlig auf.

In dem saftreichen Gewebe beiderseits des Cambiums und des Cambiums selbst befinden sich, wie schon lange bekannt, Redoxasen. Unter anderen hat auch MANSKAJA (104—107) ausgesprochen, daß sie mit der Ligninbildung zusammenhängen. Diese Fermente sind teils im Zellsaft gelöst, teils im Gewebe fixiert (66). Zur Zeit der Holzbildung und kurz davor befindet sich in demselben lebenden Gewebe in reichen Mengen das Glucosid Coniferin. Es wird durch Redoxasen nicht angegriffen. Den zur Ligninbildung benötigten Coniferylalkohol kann nur eine Glucosidase aus dem Coniferin in Freiheit setzen, aber es ist nicht gelungen, in dem Saft des Cambiums und seiner Umgebung eine wasserlösliche Glucosidase nachzuweisen.

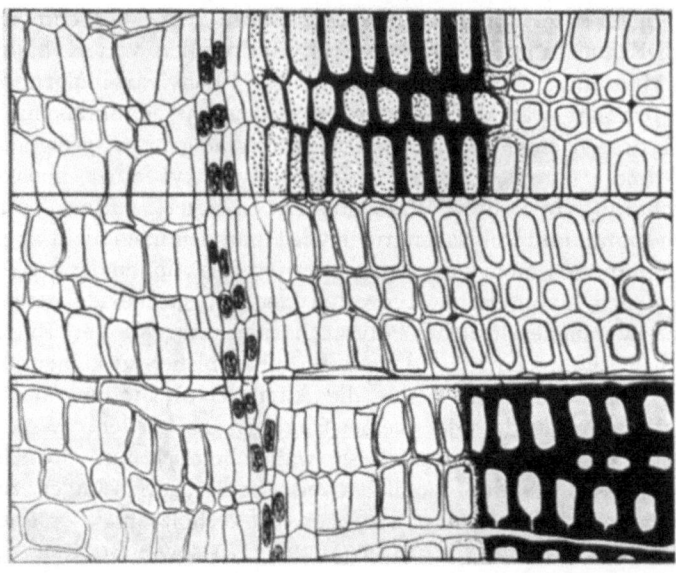

Abb. 1. Querschnitt durch das Cambium und die benachbarten Zellen einer Conifere (schematisiert). Oberer Teil mit Indicanlösung betupft, unterer Teil mit Phloroglucin-Salzsäure angefeuchtet.

Um eine zellgebundene Glucosidase festzustellen, wurde eine verdünnte Indicanlösung benutzt (66). Eine solche Lösung ist schwachgelb. Wenn Glucosidase anwesend ist, wird das Indican in Indoxyl und Glucose gespalten. Das Indoxyl verwandelt sich am Luftsauerstoff alsbald in Indigo. BEIJERINCK (17) hat auf diese Weise den Sitz des Emulsins in Mandeln untersucht. Tatsächlich trat beim Betupfen des Holzschnittes mit Indicanlösung nach einiger Zeit Indigo auf (oberer Teil der Abb. 1). Der Indigo zeigt den Sitz einer Glucosidase an, die in jenen Zellen streng fixiert ist, in denen die Holzbildung vor sich geht. Wo das Holz fertiggestellt ist (rechts im Bilde), wird keine Glucosidase mehr gefunden. Der Abfall der Glucosidase gegen das fertige Holz hin ist auffallend schroff. Folgt man den Zellreihen, in denen der Indigoeffekt auftritt und demnach Glucosidase nachweisbar ist, nach unten in das mit Phloroglucin-Salzsäure angefärbte Gebiet, so sieht man, daß

Abb. 2. Querschnitt durch *Araucaria*. Links mit Phloroglucin-Salzsäure, rechts mit Indicanlösung benetzt.

beide Effekte sich teilweise überdecken. Da, wo der Indigoeffekt am stärksten ist, läßt sich auch bereits Lignin nachweisen.

Abb. 2 (S. 71) zeigt die Aufnahmen von zwei Schnitten durch einen einjährigen Trieb von *Araucaria excelsa.* Der linke ist mit Phloroglucin-Salzsäure benetzt. Das dunkelrot angefärbte Lignin ist in lockerer Verteilung sichtbar im Mark und in sehr dichter Ablagerung im Holz. Die Verholzung reicht nicht ganz an das Cambium. In der äußeren Rinde findet sich ein unregelmäßiger, unterbrochener Ring von verholzten Zellen. Das rechte Bild zeigt einen entsprechenden Schnitt nach der Betupfung mit Indicanlösung. Innerhalb des Cambiums liegen wenige ungefärbte Zellen. Dann beginnt eine nach innen zunehmende Blaufärbung, die am fertigen Holz plötzlich abbricht. Entlang den Markstrahlen dringt die Färbung tiefer ein. Im Mark zeigt sich locker verteilte Blaufärbung. In der Rinde erscheint eine unregelmäßige blaugefärbte Zone.

Andere Coniferen, ferner z. B. die Liliacee *Cordyline congesta (66)* verhalten sich entsprechend. Verschiedene Laubhölzer zeigen den Indigoeffekt, andere nicht oder undeutlich.

Die *Verholzung* vollzieht sich demnach folgendermaßen: Unmittelbar außerhalb der fertigen Holzzellen, da, wo die Verholzung im Gange ist, befindet sich eine Glucosidase. Diese spaltet das im Zellsaft vorhandene Coniferin, setzt den Coniferylalkohol in Freiheit, der nunmehr von den Redoxasen in Lignin verwandelt wird. Wie schon erwähnt, werden dabei dieselben Durchgangsstufen durchlaufen, die auch in vitro festgestellt sind: Außer dem Coniferylalkohol selbst sind die vier sekundären Bausteine *A* bis *D* in geringer Menge im frischen Cambialsaft nachweisbar. Das abrupte Verschwinden der Glucosidase in dem fertigen Holz beruht wohl auf der Überdeckung des Ferments mit Lignin, durch die zugleich das Leben in der Zelle aufhört.

Daß im Gebiet der Verholzung eine Glucosidase postuliert und auch gefunden wurde, ist eine starke Stütze für die hier vertretenen Anschauungen von der Holzbildung und letzten Endes auch der Auffassung, daß das künstliche Lignin ein wirkliches Lignin ist.

Pflanzengewebe kann in vitro gezüchtet werden [Gautheret (73, 73a), Jacquiot (91—93a)]. In einer Gewebekultur von Carotten-cambium wird Verholzung nur andeutungsweise beobachtet; wird jedoch nach Wacek (134) der Kulturflüssigkeit Coniferin beigegeben, so vermehrt sich die Verholzung in deutlicher Weise.

Um weitere Einblicke in die Holzbildung zu gewinnen, wurde Coniferin synthetisiert, das am endständigen Kohlenstoffatom der Seitenkette radioaktiv markiert ist [(47); Kratzl (98, 99); Brown, Tanner und Stone (24)].

Hiervon wurden in der ersten Maihälfte 1—2 mg unter die Rinde eines vier-
jährigen Fichtenstämmchens im Trieb des zweiten Jahres eingeschoben
(47). Die senkrechte Schnittwunde wurde mit Baumwachs verklebt. Im
August wurde der Baum untersucht. Das radioaktive Material hatte sich
3 cm oberhalb und unterhalb des Schnitts verbreitet und war nur in der
Mitte ganz um das Stämmchen herumgelangt. Die Wundstelle war ver-
narbt, durch die Verletzung war in der Nähe dieser Stelle eine Schicht sehr

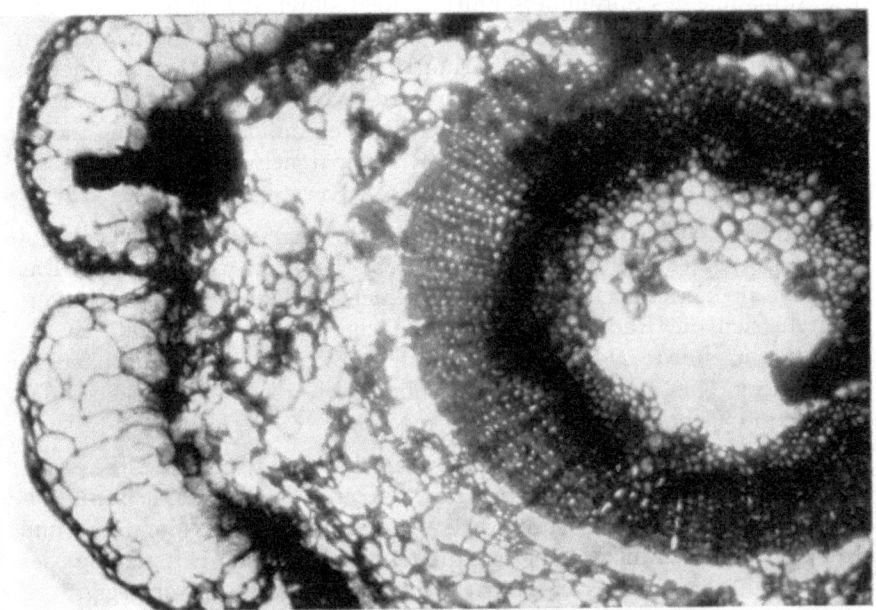

Abb. 3. Querschnitt durch das Cambialgebiet der Fichte *(Picea excelsa)* nach Zufuhr (durch die Nadeln) von
radioaktivem *D*-Coniferin.

dichter Holzzellen entstanden. Weiter oben und unten war das Gewebe
nahezu normal. Das radioaktive Material war dicht auf dem Holz des
Vorjahres fixiert, und zwar in dünner Zone, die innerhalb der ersten
zwei Wochen nach der Einlage des radioaktiven Materials gebildet sein
mußte; denn darüber lagerte sich eine dicke inaktive Holzschicht, die
in den Wochen und Monaten nach der Operation entstanden war. Die
radioaktiven Zellreihen wurden nach Möglichkeit herauspräpariert.
Wasserlösliche radioaktive Anteile waren nicht mehr oder nur noch in
Spuren vorhanden. Der radioaktive Anteil ließ sich nicht von dem
Lignin des Holzes trennen. Das radioaktive Holz wurde auf Salzsäure-
lignin verarbeitet. Es enthielt die gesamte Radioaktivität des Holzes.
Aus dem radioaktiven Lignin wurde Formaldehyd abgespalten. Er
war radioaktiv, wenn auch nicht in der erwarteten Stärke. Es gelingt

auch, die Lösung des radioaktiven Coniferins durch die Enden frischer Fichtentriebe aufnehmen zu lassen, wenn die Nadeln bis auf einen Stumpf abgeschnitten werden (*44*). Die radioaktive Substanz verbreitet sich in den benachbarten Teilen der kleinen Zweige und des Stämmchens.

Abb. 3 (S. *73*) zeigt den Querschnitt durch einen einjährigen Trieb von *Picea excelsa*. Dargestellt ist eine Radioautographie auf „stripping film". Die Aufnahme des Sproßquerschnitts erfolgt durch die dem Schnitt aufliegende belichtete und entwickelte Emulsion hindurch. Innen finden sich lose zerrissene Markzellen, darum Holzzellen des einjährigen Triebes. Bei Beginn der zweiten Wachstumsperiode wurde radioaktives Coniferin durch die Nadeln eingeführt. Über der dunklen radioaktiven Zone finden sich Zellen des zweiten Sommers. Im Rindenparenchym ist ein verholzter Sklerenchym-ring mit eingebauter radioaktiver Substanz sichtbar.

Derselbe Versuch wurde mit *L*-Coniferin angestellt, das aus radioaktivem Coniferylalkohol und *L*-Glucose bereitet worden war. Das *L*-Coniferin kann von der Glucosidase nicht zerlegt werden. Hier verbreitete sich die Radioaktivität in ähnlicher Weise wie das durch die Nadeln eingeführte *D*-Coniferin, aber die radioaktive Substanz blieb zu mindestens 90% wasserlöslich, offenbar weil die *D*-Glucosidase das *L*-Glucosid nicht zu verarbeiten vermochte.

Diese Versuche mit radioaktivem Material bestätigen die Auffassung, daß das natürliche *D*-Coniferin gespalten und der gebildete Coniferylalkohol zu Lignin verarbeitet wird. Lignin entsteht aus Coniferin und entsprechenden Glucosiden. Spekulationen über die Entstehung des Lignins, die dieser Tatsache nicht Rechnung tragen, sind abwegig.

XIII. Schlußwort.

Auch nachdem die Beziehung des Lignins zum Coniferylalkohol aus dem Stadium der Vermutung in das der Gewißheit getreten war, stand die Ligninchemie vor der Frage, ob dieser Naturstoff ein wirres, systemloses Gefüge aus Umwandlungsprodukten des Grundstoffes sei, ob das Riesenmolekül, um es drastisch auszudrücken, einem strukturlosen Schutthaufen verglichen werden muß, oder ob auch diese hochpolymere Substanz wie alle anderen einen geordneten Bauplan besitzt. Der Versuch [Erdtman (*33*); Brauns (*23a*)], auf analytischem Wege erkannte oder wahrscheinlich gemachte Einheiten in beliebiger Folge aneinanderzufügen, ist als ein Schritt auf diesem Wege anzusehen, aber er läßt kein ordnendes Prinzip erkennen. Heute beginnt sich ein solches abzuzeichnen. Der Coniferylalkohol und seine Verwandten verwandeln sich bei der Dehydrierung zunächst in die sekundären Bausteine, deren Bildung aus den primären

Bausteinen durchsichtig und verständlich ist. Die sekundären Bausteine enthalten bereits jene Gruppen, die mit der analytischen Chemie des Lignins im Einklang stehen. Die Verschweißung der sekundären Bausteine zu höheren Gebilden ist aus denselben Prinzipien zu verstehen, die von den primären Bausteinen, den Oxyzimtalkoholen, zu den sekundären Bausteinen führen. Der ordnende Gedanke, wenn man diesen teleologischen Ausdruck anwenden darf, ist erkannt und bestätigt worden. Er erklärt sowohl die Ordnung wie die Unordnung im Gefüge des Lignins, die sich in der erstaunlichen Mannigfaltigkeit der das Lignin bildenden Dehydrierungsprodukte der Oxyzimtalkohole kundgibt. Trotz aller Mannigfaltigkeit ist das Lignin ein polymerer Naturstoff wie jeder andere auch und sein Bauprinzip ist durchsichtig geworden. Die Substanz, die der Konstitutionsforschung ursprünglich weniger zugänglich erschien als irgendeine andere, steht nicht mehr abseits und ist der methodischen Behandlung durch die organische Chemie erschlossen. Die Forschung wendet sich mehr und mehr Einzelfragen zu, deren viele noch ungelöst sind.

Der große Gegenspieler des Lignins, die Cellulose, hat trotz des kristallklaren Aufbaus, der diese Substanz von allen anderen großen Naturstoffen abhebt, das Geheimnis der Entstehung noch nicht hergegeben. Beim Lignin ist dies anders. Das Zusammenspiel der Fermente bei der Verwandlung des zelleigenen Coniferins in Lignin im verholzenden Gewebe ist, für die Coniferen wenigstens, in großen Zügen aufgeklärt. Für die Laubhölzer fügt sich das eigentümliche Zusammenwirken des Coniferyl- und Sinapinalkohols dem Bilde ein, das schärfer zu umreißen die Aufgabe der Botanik sein wird.

Die Genese der primären Bausteine selbst, der Oxyzimtalkohole, bleibt ungeklärt. Während man der Genese der Glucose, des Bausteins der Cellulose, Schritt um Schritt näherkommt, liegt die Genese der Phenylpropankörper, der C_9-Gruppe, noch im Dunkeln. Daß sie ursprünglich aus den Zuckern entstehen, ist selbstverständlich, aber auf welchen Wegen oder Umwegen ist unbekannt. Die Frage nach der Entstehung der Zimtalkohole kann nur im Rahmen aller übrigen Phenylpropanderivate angegriffen und gelöst werden. Die Versuche, die Entstehung der lignin-bildenden Oxyzimtalkohole ad hoc aus anderen Holzbestandteilen abzuleiten, scheint ein müßiges Beginnen zu sein. Diese Frage steht auf einem viel breiteren Hintergrund und kann nur im Zusammenhang mit dem Problem der Bildung der Phenole in der Pflanze, insbesondere mit der C_7- und der C_{15}-Gruppe, angegriffen werden.

Naturwissenschaftliches Streben projiziert Ordnung in das Unbekannte, scheinbar Ungeordnete, Willkürliche. Es darf ausgesprochen werden, daß unser Naturstoff trotz vieler noch offener Fragen der ordnenden Erkenntnis gewonnen ist.

Literaturverzeichnis.

1. ADLER, E.: Privatmitteilung.
2. ADLER, E. and K. J. BJÖRKQVIST: Synthesis and Reactions of x-(3-Methoxy-4-hydroxyphenyl)-glycerol ("Guaiacylglycerol"). I. Preliminary Experiments. Acta Chem. Scand. **7**, 561 (1953).
3. ADLER, E., K. J. BJÖRKQVIST und S. HÄGGROTH: Über die Ursache der Farbreaktionen des Holzes. Acta Chem. Scand. **2**, 93 (1948).
4. ADLER, E. und L. ELLMER: Coniferylaldehydgruppen im Holz und in isolierten Ligninpräparaten. Acta Chem. Scand. **2**, 839 (1948).
5. ADLER, E. and B. O. LINDGREN: Some Aspects on Lignin Structure. Svensk Papperstidn. **55**, 563 (1952).
6. ADLER, E., B. O. LINDGREN and U. SAEDÉN: The β-Guacacyl Ether of α-Veratrylglycerol as a Lignin Model. Svensk Papperstidn. **55**, 245 (1952).
7. ADLER, E. and S. YLLNER: Synthesis and Reactions of α-(3-Methoxy-4-hydroxyphenyl)-glycerol ("Guaiacylglycerol"). II. Synthesis. Acta Chem. Scand. **7**, 570 (1953). III. Svensk Papperstidn. **57**, 78 (1954).
8. ASAHINA, Y.: Flechtenstoffe. Fortschr. Chem. organ. Naturstoffe **2**, 27 (1939).
8a. — Neuere Entwicklungen auf dem Gebiete der Flechtenstoffe. Fortschr. Chem. organ. Naturstoffe **8**, 207 (1951).
9. AULIN-ERDTMAN, G.: Spectrographic Contributions to Lignin Chemistry. Svensk Papperstidn. **47**, 91 (1944).
10. — Ultraviolet Spectroscopy of Lignin and Lignin Derivatives. Techn. Assoc. Pulp Paper Ind. (TAPPI) **32**, 160 (1949).
11. — Spectrographic Contributions to Lignin Chemistry. II. Svensk Papperstidn. **55**, 745 (1952).
12. — Spectrographic Contributions to Lignin Chemistry. III. Svensk Papperstidn. **56**, 91 (1953).
13. — Spectrographic Contributions to Lignin Chemistry. IV. Svensk Papperstidn. **56**, 287 (1953).
14. AULIN-ERDTMAN, G., A. BJÖRKMAN, H. ERDTMAN und S. E. HÄGGLUND: Einige Überlegungen und Modellversuche zur Sulfitierung des Lignins. Hägglund-Festschrift, Svensk Papperstidn. **50**, Sonderheft *81* (1947).
15. BAILEY, A. J.: Lignin in Douglas Fir. Composition of the Middle Lamella. Ind. Eng. Chem. Analyt. Ed. **8**, 52 (1936).
16. — The Mechano Chemical Dissection of Wood Fibers. Paper Ind., Jan. 1936.
17. BEIJERINCK, M. W.: Indigo Fermentation. Kon. Ned. Akad. Wetensch. Proc. **2**, 495 (1900); Verslagen Kon. Akad. Wetensch. Wisen Natuurk. Afd. Amsterdam Deel **8**, 572 (1900); Verzamelde Geschr. Delft, **3**, 342 (1921).
18. BLAND, D. E., G. HO and W. E. COHEN: Aromatic Aldehydes from the Oxidation of some Australian Woods and their Chromatographic Separation. Austral. J. Sci. Res., Ser. A, **3**, No. 4, 642 (1950).
18a. BRAUNS, F. E.: Lignin. Fortschr. Chem. organ. Naturstoffe **5**, 175 (1948).
19. — The Chemistry of Lignin. New York: Acad. Press. 1952.
20. — The Chemistry of Lignin, p. 51.
21. — The Chemistry of Lignin, p. 373.
22. — The Chemistry of Lignin, p. 745.
23. — Native Lignin. I. Its Isolation and Methylation. J. Amer. Chem. Soc. **61**, 2120 (1939).
23a. — Ungelöste Probleme der Ligninchemie. Das Papier **7**, 446 (1953).
24. BROWN, S. A., K. G. TANNER and J. E. STONE: Studies of Lignin Biosynthesis Using Isotopic Carbon. II. Short-Term Experiments with $C^{14}O_2$. Canad. J.Chem. **31**, 755 (1953).

24a. Bucher, H.: Die Tertiärlamelle von Holzfasern und ihre Erscheinungsformen bei Coniferen. Cellulosefabr. Attisholz, Solothurn, 1953.

24b. Bucher, H. und L. P. Widerkehr-Scherb: Morphologie und Struktur von Holzfasern. Cellulosefabr. Attisholz, Solothurn, 1947.

25. Coggeshall, N. D. and A. S. Glessner, Jr.: Ultraviolet Absorption Study of the Ionization of Substituted Phenols in Ethanol. J. Amer. Chem. Soc. 71, 3150 (1949).

26. Cousin, H. et H. Hérissey: Oxidation de l'iso-eugénol. C. R. hebd. Séances Acad. Sci. 147, 247 (1908); J. pharmac. chim. (6) 28, 93 (1908); Bull. soc. chim. France (4) 3, 1070 (1908).

27. Creighton, R. H. J., R. D. Gibbs and H. Hibbert: Studies on Lignin and Related Compounds. LXXV. Alkaline Nitrobenzene Oxidation of Plant Materials and Application to Taxonomic Classification. J. Amer. Chem. Soc. 66, 32 (1944).

28. Creighton, R. H. J. and H. Hibbert: Studies on Lignin and Related Compounds. LXXVI. Alkaline Nitrobenzene Oxidation of Corn Stalks. Isolation of p-Hydroxybenzaldehyde. J. Amer. Chem. Soc. 66, 37 (1944).

28a. DeBaun, R. M. and F. F. Nord: Investigations on Lignin and Lignification. V. Lignin of Cork. J. Amer. Chem. Soc. 73, 1358 (1951).

29. Enkvist, T. und B. Alfredsson: Spektrophotometrische Studien über das Vorkommen von Phenolhydroxyl in den Thioligninen. Svensk Papperstidn. 54, 185 (1951).

30. Erdtman, H.: Streifzüge durch die Entwicklung der Ligninchemie in den letzten Jahren. Svensk Papperstidn. 44, 243 (1941).

31. — Dehydrierungen in der Coniferylreihe. I. Dehydrodi-eugenol und Dehydrodi-isoeugenol. Biochem. Z. 258, 172 (1933).

32. — Dehydrierungen in der Coniferylreihe. II. Dehydrodi-isoeugenol. Liebigs Ann. Chem. 503, 283 (1933).

33. — The Chemical Nature of Lignin. Techn. Assoc. Pulp Paper Ind. (TAPPI) 32, 71 (1949).

34. — Chemistry of Some Heartwood Constituents of Conifers. Progr. Organ. Chemistry 1, 29 (1952).

35. Erdtman, H. und J. Gripenberg: Die Konstitution der Harzphenole und ihre biogenetischen Zusammenhänge. X. Herausspaltung des „Mittelstücks" des Pinoresinols. Acta Chem. Scand. 1, 71 (1947).

36. Flexser, L. A., L. P. Hammett and A. Dingwall: The Determination of Ionization by Ultraviolet Spectrophotometry: Its Validity and its Application to the Measurement of the Strength of very Weak Bases. J. Amer. Chem. Soc. 57, 2103 (1935).

37. Freudenberg, K.: Die Chemie der natürlichen Gerbstoffe. Berlin: Springer. 1920. SS. 7 und 126.

38. — Zur Kenntnis der Cellulose. Ber. dtsch. chem. Ges. 54, 767 (1921).

39. — Zur Kenntnis des Fichtenholzlignins. Sitz.-Ber. Heidelbg. Akad. Wiss. 1928, 19. Abhandlung.

39a. — Lignin. Fortschr. Chem. organ. Naturstoffe 2, 1 (1939).

40. — Die Grundzüge der Ligninchemie. Svensk Kem. Tidskr. 55, 201 (1943).

41. — Lignin. Fiat Rev. German Sci. 1939–1946. Biochem. 3, 159 (1948).

42. — Über ein ligninähnliches Produkt aus Coniferylalkohol. Angew. Chem. 61, 228 (1949).

43. — Die Bildung ligninähnlicher Stoffe unter physiologischen Bedingungen. Sitz.-Ber. Heidelbg. Akad. Wiss. 1949, 5. Abh.

43a. — Moderne Methoden der Pflanzenanalyse, Abschnitt Lignin. (Herausgeber K. Paech und M. V. Tracey.) Berlin-Göttingen-Heidelberg: Springer-Verlag. 1954.

44. — (unveröffentlicht).

45. Freudenberg, K. und F. Bittner: Coniferylalkohol aus Siambenzoe. Ber. dtsch. chem. Ges. **83**, 600 (1950).

46. — — Coniferyl-methyläther und andere mit dem Lignin zusammenhängende Stoffe. Ber. dtsch. chem. Ges. **85**, 86 (1952).

47. — — Versuche mit Coniferylalkohol, der radioaktiven Kohlenstoff enthält. Ber. dtsch. chem. Ges. **86**, 155 (1953).

48. Freudenberg, K. und G. Dietrich: Vergleichende Untersuchung des Fichten- und Buchenlignins. Liebigs Ann. Chem. **563**, 146 (1949).

49. Freudenberg, K. und H. Dietrich: Über das Syringaresinol, ein Dehydrierungsprodukt des Sinapinalkohols. Ber. dtsch. chem. Ges. **86**, 4 (1953).

50. — — Synthese des *d,l*-Pinoresinols und andere Versuche im Zusammenhang mit dem Lignin. Ber. dtsch. chem. Ges. **86**, 1157 (1953).

51. Freudenberg, K., H. Dietrich und W. Siebert: Ultrarotspektren ligninverwandter Stoffe. Ber. dtsch. chem. Ges. **84**, 961 (1951).

52. Freudenberg, K. und R. Dillenburg: Coniferylaldehyd und Sinapinalkohol. Ber. dtsch. chem. Ges. **84**, 67 (1951).

53. Freudenberg, K., K. Engler, E. Flickinger, A. Sobeck und F. Klink: Der Abbau des Fichten-Lignins zu Phenolcarbonsäuren. Ber. dtsch. chem. Ges. **71**, 1810 (1938).

54. Freudenberg, K. und G. Gehrke: *p*-Cumaralkohol und sein Dehydrierungspolymerisat. Ber. dtsch. chem. Ges. **84**, 433 (1951).

55. Freudenberg, K. und W. Heel: Dioxy- und Trioxy-zimtalkohol. Ber. dtsch. chem. Ges. **86**, 190 (1953).

56. Freudenberg, K. und W. Heimberger: Die biochemische Synthese ligninartiger Stoffe. Ber. dtsch. chem. Ges. **83**, 519 (1950).

57. Freudenberg, K. und H. H. Hübner: Oxyzimtalkohole und ihre Dehydrierungspolymerisate. Ber. dtsch. chem. Ges. **85**, 1181 (1952).

58. Freudenberg, K., A. Janson, E. Knopf und A. Haag: Zur Kenntnis des Lignins. Ber. dtsch. chem. Ges. **69**, 1415 (1936).

59. Freudenberg, K. und R. Kraft: Methyliertes Fichtenlignin und seine künstliche Nachbildung. Ber. dtsch. chem. Ges. **83**, 530 (1950).

60. Freudenberg, K., R. Kraft und W. Heimberger: Über den Sinapinalkohol, den Coniferylalkohol und ihre Dehydrierungspolymerisate. Ber. dtsch. chem. Ges. **84**, 472 (1951).

60a. Freudenberg, K., W. Lautsch und K. Engler: Die Bildung von Vanillin aus Fichtenholz. Ber. dtsch. chem. Ges. **73**, 167 (1940).

61. Freudenberg, K. und H. G. Müller: Synthetische Versuche im Zusammenhang mit dem Lignin. Liebigs Ann. Chem. **584**, 40 (1953).

62. Freudenberg, K. und E. Plankenhorn: Über Essigsäure-Lignin. Ber. dtsch. chem. Ges. **75**, 857 (1942).

63. Freudenberg, K. und Th. Ploetz: Über die Verschiedenheit von Cellulase und Lichenase. Z. physiol. Chem. (Hoppe-Seyler) **259**, 19 (1939).

64. — — Über enzymatische Abbauversuche an Holz. Holz als Roh- und Werkstoff **3**, 105 (1940).

65. Freudenberg, K. und D. Rasenack: *d,l*-Pinoresinol, ein weiteres Zwischenprodukt der Ligninbildung. Ber. dtsch. chem. Ges. **86**, 755 (1953).

65a. Freudenberg, K., M. Reichart und L. Knof: Naturwiss. 1954 (im Druck). Lösliches Lignin enthält weniger Phenylcumaran-einheiten als andere Ligninpräparate (Zusatz bei der Korrektur).

66. Freudenberg, K., H. Reznik, H. Boesenberg und D. Rasenack: Das an der Verholzung beteiligte Fermentsystem. Ber. dtsch. chem. Ges. **85**, 641 (1952).

67. FREUDENBERG, K. und H. RICHTZENHAIN: Die Konstitution des Dehydrodi-isoeugenols und seine Bedeutung für die Chemie des Lignins. Liebigs Ann. Chem. **552,** 126 (1942).

68. — — Enzymatische Versuche zur Entstehung des Lignins. Ber. dtsch. chem. Ges. **76,** 997 (1943).

69. FREUDENBERG, K., W. SIEBERT, W. HEIMBERGER und R. KRAFT: Ultrarot-spektren von Lignin und ligninähnlichen Stoffen. Ber. dtsch. chem. Ges. **83,** 533 (1950).

70. FREUDENBERG, K. und H. WALCH: Die Phenolgruppen im Lignin. Ber. dtsch. chem. Ges. **76,** 305 (1943).

71. FREUDENBERG, K. und G. WILKE: Modelle zur Chemie des Lignins, insbesondere der Formaldehyd abspaltenden Gruppen. Ber. dtsch. chem. Ges. **85,** 78 (1952).

72. FREUDENBERG, K., H. ZOCHER und W. DÜRR: Weitere Versuche mit Lignin. Ber. dtsch. chem. Ges. **62,** 1814 (1929).

72a. FREY-WYSSLING, A. und K. MÜHLETHALER: The Fine Structure of Cellulose. Fortschr. Chem. organ. Naturstoffe **8,** 1 (1951).

73. GAUTHERET, R. J.: Plant Tissue Culture. Endeavour **7,** 75 (1948).

73a. — Remarques sur les besoins nutritifs des cultures de tissus de *Salix caprea.* C. R. Séances Soc. Biol. **144,** 173 (1950).

74. GIERER, J., sowie E. ADLER und J. GIERER: Vorträge, XIII. Int. Kongr. f. reine und angew. Chem. Stockholm. 1953.

75. GOLDSCHMIDT, O.: The Effect of Alkali and Strong Acid on the Ultraviolet Absorption Spectrum of Lignin and Related Compounds. J. Amer. Chem. Soc. **75,** 3780 (1953).

76. GORIS, M.: Sur un nouveau principe cristallisé de la Kola fraîche. C. R. hebd. Séances Acad. Sci. **144,** 1162 (1907).

77. HÄGGLUND, E.: Chemistry of Wood. New York: Acad. Press. 1951.

78. — Chemistry of Wood, pp. 215 and 415.

79. HÄGGLUND, E. und T. JOHNSON: Über die Sulfonierung des Fichtenholzlignins. I. Biochem. Z. **202,** 439 (1928).

80. HARRIS, E. E.: A Lignin-Carbohydrate Bond in Wood. Techn. Assoc. Pulp Paper Ind. (TAPPI) **36,** 402 (1953).

81. HEDÉN, S. und B. HOLMBERG: Bisulfitkok met aromatiska alkoholer. Svensk Kem. Tidskr. **48,** 207 (1936).

82. HIBBERT, H.: In: F. E. BRAUNS, The Chemistry of Lignin, p. 467 ff. New York: Acad. Press. 1952.

83. HODGE, H. J. and A. B. WARDROP: An Electron Microscopic Investigation of the Cell Wall Organization of Conifer Tracheids and Conifer Cambium. Austral. J. Sci. Res., Ser. B, **3,** 265 (1950).

84. HOLMBERG, B.: Die Mercaptolyse des Fichtenholzes. Ing. Vetenskapsakad. Handl. No. 103 (1930).

85. — Aromatische Alkohole und Thioglykolsäure. J. prakt. Chem. (2) **141,** 93 (1934).

86. — Thioglykolsäure als Ligninreagens. Ing. Vetenskapsakad. Handl. No. 131 (1934).

87. — Die Alkoholyseprodukte des Fichtenholzes. Svensk Papperstidn. **39,** Sonder-Nr. 113 (1936).

88. — Lignin und Thioglykolsäure. Österr. Chem.-Ztg. **43,** 152 (1940).

89. — Om granens lignin. Finska kemist samf. Meddelanden **54,** 124 (1945).

89a. — Peter Klasons ligninchemiska arbeten. Jahrbuch d. Kgl. Schwed. Akad. d. Wiss. für 1953, S. 338.

90. HOLMBERG, B. und M. GRALÉN: Die Stöchiometrie des Fichtenlignins. Ing. Vetenskapsakad. Handl. No. 162 (1942).

91. JACQUIOT, C.: Sur la culture *in vitro* de tissu cambial de Châtaignier (*Castanea vesca* GAERTN.). C. R. hebd. Séances Acad. Sci. **231**, 1080 (1950).

92. — Sur les phénomènes d'histogenese observés dans des cultures *in vitro* de tissu cambial de Chênes (*Quercus sessiflora* SM., *Quercus pedunculata*, EHR., *Q. suber* L.). C. R. hebd. Séances Acad. Sci. **234**, 1468 (1952).

93. JACQUIOT C.: Sur les anomalies de la différenciation et de la lignification du tissu cambial de divers arbres cultivés *in vitro*. Bull. assoc. tech. ind. papetière **6**, 13 (1952)

93 a. — Nouvelles recherches sur les processus de liginification dans les tissus d'arbres cultivés *in vitro*. Bull. assoc. tech. ind. papetière **7**, 23 (1953).

94. KLASON, P.: Beitrag zur Konstitution des Fichtenlignins. III. Ber. dtsch. chem. Ges. **56**, 300 (1923).

95. — Beiträge zur Konstitution des Fichtenlignins. VIII. Untersuchung des Nahrungssaftes der Fichte. Ber. dtsch. chem. Ges. **62**, 635 (1929).

95 a. — Beiträge zur Konstitution des Fichtenlignins. IX. Bemerkungen zu FREUDEN-BERGs Mitt. 11 über Lignin und Cellulose. Ber. dtsch. chem. Ges. **62**, 2523 (1929).

96. KRATZL, K.: Neue Synthesen und Reaktionen fettaromatischer Sulfosäuren. Österr. Chem.-Ztg. **49**, 143 (1948).

97. — Über die Konstitution der Ligninsulfosäure (vorl. Mitt.). Monatsh. Chem. **78**, 173 (1948).

97 a. — Zur Bromierung der Ligninsulfosäure und ihrer Modellsubstanzen (kurze Mitt.). Monatsh. Chem. **80**, 437 (1949).

98. KRATZL, K. und G. BILLEK: Die Synthese des mit C^{14} markierten Coniferins. Monatsh. Chem. **84**, 406 (1953).

99. — — Synthese von mit C^{14} markierten Ligninbausteinen. Holzforsch. **7**, 66 (1953).

99 a. KUDZIN, S. F., R. M. DeBAUN and F. F. NORD: Investigations on Lignin and Lignification. VI. The Comparative Evaluation of Native Lignins. J. Amer. Chem. Soc. **73**, 4615 (1951).

99 b. KUDZIN, S. F. and F. F. NORD: Investigations on Lignin and Lignification. IV. Studies on Hardwood Lignin. J. Amer. Chem. Soc. **73**, 690 (1951).

99 c. — — Investigations on Lignin and Lignification. VII. Characterization of Enzymatically Liberated Hardwood Lignins. J. Amer. Chem. Soc. **73**, 4619 (1951).

100. LANGE, P. W.: Über Natur und Verteilung des Lignins im Fichtenholz. Svensk Papperstidn. **47**, 262 (1944).

101. — Ultraviolet Absorption and Dichroism of Lignin in Wood.: Svensk Pappenstidn. **48**, 241 (1945).

102. LINDGREN, B. O.: The Sulfonatable Groups of Lignin. Svensk Papperstidn. **55**, 78 (1952).

103. LINDGREN, B. O. and U. SAEDÉN: Pinoresinol as a Lignin Model. Acta Chem. Scand. **6**, 91 (1952).

104. MANSKAJA, S. M.: Die Bildung des Lignins in den Pflanzen. Uspechi Sowremennoi Biologii **23**, 203 (1947) [Chem. Zbl. **1948** II, 745].

105. — Teilnahme von Oxydasen an der Bildung des Lignins. Doklady Chimii Nauk **62**, 369 (1948) [Chem. Abstr. **43**, 2282 (1949)].

106. MANSKAJA, S. M. und M. S. BARDINSKAJA: Über die Bildungsweise des Lignins in der Holzfaser. Biochimia **17**, 711 (1952) [Chem. Abstr. **47**, 5115 (1953)].

107. MANSKAJA, S. M. und M. N. KOTSCHNEVA: Das Lignin verschiedener Pflanzengruppen. Doklady Chimii Nauk **62**, 505 (1948) [Chem. Abstr. **43**, 2282 (1949)].

108. MAYER, W.: Dehydro-digallussäure. Liebigs Ann. Chem. **578**, 34 (1952).

109. MORTON, R. A. and A. L. STUBBS: Absorption Spectra of Hydroxy-aldehydes. Hydroxy-ketones, and their Methyl Ethers. J. Chem. Soc. (London) **1940**, 1347.

110. NORD, F. F. and J. C. VITUCCI: On the Mechanism of Enzyme Action. XXXI. The Mechanism of Methyl-*p*-methoxycinnamate Formation by *Lentinus lepideus* and its Significance in Lignification. Arch. Biochemistry 15, 465 (1947).

111. PEW, J. C.: Structural Aspects of the Color Reaction of Lignin with Strong Acids. J. Amer. Chem. Soc. 74, 2850 (1952).

112. PLOETZ, TH.: Über einige Enzyme des Hausschwamms *(Merulius lacrimans)*. 2. Mitt. über den enzymatischen Abbau polymerer Kohlenhydrate. Z. physiol. Chem. (Hoppe-Seyler) 261, 183 (1939).

113. — Über das Verhalten von Lindenholz gegen Äthylendiaminkupferoxyd-Lösung und den enzymatischen Abbau der Hauptfraktionen. 3. Mitt. über den enzymatischen Abbau polymerer Kohlenhydrate. Ber. dtsch. chem. Ges. 72 1885 (1939).

114. — Vergleichender enzymatischer Abbau einiger isolierter Holzbestandteile mit Schneckenferment. 4. Mitt. über den enzymatischen Abbau polymerer Kohlenhydrate. Ber. dtsch. chem. Ges. 73, 57 (1940).

115. — Enzymatischer Abbau einiger nativer ligninhaltiger Materialien. 5. Mitt. über den enzymatischen Abbau polymerer Kohlenhydrate. Ber. dtsch. chem. Ges. 73, 61 (1940).

116. — Über den Bindungszustand des Lignins im Holz. 6. Mitt. über den enzymatischen Abbau polymerer Kohlenhydrate. Ber. dtsch. chem. Ges. 73, 74 (1940).

117. — Weitere Fraktionierungsversuche an Lindenholz und enzymatischer Abbau der Fraktionen. 7. Mitt. über den enzymatischen Abbau polymerer Kohlenhydrate. Ber. dtsch. chem. Ges. 73, 790 (1940).

118. — Beiträge zur Kenntnis des Baues der verholzten Faser. Sitz.-Ber. Heidelbg. Akad. Wiss. 1940, 10. Abh.

119. REINITZER, F.: Untersuchungen über das Olivenharz. Monatsh. Chem. 45, 87 (1924).

120. RICHTZENHAIN, H.: Enzymatische Versuche zur Entstehung des Lignins. 2. Mitt.: Die Dehydrierung des 5-Methyl-pyrogallol-1,3-dimethyläthers. Ber. dtsch. chem. Ges. 77, 409 (1944).

121. — Enzymatische Versuche zur Entstehung des Lignins. 3. Mitt.: Die Dehydrierung des 5-Allyl-pyrogallol-1,3-dimethyläthers. Ber. dtsch. chem. Ges. 81, 260 (1948).

121a. — Enzymatische Versuche zur Entstehung des Lignins. 4. Mitt.: Dehydrierungen in der Guajacolreihe. Ber. dtsch. chem. Ges. 82, 447 (1949).

122. — Zur Konstitution des Fichtenlignins. Acta Chem. Scand. 4, 589 (1950).

123. — Die phenolischen Gruppen des Fichtenholzes. Ber. dtsch. chem. Ges. 83, 488 (1950).

124. — Über den oxydativen Abbau von methylierten Ligninpräparaten. Svensk Papperstidn. 53, 644 (1950).

125. RICHTZENHAIN, H. und B. ALFREDSSON: Über den Abbau von Lignin und Ligninmodellsubstanzen mit Hypochlorit. I. Acta Chem. Scand. 7, 1177 (1953).

126. SANGER, F.: The Free Amino Groups of Insulin. Biochemic. J. 39, 507 (1945).

127. SCHORYGINA, N. N. und T. JA. KEFELI: Die Spaltung des Lignins durch metallisches Natrium in flüssigem Ammoniak. J. Obschtschei Chimii 20, (82) 1199 (1950) [Chem. Zbl. 1951 II, 233].

128. SCHORYGINA, N. N., T. JA. KEFELI und A. F. SSEMETSCHKINA: Die Spaltung des Lignins durch metallisches Natrium in flüssigem Ammoniak. J. Obschtschei Chimii 19 (81), 1558 (1949) [Chem. Zbl. 1950 I, 728]; Doklady Acad. Nauk 64, 689 (1949) [Chem. Zbl. 1950 I, 1366].

128a. SCHUBERT, W. J. and F. F. NORD: Investigations on Lignin and Lignification. I. Studies on Softwood Lignin. J. Amer. Chem. Soc. 72, 977 (1950).

128b. — — Investigations on Lignin and Lignification. II. The Characterization of Enzymatically Liberated Lignin. J. Amer. Chem. Soc. 72, 3835 (1950).

128c. SCHUBERT, W. J. and F. F. NORD: A Methoxy-containing Lignin-like Component of the Mold *Trametes Pine.* J. Amer. Chem. Soc. **72**, 5338 (1950).
128d. SCHUBERT, W. J., A. PASSANNANTE, G. DE STEVENS, M. BIER and F. F. NORD: Investigations on Lignin and Lignification. XIII. Electrophoresis of Native and Enzymatically Liberated Lignins. J. Amer. Chem. Soc. **72**, 1969 (1953).
129. SCHÜTZ, FR. und W. KNACKSTEDT: Holzaufschluß mit Salzsäure oder Chloriden als Katalysator in essigsaurer Lösung. Cellulosechemie **20**, 15 (1942).
129a. STEVENS, G. DE and F. F. NORD: Investigations on Lignin and Lignification. IX. The Relationship between the Action of Brown Rot Fungi, Cellulose Degradation and Lignin Composition in Bagasse. J. Amer. Chem. Soc. **74**, 3326 (1952).
129b. — — Investigations on Lignin and Lignification. X. The Isolation and Characterization of the Native Lignin from Kiri Wood. J. Amer. Chem. Soc. **74**, 3447 (1952).
129c. — — Investigations on Lignin and Lignification. XI. Structural Studies on Bagasse Native Lignin. J. Amer. Chem. Soc. **75**, 305 (1953).
129d. — — Investigations on Lignin and Lignification. VIII. Isolation and Characterization of Bagasse Native Lignin. J. Amer. Chem. Soc. **73**, 4622 (1951).
130. STUMPF, W. und K. FREUDENBERG: Lösliches Lignin aus Fichten- und Buchenholz. Angew. Chem. **62**, 537 (1950).
131. STUMPF, W., F. WEYGAND und O. A. GROSSKINSKY: Synthese von Dioxan [14C] und Anwendung als Extraktionsmittel für lösliches Lignin. Ber. dtsch. chem. Ges. **86**, 1391 (1953).
132. TIEMANN, F. und B. MENDELSOHN: Zur Kenntnis der Bestandteile des Holzteerkreosots. Ber. dtsch. chem. Ges. **8**, 1136, Anmerk. 1139 (1875).
133. TSCHUDAKOFF, M. J.: Natürliches Lignin aus destructiv zersetztem Holz. Priklad. Chimii **22**, 392 (1949).
134. WACEK, A. V., O. HÄRTEL und S. MERALLA: Über den Einfluß von Coniferinzusatz auf die Verholzung von Karottengewebe bei Kultur *in vitro.* Holzforsch. **7**, 58 (1953).
135. WILLSTÄTTER, R. und L. ZECHMEISTER: Zur Kenntnis der Hydrolyse von Cellulose. I. Ber. dtsch. chem. Ges. **46**, 2401 (1913).
135a. WISE, L. E. und E. C. JAHN: Wood Chemistry. New York: Reinhold Publ. Co. 1952.
136. ZAHN, H. und A. WÜRZ: Zur quantitativen Bestimmung von Phenolen mit 2,4-Dinitrofluorbenzol. Z. analyt. Chem. **134**, 183 (1951).
137. ZIEGLER, E. und K. GARTLER: Über Indophenole. II. Mitt. Monatsh. Chem. **80**, 759 (1949).
138. ZIEGLER, E. und G. SNATZKE: Spaltungen mittels Diazoniumverbindungen. XIII. Mitt. über Phenolderivate. Monatsh. Chem. **84**, 278 (1953).

(Eingelaufen am 28. Dezember 1953.)

Probleme und neuere Ergebnisse in der Vitamin D-Chemie.

Von **H. H. Inhoffen** und **K. Brückner**, Braunschweig.

I. Präcalciferol, ein neues Isomeres in der Reihe der Bestrahlungsprodukte des Ergosterins.

Wie wir auf Grund der umfassenden Arbeiten von Windaus und seiner Schule wissen, verläuft der photochemische Übergang vom Ergosterin (I) zum Vitamin D_2 (Calciferol) (IV) über Lumisterin (II) und Tachysterin (III), wobei in jedem Falle alle diese Stufen durchschritten werden müssen. Während bisher angenommen wurde, daß diese Bildung des Vitamins D_2 ein rein photochemischer Vorgang und nur abhängig von Bestrahlungsart und -dauer sei, haben Velluz und Mitarbeiter (*80, 81*) in neuerer Zeit festgestellt, daß auch die thermischen Bedin-

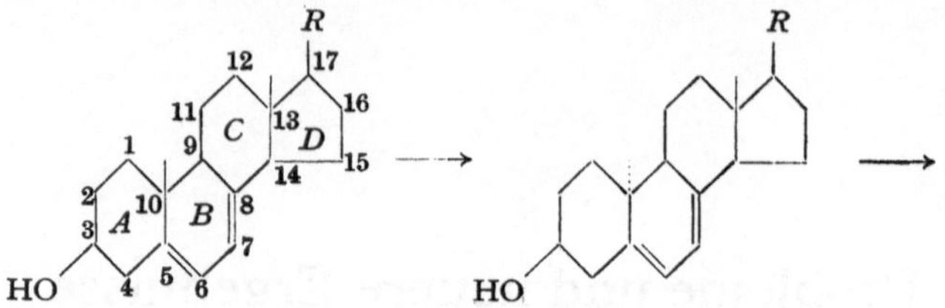

(I.) Ergosterin.   (II.) Lumisterin.

$$R = -CH-CH=CH-CH-CH(CH_3)_2$$
$$\quad\quad\;|\quad\quad\quad\quad\;|$$
$$\quad\quad CH_3\quad\quad\quad CH_3$$

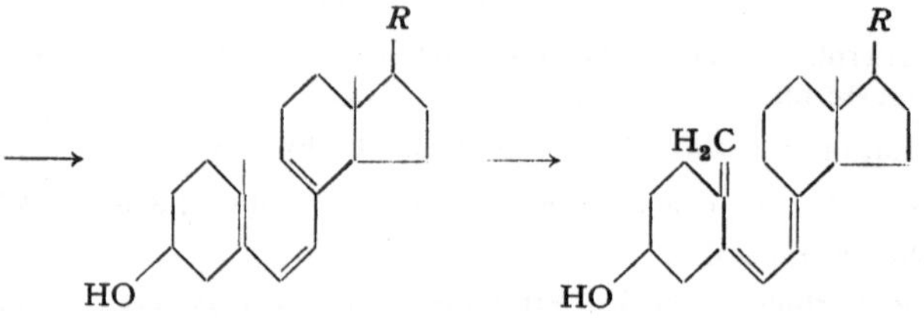

(III.) Tachysterin.   (IV.) Vitamin D₂.

gungen von Einfluß sind, und daß es ein *Präcalciferol* gibt, welches nicht durch Belichtung, wohl aber durch gelindes Erwärmen in Vitamin D₂ übergeht. Den Ausgangspunkt ihrer Untersuchungen bildete die Beobachtung, daß rohe Bestrahlungsprodukte des Ergosterins bei langem Stehen oder schneller bei mäßigem Erwärmen (bis zu 100°) ihr Drehungsvermögen ständig bis zu einem bestimmten positiven Grenzwert ändern. Schon Windaus und Auhagen (90) hatten dies festgestellt, jedoch nicht näher untersucht. Velluz gibt an, daß es sich bei dem hier entstehenden Produkt um Vitamin D₂ handelt.

In Verfolg dieser Beobachtungen bestrahlte Velluz ätherische oder benzolische Ergosterinlösungen bei gewöhnlicher Temperatur und nahm die Aufarbeitung des rohen Bestrahlungsproduktes sowie die Abtrennung des nicht umgesetzten Ergosterins unter peinlicher Vermeidung jeder Temperaturerhöhung vor. Wurde das dabei gewonnene Rohprodukt anschließend mehrere Stunden in benzolischer Lösung bei Ausschluß von Licht unter Rückfluß gekocht, so ließ sich eine Erhöhung des Vitamin D-Gehalts im Rohöl auf das Drei- bis Vierfache feststellen, z. B. von 11% auf 36%.

Daraus ergab sich der Schluß, daß in der Reihe der Übergangsstufen vom Ergosterin zum Vitamin D₂, wie schon erwähnt, ein weiteres Produkt vorhanden sein muß, das sich lediglich durch Erwärmen in Vitamin D₂ überführen läßt. Da anderseits die Isomerisierung des Tachysterins eindeutig Photonen erfordert (104), kann der neue Stoff nur zwischen Tachysterin und Vitamin D₂ eingeordnet werden. Dies setzt allerdings die noch nicht bewiesene, jedoch sehr wahrscheinliche Tatsache voraus, daß Präcalciferol aus Tachysterin und nicht direkt aus einem seiner Vorprodukte gebildet wird. Später gelang es dann VELLUZ und AMIARD (76), das Präcalciferol als 3,5-Dinitrobenzoesäureester nach einer chromatographischen Trennung in reiner Form zu gewinnen. Die Darstellung verlief dabei im einzelnen wie folgt.

50 g Ergosterin wurden in 4000 ccm Äther mit dem UV-Licht einer Magnesiumfunkenstrecke bestrahlt, danach die ätherische Lösung bei 10° zur Hälfte eingeengt und der größte Teil des unveränderten Ergosterins abgetrennt. Weiteres Einengen und Zugabe von 300 ccm Methanol ergaben eine zweite Kristallisation von Ergosterin. Dann wurde die methanolische Lösung bei 30° abdestilliert, der Rückstand mehrfach mit Benzol bei Raumtemperatur eingeengt und die danach erhaltenen 19,4 g Öl in 60 ccm Benzol und 20 ccm Pyridin gelöst, bei 10—15° mit 30 g 3,5-Dinitrobenzoylchlorid in 45 ccm Benzol versetzt und 1 Stunde bei 15—20° stehengelassen. Nach üblicher Aufarbeitung bei Temperaturen bis maximal 20° mußte das Rohprodukt einer chromatographischen Trennung unterworfen werden. Dazu wurde es in 400 ccm Petroläther und 40 ccm Benzol gelöst, auf 300 g neutrales Aluminiumoxyd aufgezogen und mit Petroläther-Benzol (9 : 1) so lange entwickelt, bis die gelbe Farbe den Boden der Kolonne erreichte. Die unterste Zone wurde nun mit Äther eluiert (13,1 g) und mit Methyl-äthylketon und Alkohol versetzt. Dabei kristallisierten 8,5 g Präcalciferol-3,5-dinitrobenzoesäureester aus. Durch Nach-chromatographie der im ersten Chromatogramm erhaltenen Mischfraktion ließen sich noch einmal 1,5 g kristallisiertes Produkt gewinnen.

Das Dinitrobenzoat des Präcalciferols kristallisiert im Gegensatz zum gleichen Derivat des Vitamins D_2 in feinen blaßgelben Nadeln vom Schmelzpunkt 103—104° und mit einer Drehung von $[\alpha]_D = + 30°$ in Benzol ($+ 45°$ in Chloroform). (Vitamin D_2-3,5-dinitrobenzoat: Schmelzp. 158—159°, $[\alpha]_D = + 57°$ in Benzol, $+ 88°$ in Chloroform).

Obwohl Präcalciferol sich durch Analyse und Molekulargewichtsbestimmung als Isomeres des Vitamins D_2 erwies, unterscheidet es sich durch die ebengenannten Daten seines Esters von allen bisher bekannten Isomeren. Durch Verseifung bei Raumtemperatur ließ sich daraus das freie Präcalciferol als Öl erhalten, gekennzeichnet durch seine Drehung von $[\alpha]_D = + 43°$ in Benzol im Gegensatz zu $[\alpha]_D = + 85°$ für Vitamin D_2.

Wird sowohl der Ester als auch das freie Präcalciferol unter Ausschluß des Lichts in benzolischer Lösung gelinde erwärmt (60°), so steigt in beiden Fällen die Drehung bis zu einem Grenzwert an ($+ 50°$ bzw. $+ 70°$), und es lassen sich danach 65% bzw. 75% Vitamin D_2 als 3,5-Dinitrobenzoat gewinnen.

Überraschend war nun die ebenfalls von VELLUZ (77) beobachtete Tatsache, daß bei gleichem Erwärmen sowohl von Vitamin D_2-3,5-dinitrobenzoat als auch von Vitamin D_2 selbst die Drehung bis auf einen Grenzwert sinkt, der im Falle des Esters dem vorher erwähnten nahezu gleicht ($+ 51°$ gegenüber $+ 50°$), im Falle des freien Vitamins D_2 — wahrscheinlich infolge störender Oxydationsreaktionen — dem obigen Wert nahekommt ($+ 77°$ gegenüber $+ 70°$). Die daraus abgeleitete Vermutung, daß sich beim Erwärmen zwischen Präcalciferol und Vitamin D_2

bzw. zwischen ihren Estern ein Gleichgewicht einstellt, ließ sich durch Isolierung von Präcalciferol-*m*-dinitrobenzoat aus längere Zeit erhitzten Lösungen von Vitamin D_2 oder seines Esters bestätigen. So konnten z. B. aus 100 g Vitamin D_2-*m*-dinitrobenzoat nach achtstündigem Kochen in Benzol, anschließender Chromatographie und fraktionierter Kristallisation 12 g Präcalciferol-*m*-dinitrobenzoat gewonnen werden. Beide Körper stehen also tatsächlich in einem Gleichgewicht, das weit auf der Seite des Vitamins D_2 als dem stabileren der beiden Isomeren liegt.

Wie weitere Untersuchungen von VELLUZ (*78*) ergaben, gelten vollkommen analoge Verhältnisse für Vitamin D_3 und das bisher unbekannte Präcalciferol D_3, dessen Dinitrobenzoat ebenfalls in feinen Nadeln vom Schmelzpunkt 110—111° und $[\alpha]_D = +38,5°$ in Benzol ($+52°$ in Chloroform) kristallisiert. Das nach Verseifung erhaltene Präcalciferol D_3 besitzt eine Drehung von $[\alpha]_D = +40°$.

Bei der Diskussion der *Struktur* dieses interessanten Stoffes kommen VELLUZ und Mitarbeiter (*79*) zu folgenden Ergebnissen: Das Präcalciferol kann auf Grund von Molekulargewichtsbestimmungen weder ein Polymeres sein, noch kann es die 3-epi-Verbindung des Vitamins D_2 darstellen, da für eine solche Epimerisierung erfahrungsgemäß wesentlich stärkere Reaktionsbedingungen notwendig sind. Auch eine sterische Umlagerung der Methylgruppe an $C_{(13)}$ kann ausgeschlossen werden, da sie ohne Zweifel wie im Falle des Oestrons (*16, 14*) die Einwirkung von UV-Licht notwendig machen dürfte. Die anfangs geäußerte Annahme einer hypothetischen Cyclobuten-Struktur wurde gleichfalls fallengelassen, da sowohl Hydrierung als auch Benzopersäure-titration die Anwesenheit von vier Doppelbindungen im Molekül bewiesen. Für ein konjugiertes Trien-System im Molekül sprach schließlich die UV-Absorption, deren Maximum wie beim Vitamin D_2 bei 265 mμ liegt, aber eine wesentlich niedrigere molare Extinktion aufweist, $\varepsilon = 9000$. VELLUZ schloß daraus, daß der Molekülaufbau dem der ring-offenen Isomeren entspricht und diskutierte abschließend zwei mögliche Strukturen, ohne einen Beweis für sie zu geben.

Die eine Deutung setzt voraus, daß Vitamin D_2 in den zwei Formen (IV) und (V) auftreten kann. Dies wäre ein Beispiel der allgemein als „Atropisomerie" benannten Erscheinung, nach der sich ein Molekül in Bezug auf eine Einfachbindung in bestimmten räumlichen Lagen stabilisiert. Im konjugierten System resultieren daraus eine *cis*- bzw. *trans*-Konfiguration, bezeichnet als „s-*cis*" und „s-*trans*". Im vorliegenden Fall sollte das instabilere Präcalciferol möglicherweise die s-*cis*-Form darstellen (s. S. 87). Die zweite Möglichkeit wird in der Struktur (VI) gesehen, die einen Zwischenzustand zwischen Tachysterin (III) und Vitamin D_2 (IV) darstellt, bei der also schon eine Verlagerung des Trien-Systems eingetreten, die spätere semicyclische Methylen-Doppelbindung jedoch noch im Ring A von $C_{(10)}$ nach $C_{(1)}$ gelagert wäre. Dieser Anschauung widerspricht die relativ leicht verlaufende Rückreaktion, die nach den bisherigen Erfahrungen eines sauren oder basischen Katalysators bedürfte. Darüber

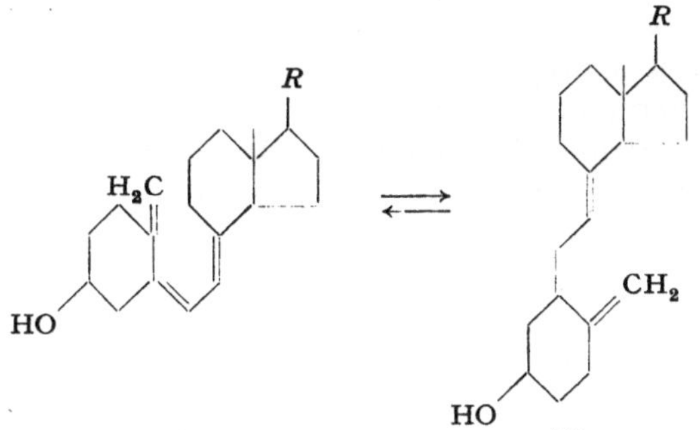

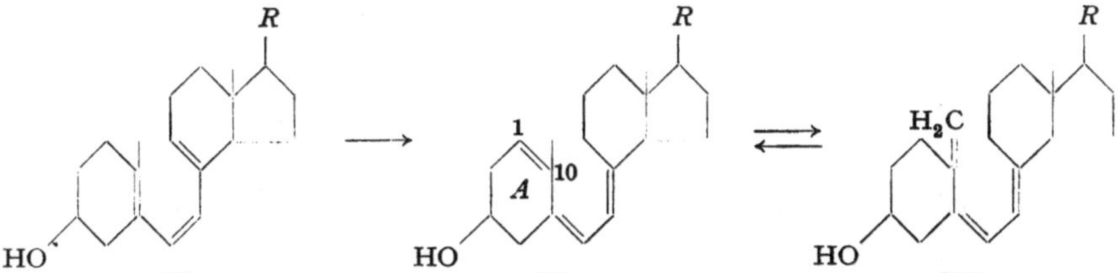

hinaus müßte bei einer solchen Isomerisierung nach unserer Meinung wenigstens eine geringe Verlagerung des Absorptionsmaximums zu erwarten sein. Dies ist jedoch, wie schon erwähnt wurde, nicht der Fall. Ist daher mit Angabe der exakten experimentellen Unterlagen die Existenz eines Präcalciferols gesichert, so stellt die theoretische Deutung seiner Struktur ein völlig offenes Kapitel der Vitamin D-Chemie dar. Es erscheint nicht ausgeschlossen, daß sich seine Aufklärung befruchtend auf die Frage der Feinstruktur aller D-Isomeren auszuwirken vermag.

II. Konstitution des Vitamins D₂ und des Tachysterins.

Nachdem im Jahre 1936 die Konstitution des Vitamins D_2 (94) nach zahlreichen Studien insbesondere von WINDAUS in der heute gebräuchlichen Form festgelegt werden konnte, blieb nur noch die Frage der *geometrischen Isomerie* an den beiden Doppelbindungen offen. Nur wenige Arbeiten beschäftigen sich seitdem mit dieser Frage seiner Feinstruktur, wie auch der des Tachysterins, die vor allem für den Versuch eines synthetischen Aufbaus beider Stoffe von großer Bedeutung ist. Die heute gebräuchliche Formel des Tachysterins war im Jahre 1939 als Resultat längerer Untersuchungen festgelegt worden (28, 88), ohne z. B. besonders für die Lage der Doppelbindung zwischen $C_{(8)}$—$C_{(9)}$ einen endgültigen Beweis zu besitzen.

Für beide Verbindungen lassen sich unter Einbeziehung aller bis dahin geführten Strukturbeweise noch folgende Formen aufstellen, die durch die Äthylen-isomerie an den Doppelbindungen (unter Außeracht-lassung eventuell vorhandener Atropisomerer) gegeben werden.

Für Tachysterin (III.) und (VII.):

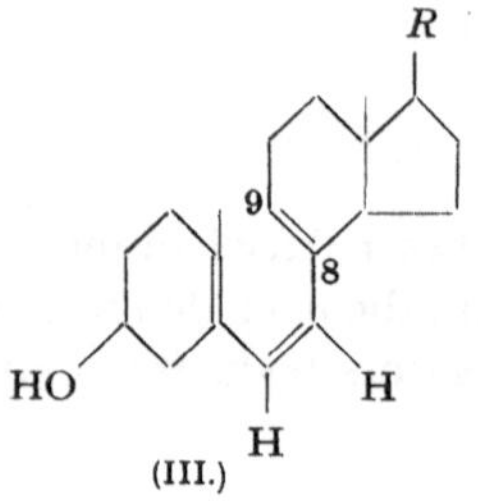

(III.) 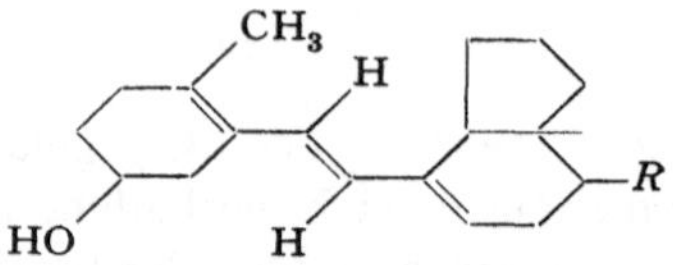 (VII.)

Für Vitamin D_2 (IV.), (V.) und (VIII.) bis (X.):

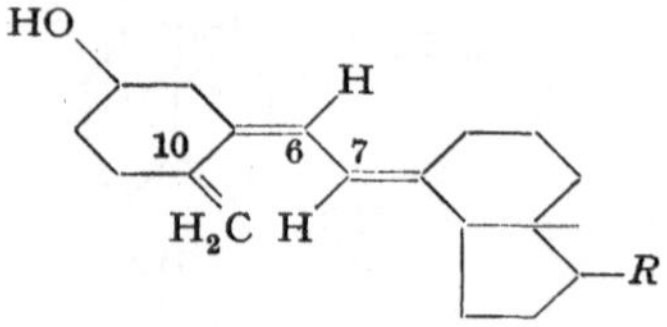

(V.) 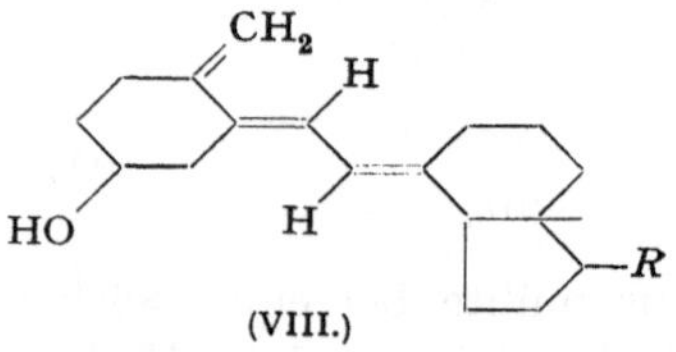 (VIII.)

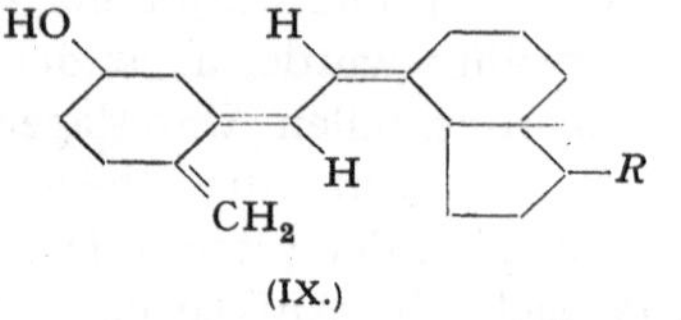

(IX.) 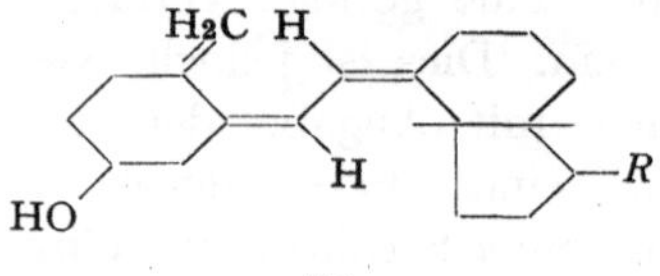 (X.)

Betrachten wir zunächst das *Tachysterin*, so sprachen sich Koch (*41*) und Dimroth (*21*) auf Grund von Vergleichen der Extinktionen bekannter *cis-trans*-Isomerer mit derjenigen des Tachysterin-Maximums für eine *cis*-Struktur desselben aus. Die *cis*-Isomeren zeigen auf Grund sterischer Hinderung stets die geringere Extinktion und zumeist eine ins Kurzwellige verschobene Absorption. Auch beim Tachysterin sprechen diese Daten (281 mμ, $\varepsilon = 24600$) für eine *cis*-Form. So zeigt sein einfachstes Modell (XI) mit eindeutiger *trans*-Verknüpfung (*12*) eine molare Extinktion von 42600 bei 269 mμ. Außerdem tritt bei zwei Stoffen mit eindeutiger *trans*-Verknüpfung, (XII, XIII), die im Verlauf dieses Berichtes noch beschrieben werden (S. 99, 109), bei gleichbleibender bzw. nur um eins erhöhter Substituentenzahl am chromophoren System eine Verschiebung des Maximums bis 288 bzw. 290 mμ und eine Erhöhung

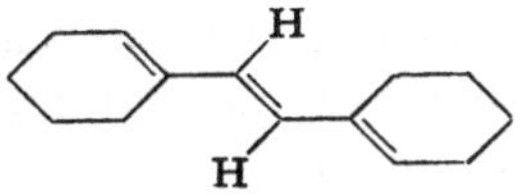

CH$_3$

H

HO

H H

R

(XI.) 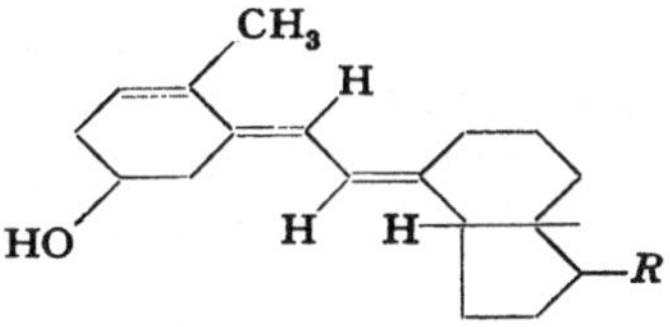 (XII.) Isovitamin D$_2$.

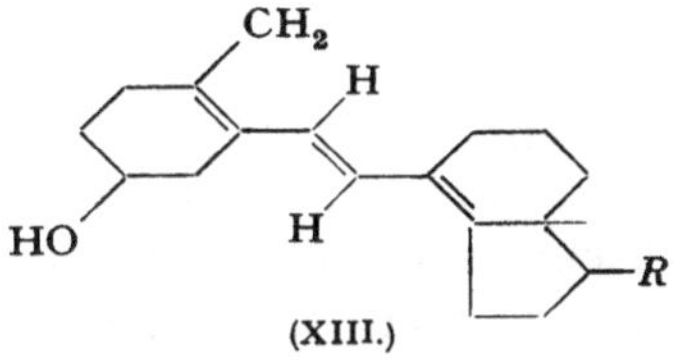

CH$_2$

H

HO

H

R

(XIII.)

der Extinktion auf 41 800 ein. Es erscheint daher genügend gerecht-fertigt, dem Tachysterin die bisher übliche *cis*-Struktur zuzuschreiben, die einer Ringöffnung des Lumisterins ohne anschließende *cis* → *trans*-Umlagerung entspricht.

Wie später noch diskutiert werden wird, ist es wahrscheinlich, daß Pyrotachysterin (*104*) die *trans*-Form des Tachysterins darstellt. Die dabei auftretende Verschiebung der Maxima auf 265, 289 und 300 mμ würde mit einer Aufhebung der sterischen Hinderung bei diesem Über-gang gut übereinstimmen.

Von den vier für *Vitamin D*$_2$ auf Grund der Äthylen-isomerie möglichen Strukturen spricht der leichte thermische Ringschluß, wie Modellbetrach-tungen zeigen, für Formel (V) bzw. noch besser für deren atropisomere s-*cis*-Form (IV), welche die bisher übliche Schreibweise darstellt. Besonders für letztere spricht das UV-Spektrum des Vitamins. Dieses weist ein äußerst kurzwelliges Maximum von überraschend niedriger Extinktion auf (265 mμ, $\varepsilon = 18300$), das seine Erklärung in der starken sterischen Hinderung dieser Form findet. Den gleichen Schluß zogen schon KOCH und DIMROTH im Rahmen der bereits erwähnten Arbeiten. Wenn diese Struktur des Vitamins D$_2$ allgemein mit ,,*cis*'' bezeichnet wird, so muß man darunter diejenige Form verstehen, bei der einmal unter Einbeziehung der freien Drehbarkeit um die Achse C$_{(6)}$—C$_{(7)}$ die Methylengruppe an C$_{(10)}$ dem ,,Ring'' *B* zugewendet werden kann und darüber hinaus das Molekül in seiner atropisomeren s-*cis*-Struktur vorliegt.

Einen direkten Beweis für die Formel (V) des Vitamins D$_2$ leistete die von CROWFOOT und DUNITZ (*17*) durchgeführte Röntgenstruktur-analyse des kristallinen Vitamin D$_2$-4-jod-5-nitro-benzoats. Dabei kommen die Autoren zu dem Ergebnis, daß hier das Molekül nicht in der für gewöhnlich beschriebenen ,,aufgerollten'', sondern in der gestreckten atropisomeren s-*trans*-Struktur (V) vorliegt. Diese entspricht der bisher üblichen Schreib-weise (IV), nur ist darin eine Hälfte der Molekel um die frei bewegliche

Achse $C_{(6)}$—$C_{(7)}$ um 180° gedreht. Darüber hinaus bestätigt die Untersuchung alle bisher auf chemischem Wege bewiesenen Strukturelemente.

So beschränkt sich das Problem der Feinstruktur des Vitamins D_2 nach Ausschluß der Strukturen (VIII) bis (X) nur noch auf die Frage, ob immer eine feste Zuordnung zu einer der beiden im Verhältnis der Atropisomerie stehenden Formen (IV) und (V) möglich ist oder ob sich für gewöhnlich ein Gleichgewicht zwischen den beiden Formen einstellt und eine Unterscheidung in reiner Form nicht möglich ist. CROWFOOT und DUNITZ halten es für möglich, daß die Festlegung der gestreckten Form (V) im speziellen Fall des kristallinen 4-Jod-5-nitrobenzoats verwirklicht, dagegen vielleicht in anderen Derivaten bzw. im Vitamin D_2 selbst Form (IV) mehr bevorzugt ist. Jedoch liegen darüber keine weiteren Untersuchungen vor.

Anderseits sehen VELLUZ und Mitarbeiter (79), wie schon erwähnt, in der vorliegenden Atropisomerie eine mögliche Erklärung für das Gleichgewicht Vitamin $D_2 \rightleftarrows$ Präcalciferol. Dabei wird dem instabilen Präcalciferol die s-cis-Konfiguration zugeschrieben. Wir möchten dagegen umgekehrt für möglich halten, daß gerade der cis-Form die größere Stabilität zuzuschreiben ist, da die 6 π-Elektronen darin einen Stabilisierungskomplex bilden könnten. Derartige Stabilisierungskomplexe nimmt man heute als Vorstufen im Verlauf von Diensynthesen an (32). Von welcher Seite aus man heute auch diese Frage der letzten Feinheit in der Struktur des Vitamins D_2 betrachtet, zur Zeit läßt sich keine endgültige Entscheidung geben. Nach unserer Meinung erscheint die quantitative Festlegung des Vitamins D bzw. gegebenenfalls auch des Präcalciferols besonders in der s-trans-Form zumindest in Lösung unwahrscheinlich, da kein Grund für ihre ausschließliche Stabilisierung angegeben werden kann.

III. Zusammenfassung neuerer Einzelergebnisse aus der Chemie der Vitamine und Provitamine D.

1. Neue Verbindungen des Vitamins D_2.

In bezug auf die Darstellung neuer Derivate oder Abwandlungsprodukte der Vitamine D und ihrer Isomeren ist seit Einstellung der Arbeiten seitens des Göttinger Instituts (1943) kaum noch eine Bereicherung erfolgt. Lediglich in einigen Patentschriften wird die Darstellung von mehr technisch interessierenden Derivaten beschrieben. Es sind dies unter anderem Doppelverbindungen der Vitamine D_2 und D_3 mit Cholesterin, Cholestanol und ähnlichen Verbindungen (34), die eine Stabilisierung derselben bewirken, sowie verschiedene bisher unbekannte kristalline Ester (63), wie Benzoat, Acetat, Propionat, Oleat usw. und der Methyl-bzw. Trityläther des Vitamins D_2 (55, 7). Größeres wissenschaftliches

Interesse kann ein ionisches, jodhaltiges Derivat des Vitamins D_2 beanspruchen (87) (XIV), das sich mit p-Dimethylamino-benzaldehyd in Benzol und Essigsäure über ein Perchlorat durch anschließende Behandlung mit Pyridin und Jod darstellen läßt. Wir möchten hierbei auch die chinoide Form (XIVa) berücksichtigen.

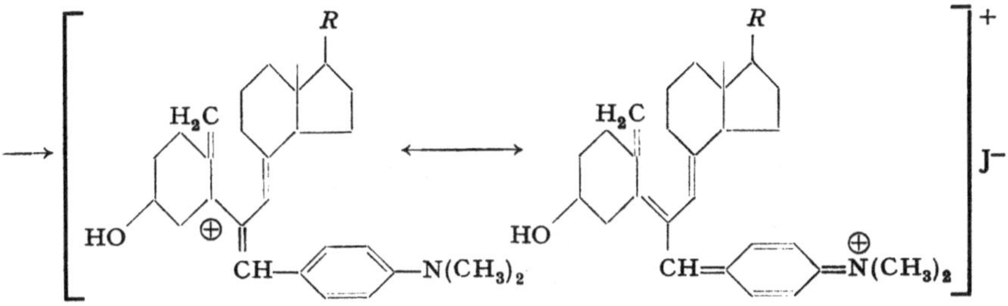

(XIV.)   (XIVa.)

Interesse bietet ferner die Tatsache, daß Vitamin D_2 durch Natrium/Alkohol-Hydrierung bei Verwendung von Alkoholen mit mehr als drei C-Atomen nicht wie üblich Dihydro-vitamin D_2 (99), sondern ausschließlich Dihydro-tachysterin liefert (88).

2. Neuere Methoden zur Darstellung von Provitaminen und Vitaminen D.

Während die Gewinnung des Provitamins D_2, des Ergosterins, noch heute aus Hefe oder Schimmelpilzen erfolgt, gibt es für 7-Dehydrocholesterin (XX), das Provitamin D_3, kein ausreichendes natürliches Vorkommen. Aus diesem Grunde sind zu seiner Darstellung, ausgehend vom Cholesterin, eine Vielzahl von Methoden ausgearbeitet worden. Dabei ist das alte Verfahren von WINDAUS (97), das vom Cholesterinacetat (XV) über die 7-Keto-Verbindung (XVI) nach Reduktion derselben zum 3,7-Dibenzoat (XVIII) führt und aus diesem durch Abspaltung von Benzoesäure und anschließende Verseifung das gewünschte 7-Dehydrocholesterin (XX) liefert, in seinen einzelnen Stufen teilweise bedeutend verbessert worden. So gelingt die Oxydation zum 7-Keton, die früher durch Chromsäure mit einer Ausbeute von 28% erreicht wurde, heute nach dem Verfahren von OPPENAUER (60) mit tertiärem Butylchromat mit 92%; die früher nach MEERWEIN-PONNDORF durchgeführte Reduktion und anschließende Überführung in das Dibenzoat nach FIESER durch Lithiumaluminiumhydrid mit 59%, gegenüber früher 29%. Schließlich konnte die ursprünglich thermisch bewirkte Abspaltung der Benzoesäure (58%) durch HASLEWOOD (29, vgl. 13) mittels Kochen in Dimethylanilin auf 69% gesteigert werden.

Im Verlauf einer neuen Darstellung wird Cholesterinacetat nach ZIEGLER mit N-Bromsuccinimid unter Belichtung zur 7-Brom-Ver-

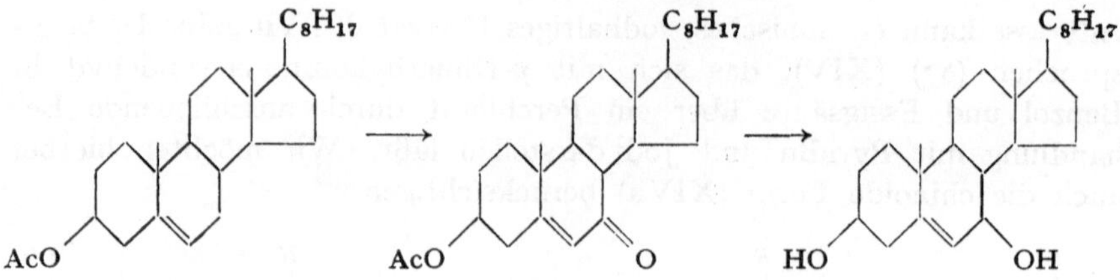

(XV.) Cholesterin-acetat.   (XVI.)   (XVII.)

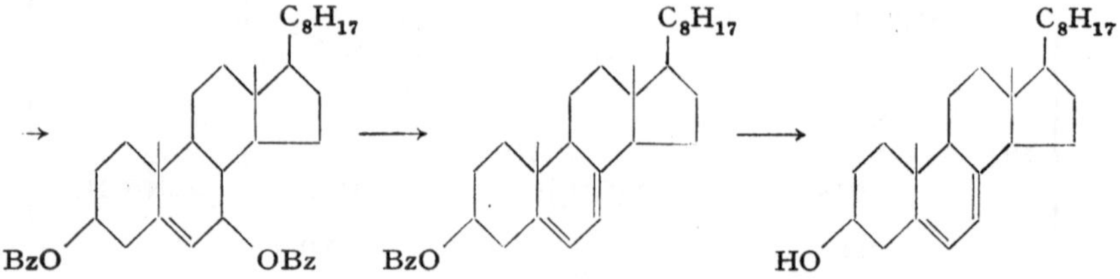

(XVIII.) (Bz = C₆H₅ . CO—)   (XIX.)   (XX.) 7-Dehydro-cholesterin.

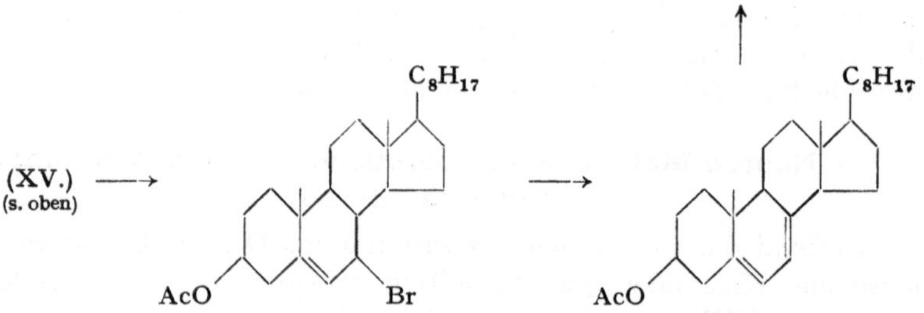

(XV.) ——→
(s. oben)

(XXI.)   (XXII.)

bindung (XXI) umgesetzt, danach mit Dimethylanilin Bromwasser-
stoff abgespalten (XXII) und schließlich an $C_{(3)}$ verseift (XX) (Ausbeute
insgesamt 23%) (*31, 11, 68, 82, 85, 40*). Die Bromierung von Cholesterinacetat gelingt nach einem Verfahren von Schaltegger (*72*) auch mit
freiem Brom, wenn man in Tetrachlorkohlenstoff unter Einwirkung von
Licht (410 mμ) arbeitet (Ausbeute 55—65%).

Vom biologischen Standpunkt aus ist die von Mazza und Migliardi (*52*)
aufgefundene, von Sah (*57, 71*) genauer erforschte *Dehydrierung des
Cholesterins* durch Chinone interessant. Dabei gaben 1,4-Chinone beim
6—8stündigen Erhitzen von Cholesterinacetat auf 120—135° Provitamin D$_3$. 1,2-Chinone zeigten nur schwache Wirkung, während aliphatische und aromatische Diketone unwirksam waren. In diesem
Zusammenhang scheint auch die Mitteilung von Banchetti (*5*) interessant, wonach Lösungen von Cholesterin bzw. seinem Acetat mit Benzo-
phenon, die monatelang dem Sonnenlicht ausgesetzt worden waren, bis
zu 6% Vitamin D$_3$ (bezogen auf Cholesterin) enthielten. (Acetophenon
gab schlechte bzw. unklare Resultate.)

Die *Aktivierung der Provitamine* erfolgt fast durchweg mit Quecksilberdampflampen oder einer Magnesiumfunkenstrecke. In einer Vielzahl von Patenten wurde ferner die Aktivierung mittels Röntgen-, α,β,γ-Strahlen, Gleichstrom oder Wechselstrom beschrieben. Genauer bekannt geworden und technische Bedeutung erlangt hat darunter nur der sog. WHITTIER-Prozeß (*18, 89*). Dabei wird am Boden eines evakuierten vertikalen Rohres Indium zum Schmelzen gebracht und kurz über dem Metall Hochspannung angelegt. Auf die Oberfläche des Indiums wird nun Ergosterin aufgetragen, welches verdampft und im elektrischen Feld zu Vitamin D_2 umgewandelt wird. Eine Analyse des Reaktionsproduktes hat ergeben, daß es aus Vitamin D_2 (30%), Neo-ergosterin und einem bisher unbekannten Steroid besteht.

3. Neue Vitamine D und Beiträge zum Zusammenhang zwischen Konstitution und physiologischer Wirkung.

Nach Abschluß der Untersuchungen des Göttinger Arbeitskreises waren Vitamine D mit folgenden *Seitenketten* bekannt:

D_2 $\mathrm{HC{-}CH{=}CH{-}CH{-}CH}$ mit $\mathrm{CH_3}$ (oben), $\mathrm{CH_3}$ (unten am HC), $\mathrm{CH_3}$ und $\mathrm{CH_3}$ (am endständigen CH)   (*79, 3*)

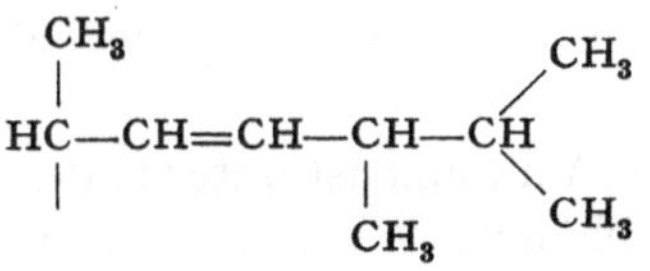

D_3 $\mathrm{HC{-}CH_2{-}CH_2{-}CH_2{-}CH}$ mit $\mathrm{CH_3}$ (oben und unten am HC), $\mathrm{CH_3}$, $\mathrm{CH_3}$ (am endständigen CH)   (*73, 102*)

D_4 $\mathrm{HC{-}CH_2{-}CH_2{-}\overset{(24)}{CH}{-}CH}$ mit $\mathrm{CH_3}$ (oben und unten am HC), $\mathrm{CH_3}$ (am (24)-CH), $\mathrm{CH_3}$, $\mathrm{CH_3}$ (am endständigen CH)   (*96, 103, 95*)

D_5 $\mathrm{HC{-}CH_2{-}CH_2{-}CH{-}CH}$ mit $\mathrm{CH_3}$ (oben am HC), $\mathrm{C_2H_5}$ (unten), $\mathrm{CH_3}$, $\mathrm{CH_3}$ (am endständigen CH)   (*105*)

D_6 $\mathrm{HC{-}CH{=}CH{-}CH{-}CH}$ mit $\mathrm{CH_3}$ (oben am HC), $\mathrm{C_2H_5}$ (unten), $\mathrm{CH_3}$, $\mathrm{CH_3}$ (am endständigen CH)   (*30, 50*)

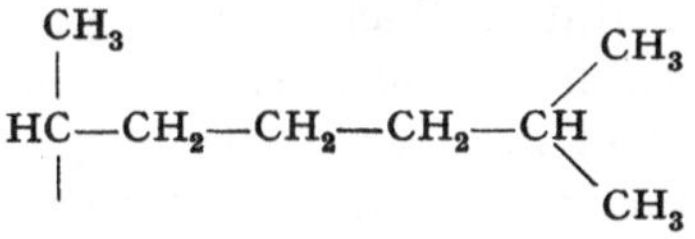

ferner:

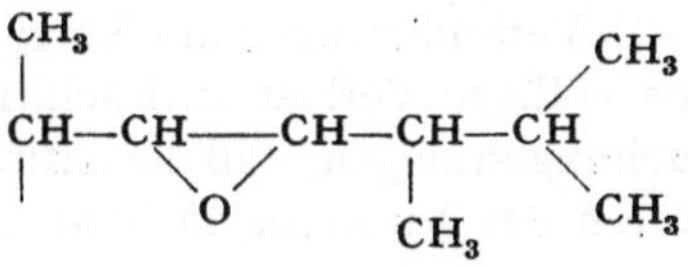

$\mathrm{CH{-}CH{-}CH{-}CH{-}CH}$ mit $\mathrm{CH_3}$ (oben), O-Brücke, $\mathrm{CH_3}$, $\mathrm{CH_3}$, $\mathrm{CH_3}$ (*25*)

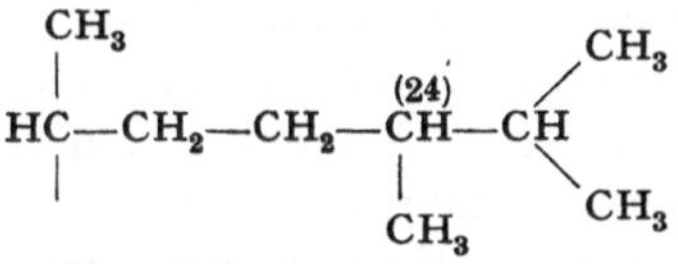

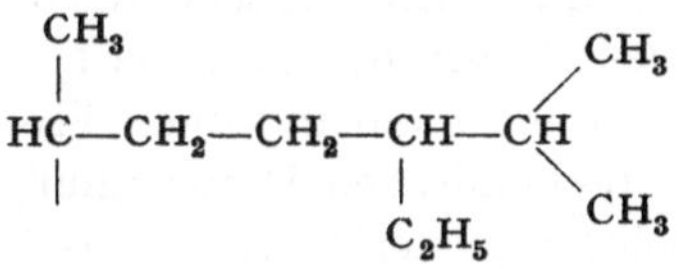

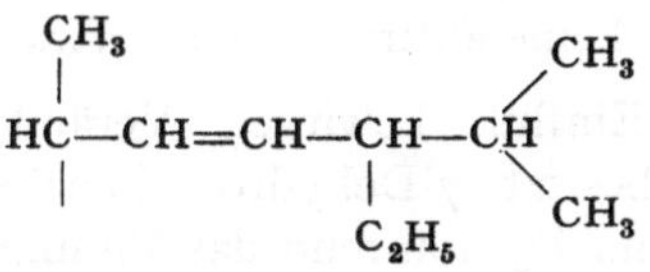

Darüber hinaus konnte von Ruigh (70) Vitamin D_7 über das Dehydro-campesterin dargestellt werden.

$$D_7 \qquad \underset{\displaystyle |}{\overset{\displaystyle CH_3}{\overset{\displaystyle |}{HC}}}-CH_2-CH_2-\underset{(24)}{\overset{\displaystyle CH_3}{\overset{\displaystyle |}{CH}}}-CH\begin{matrix} \nearrow CH_3 \\ \searrow CH_3 \end{matrix} \qquad (70)$$

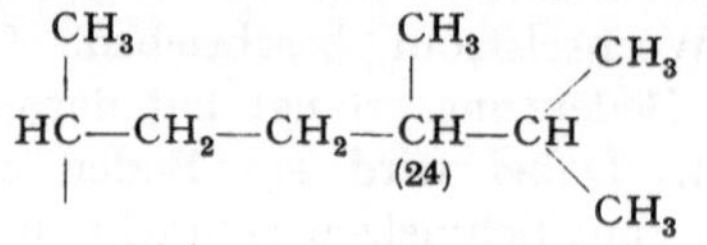

Ein weiteres Provitamin, das $\Delta^{5,\,7}$-Nor-cholestadien-3β-ol, wurde von Alberti, Camerino und Mamoli (1) gewonnen, jedoch nicht bestrahlt und auf seine Provitamin-Wirksamkeit geprüft.

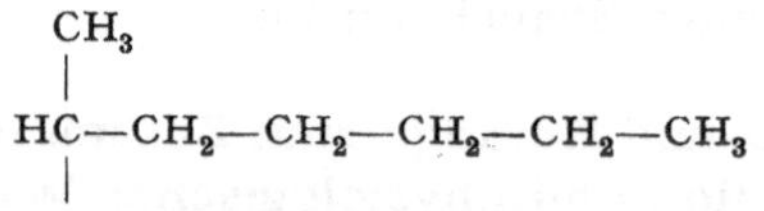

$$\underset{\displaystyle |}{\overset{\displaystyle CH_3}{\overset{\displaystyle |}{HC}}}-CH_2-CH_2-CH_2-CH_2-CH_3 \qquad (1)$$

Van der Vliet (83, 84) gelang die Isolierung eines Provitamin-Gemisches aus Muscheln, in dem neben 7-Dehydro-cholesterin und Ergosterin $\Delta^{5,\,7,\,22}$-Cholestatrien-3β-ol und ein unbekanntes Provitamin enthalten waren.

Die *Wirkung* der einzelnen Vitamine ist unter anderem sehr stark vom Charakter der Seitenkette abhängig. Wie schon länger bekannt ist, sind D_2 und D_3 gegenüber der Ratten-Rachitis von gleicher Wirksamkeit. Gegenüber der Hühnchen-Rachitis jedoch ist D_3 wesentlich stärker wirksam als D_2. Vitamin D_4 steht zwischen D_2 und D_3. Während die eine Äthylgruppe tragenden Vitamine D_5 und D_6 praktisch gar keine Wirkung mehr besitzen, zeigt D_7 wieder ein Zehntel der Wirksamkeit von D_2. Der Vergleich zum Vitamin D_4 zeigt hier, daß die auf Stereoisomerie an $C_{(24)}$ beruhenden Aktivitätsunterschiede bedeutender sind als das Vorhandensein bzw. Fehlen einer Doppelbindung in der Seitenkette. Daß die Anwesenheit der Seitenkette für die antirachitische Wirksamkeit notwendig ist, zeigt die Bestrahlung des $\Delta^{5,\,7}$-Androstadien-3,17-diols (15) und des $\Delta^{5,\,7}$-Androstadien-3-ols (58), die zu unwirksamen Produkten führt, obwohl der Charakter der photochemischen Umwandlung in Zeit und Art völlig der gleiche ist wie beim Ergosterin. Auch das Vorhandensein der Gallensäure-Seitenkette läßt keine antirachitische Wirkung auftreten (30).

Einen gleich starken Einfluß bewirken Veränderungen an der $C_{(3)}$-Hydroxylgruppe. So besitzt 7-Dehydro-epi-cholesterin (100) nur $^1/_{20}$ der Aktivität von Vitamin D_3, während das Vitamin D_2-Keton überhaupt keine mehr zeigt (91). Veränderungen am Ringgerüst und im Triensystem rufen ebenfalls den völligen Verlust antirachitischer Wirksamkeit hervor. Alle diese Untersuchungen zeigen, daß die antirachitische Wirkung sehr eng an die Konstitution der Vitamine D_2 und D_3 gebunden ist.

IV. Ozon-Abbau des Vitamins D₂; zugleich ein Beitrag zur Stereochemie der Steroide.

Durch zahlreiche Arbeiten konnte bei der überwiegenden Mehrzahl der Steroide die exakte sterische Einordnung der Substituenten und Seitenketten in bezug auf ihre Lage zu den beiden angulären Methylgruppen vorgenommen werden. Unbekannt ist bis heute aber immer noch die absolute Konfiguration und die Zuordnung zum *DL*-System des Glycerinaldehyds und der Zucker. Die Festlegung der angulären Methylgruppen als β-ständig, d. h. oberhalb der Ringebene liegend, war bekanntlich seinerzeit willkürlich erfolgt.

Einen ersten Versuch dieser Zuordnung unternahmen BERGSTRÖM, LARDON und REICHSTEIN (*8, 7, 48, 49*). Sie ozonisierten Vitamin D₂-methyläther (IVa) und spalteten das erhaltene Ozonid oxydativ mit Kaliumpermanganat in Aceton auf. Es gelang ihnen, nach schwieriger Trennung (am besten mittels Verteilungschromatographie) eine optisch-aktive (—)β-Methoxy-adipinsäure zu gewinnen, die über das Bariumsalz gereinigt und als Benzylthiuroniumsalz und Dianilid charakterisiert wurde. Nach den bisherigen Vorstellungen wäre für sie Struktur (XXIII) zu erwarten gewesen.

Daneben konnte aus Hydrochinon-monomethyläther durch Hydrierung und anschließende Oxydation eine synthetische β-*DL*-Methoxy-adipinsäure gewonnen werden und das Racemat mit Strychnin in die beiden optisch-aktiven Formen aufgespalten und beide als Benzylthiuroniumsalze und als Dianilide charakterisiert werden. Dabei zeigte die (—)β-Methoxy-adipinsäure in allen Daten weitgehende Übereinstimmung mit der aus Vitamin D₂ gewonnenen Säure. [Die (+)β-Methoxy-adipinsäure kristallisiert schwerer und läßt sich aus den Mutterlaugen gewinnen.]

(—)β-Methoxy-adipinsäure mußte nun in eindeutiger Weise mit einem Stoff bekannter Konfiguration verknüpft werden. Als Ausgangspunkt dafür diente die von FREUDENBERG und BRAUNS (*27*) in ihrer Konfiguration bewiesene *L*-Äpfelsäure (XXIV), die über den Dimethylester (XXV) mit Lithiumaluminiumhydrid in das Diol (XXVI) übergeführt wurde, das sich über das Diacetat reinigen ließ. Nach Umsatz zum Ditosylat und zum Dijodid (XXVIII) (in Aceton mit Natriumjodid) wurde das Dinitril (XXIX) gewonnen, das zur Methoxy-adipinsäure (XXIII) verseift und als Dianilid der schwer kristallisierenden (+)β-Methoxy-adipinsäure charakterisiert werden konnte.

Die aus *L*-Äpfelsäure synthetisierte, in ihrer Struktur eindeutige (+)β-Methoxy-adipinsäure ist also *nicht* identisch mit der aus Vitamin D₂ gewonnenen Säure, der infolgedessen die Konfiguration (XXX) zugeschrieben werden muß. Dies würde aber bedeuten, daß im Vitamin D₂ die OH-Gruppe im Gegensatz zur bisherigen Annahme nach „unten" gerichtet

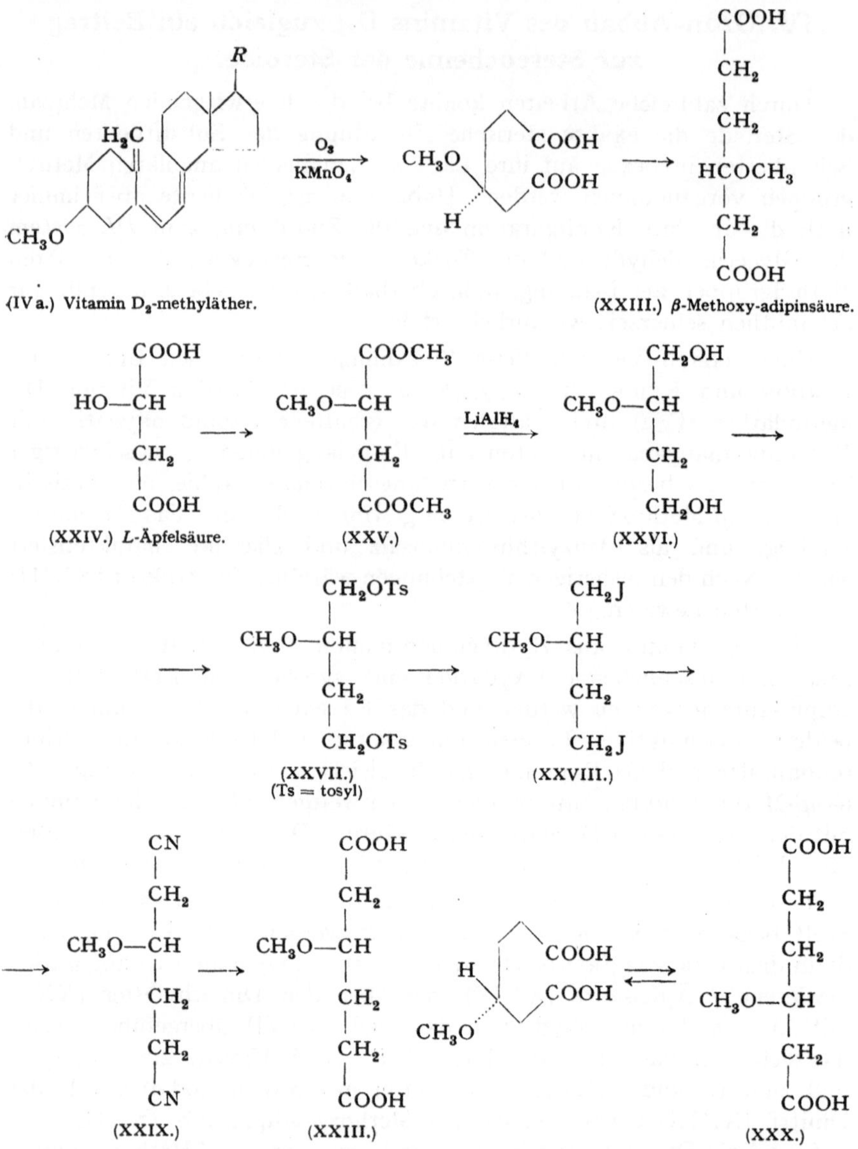

(IVa.) Vitamin D_2-methyläther.   (XXIII.) β-Methoxy-adipinsäure.

(XXIV.) L-Äpfelsäure.   (XXV.)   (XXVI.)

(XXVII.)   (XXVIII.)
(Ts = tosyl)

(XXIX.)   (XXIII.)   (XXX.)

sein muß und — in Übertragung auf alle Steroide — die bisher gebräuchlichen Formeln durch ihre Spiegelbilder zu ersetzen wären, wenn in der
Zuckerreihe die absolute Konfiguration mit den seinerzeit von E. FISCHER
willkürlich festgelegten Formelbildern übereinstimmt. Dies ist, wie
W. KUHN (*47*) und BIJVOET (*10, 62*) zeigen konnten, der Fall. Die sich
daraus ergebende Konsequenz der Umstellung aller Steroidformeln zieht

REICHSTEIN aber nicht, da noch gewisse kleine Unsicherheiten vorhanden sind. Diese bestehen vor allem in den sehr geringen Drehungswerten der Dianilide: $+2,7° \pm 0,7°$ bei der durch Spaltung des Racemats erhaltenen Säure, $+3,6° \pm 1,5°$ bei der durch Synthese gewonnenen Säure. Darüber hinaus verläuft die Darstellung der optisch aktiven β-Methoxy-adipinsäure über mehrere nichtkristalline Stufen.

Was die sterische Anordnung der *Hydroxyl-Gruppe* im Vitamin D_2 anbetrifft, so dürfte kaum daran zu zweifeln sein, daß ihre Lage die gleiche ist wie im Ergosterin. Außer der röntgenographischen Messung von CROWFOOT liegt noch die folgende Reaktionsfolge aus dem WINDAUS-schen Institut vor: Vitamin D_2 liefert beim einfachen Erhitzen auf 180° das Gemisch der Produkte Pyro-calciferol und Isopyro-Vitamin D; in diesen ist die 9,10-Bindung wiederhergestellt. Wird nun die Asymmetrie am $C_{(9)}$ durch Dehydrierung mittels Mercuriacetat beseitigt, so erhält man schließlich Dehydro-lumisterin und Dehydro-ergosterin. Die Rückgewinnung von Dehydro-ergosterin aus Ergosterin unter den vorgenannten Bedingungen darf als überzeugendes Argument für die unveränderte Stellung der OH-Gruppe angesehen werden.

V. Isomerisierung des Vitamins D_2.

Im Verlauf eigener Versuche (*37*) stellten wir fest, daß sich Vitamin D_2 sowohl mit Bortrifluorid-ätherat als auch Phosphorsäure zu einem einheitlichen neuen Isomeren umlagern läßt. Es zeichnet sich durch ein charakteristisches Hauptmaximum bei 290 mμ von außerordentlicher Intensität aus ($\varepsilon = 41800$) (Nebenmaxima bei 280 und 302 mμ). Wir bezeichneten diese neue Verbindung aus Gründen, die später noch besprochen werden, als iso-Tachysterin (XIII; Formel auf S. 99).

Ähnliche spektrale Veränderungen hatten schon THIBAUDET (*74*) und MEUNIER (*54, 65*) nach Einwirkung von Antimontrichlorid in einer Lösung von Dichloräthan beobachtet, das mit 1% Glycerin-dichlorhydrin oder 0,02 bis 0,2% Acetylchlorid versetzt war.

Nach Eliminierung des Katalysators wurden in allen Fällen Produkte erhalten, deren Maxima ähnlich wie im iso-Tachysterin liegen, diesen jedoch nicht völlig gleichen (280, 292 und 305 mμ). Trotzdem möchten wir annehmen, daß es sich immer um den gleichen Stoff handelt und die geringe Verschiedenheit der Spektren nur auf ungenauer Messung bzw. ungenügender Reinheit der Substanzen beruht. MEUNIER gibt darüber hinaus die Daten eines kristallinen 4-Methyl-3,5-dinitrobenzoesäureesters seines Umlagerungsproduktes an, dessen Schmelzpunkt dem des von uns gewonnenen iso-Tachysterin-4-methyl-3,5-dinitrobenzoats sehr ähnlich ist (123—125° gegenüber 126—127,5°). Jedoch weichen alle weiteren Angaben über das Verhalten dieses Stoffes, die ohne exakte Beschreibung der experimentellen Bedingungen gemacht wurden, sowie die theoretische Auswertung und möglichen Rückschlüsse auf seine Struktur so weit von unseren Ergebnissen ab, daß es sich dabei entweder tatsächlich um eine andere Verbindung handeln muß, wahrscheinlicher aber diese Ergebnisse einer generellen Revision bedürfen.

Unter den wenigen bisher bekannten ring-offenen Steroiden vom Typus der Bestrahlungsprodukte des Ergosterins stimmt nur das Pyrotachysterin (*104*) in der Lage des Hauptmaximums der UV-Absorption mit dem von uns neu gewonnenen Isomeren nahezu überein. Jedoch ist dort hauptsächlich die Lage der Nebenmaxima eine andere (265, *289* und 300 mμ) und damit — wie auch im Hinblick auf das unterschiedliche Drehungsvermögen — eine Identität beider Produkte unwahrscheinlich.

Wie schon die relativ langwellige Lage seines UV-Maximums zeigt, konnte es sich beim iso-Tachysterin um kein ring-geschlossenes Isomeres handeln. Während der Umlagerung mußte also entweder nur eine Verlagerung der semicyclischen Doppelbindung $C_{(10)}$—$C_{(19)}$ in den Ring *A* nach $C_{(1)}$—$C_{(10)}$ zu der bisher unbekannten $\Delta^{1,\,10-5,\,6-7,\,8}$-Trien-konfiguration (VI), oder eine Verschiebung des gesamten Trien-systems stattgefunden haben.

Bei einfacher Verlagerung nur der einen Doppelbindung hätte sich beim Ozonabbau die von Windaus (*94*) bei der Konstitutionsaufklärung des Vitamins D$_2$ (IV, S. 84) aufgefundene C_{13}-Abbauketosäure (XXXI)

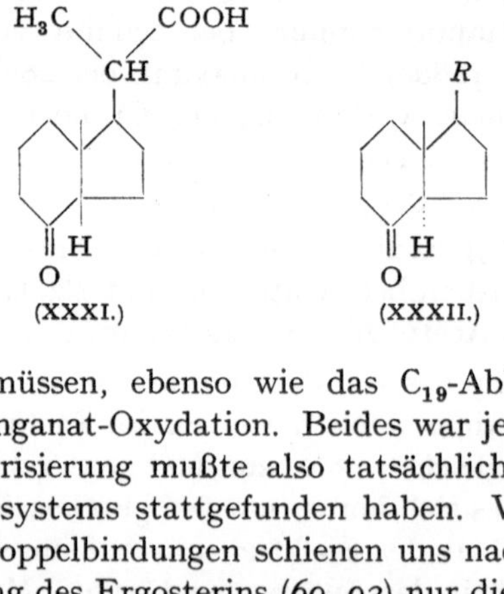

(XXXI.)   (XXXII.)

nachweisen lassen müssen, ebenso wie das C_{19}-Abbauketon (XXXII) durch Kaliumpermanganat-Oxydation. Beides war jedoch nicht der Fall. Während der Isomerisierung mußte also tatsächlich eine Verschiebung des gesamten Trien-systems stattgefunden haben. Von allen möglichen Wanderungen der Doppelbindungen schienen uns nach den Erfahrungen bei der Isomerisierung des Ergosterins (*69*, *93*) nur die als wahrscheinlich, die gemäß der Reihenfolge der C-Atome 5, 6, 7, 8, (9), 14, 15 verlaufen. Wenn man beachtet, daß Bortrifluorid auch *cis-trans*-Umlagerungen an Doppelbindungen katalysiert (*64*), ergeben sich somit folgende Möglichkeiten für die Konstitution des iso-Tachysterins, wenn man atropisomere Formen unberücksichtigt läßt. Bei einfacher Wanderung:

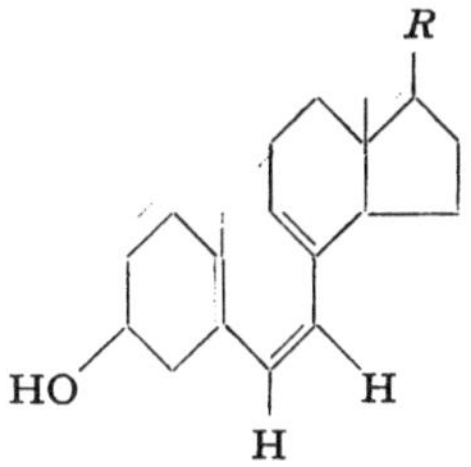

(III.) Tachysterin.

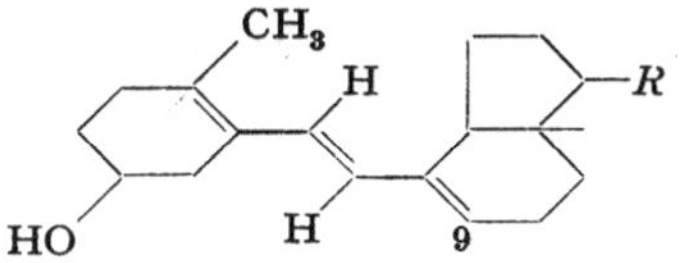

(VII.) Pyrotachysterin (?).

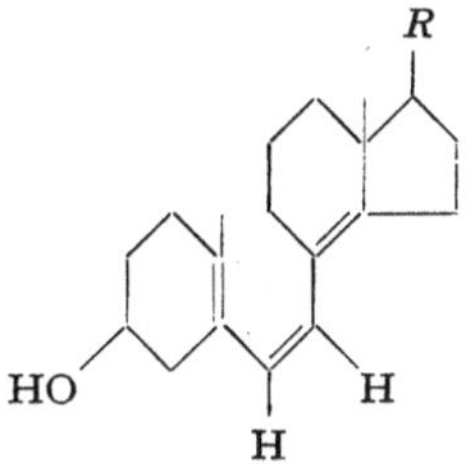

(XXXIII.)

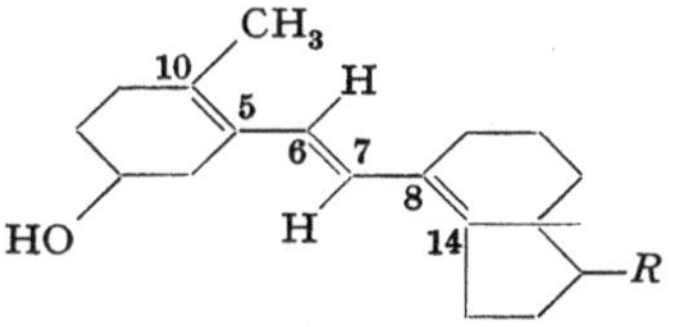

(XIII.) iso-Tachysterin.

Darunter scheiden (III) und (VII) aus, weil sie dem Tachysterin bzw. Pyro-tachysterin (?) entsprechen, und außerdem eine Wanderung des Trien-systems nach dem sekundären $C_{(9)}$ weniger wahrscheinlich ist als nach $C_{(14)}$. Ganz allgemein wird sich während der Isomerisierung mit Bortrifluorid-ätherat mit großer Wahrscheinlichkeit die energetisch günstigste Konfiguration einstellen, d. h. diejenige geringster sterischer Hinderung und maximaler coplanarer Ausrichtung des chromophoren Systems. Unter diesem Gesichtspunkt ist von den beiden Strukturen (XXXIII) und (XIII) die all-*trans*-Konfiguration in (XIII) zweifellos gegenüber der anderen bevorzugt. Die vier möglichen $\Delta^{5,\,6-7,\,8-14,\,15}$-Trien-strukturen (XXXIV)—(XXXVII) werden als unwahrscheinlich erachtet. Der Grund dafür liegt in der überraschenden Ähnlichkeit der Spektren von iso-Tachysterin mit dem eines später durch Partialsynthese gewonnenen und in seiner Konstitution gesicherten iso-Vitamins D_2 (XII, S. 109), das zweifellos eine all-*trans*-Verknüpfung des konjugierten Systems besitzt. Jede der Formeln (XXXIV)—(XXXVII) zeigt jedoch infolge Einbau einer Doppelbindung in den Ring D eine Störung dieser all-trans-Konfiguration, die sich wahrscheinlich in einer größeren spektralen Verschiedenheit bemerkbar machen müßte. Der chemische Strukturbeweis bereitet größere Schwierigkeiten und muß noch gegeben werden.

iso-Tachysterin besitzt also mit großer Wahrscheinlichkeit die Konstitution (XIII), bei der die drei Doppelbindungen in linearer Verknüpfung entsprechend einer all-*trans*-Konfiguration bei langkettigen Polyenen zwischen den C-Atomen 10, 5—6, 7—8, 14 liegen. Mit dieser Struktur steht vor allem die außerordentlich starke Rotverschiebung des Ab-

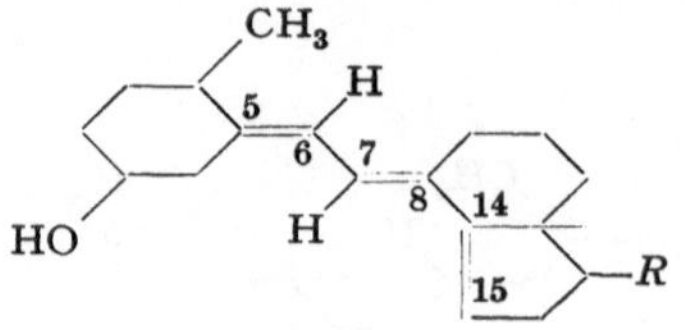

(XXXIV.) 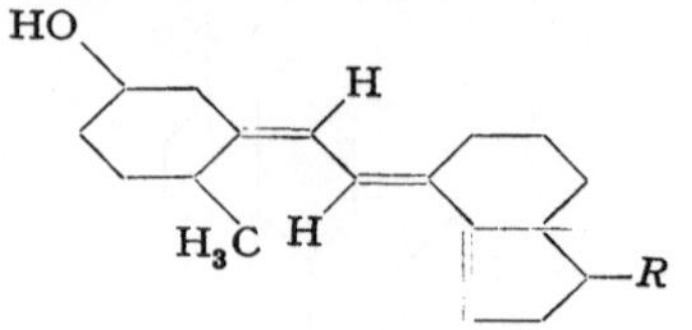   (XXXV.)

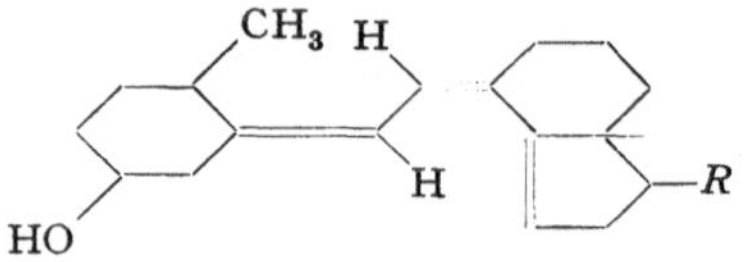

(XXXVI.) 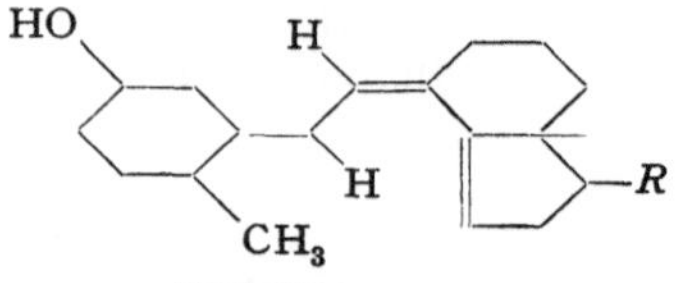   (XXXVII.)

sorptionsmaximums um 25 mμ während der Isomerisierung in Einklang. Diese ließe sich nicht allein durch den Substituenten-zuwachs am chromophoren System von vier auf sechs erklären, wohl aber dann, wenn man die Aufhebung der starken sterischen Hinderung beim Übergang von Vitamin D$_2$ in das völlig freie iso-Tachysterin einbezieht.

Bezüglich der Benennung der neu gewonnenen Verbindung hielten wir uns an folgende Richtlinie: Verbindungen mit Doppelbindungen zwischen den C-Atomen 5,6 und 7,8 sollten als Isomere des Vitamins D und solche mit einer Doppelbindung zwischen den C-Atomen 6 und 7 als Isomere des Tachysterins angesprochen werden. Es sollte jedoch offen bleiben, ob die weitere Entwicklung der Chemie der ring-offenen ungesättigten Steroide eines Tages die Einführung einer exakteren Nomenklatur notwendig macht.

Es erscheint daher schon jetzt angebracht, einen *Nomenklaturvorschlag* heranzuziehen (*85a*), nach dem bei ring-offenen Steroiden die geöffnete Bindung durch die Vorsilbe „seco" gekennzeichnet und im übrigen der Name des entsprechenden ringgeschloßenen Grundkörpers zugrunde gelegt wird. Danach kann die als iso-Tachysterin bezeichnete Verbindung auch 9,10-seco-Δ[10, 5—6, 7—8, 14—22, 23]-Ergostatetraen-3β-ol genannt werden.

iso-Tachysterin ließ sich aus den Rohprodukten der mit Bortrifluorid in Benzol katalysierten Umlagerung leicht durch Hochvakuumdestillation abtrennen und reinigen. Bei Verwendung von Phosphorsäure in Eisessig gelang dies trotz besserer Rohausbeute jedoch nicht, da dort die auftretenden Nebenprodukte nicht polymerisiert waren. Die reine Substanz stellt ein schwach gelblich gefärbtes Glas dar, das bisher analog dem Tachysterin nicht kristallin erhalten werden konnte.

Jedoch kristallisiert sein 4-Methyl-3,5-dinitrobenzoesäureester in feinen Nadeln (Schmelzp. 126—127,5°; $[\alpha]_D = +55,6°$ in Chloroform) und liefert nach der Verseifung den Alkohol wieder in öliger Form zurück. Das so gereinigte iso-Tachysterin zeigt eine Drehung von —71° und folgende Maxima der UV-Absorption: 280 mμ, $\varepsilon = 32000$; 290 mμ, $\varepsilon = 41800$; 302 mμ, $\varepsilon = 31200$.

iso-Tachysterin erwies sich als thermisch sehr stabil; selbst nach achtstündigem Erhitzen im Hochvakuum auf 230° wurde noch keinerlei Veränderung festgestellt. Bestrahlung mit kontinuierlichem UV-Licht (bis 220 mμ) aber führte innerhalb kurzer Zeit zu völligem Abbau seines charakteristischen Spektrums, jedoch ließ sich bislang kein Vitamin D$_2$ im Bestrahlungsprodukt chemisch nachweisen.

Im Verlauf einer Alkoholat-oxydation mit Aluminium-iso-propylat/ Cyclohexanon in Toluol ließ sich iso-Tachysterin zum Keton (XXXVIII) oxydieren, das nach sorgfältiger Reinigung zur Kristallisation gebracht werden konnte (Schmelzpunkt 79—83°; $\lambda_{max.}$ 280 mμ, $\varepsilon = 34400$; 290 mμ, $\varepsilon = 44900$; 302 mμ, $\varepsilon = 32900$) und ein schön kristallisierendes Semicarbazon lieferte (Schmelzpunkt 182—184°, $[\alpha]_D = -17°$). Die unveränderte Lage der UV-Absorptionsmaxima sowohl im Keton als auch in seinem Semicarbazon (280 mμ, $\varepsilon = 36800$; 290 mμ, $\varepsilon = 46200$; 302 mμ, $\varepsilon = 33800$) bewies die gleiche Stellung der Doppelbindungen nach der Oxydation.

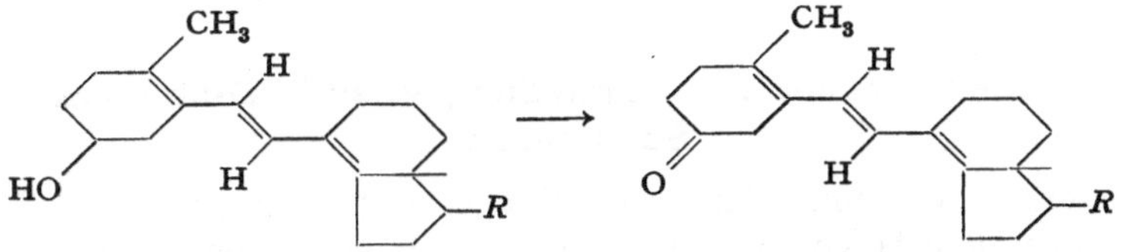

(XIII.) iso-Tachysterin.   (XXXVIII.)

Die Spaltung des Semicarbazons von (XXXVIII) mittels Oxalsäure führt zu einem öligen isomeren Keton, das eine außerordentlich intensive UV-Absorption bei 324 mμ aufweist. Sein gut kristallisierendes Semicarbazon schmilzt bei 211—212° und zeigt folgende Maxima der UV-Absorption: 320 mμ, $\varepsilon = 48000$; 334 mμ, $\varepsilon = 72000$; 351 mμ, $\varepsilon = 65500$. Ihre Lage und Intensität zeigt, daß es sich bei diesem Produkt um das völlig durchkonjugierte Keton (XXXVIIIa) handeln muß. Gemäß dem erwähnten Nomenklaturvorschlag ist es als 9,10-seco-$\Delta^{4,\,5-6,\,7-8,\,14-22,\,23}$-Ergostatetraen-3-on zu bezeichnen.

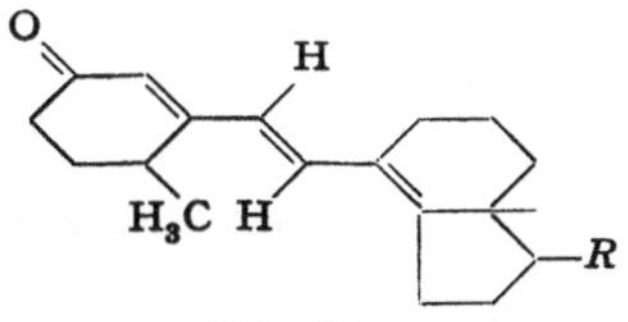

(XXXVIIIa.)

Wie inzwischen auf Grund weiterer Arbeiten zu erkennen ist und später noch beschrieben wird, stellt iso-Tachysterin nicht nur für Vitamin D_2, sondern auch für Tachysterin und das später beschriebene iso-Vitamin D_2 (S. 109) einen Endzustand chemischer Isomerisierung dar, ja es erscheint sogar als möglich, daß sich alle denkbaren 9,10-seco-Trien-systeme in dieser Richtung stabilisieren. Auf Grund seines äußerst charakteristischen Spektrums vermag es selbst in sehr unreinen Gemischen als qualitativer Nachweis für die Anwesenheit 9,10-ring-offener Triene zu dienen.

VI. Synthetische Versuche in der Vitamin D-Reihe.

Unter allen in ihrer Struktur bekannten Vitaminen stellen die D-Vitamine zusammen mit ihren Vorstufen bis auf den heutigen Tag die einzigen dar, deren Total- oder sogar Partial-synthese aus einzelnen Bausteinen noch nicht geglückt ist. Scheint dies auf den ersten Blick hin überraschend, so sollen am Schluß dieses Kapitels, nach Betrachtung aller bisherigen Ansätze in dieser Richtung, die Schwierigkeiten einer solchen Synthese zusammenfassend betrachtet werden.

1. Modellversuche zur Darstellung Vitamin D-ähnlicher Substanzen.

Als erstes Vitamin D-ähnliches Modell wurde im Jahre 1938 von BURKHARDT und HINDLEY (*12*) das α,β-Di-Δ^1-cyclohexenyl-äthylen (XI) dargestellt, indem sie zunächst Cyclohexanon und Acetylen unter der katalytischen Wirkung von Natrium-tert.-butylat zum Äthinyl-cyclo-hexanol kondensierten und dieses mit Kalium-tert.-butylat in siedendem

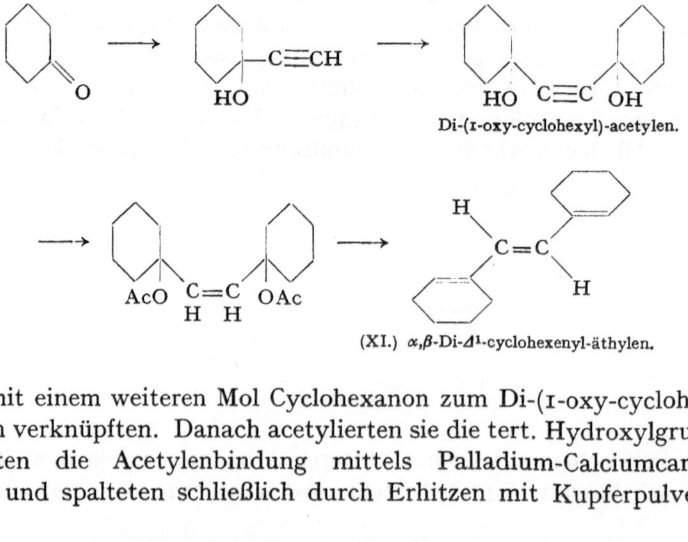

Di-(1-oxy-cyclohexyl)-acetylen.

(XI.) α,β-Di-Δ^1-cyclohexenyl-äthylen.

Äther mit einem weiteren Mol Cyclohexanon zum Di-(1-oxy-cyclohexyl)-acetylen verknüpften. Danach acetylierten sie die tert. Hydroxylgruppen, hydrierten die Acetylenbindung mittels Palladium-Calciumcarbonat partiell und spalteten schließlich durch Erhitzen mit Kupferpulver auf

145—160° zwei Mol Essigsäure ab. (Variationen in den einzelnen Stufen führten zu Mißerfolgen.) Damit war zum ersten Male das einfachste Analogon des Tachysterin-triensystems dargestellt worden, das vor allen Dingen in bezug auf sein spektrales Verhalten im Vergleich zum Tachysterin interessant war (s. S. 88 und 112), 259,5, *269* und 281 mμ; $\varepsilon = 42\,600$).

In Fortführung solcher Versuche zur Darstellung Vitamin D-ähnlicher Substanzen beschritten zwei Arbeitskreise gleichzeitig einen anderen Weg, indem sie Cyclohexyliden-acetaldehyd im Verlauf einer Aldolkondensation mit geeignet erscheinenden Ring *A*-Bausteinen umsetzten.

So kondensierte DIMROTH (*19, 20*) Cyclohexyliden-acetaldehyd in wäßriger 1proz. Natronlauge mit Cyclohexanon und erhielt ein kristallines, luftempfindliches Keton (XXXIX) (Tabelle 1). Bei anschließender GRIGNARD-Reaktion wurde ein luftempfindliches Öl erhalten, bei dem zum größten Teil schon Wasserabspaltung in den Ring, also nicht unter Bildung einer semicyclischen Methylengruppe, stattgefunden hatte (278 mμ,

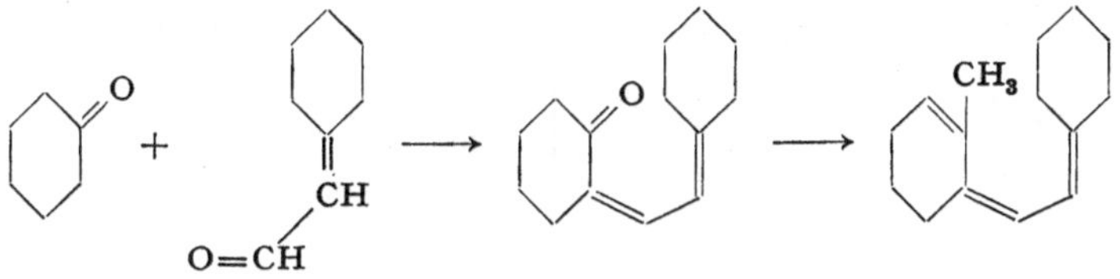

Cyclohexyliden-
acetaldehyd.   (XXXIX.)   (XL).

$\varepsilon = 22\,800$) (XL). Ein Versuch der Einführung dieser Methylengruppe nach WALLACH mit Bromessigester und Zink versagte, ließ vielmehr auch hier die Wasserabspaltung in den Ring laufen, ersichtlich am Spektrum des Reaktionsprodukts (XL), das ein charakteristisches Maximum außerordentlicher Intensität besitzt (280 mμ, $\varepsilon = 37\,800$).

ALDERSLEY und Mitarbeiter (*2*) verwandten zur Kondensation *p*-Acetoxy-cyclohexanon, das sie mit Cyclohexyliden-acetaldehyd ebenfalls in

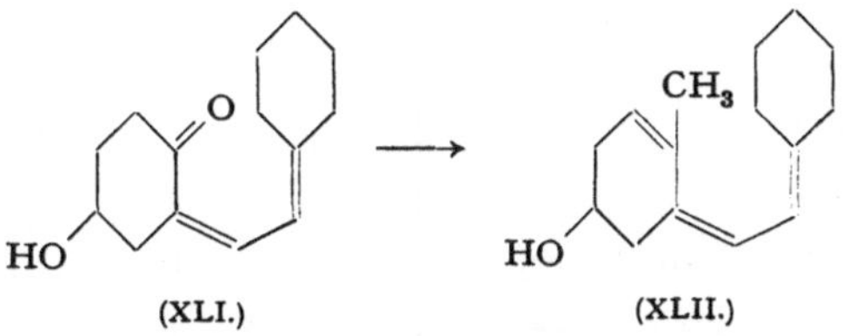

(XLI.)   (XLII.)

wäßriger Natronlauge zum Oxy-keton (XLI) kondensierten. (306 mμ, $\varepsilon = 10\,500$, in Alkohol). Der Versuch, die Ketogruppe in gleicher Weise, wie eben beschrieben, durch eine Methylengruppe zu ersetzen, scheiterte

Tabelle 1. Modellsubstanzen der Vitamin D-Reihe und ihre Spektra.

Formel	λ_{max} (mμ)	ε	Formel	λ_{max} (mμ)	ε
			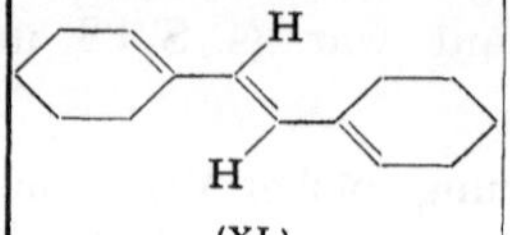 (XI.)	269	42 600
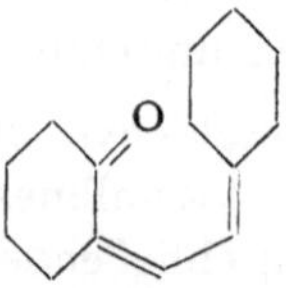 (XXXIX.)	(in Äther) 297	25 300	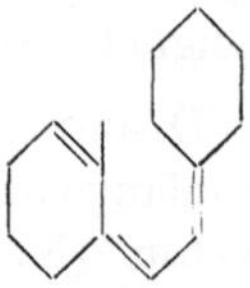 (XL.)	280	37 800
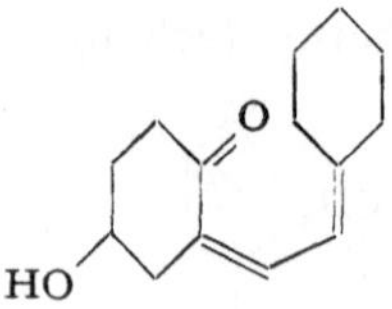 (XLI.)	(in Alkohol) 309	23 200	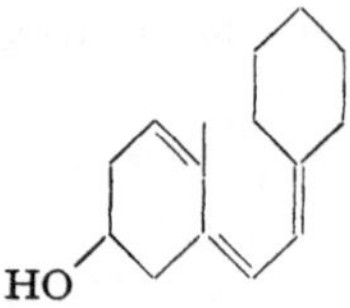 (XLII.)	280	40 000
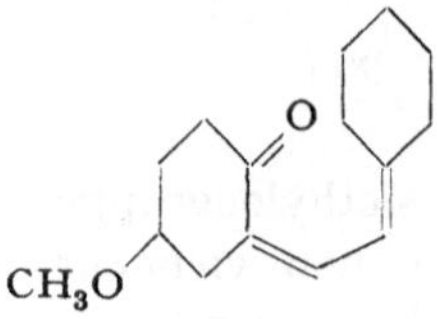 (XLIII.)	(in Äther) 297 (in Alkohol) 306—7	25 300 24 500			
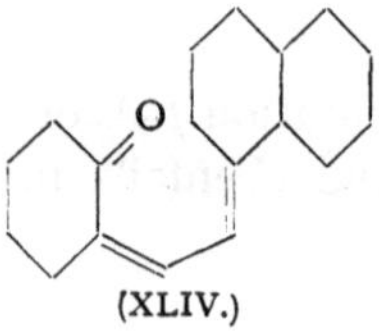 (XLIV.)	(in Äther) 300	27 300			
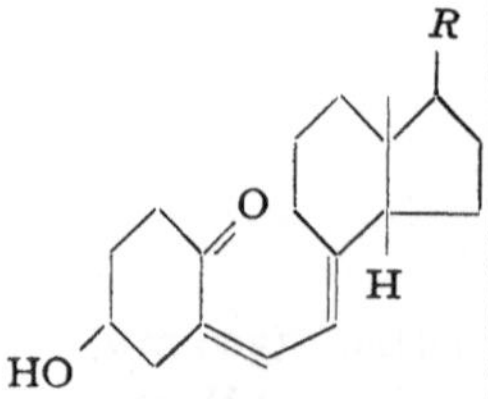 (LIX.)	(in Äther) 302	24 800	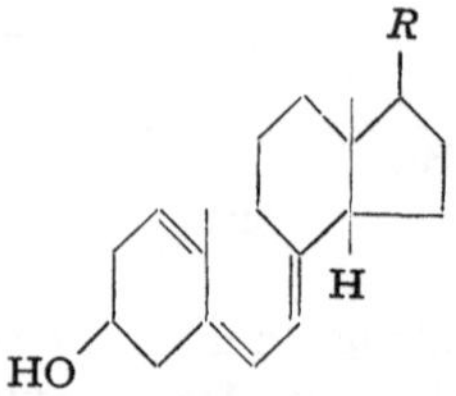 (XII.)	288	41 800

jedoch auch hier und führte zu einem Produkt (XLII), das bei 280 mμ absorbierte ($\varepsilon = 40\,000$).

DIMROTH und JONSSON (23) erweiterten diese Serie von Kondensationsversuchen, indem sie einmal Cyclohexyliden-acetaldehyd mit p-Methoxycyclohexanon, zum anderen Decahydro-naphthyliden-acetaldehyd mit Cyclohexanon kondensierten und dabei die Ketone (XLIII) und (XLIV) isolierten.

Einen Vergleich all dieser Modellsubstanzen in bezug auf ihre spektralen Eigenschaften zeigt *Tabelle 1* (die Verbindungen sind darin ohne Berücksichtigung der geometrischen Isomerieverhältnisse aufgezeichnet).

Eine weitere Gruppe von Modellreaktionen sollte nicht direkt zum Vitamin D_2 führen, sondern zu Verbindungen vom Typus des Ergosterins mit einer $\Delta^{5,\,7}$-Dien-Konfiguration.

So kondensierten 1938 HUBER (33) 2-Methyl-1-acetyl-cyclohexen-1 mit Cyclohexanon, später DIMROTH (22) mit Decalon und schließlich BAGCHI und BANERJEE (4) mit 8-Methyl-hydrindan-4-on. Die unter dem katalytischen Einfluß von Alkoholaten bewerkstelligten Kondensationen sollten gemäß der von RAPSON und ROBINSON (67) durchgeführten Reaktion zwischen 1-Acetyl-cyclohexen und Cyclohexanon analog dem Kondensationsprodukt (XLV) Verbindungen liefern, die gegebenenfalls durch Reduktion und Wasserabspaltung in den Ergosterintypus zu überführen gewesen wären. Durch Arbeiten von TURNER und VOITLE (75) einerseits und JOHNSON (39) anderseits ist jedoch gezeigt worden, daß kein Ringschluß eingetreten sein konnte. Die Anwesenheit der Methylgruppe behindert augenscheinlich die MICHAEL-Addition und läßt mit der Aldolkondensation die Reaktion beim ring-offenen Keton abbrechen.

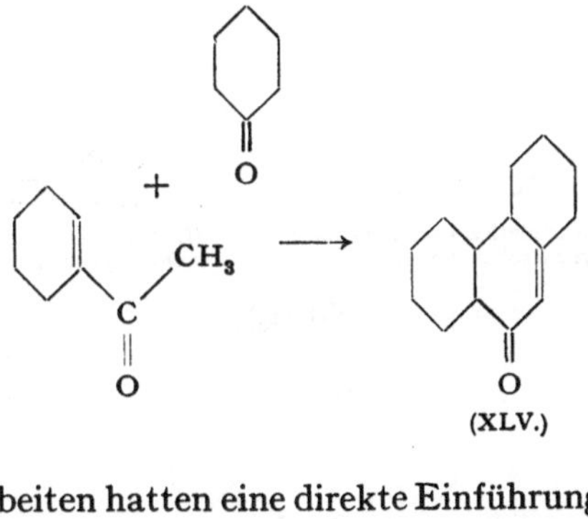

(XLV.)

Zwei andere Arbeiten hatten eine direkte Einführung der semicyclischen Methylengruppe zum Ziel.

So gelang 1939 MILAS und ALDERSON (56) die Darstellung des einfachsten Modells eines Triens mit semicyclischer Methylengruppe, indem sie 2-Dimethylamino-methyl-cyclohexanon (XLVI) mit Allylbromid umsetzten (XLVII), dann über die tertiäre Bromverbindung durch

Alkalibehandlung das Dien (XLIX) erhielten und schließlich durch erschöpfende Methylierung und Hofmannschen Abbau der Mannich-Base zum 3-(2'-Methylen-cyclohexyliden-1')-propen-1 (L) gelangten, das in seinen spektralen Eigenschaften dem Vitamin D_2 sehr ähnlich ist (255 mμ, $\varepsilon = 19000$). Das Vorliegen einer semicyclischen Methylengruppe wurde allerdings nicht durch Ozon-Abbau bewiesen.

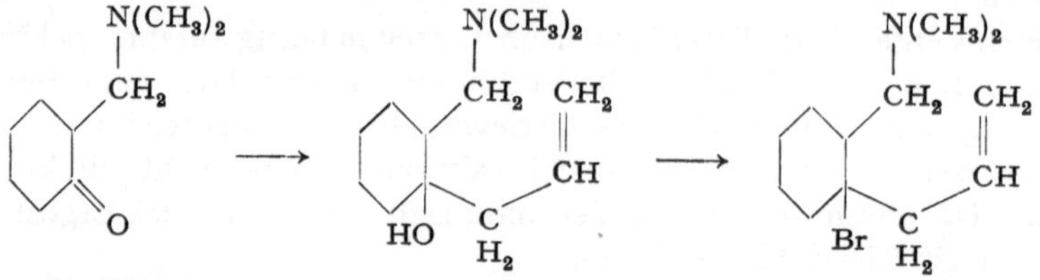

(XLVI.) 2-Dimethylamino-methyl-cyclohexanon.   (XLVII.)   (XLVIII.)

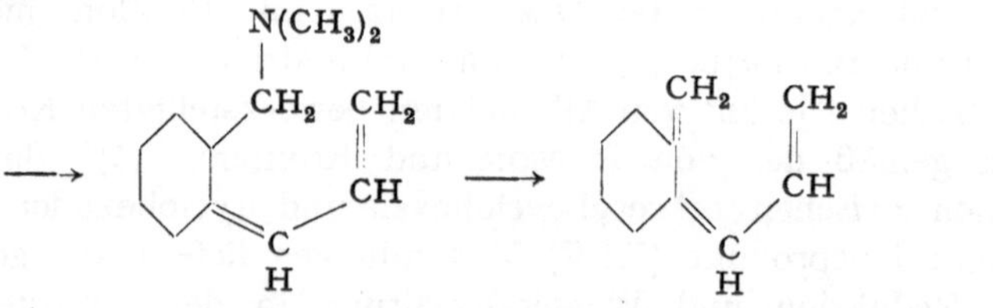

(XLIX.)   (L.) 3-(2'-Methylen-cyclohexyliden-1')-propen-1.

Eine Erweiterung dieser Versuche strebten Raphael und Sond-heimer (66) an, indem sie 2-Dimethylamino-methyl-cyclohexanon-1 mit 1-Äthinyl-cyclohexanol-1 umsetzten. Jedoch gelang der Hofmannsche

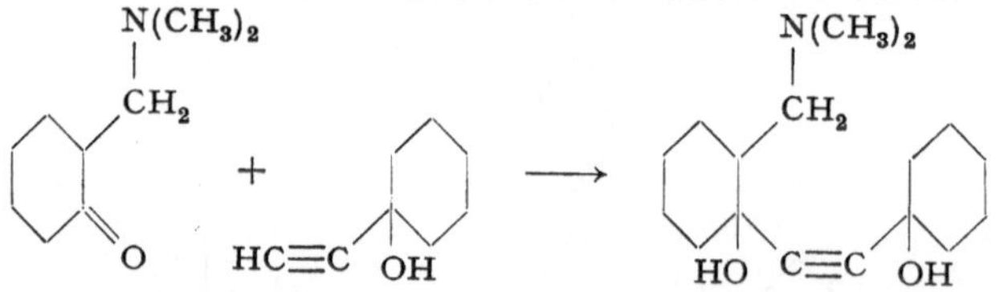

(XLVI.) 2-Dimethylamino-methyl-cyclohexanon-1   (LI.) 1-(2-Dimethylamino-methyl-1-oxy-
und 1-Äthinyl-cyclohexanol-1.   cyclohexyl)-2-(1-oxy-cyclohexyl)-acetylen.

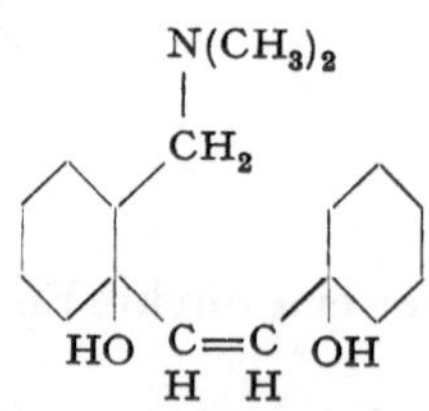

(LII.)

Abbau weder in dem dabei gewonnenen 1-(2-Dimethylamino-methyl-1-oxy-cyclohexyl)-2-(1-oxy-cyclohexyl)-acetylen (LI), noch in dem entsprechenden Äthylenderivat (LII). In jedem Fall führte diese Reaktion zu Polymeren. Auch eine Variation des vorgegebenen Weges durch Kondensation mit Äthinyl-cyclohexen-1 (LIII) beseitigte die Schwierigkeiten nicht, vielmehr war hier das letzte faßbare Produkt das 1-(2-Dimethylamino-methyl-1-oxy-cyclohexyl)-2-(cyclohexen-1-yl)-acetylen (LIV) bzw. die entsprechende Äthylenverbindung (LV).

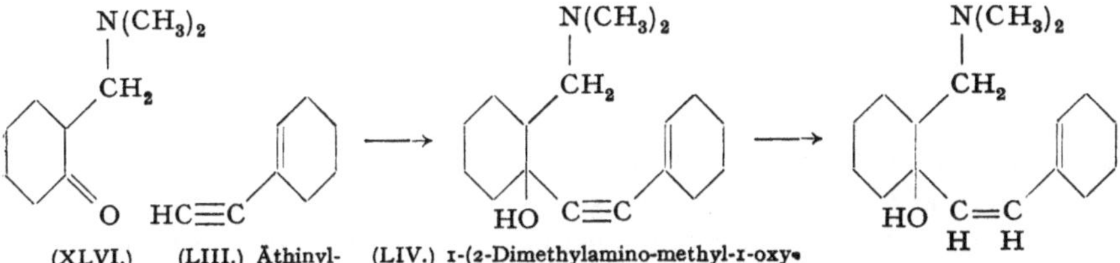

(XLVI.)   (LIII.) Äthinyl-cyclohexen-1.   (LIV.) 1-(2-Dimethylamino-methyl-1-oxy-cyclohexyl)-2-(cyclohexen-1-yl)-acetylen.   (LV.)

Die Tatsache, daß trotz der vorher eindeutig angegebenen Arbeitshypothese auch hier keine weiteren Ergebnisse veröffentlicht worden sind, mag die großen Schwierigkeiten, die einer direkten Einführung der Methylengruppe in Vitamin D-ähnlichen Modellen entgegenstehen, kennzeichnen.

Aus unveröffentlichten Versuchen (26) geht darüber hinaus hervor, daß in Bausteinen für den späteren Ring A, die nicht nur eine kupplungsfähige Ketogruppe am späteren $C_{(5)}$, sondern auch eine Sauerstofffunktion am späteren $C_{(3)}$ tragen, entweder wie im Fall von Enoläthern (LVI) bereits die Darstellung der MANNICH-Base unmöglich ist oder aber bei $C_{(3)}$-Estern oder -Äthern (LVII) diese Einführung stets mit einem

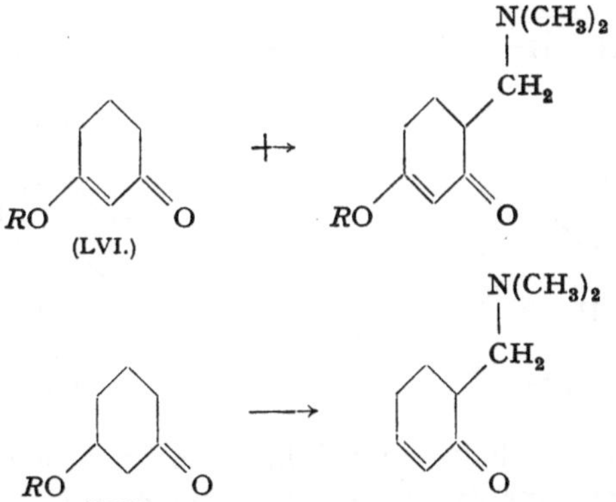

(LVI.)

(LVII.)

Verlust der 3-Acyl- oder Alkoxylgruppe verbunden ist. Damit werden diese Bausteine aber wertlos, wenn die Synthesen den Rahmen reiner Modellversuche überschreiten sollen.

2. Partialsynthese von zwei neuen Isomeren des Vitamins D$_2$.

Während die bisher geschilderten Arbeiten nicht über den Rahmen reiner Modellversuche hinausgingen, wurde im Verlauf eigener Synthesen versucht, ganz allgemein in das Studium ring-offener Vitamin-D-Isomerer, von denen bisher nur Tachysterin und Pyrotachysterin bekannt waren, einzudringen. Um dabei in jedem Fall direkte Vergleichsmöglichkeiten mit diesen Stoffen zu besitzen, verwandten wir als Baustein für das C,D-Ringsystem den C$_{21}$-Abbaualdehyd (*101*) bzw. das C$_{19}$-Abbauketon des Vitamins D$_2$ (*24, 94*). In der ersten Reaktionsfolge wurde *p*-Acetoxy-cyclohexanon mit dem C$_{21}$-Abbaualdehyd (LVIII) unter der katalytischen Wirkung von Natriumäthylat in alkoholischer Lösung in das Kondensationsprodukt (LIX) übergeführt (30%) und chromatographisch gereinigt. Die Dikondensation ließ sich durch genügenden Überschuß von *p*-Acetoxy-cyclohexanon bis auf ein geringes Ausmaß zurückdrängen. Während der Reaktion mußte mit einer Umlagerung des C,D-Hydrindan-systems von *trans* in *cis* gerechnet werden, wie sie für das Abbauketon des Vitamins D$_2$ schon beschrieben wurde (*24*). Das kristalline Keton entspricht

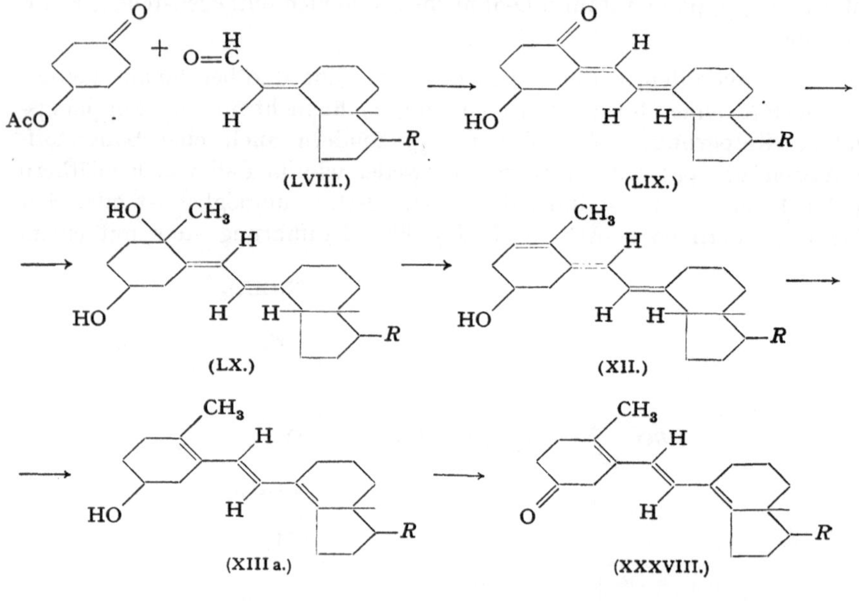

in seinem spektralen Verhalten den früher beschriebenen Modellketonen (S. 104) und besteht ohne Zweifel aus den beiden an C$_{(3)}$ epimeren Alkoholen.

Im weiteren Verlauf der Synthese ließ sich durch Umsatz mit Lithium-methyl das Diol (LX) gewinnen (78%), das trotz des möglichen Vorhandenseins von vier Isomeren ausgezeichnet kristallisiert. In seinem spektralen

Verhalten entspricht es völlig dem Dihydro-tachysterin (*88*) und besitzt drei Maxima außerordentlich starker Intensität (242 mμ, $\varepsilon = 30200$; 250 mμ, $\varepsilon = 35400$; 260 mμ, $\varepsilon = 23700$).

Die Wasserabspaltung aus diesem Produkt erfolgte schon während der Hochvakuumdestillation (85—90%) und ergab ein isomeres Vitamin D$_2$ (XII), das im Charakter des UV-Spektrums bis auf eine Verschiebung aller Maxima um 2 mμ in das Kurzwellige völlig dem des vorher beschriebenen iso-Tachysterins (XIII, S. 99) gleicht. Es liegt als farbloses Glas vor, das nicht zur Kristallisation gebracht werden konnte. Es gelang auch nicht — wahrscheinlich infolge des Vorhandenseins der beiden C$_{(3)}$-Epimeren — einen kristallinen Ester daraus zu erhalten. In bezug auf die Lage der Doppelbindungen konnte für den von uns als iso-Vitamin D$_2$ bezeichneten Stoff (S. 100) die Formel (XII) schon auf Grund des eindeutigen Syntheseweges als gesichert angesehen werden. Sie ließ sich aber noch zusätzlich durch Ozonabbau bestätigen, bei dem wohl die Abbauketosäure (XXXI, S. 98), aber kein Formaldehyd nachgewiesen werden konnte.

In bezug auf die Feinstruktur des neuen Isomeren sind (unter Nichtbeachtung eventueller atropisomerer Strukturen) für das vorliegende

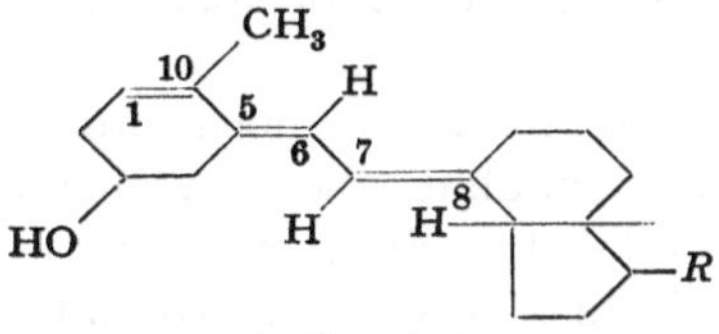

(XII.) iso-Vitamin D$_2$. 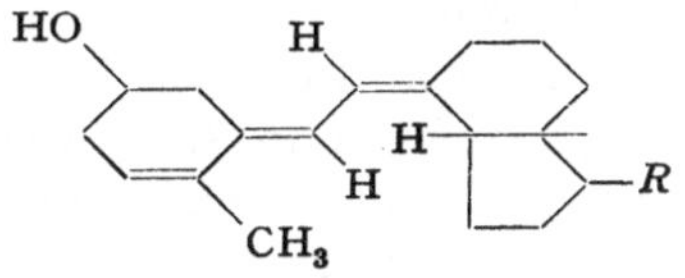 (LXI.)

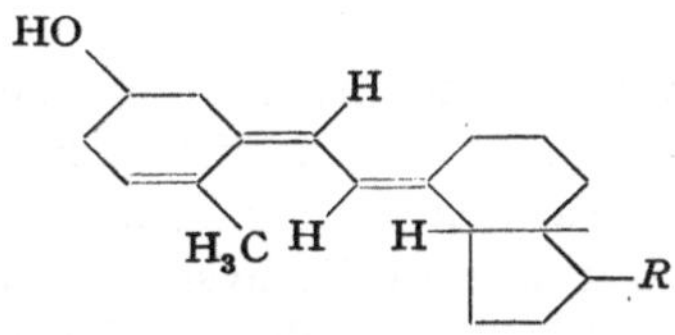

(LXII.) 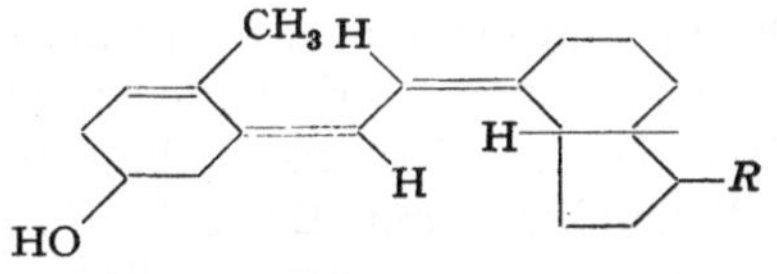 (LXIII.)

Trien-system an den beiden semicyclischen Doppelbindungen vier stereoisomere Formen möglich (XII, LXI bis LXIII). Zweifellos wird sich davon wieder die energetisch begünstigste Konfiguration einstellen, da dies die einzelnen Reaktionen der Kondensation und der Wasserabspaltung ermöglichen. Von diesen vier Formen scheiden (LXII) und (LXIII) auf Grund der für Carotinoide von PAULING (*61*) und ZECHMEISTER (*106*) aufgestellten Regel wahrscheinlich aus, wonach das System *A* gegenüber *B* weniger stabil ist, da CH$_3$ und H sich einander sterisch behindern.

Zwischen den Strukturen (XII) und (LXI) läßt sich nicht ohne weiteres entscheiden, da beide eine all-*trans*-Verknüpfung mit gestreckter und coplanarer Lage des konjugierten Systems besitzen. Trotzdem bevorzugen wir Struktur (XII), da sie im Gegensatz zur anderen in bezug auf die Drehbarkeit um die Bindung $C_{(6)}$—$C_{(7)}$ völlig frei ist.

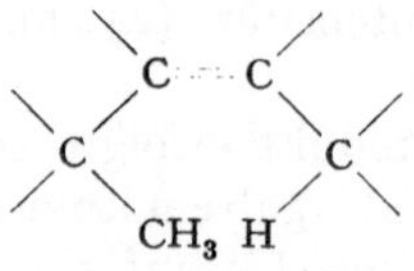

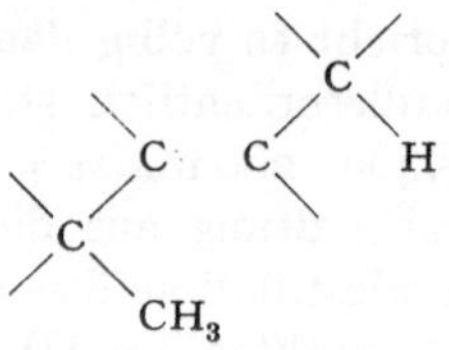

Dem partialsynthetisch gewonnenen iso-Vitamin D_2 wird also die Konfiguration (XII) zugeschrieben, die wie das iso-Tachysterin (XIII) eine all-*trans*-Verknüpfung besitzt. Nach dem vorerwähnten Nomenklaturvorschlag (S. 100) kann es auch als 9,10-seco-$\varDelta^{1,\ 10-5,\ 6-7,\ 8-22,\ 23}$-14-allo-Ergostatetraen-3α,β-ol bezeichnet werden. Interessant ist der Vergleich der UV-Spektren der beiden neuen Isomeren. Zeigen doch beide Produkte bis auf eine Verschiebung aller Maxima um 2 mμ in ihrem Charakter trotz verschiedener Substituenten-Zahl (5 bzw. 6) nahezu identische Spektren. Der Verlauf der Absorptionskurve insgesamt scheint durch das vollkommen freie all-*trans*-System der Doppelbindungen bestimmt zu werden.

Im Hinblick auf das in seiner Feinstruktur noch nicht geklärte Pyrotachysterin sind, wie schon erwähnt, die vorstehend beschriebenen Ergebnisse auf Grund des ähnlichen Spektrums als Beweismoment dafür anzusprechen, daß auch hier das ungestörte all-*trans*-System entsprechend der bereits diskutierten Struktur (VII, S. 88) vorliegt.

Die Vermutung, daß wir im iso-Tachysterin einen Endzustand chemischer Umlagerung der ring-offenen Triene vor uns haben, ließ sich auch im Falle des iso-Vitamins D_2 bestätigen. Tatsächlich konnte auch hier nach Behandlung mit Bortrifluorid-ätherat das iso-Tachysterin, allerdings als Gemisch seiner $C_{(3)}$-Epimeren (XIIIa), in gleicher Reinheit wie bei der Umlagerung des Vitamins D_2 gewonnen werden. Für einen Identitätsbeweis mußte man noch das Asymmetriezentrum an $C_{(3)}$ beseitigen, was sich durch Alkoholat-oxydation zum Keton erreichen ließ. Sein Semicarbazon erwies sich durch Schmelzpunkt, Misch-schmelzpunkt, Drehungsvermögen und UV-Spektrum als identisch mit dem des schon beschriebenen iso-Tachysterin-ketons (XXXVIII, S. 101).

Um ein möglichst vollständiges Bild zu erhalten, wurden parallel zu der eben beschriebenen Synthese andere Versuche zur Darstellung 9,10-ring-offener Triene unternommen (38). So konnte iso-Tachysterinmethyläther in großer Anreicherung auf dem folgenden Wege gewonnen werden.

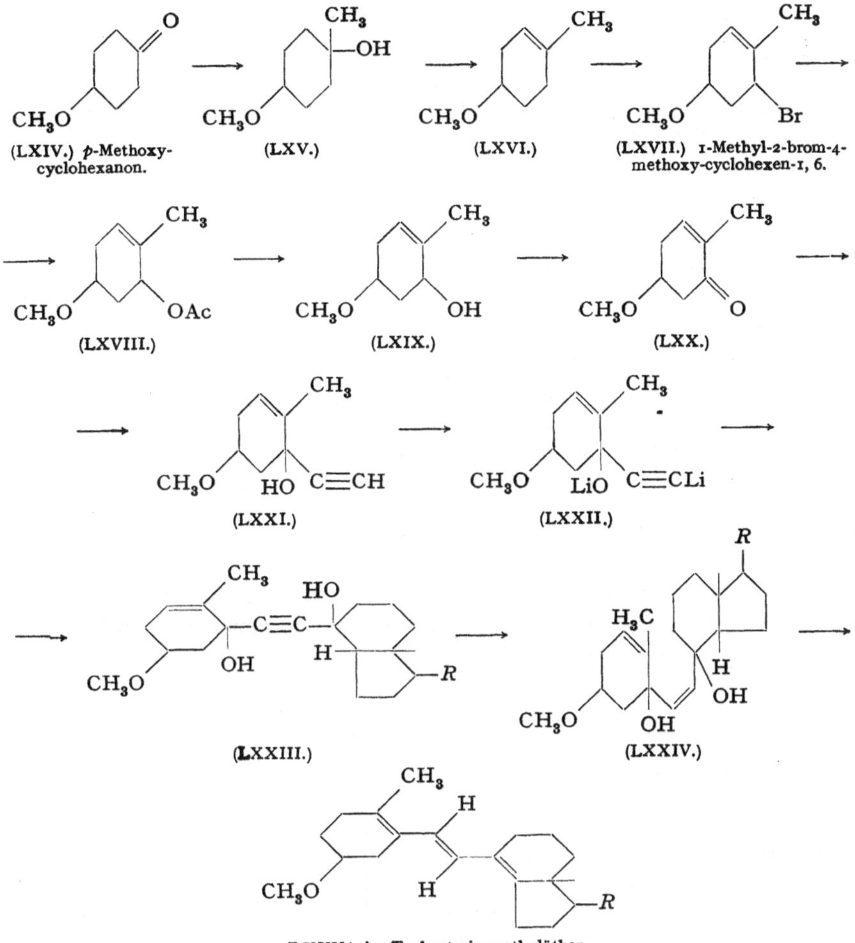

(LXIV.) p-Methoxy-cyclohexanon.

(LXV.)

(LXVI.)

(LXVII.) 1-Methyl-2-brom-4-methoxy-cyclohexen-1, 6.

(LXVIII.)

(LXIX.)

(LXX.)

(LXXI.)

(LXXII.)

(LXXIII.)

(LXXIV.)

(LXXV.) iso-Tachysterin-methyläther.

Ausgehend vom p-Methoxy-cyclohexanon (LXIV), das nach den in der Literatur gegebenen Vorschriften (53) dargestellt wurde, ließ sich über Grignardierung, Wasserabspaltung und ZIEGLER-Bromierung das 1-Methyl-2-brom-4-methoxy-cyclohexen-1,6 (LXVII) gewinnen. Von dort führten Umsatz mit Kaliumacetat und Verseifung sowie schließlich Oxydation mit Braunstein oder tert. Butylchromat zum 1-Methyl-2-keto-4-methoxy-cyclohexen-1,6 (LXX), das für die kommende Synthese als Ring A-Baustein diente. Es ist sowohl gegen Lauge als auch Säure unbeständig und geht unter Methanolabspaltung in o-Kresol über. Diese große Aromatisierungstendenz ließ auch die folgende Umsetzung zur Äthinylverbindung (LXXI) mit Lithiumacetylid in flüssigem Ammoniak nur in unbefriedigenden Ausbeuten verlaufen. Die Äthinylverbindung

wurde nun mit Lithiummethyl bzw. Lithiumphenyl in das reaktionsfähige metallorganische Derivat übergeführt und mit dem $C_{(19)}$-Abbau-
keton des Vitamins D_2 in Äther umgesetzt (94%, unter Berücksichtigung
der rückgewonnenen Ausgangsstoffe). Die Reinigung des Kondensations-
produktes (LXXIII) erfolgte durch Chromatographie. Das dabei gewonnene Öl ließ sich mit Lindlar-Katalysator oder mit chinolin-vergifteter
5proz. Palladiumkohle partiell hydrieren. Einwirkung von Phosphor-
jodür auf dieses Hydrierungsprodukt nach R. Kuhn (42—46) lieferte
das Endprodukt (LXXV) der Partialsynthese. Es zeigte im UV-Spektrum
bei gleichem Verlauf der Absorptionskurve wie beim iso-Tachysterin die
drei charakteristischen Maxima bei 280, 290 und 302 $m\mu$ und stellt somit
den Methyläther desselben dar.

Es stabilisiert sich also auch hier das π-Elektronensystem in der sterisch und
energetisch am meisten begünstigten Lage. Der iso-Tachysterin-methyläther ließ
sich durch Chromatographie und Destillation bis zu einem Gehalt von 64% an-
reichern (280 $m\mu$, $\varepsilon = 22\,400$; 290 $m\mu$, $\varepsilon = 27\,200$; 302 $m\mu$, $\varepsilon = 20\,500$). Die Reinigung
gestaltete sich hier schwierig, da keines der unmittelbaren Vorprodukte kristallin
zu erhalten war. Mit der prinzipiellen Klärung des Reaktionsverlaufes zum iso-
Tachysterin konnte dieser Weg jedoch ohne weitere Versuche abgebrochen werden.

Zum Abschluß dieser Betrachtungen soll eine Zusammenstellung
der spektralen Daten der im Rahmen dieses Berichtes interessierenden
Vitamin D_2-Isomeren gegeben werden (*Tabelle 2*).

Tabelle 2. Einige spektroskopische Daten in der Vitamin D_2-Reihe.

	Lage der Maxima (mμ)			ε
Vitamin D_2		265		18 300
Ergosterin (98)	270	280		11 500
Lumisterin (92)..........................	265	280		8 600
Tachysterin		281		24 600
Pyrotachysterin	265	289	300	23 700
iso-Vitamin D_2	278	288	300	41 800
iso-Tachysterin	280	290	302	41 800

3. Syntheseversuche des C, D-Hydrindan-Ringsystems.

Es sollen auch die Versuche erwähnt werden, die die Darstellung
geeigneter C, D-Bausteine beschreiben. Diese müssen zweifellos folgende
Bedingungen erfüllen: a) es muß das fertige Hydrindansystem vorliegen,
da spätere Umwandlungen daran in Gegenwart des empfindlichen
Triensystems als aussichtslos erscheinen; b) es müssen sowohl in 1- als
auch 4-Stellung Sauerstoff-Funktionen enthalten sein, die ein Anknüpfen
der beidseitigen Reste ermöglichen.

Diesen Bedingungen wurde bis vor kurzem nach mehreren vergeb-
lichen Versuchen (*27a, 73a*) nur eine Arbeit gerecht: die von Banerjee und Shafer (*6*) beschriebene interessante Synthese des 8-

Methyl-hydrindan-1,4-dions. Dabei wird γ-Acetyl-buttersäureäthyl-
ester, der aus Acetessigester in 65% Ausbeute gewonnen werden kann,
mit Cyanessigester umgesetzt und an das Reaktionsprodukt (LXXVII)

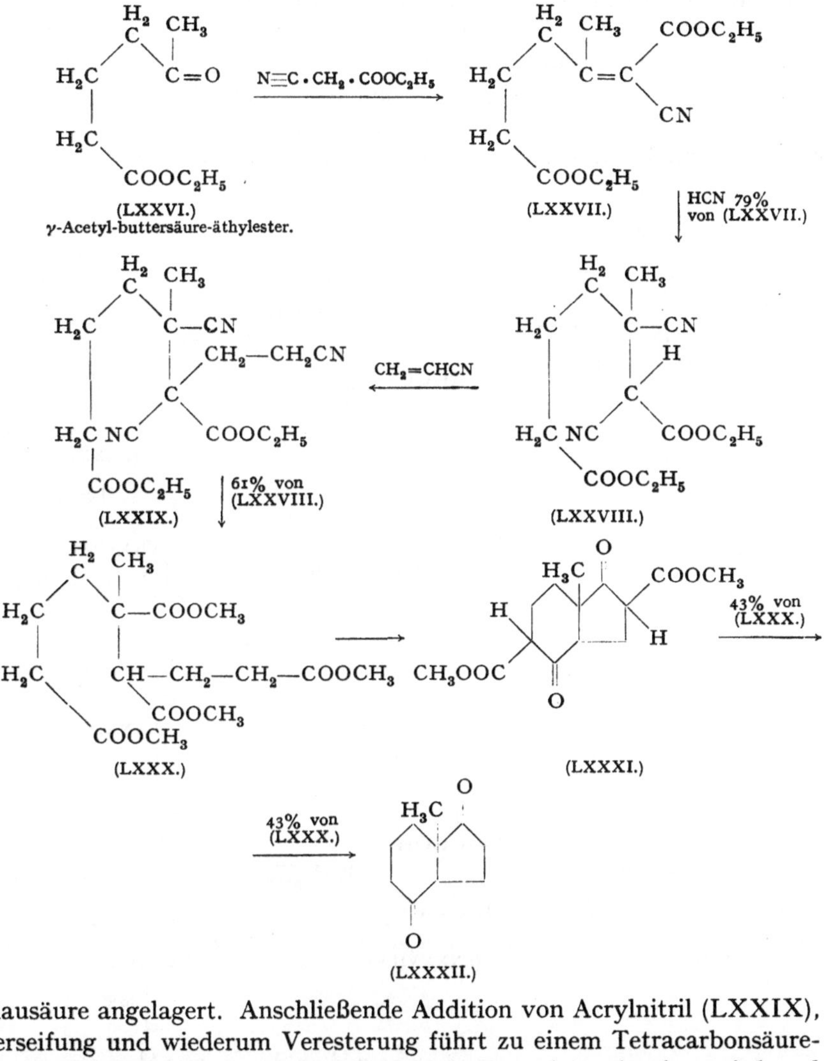

(LXXVI.)
γ-Acetyl-buttersäure-äthylester.

$N\equiv C \cdot CH_2 \cdot COOC_2H_5$

(LXXVII.)

HCN 79%
von (LXXVII.)

(LXXIX.)

61% von
(LXXVIII.)

$CH_2=CHCN$

(LXXVIII.)

(LXXX.)

43% von
(LXXX.)

(LXXXI.)

43% von
(LXXX.)

(LXXXII.)

Blausäure angelagert. Anschließende Addition von Acrylnitril (LXXIX),
Verseifung und wiederum Veresterung führt zu einem Tetracarbonsäure-
ester (LXXX), der mit Natriumhydrid in Benzol kondensiert wird und
dabei als Hauptprodukt den 8-Methyl-hydrindan-1,4-dion-2,5-dicarbon-
säure-methylester (LXXXI) und nach dessen Verseifung und Decarboxylierung das 8-Methyl-hydrindan-1,4-dion (LXXXII) liefert.

Wie die Verfasser mitteilen, besitzt die an $C_{(1)}$ stehende Ketogruppe
gegenüber der anderen eine stark herabgesetzte Reaktionsfähigkeit, so

daß dieses Produkt für weitere Synthesen, die ja einer partiellen Reaktionsfähigkeit bedürfen, als geeignet anzusprechen wäre. Allerdings besitzt es das *C,D-cis*-Hydrindangerüst, das nicht der natürlichen *trans*-Konfiguration entspricht (wahrscheinlich als Racemat).

Eine andere Darstellung der gleichen Verbindung gelang in neuester Zeit (*37a*). Dabei wurde zunächst nach Vorschriften von NENITZESCU und CIORANESCU (*58a*) sowie RABE und RAHM (*64a, 64b*) die 3-Methylcyclohexen-2-on-1-β-propionsäure-2 dargestellt. Ausgehend von Acetessigester und Paraformaldehyd ließ sich durch Einwirkung von Piperi-

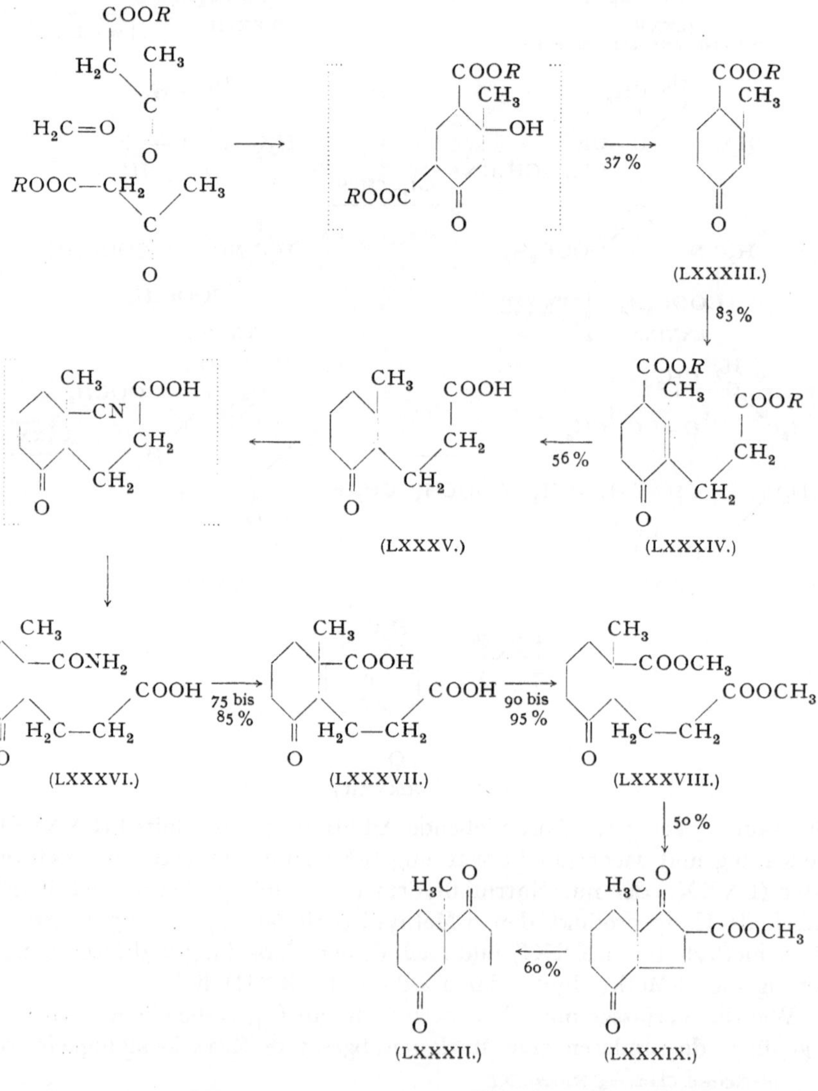

$COOR$
H_2C CH_3
C
$H_2C=O$
O
$ROOC-CH_2$ CH_3
C
O

$\longrightarrow$

$COOR$
CH_3
$-OH$
$ROOC$
O

$\xrightarrow{37\%}$

$COOR$
CH_3
O
(LXXXIII.)

$\downarrow 83\%$

CH_3 $COOH$
$-CN$
CH_2
CH_2
O

$\longleftarrow$

CH_3 $COOH$
CH_2
CH_2
O
(LXXXV.)

$\xleftarrow{56\%}$

$COOR$
CH_3 $COOR$
CH_2
CH_2
O
(LXXXIV.)

$\downarrow$

CH_3
$-CONH_2$
$COOH$
H_2C-CH_2
O
(LXXXVI.)

$\xrightarrow[85\%]{75\ bis}$

CH_3
$-COOH$
$COOH$
H_2C-CH_2
O
(LXXXVII.)

$\xrightarrow[95\%]{90\ bis}$

CH_3
$-COOCH_3$
$COOCH_3$
H_2C-CH_2
O
(LXXXVIII.)

$\downarrow 50\%$

H_3C O
$-COOCH_3$
O
(LXXXIX.)

$\xleftarrow{60\%}$

H_3C O
O
(LXXXII.)

din der sogenannte HAGEMANN-Ester (LXXXIII) gewinnen und dieser durch Kondensation mit β-Chlor-propionsäure-äthylester mittels Natriumäthylat (LXXXIV) sowie nach anschließender saurer oder alkalischer Verseifung bei gleichzeitiger Decarboxylierung in die kristalline ungesättigte Ketosäure (LXXXV) überführen.

An die Doppelbindung derselben konnte in methanolisch-wäßriger Kaliumcyanid-Lösung Blausäure angelagert und das bei der Aufarbeitung aus dem Nitril entstandene Carbonsäureamid (LXXXVI) alkalisch zur kristallinen Ketodicarbonsäure (LXXXVII) verseift werden. Nach anschließender Veresterung mit Diazomethan wurde der Keto-dicarbonsäureester (LXXXVIII) durch Kochen seiner benzolischen Lösung in Gegenwart von Natriummethylat einer DIECKMANN-Kondensation unterworfen. Der dabei entstandene 8-Methyl-hydrindan-1,4-dion-2-carbonsäure-methylester (LXXXIX) lieferte nach Verseifung und Decarboxylierung mittels 30%iger Schwefelsäure das gewünschte 8-Methyl-hydrindan-1,4-dion (LXXXII), wahrscheinlich ebenfalls als Racemat.*

Nach Betrachtung aller Modellversuche und Partialsynthesen von Vitamin D-ähnlichen Stoffen läßt sich sagen, daß hauptsächlich drei Voraussetzungen eine Synthese des Vitamins D_2 problematisch erscheinen lassen: 1. Die Schwierigkeit der direkten Einführung der semicyclischen Methylengruppe. 2. Die Frage der geometrischen Isomerien an den Doppelbindungen verbunden mit der Erkenntnis, daß sowohl Vitamin D_2 als auch Tachysterin in der weniger stabilen *cis*-Form vorliegen. 3. Der Einbau des weniger stabilen *trans*-Hydrindansystems. Hinzu kommt die Empfindlichkeit all dieser Stoffe, das zum Teil geringe Kristallisationsvermögen und ihre strukturelle Ähnlichkeit, die eine Reinigung oft schwierig, wenn nicht unmöglich machen können.

So einfach die für eine Synthese des Vitamins D_2 nötigen Konzeptionen oftmals aussehen mögen, erscheint sie wie auch die Synthese des Tachysterins in bezug auf Feinheiten der Struktur als ein schwieriges, aber zugleich äußerst interessantes Problem der synthetischen Chemie.

VII. Photodehydro-ergosterin.

Außer den Bestrahlungsprodukten des Ergosterins haben WINDAUS und seine Mitarbeiter eine Verbindung beschrieben, die bei der Ultraviolett-Bestrahlung des Dehydro-ergosterins entsteht, das sich vom Ergosterin lediglich durch eine vierte Doppelbindung unterscheidet, die beim Behandeln mit Mercuriacetat in 9,11-Stellung gebildet wird (XC).

Interessanterweise wird dieses Dehydro-ergosterin gleichfalls durch UV-Licht verändert, wobei zum Unterschied vom Ergosterin nur *ein* Photoprodukt gefaßt werden konnte, nämlich das Photodehydro-ergo-

---

* Ein Vergleich der beiden Präparate konnte noch nicht durchgeführt werden.

sterin (*98*). Über seine Konstitution ist bisher nur soviel bekannt, daß es sich um ein Isomeres des Dehydro-ergosterins der Summmenformel $C_{28}H_{42}O$ handelt, und daß der neue Stoff gleichfalls vier Doppelbindungen aufweist wie die Ausgangssubstanz. Hieraus ist zu folgern, daß ein tetracyclisches Ringsystem vorliegt, während Tachysterin und Vitamin D bekanntlich tricyclische Ringsysteme besitzen, entstanden durch Ring-Öffnung unter Hinzutritt einer weiteren Doppelbindung.

Nun haben vor kurzem NES und MOSETTIG (*59*) eine neuartige und höchst bemerkenswerte Umlagerung beschrieben, die das Dehydro-ergosterin unter der Einwirkung von Chlorwasserstoffsäure in Chloroform-lösung erleidet, d. h. bei derjenigen Reaktionsbedingung, die WINDAUS zur Isomerisierung des Ergosterins angewandt hat. Man darf wohl annehmen, daß NES und MOSETTIG ähnliche Doppelbindungsverschie-bungen am Dehydro-ergosterin vornehmen wollten. Aber der Reaktions-verlauf ist hier ein grundsätzlich anderer. Wie NES und MOSETTIG gezeigt haben, wird hierbei die $C_{(10)}$-$C_{(1)}$-Bindung geöffnet und ein neuer Ring $C_{(1)}$—$C_{(6)}$ geschlossen, wobei unter Aromatisierung des Ringes *B* ein Anthracenderivat erhalten wird, (XC) → (XCI):

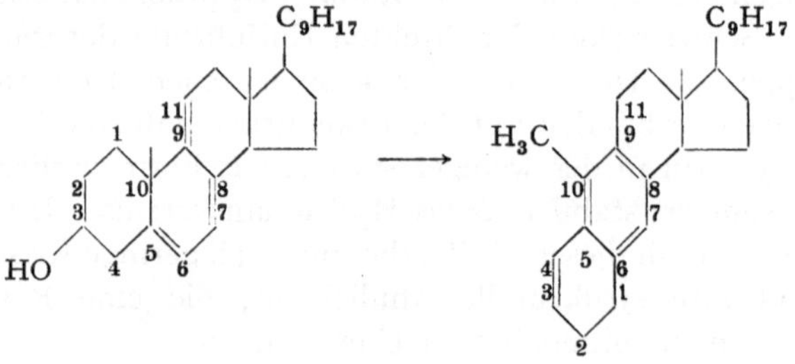

Wie ist diese Reaktion zu verstehen? Während beim Ergosterin das Doppelbindungssystem 5,6—7,8 nach der STAUDINGER-SCHMIDTschen Doppelbindungsregel die Bindung 9—10 lockert, wird im Dehydro-ergosterin unter dem unmittelbaren Einfluß der Doppelbindungen 5,6 und 9,11 die Bindung $C_{(10)}$—$C_{(1)}$ so geschwächt, daß durch Protonen-Katalyse Sprengung eintritt.

Wir haben also zwei bemerkenswerte Ringsprengungen vor uns: Die Öffnung der 9,10-Bindung im Ergosterin bzw. Lumisterin durch Ultraviolett-Bestrahlung und die Öffnung der 10,1-Bindung im Dehydro-ergosterin unter der Einwirkung von HCl in Chloroform.

Es erscheint daher vorerst plausibel anzunehmen, daß auch bei der Ultraviolett-Bestrahlung des Dehydro-ergosterins die 10,1-Bindung geöffnet wird. Da das hierbei gebildete Photodehydro-ergosterin, wie schon erwähnt, vier Doppelbindungen aufweist, müßten wir weiterhin folgern, daß sich auch in diesem Fall ein neuer Ring, möglicherweise gleichfalls zwischen $C_{(1)}$ und $C_{(6)}$, gebildet hat (*36*). Es darf in diesem Zusammenhang daran erinnert werden, daß bei der Überbestrahlung des Ergosterins bzw. der nachfolgenden Bestrahlungsprodukte die sog.

(I.) Ergosterin. ⟶
 (S. 84.)

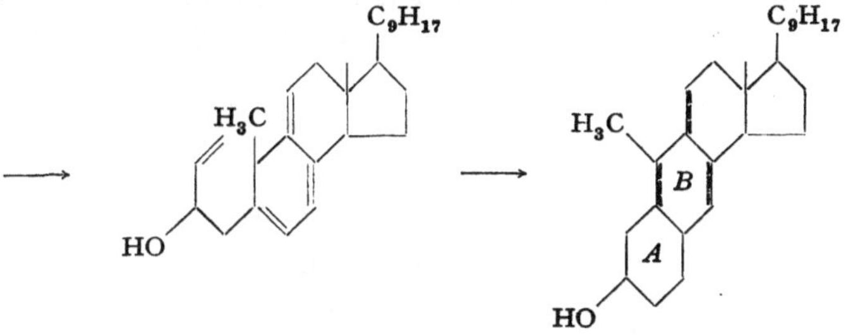

(XCII.) (?)

Suprasterine erhalten werden, wobei in einem Fall ebenfalls mit der Bildung einer neuen Kohlenstoff-Kohlenstoff-Bindung zu rechnen ist. Man könnte hierbei an folgende Übergänge denken (*35*):

(IV.) Vitamin D$_2$. ⟶
 (S. 84.)

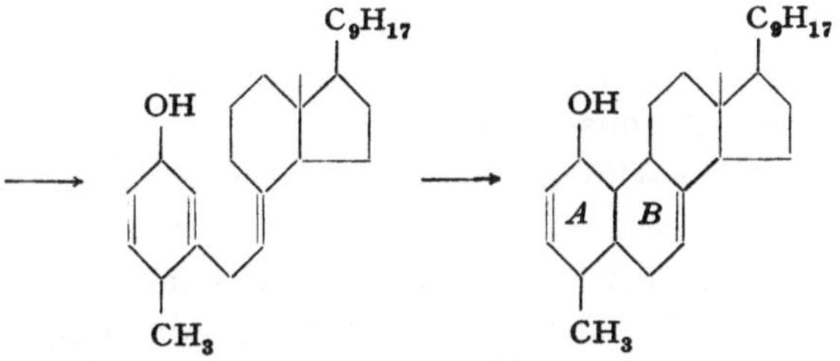

Suprasterin I (?).

(XCIII.)

Die vierte Variante wäre eine Untersuchung der Frage, ob bei der Chlorwasserstoff-Behandlung des Ergosterins und Lumisterins neben der Bildung der Iso-ergosterine auch zum Teil die 9,10-Bindung geöffnet und gleichfalls ein neuer Ring zwischen $C_{(4)}$ und $C_{(9)}$ geschlossen wird (*36*).

Derartige chemische Übergänge in (XCII), (XCIII) und (XCIV) sind auch von Interesse für physiologische bzw. pathologische Zellvorgänge. Es handelt sich nämlich hierbei um abgewandelte Steroide, die nicht mehr das ursprüngliche Kohlenstoffskelett und damit

(I.) Ergosterin.
(II.) Lumisterin. ⟶
 (S. 84.)

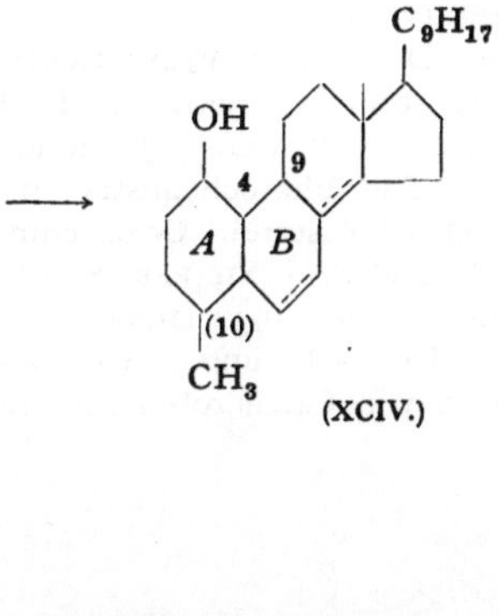

(XCIV.)

zusammenhängend auch nicht mehr die anguläre Methylgruppe an dem ehemaligen quartären C-Atom 10 besitzen, und bei denen infolgedessen durch einfachen Wasserstoffentzug relativ leicht die Ringe *A* und *B* aromatisch werden können. Die resultierenden Derivate des Anthracens bzw. Phenanthrens könnten — falls sie auch in vivo im Verlauf pathologischer Stoffwechselprozesse entstünden — direkt oder nach weiterer Umwandlung Veranlassung zur Entstehung bösartiger Geschwülste geben (*35*).

VIII. Schlußwort.

Die vorstehende Zusammenfassung der neueren Ergebnisse in der Vitamin D-Chemie läßt erkennen, daß auch hier wie bei den eigentlichen Steroiden mit den hauptsächlich von WINDAUS nachgelassenen Problemen eine Intensivierung der Forschungen zu erwarten ist.

Es sei zum Schluß auf einige der noch offenen Fragen hingewiesen:

1. Partial- und Total-Synthese von Vitamin D und Tachysterin bzw. von seitenkettenlosen Analoga.

2. Konstitution der Suprasterine.

3. Konstitution des Photodehydro-ergosterins im Licht der NES-MOSETTIG-Umlagerung.

4. Ring-Öffnung zwischen den Kohlenstoffatomen 9 und 10 bei Steroiden mit rein chemischen Mitteln.

5. Untersuchung aller Möglichkeiten, aus Steroiden durch Umlagerung im Ringsystem zu neuen Verbindungen zu gelangen, die durch Wasserstoffentzug in cancerogene Stoffe überzugehen vermögen.

6. Weitere Untersuchung des Präcalciferols von VELLUZ.

Literaturverzeichnis.

1. ALBERTI, C. G., B. CAMERINO und L. MAMOLI: Ein neues Provitamin D: das $\Delta^{5,7}$-Norcholestadien-3β-ol. Helv. Chim. Acta **32**, 2038 (1949).

2. ALDERSLEY, J. B., G. N. BURKHARDT, A. E. GILLAM and N. C. HINDLEY: The Synthesis of Compounds Related to the Antirachitic Vitamins. II. J. Chem Soc. (London) **1940**, 10.

3. ASKEW, F. A., R. B. BOURDILLON, H. M. BRUCE, R. K. CALLOW, J. ST. L. PHILPOT and T. A. WEBSTER: Crystalline Vitamin D. Proc. Roy. Soc. (London) Ser. B, **109**, 488 (1932).

4. BAGCHI, P. and D K. BANERJEE: Synthetic Investigations on Sterols, Bile Acids, Hormones, etc. IV. Synthesis of 9-Keto-2 : 13-Dimethyl-Δ^{10}-1 : 2-*Cyclo*-pentano-perhydrophenanthrene. J. Indian Chem. Soc. **23**, 397 (1946).

5. BANCHETTI, A.: The Photochemistry of Cholesterol. II. Photosynthesis of Vitamin D_3 from Cholesterol. Gazz. chim. ital. **78**, 404 (1948).

6. BANERJEE, D. K. and P. R. SHAFER: Synthesis of 8-Methylhydrindan-1,4-dione. J. Amer. Chem. Soc. **72**, 1931 (1950).

7. BERGSTRÖM, S.: Die Isolierung von (—)-β-Methoxy-adipinsäure aus den Oxydationsprodukten von Calciferol-methyläther. Helv. Chim. Acta **32**, 3 (1949).

8. BERGSTRÖM, S., A. LARDON und T. REICHSTEIN: Die Isolierung von (—)-β-Methoxy-adipinsäure aus den Oxydationsprodukten von Calciferol-methyläther. 2. Mitt. Steroide, 3. Mitt. Helv. Chim. Acta **32**, 1617 (1949).

9. BERNSTEIN, S., L. J. BINOVI, L. DORFMAN, K. J. SAX and Y. SUBBAROW: 7-Dehydrocholesterol. J. Organ. Chem. (U. S. A.) **14**, 433 (1949).

10. BIJVOET, J. M., A. F. PEERDEMAN and A. J. VAN BOMMEL: Determination of the Absolute Configuration of Optically Active Compounds by Means of X-Rays. Nature (London) **168**, 271 (1951).

11. BUISMAN, J. A. K., W. STEVENS and J. VAN DER VLIET: Investigations on Sterols. A New Synthesis of 7-Dehydrocholesterol (Provitamin D). Rec. trav. chim. Pays-Bas **66**, 83 (1947).

12. BURKHARDT, G. N. and N. C. HINDLEY: The Synthesis of Compounds Related to the Antirachitic Vitamins. $\alpha\beta$-Di-Δ^1-cyclohexenylethylene. J. Chem. Soc. (London) **1938**, 987.

13. BUSER, W.: γ-Cholestenon und epi-γ-Cholestenon. Helv. Chim. Acta **30**, 1379 (1947).

14. BUTENANDT, A., W. FRIEDRICH und L. POSCHMANN: Über Lumi-oestron. II. Mitt.: Die Bestrahlung von Oestron mit monochromatischem Ultraviolett-Licht. Ber. dtsch. chem. Ges. **75**, 1931 (1942).

15. BUTENANDT, A., E. HAUSMANN und J. PALAND: Über $\Delta^{5,7}$-Androstadiendiol-(3.17). Ber. dtsch. chem. Ges. **71**, 1316 (1938).

16. BUTENANDT, A., A. WOLFF und P. KARLSON: Über Lumi-oestron. Ber. dtsch. chem. Ges. **74**, 1308 (1941).

17. CROWFOOT, D. and J. D. DUNITZ: Structure of Calciferol. Nature (London) **162**, 608 (1948).

18. DASLER, W. and C. D. BAUER: Electrically Activated Ergosterol (Whittier Process). J. Biol. Chem. **167**, 581 (1947).

19. DIMROTH, K.: Synthetische Versuche zur Darstellung der antirachitischen Vitamine. I. Mitt.: Synthese α,β-ungesättigter Alkohole und Aldehyde mit semicyclischer Doppelbindung. Ber. dtsch. chem. Ges. **71**, 1333 (1938).

20. — Synthetische Versuche zur Darstellung der antirachitischen Vitamine. II. Mitt.: Kondensation von Cyclohexyliden-acetaldehyd mit Cyclohexanon. Ber. dtsch. chem. Ges. **71**, 1346 (1938).

21. — Diskussionsbemerkung zum Vortrag von H. FROMHERZ, L. THALER und G. WOLF: Lichtabsorption und Konstitution einiger polycyclischer Kohlenwasserstoffe. Z. Elektrochem. **49**, 398 (1943).

22. — Über die Synthese von Modellen ungesättigter Steroide vom Typ des Ergosterins. Angew. Chem. **59**, 215 (1947).

23. DIMROTH, K. und H. JONSSON: Synthetische Versuche zur Darstellung der antirachitischen Vitamine. III. Mitt. Ber. dtsch. chem. Ges. **71**, 2658 (1938).

24. — — Die sterische Verknüpfung der Ringe C und D bei den Steroiden. Ber. dtsch. chem. Ges. **74**, 520 (1941).

25. DIMROTH, K. und J. PALAND: Über die Ultraviolett-Bestrahlung des $\Delta^{5,7}$-Androstadien-diols-(3.17). Ber. dtsch. chem. Ges. **72**, 187 (1939).

26. ERDMANN, D.: Dipl. Arbeit, Techn. Hochschule Braunschweig. 1952.

27. FREUDENBERG, K. und F. BRAUNS: Die Konfiguration der einfachen α-Oxysäuren. Ber. dtsch. chem. Ges. **55**, 1339 (1922).

27a. GOLDBERG, M. W., F. HUNZIKER, J. R. BILLETER und H. R. ROSENBERG: Über Steroide und Sexualhormone. 135. Mitt. Versuche zur Herstellung von 8-Methyl-hydrindan-1,4-dion. Helv. Chim. Acta **30**, 200 (1947).

28. GRUNDMANN, W.: Beitrag zur Konstitutionsermittlung des Tachysterins. Z. physiol. Chem. (Hoppe-Seyler) **252**, 151 (1938).

29. Haslewood, G. A. D.: Derivates of cis-3-Hydroxy-Δ^5-cholenic Acid. J. Chem. Soc. (London) 1938, 224.

30. — The Action of Light on Substances Related to Ergosterol. Biochemic. J. 33, 454 (1939).

31. Henbest, H. B., E. R. H. Jones, A. E. Bide, R. W. Peevers and P. A. Wilkinson: A New Route to 7-Dehydrocholesterol, Provitamin D_3. Nature (London) 158, 169 (1946).

32. Henecka, H.: Der Mechanismus der Dien-Synthese. Z. Naturforsch. 4b, 15 (1949).

33. Huber, W.: Über einige synthetische Versuche an Polyterpenen. Ber. dtsch. chem. Ges. 71, 725 (1938).

34. I. G. Farbenindustrie A. G.: Doppelverbindungen des Vitamins D_3 und D_2. Franz. Pat. 860509 (1939).

35. Inhoffen, H. H.: The Relationship of Natural Steroids to Carcinogenic Aromatic Compounds. Progr. Organ. Chem. 2, 151 (1953); London: Butterworths Scientific Publ.

36. — Untersuchungen an Steroiden. Naturwiss. 40, 455 (1953).

37. Inhoffen, H. H., K. Brückner und R. Gründel: Studien in der Vitamin D-Reihe: Umlagerung des Vitamins D_2 zu einem iso-Tachysterin und Partialsynthese eines iso-Vitamins D_2. Chem. Ber. 87, 1 (1954).

37a. Inhoffen, H. H. und E. Prinz: Chem. Ber. (1954) (im Druck).

38. Inhoffen, H. H. und K. Weissermel: Studien in der Vitamin D-Reihe. II. Partialsynthese des iso-Tachysterinmethyläthers. Chem. Ber. 87, 187 (1954).

39. Johnson, W. S., J. Szmuszkovicz and M. Miller: The Condensation of 1-Methyl-2-acetylcyclohexene with 1-Decalone and with 1,5-Decalin-dione. J. Amer. Chem. Soc. 72, 3726 (1950).

40. Jones, E. R. H. and R. W. Peevers (and Glaxo Laboratories Ltd.): 7-Dehydrosterols. Brit. Pat. 574432 (1946).

41. Koch, H. P.: The Light Absorption of Geometrical Isomerides and the Structure of Vitamin-D. Chem. and Ind. 61, 273 (1942).

42. Kuhn, R. und G. Platzer: Über Kumulene. III. Ber. dtsch. chem. Ges. 73, 1410 (1940).

43. Kuhn, R. und K. Wallenfels: Über Kumulene. I. Synthese von Tetraphenyl-hexapentaen und Di-biphenylen-hexapentaen. Ber. dtsch. chem. Ges. 71, 783 (1938).

44. — — Über Kumulene. II. Eine wesentliche Verbesserung des Darstellungs-Verfahrens. Ber. dtsch. chem. Ges. 71, 1510 (1938).

45. — — Synthese von Polyenen mit Hilfe von Acetylen und Diacetylen. Ber. dtsch. chem. Ges. 71, 1889 (1938).

46. Kuhn, R. und A. Winterstein: Über konjugierte Doppelbindungen. I. Synthese von Diphenyl-polyenen. Helv. Chim. Acta 11, 87 (1928).

47. Kuhn, W.: Das Problem der absoluten Konfiguration des tetraedrischen Kohlenstoffatoms. Z. Elektrochem. 56, 506 (1952).

48. Lardon, A. und T. Reichstein: Synthese der optisch aktiven β-Methoxy-adipinsäure. Steroide, 2. Mitt. Helv. Chim. Acta 32, 1613 (1949).

49. — — Die Konfiguration der optisch aktiven β-Methoxy-adipinsäuren und des Calciferols. Vorl. Mitt. Steroide, 4. Mitt. Helv. Chim. Acta 32, 2003 (1949).

50. Linsert, O.: Über das 7-Dehydro-stigmasterin. Z. physiol. Chem. (Hoppe-Seyler) 241, 125 (1936).

51. Linsert, O. (und I. G. Farbenindustrie A. G.): Physiologisch hochwirksames Produkt durch Hydrieren von bestrahltem Ergosterin bzw. Vitamin D_2... D. R. Pat. 730017 (1938).

52. MAZZA, F. P. und C. MIGLIARDI: Die Dehydrierung von Cholesterin zu 7-Dehydrocholesterin. Quad. Nutriz. (Turin) **8**, 86 (1941) [Chem. Zbl. **1942** I, 2016].

53. MERCK, Firma E.: Verfahren zur Darstellung von Monoalkyläthern aus Dialkyläthern zweiwertiger Phenole. 2. Zusatz zum D. R. Pat. 78910 (1894).

54. MEUNIER, P. et A. VINET: Chromatographie et mésomérie. Adsorption et résonance. Paris: Masson et Cie. 1947.

55. MILAS, N. A. (and Research Corp., New York): Alkyl Ethers of Vitamin D. U. S. Pat. 2410893 (1946).

56. MILAS, N. A. and W. L. ALDERSON, Jr.: Studies in the Synthesis of the Antirachitic Vitamins. I. The Synthesis of 3-[2'-Methylenecyclohexylidene-1']-propene-1. J. Amer. Chem. Soc. **61**, 2534 (1939).

57. MILAS, N. A. and R. HEGGIE (and Research Corp., New York): Herstellung synthetischer antirachitischer Provitamine und Vitamine. U. S. Pat. 2260085 (1938).

58. MILAS, N. A. and C. R. MILONE: Studies in the Synthesis of the Antirachitic Vitamins. II. Δ^5-Androstenol-3 and its Partial Dehydrogenation with Benzoquinone[1, 2]. J. Amer. Chem. Soc. **68**, 738 (1946).

58a. NENITZESCU, C. D. und E. CIORANESCU: Durch Aluminiumchlorid katalysierte Reaktionen. XXIII. Mitt. Versuche zur Synthese von Verbindungen mit Steringerüst. Ber. dtsch. chem. Ges. **75**, 1765 (1942).

59. NES, W. R. and E. MOSETTIG: The Rearrangement of Dehydroergosteryl Acetate to a s-Octahydroanthracene Derivative. J. Amer. Chem. Soc. **75**, 2787 (1953).

60. OPPENAUER, R. V. and H. OBERRAUCH: A New Oxidant: Tertiary Butyl Chromate. Ann. asoc. quím. argentina **37**, 246 (1949).

61. PAULING, L.: Recent Work on the Configuration and Electronic Structure of Molecules; with some Applications to Natural Products. Fortschr. Chem. organ. Naturstoffe **3**, 203 (1939).

62. PEERDEMAN, A. F., A. J. VAN BOMMEL and J. M. BIJVOET: Determination of Absolute Configuration of Optically Active Compounds by Means of X-Rays. Kon. Ned. Akad. Wetensch. Proc. Ser. B, **54**, 16 (1951).

63. PENAU, H. and G. HAGEMANN (and Usines Chimiques des Laboratoires Français): Calciferol Esters. U. S. Pat. 2484526 (1949).

64. PRICE, C. C. and M. MEISTER: cis-trans-Isomerization with Boron Fluoride. J. Amer. Chem. Soc. **61**, 1595 (1939).

64a. RABE, P. und F. RAHM: Über Kondensationsprodukte aus Acetessigester und Formaldehyd. Liebigs Ann. Chem. **332**, 1 (1904).

64b. — — Über die Konstitution des sogenannten Hagemannschen Esters. Ber. dtsch. chem. Ges. **38**, 969 (1905).

65. RAOUL, Y., J. CHOPIN, P. MEUNIER et N. LE BOULCH: Préparation d'un tachystérol se transformant spontanément en vitamine D à l'aide d'un réactif ionisant transparent. C. R. hebd. Séances Acad. Sci. **228**, 1064 (1949).

66. RAPHAEL, A. and F. SONDHEIMER: The Condensation of 2-Dimethyl-aminomethylcyclohexan-1-one with Acetylenic Compounds. J. Chem. Soc. (London) **1950**, 3185.

67. RAPSON, W. S. and R. ROBINSON: Experiments on the Synthesis of Substances related to the Sterols. II. A New General Method for the Synthesis of Substituted Cyclohexenones. J. Chem. Soc. (London) **1935**, 1285.

68. REDEL, J. et B. GAUTHIER: Contribution à la synthèse du déhydro-7-cholestérol. Bull. soc. chim. France **1948**, 607.

69. REINDEL, F., E. WALTER und H. RAUCH: Über das Ergosterin der Hefe. I. Liebigs Ann. Chem. **452**, 42 (1927).

70. RUIGH, W. L.: 7-Dehydrocampesterol, a New Provitamin D. J. Amer. Chem. Soc. **64**, 1900 (1942).

71. SAH, P. P. T.: Chinones as Reagents for the Dehydrogenation of Cholesteryl Acetate to Pro-vitamin D. Rec. trav. chim. Pays-Bas **59**, 454 (1940).

72. SCHALTEGGER, H.: Die direkte Halogenierung des Cholesterins in der Allylstellung (C_7-Atom). Experientia **5**, 321 (1949).

73. SCHENCK, FR.: Über das kristallisierte Vitamin D_3. Naturwiss. **25**, 159 (1937).

73a. SCHWENK, E. and E. BLOCH: Preparation of β-(2-Methyl-6-oxo-1-cyclohexen-1-yl)-propionic Acid. J. Amer. Chem. Soc. **64**, 3050 (1942).

74. THIBAUDET, G.: Chromatographie et mésomérie: du calciférol au tachystérol. C. R. hebd. Séances Acad. Sci. **220**, 751 (1945).

75. TURNER, R. B. and D. M. VOITLE: The Reaction of 1-Acetyl-2-methyl-cyclohexene with Cyclohexanone. J. Amer. Chem. Soc. **72**, 4166 (1950).

76. VELLUZ, L. et G. AMIARD: Le précalciférol. C. R. hebd. Séances Acad. Sci. **228**, 692 (1949).

77. — — Equilibre de réaction entre précalciférol et calciférol. C. R. hebd. Séances Acad. Sci. **228**, 853 (1949).

78. — — Nouveau précurseur de la vitamine D_3. C. R. hebd. Séances Acad. Sci. **228**, 1037 (1949).

79. VELLUZ, L., G. AMIARD et A. PETIT: Le précalciférol. — Ses relations d'équilibre avec le calciférol. Bull. soc. chim. France **1949**, 501.

80. VELLUZ, L., A. PETIT et G. AMIARD: Sur un stade non photochimique dans la formation des calciférols. Essais d'interprétation. Bull. soc. chim. France **1948**, 1115.

81. VELLUZ, L., A. PETIT, G. MICHEL et G. ROUSSEAU: Sur un stade non photochimique dans la formation des calciférols. C. R. hebd. Séances Acad. Sci. **226**, 1287 (1948).

82. VLIET, J. VAN DER: Die Synthese von Provitamin D_3. Chem. Weekbl. **44**, 692 (1948).

83. — Investigations on Sterols. II. Vitamin-D_2 and -D_3 in Irradiated Sterol from the Mussel *(Mytilus edulis)*. Rec. trav. chim. Pays-Bas **67**, 246 (1948).

84. — Investigations on Sterols. III. The Provitamins-D from the Mussel *(Mytilus edulis)*. Rec. trav. chim. Pays-Bas **67**, 265 (1948).

85. VLIET, J. VAN DER and W. STEVENS: 7-Dehydro-sterol. U. S. Pat. 2441091 (1948).

85a. Vorschläge zur Nomenklatur der Steroide. Helv. Chim. Acta **34**, 1680 (1951).

86. WALLACH, O.: Zur Kenntnis der Terpene und der ätherischen Öle. Liebigs Ann. Chem. **365**, 255 (1909).

87. WANDER, A., A. G.: Ionische Derivate einer Polyhydrophenanthrenverbindung. Schweiz. Pat. 251384 (1948).

88. WERDER, F. V.: Über Dihydro-tachysterin. Z. physiol. Chem. (Hoppe-Seyler) **260**, 119 (1939).

89. WHITTIER, C. C. (and Nutrition Research Laboratories Inc.): Vitamin D. U. S. Pat. 2106779—80 (1934) and 2106781—82 (1935).

90. WINDAUS, A. und E. AUHAGEN: Über die Beständigkeit des bestrahlten Ergosterins. Z. physiol. Chem. (Hoppe-Seyler) **196**, 108 (1931).

91. WINDAUS, A. und K. BUCHHOLZ: Über ein Keton des Vitamins D_2. Z. physiol. Chem. (Hoppe-Seyler) **256**, 273 (1938).

92. WINDAUS, A., K. DITHMAR und E. FERNHOLZ: Über das Lumisterin. Liebigs Ann. Chem. **493**, 259 (1932).

93. WINDAUS, A., K. DITHMAR, H. MURKE und F. SUCKFÜLL: Weitere Untersuchungen über die Isomerisierung des Ergosterins und seiner Derivate. Liebigs Ann. Chem. **488**, 91 (1931).

94. WINDAUS, A. und W. GRUNDMANN: Über die Konstitution des Vitamins D_2. II. Liebigs Ann. Chem. **524**, 295 (1936).

95. WINDAUS, A. und B. GÜNTZEL: Über einige Bestrahlungsprodukte des 22-Dihydro-ergosterins. Liebigs Ann. Chem. **538**, 120 (1939).

96. WINDAUS, A. und R. LANGER: Über das 22-Dihydro-ergosterin. Liebigs Ann. Chem. **508**, 105 (1934).

97. WINDAUS, A., H. LETTRÉ und FR. SCHENCK: Über das 7-Dehydro-cholesterin. Liebigs Ann. Chem. **520**, 98 (1935).

98. WINDAUS, A. und O. LINSERT: Über die Ultraviolett-Bestrahlung des Dehydroergosterins. Liebigs Ann. Chem. **465**, 148 (1928).

99. WINDAUS, A., O. LINSERT, A. LÜTTRINGHAUS und G. WEIDLICH: Über das kristallisierte Vitamin D_2. Liebigs Ann. Chem. **492**, 226 (1932).

100. WINDAUS, A. und J. NAGGATZ: Über das 7-Dehydro-epi-cholesterin. Liebigs Ann. Chem. **542**, 204 (1939).

101. WINDAUS, A. und U. RIEMANN: Über die Einwirkung von Bleitetra-acetat auf einige Sterinderivate. Z. physiol. Chem. (Hoppe-Seyler) **274**, 206 (1942).

102. WINDAUS, A., FR. SCHENCK und F. v. WERDER: Über das antirachitisch wirksame Bestrahlungsprodukt aus 7-Dehydro-cholesterin. Z. physiol. Chem. (Hoppe-Seyler) **241**, 100 (1936).

103. WINDAUS, A. und G. TRAUTMANN: Über das kristallisierte Vitamin D_4. Z. physiol. Chem. (Hoppe-Seyler) **247**, 185 (1937).

104. WINDAUS, A., F. v. WERDER und A. LÜTTRINGHAUS: Über das Tachysterin. Liebigs Ann. Chem. **499**, 188 (1932).

105. WUNDERLICH, W.: Über das 7-Dehydro-sitosterin. Z. physiol. Chem. (Hoppe-Seyler) **241**, 116 (1936).

106. ZECHMEISTER, L.: *cis-trans* Isomerization and Stereochemistry of Carotenoids and Diphenylpolyenes. Chem. Rev. **34**, 267 (1944).

(Eingelaufen am 15. Dezember 1953.)

Natürlich vorkommende Chromone.

(Mit Anhang über weitere Eleutherine-Inhaltsstoffe.)

Von H. SCHMID, Zürich.

Inhaltsübersicht.

I. Einleitung.

Die natürlichen, d. h. von Pflanzen aufgebauten Chromone,
leiten sich vom γ-Benzopyron ab. Zum Unterschied von den im
Pflanzenreich weitverbreiteten Flavonen und ihren im Pyronring
hydrierten und oxydierten Abkömmlingen, die am $C_{(2)}$ einen Arylrest
tragen, besitzen die natürlichen Chromone an diesem Kohlenstoffatom
eine Methyl- (in einem Fall eine Oxymethyl-) Gruppe. Während aber
die flavanoiden Pflanzenstoffe schon lange bekannt sind und in analytischer
und synthetischer Hinsicht eine gründliche
Bearbeitung erfuhren, haben sich die natürlichen
Chromone dem Zugriff des Chemikers lange Zeit
entzogen. Zwar beschäftigten sich schon um das
Jahr 1880 Mustapha (*83*) und Malosse (*70*) mit
den Früchten von *Ammi visnaga*, einer in der
ägyptischen Volksmedizin seit altersher ge-
schätzten Droge, die, wie man heute weiß, reich

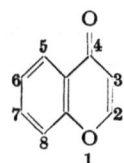

Numerierung des Chromonringes.

an Chromonen ist, ohne allerdings praktisch verwertbare Resultate zu
erzielen. 50 Jahre später isolierte Samaan (*104*) sowie Fantl und
Salem (*31*) aus derselben Droge eine Reihe von kristallisierten Stoffen,
unter ihnen das Khellin, Visnagin und Khellolglucosid. Die exakte
Konstitutionsaufklärung dieser Inhaltsstoffe und damit ihre Erkennung
als Chromone verdankt man Arbeiten von Späth und Gruber (*149—151*)
aus den Jahren 1938—1941.

Synthetisch sind allerdings die Chromone, hauptsächlich durch die
Untersuchungen von Kostanecki und seiner Schule, schon zu Beginn
dieses Jahrhunderts erschlossen worden.

Bis heute sind 11 Pflanzenstoffe als Chromone identifiziert worden.
Wenn trotz der Fülle der bisher ausgeführten pflanzenchemischen Arbeiten
eine so geringe Zahl von Chromonen isoliert wurde, so dürfte dies nicht
auf ihr seltenes Vorkommen, sondern auf Schwierigkeiten, die ihr Nachweis
und ihre Gewinnung bereiten, zurückzuführen sein. Die natürlichen
(meist farblosen) Chromone zeigen, im Gegensatz zu vielen anderen
Klassen organischer Naturstoffe, keine für eine spezifische Isolierung
geeignete Reaktionen. Einzig die gelborange gefärbten Oxoniumsalze

mit konz. Mineralsäuren würden sich vielleicht manchmal hierfür eignen (*30*). Empfindliche Chromone (z. B. Peucenin) können dabei aber verändert werden. Es ist jedenfalls bezeichnend, daß die natürlichen Chromone ihre Entdeckung nicht ihren chemischen oder physikalischen, sondern ihren pharmakologischen Eigenschaften verdanken. Khellin wird heute mit gutem Erfolg zur therapeutischen Behandlung von Koronarsclerosen, wie *Angina pectoris*, angewendet.

II. Pflanzlicher Ursprung und Isolierung der Chromone.

In der *Tabelle 1* sind die Pflanzen zusammen mit den aus ihnen gewonnenen Chromonen aufgeführt. Alle Pflanzen gehören höchst differenzierten, stark entwickelten Familien an.

Tabelle 1. Aus Pflanzen isolierte Chromone.

Ammi visnaga (L.) Lam., Früchte	Khellin (Visammin)
	Visnagin (Visnagidin)
	Visamminol
	Khellinol
	Khellolglucosid (Khellinin)
Peucedanum Ostruthium (Koch), Rhizom	Peucenin
Eugenia caryophyllata (L.) Thumbg., Nelken	Eugenin
	Eugenitin
	Isoeugenitin
	Isoeugenitol
Eleutherine bulbosa (Mill.) Urb., Knollen	Eleutherinol

Ammi visnaga (L.) Lam., eine Dicotyledone aus der Familie der Umbelliferen, kommt hauptsächlich in Ägypten, vor allem im Nildelta, in Palästina, Mesopotamien und Madeira vor. Sie wird von den Eingeborenen als „Chellah", „Khilla", „Khella" oder „Gazor Shitani" bezeichnet. Dekokte und Tinkturen aus ihren Früchten spielen eine wichtige Rolle in der ägyptischen Volksmedizin (vgl. S. 154). Khellin kommt in der Droge zu etwa 1%, Visnagin zu 0,1% und Khellolglucosid zu etwa 0,3% vor. Neben den Chromonen sind aus dieser Pflanze von Samaan (*104*) außer Fetten (*43*) noch vier weitere Stoffe, unter ihnen das amorphe Visnagan, extrahiert, aber nur ungenügend charakterisiert worden. Visnagan, welches das Khellin in seiner pharmakologischen Wirkung übertrifft, wurde kürzlich kristallin erhalten (*143*). Es scheint aber kein Chromon zu sein. Ferner kommt in *Ammi visnaga* noch ein Cumarin, Khellakton (*44, 71*) vor, dessen Konstitution noch nicht feststeht.

Ammi visnaga L. wird häufig mit der in Ägypten heimischen, ihr sehr nahe verwandten *Ammi majus* L. (Umbelliferae) verwechselt. Diese Pflanze enthält aber keine Chromone, sondern die Furocumarine Xanthotoxin (*136, 139*), Imperatorin (*136, 139*) und Bergapten (*29*).

Peucedanum Ostruthium KOCH, manchmal auch als *Imperatoria Ostruthium* L. bezeichnet, gehört ebenfalls zur Familie der Umbelliferen und ist nahe verwandt mit *Ammi visnaga* L. Ihr Rhizom, die Meisterwurz, wird noch heute medizinisch verwendet. Aus dem Rhizom konnten neben einer größeren Zahl von Cumarinen, nämlich Osthol, Ostruthin, den Furocumarinen Imperatorin, Isoimperatorin, Oxypeucedanin und Ostruthol, das Chromon Peucenin (0,03%) gewonnen werden [siehe die Zusammenfassungen von SPÄTH (*145*, *146*) und DEAN (*25*)].

Eugenia caryophyllata (L.) THUMBG., die wild wachsende Gewürznelke, gehört zur Familie der Myrtaceen und ist den Umbelliferen verwandt. Ihre Heimat ist die Molukkeninsel Makian. Ihre Nelken zeichnen sich gegenüber denjenigen der kultivierten Form durch einen geringeren Gehalt an essentiellen Ölen und durch ihren Reichtum an Chromonen aus. Ein weiterer Inhaltsstoff, das Diketon Eugenon (*126*), scheint mit den Chromonen in einem nahen biogenetischen Zusammenhang zu stehen.

Eleutherine bulbosa (MILL.) URB. ist die bisher einzige Monocotyledone unter den Pflanzen, aus denen Chromone extrahiert wurden. Sie gehört der Familie der Iridacaeen an, stammt aus dem tropischen Amerika und wird auf Java kultiviert. Die Knollen sollen in der javanischen Volksmedizin eine gewisse Bedeutung erlangt haben. Das aus ihr gewonnene Chromon Eleutherinol (*27*) stellt das erste natürliche Naphthopyron dar und ist damit verwandt mit den anderen von Naphthalin sich ableitenden Inhaltsstoffen dieser Pflanze (vgl. S. 157).

Zur *Isolierung der Chromone* wurde das getrocknete Pflanzenmaterial meistens mit Äther oder Aceton extrahiert und dieser Extrakt durch fraktionierte Kristallisation weiter aufgearbeitet. Die Reinigung geschah durch Umlösen aus verschiedenen Lösungsmitteln, durch Hochvakuumsublimation oder Chromatographie an neutralem, schwach aktivem Aluminiumoxyd oder Magnesol (*149*, *150*, *14*). Für die Papierchromatographie eignet sich das System Isopropanol-Wasser 1 : 3—1 : 6. Gemische aus Chromonen, die phenolische Hydroxylgruppen besitzen, ließen sich vorteilhaft durch fraktioniertes Ausschütteln ihrer Äther-Chloroform-Lösung mit Soda und wäßriger Lauge steigender Konzentration auftrennen (*147*, *119*). Khellolglucosid gewann man aus dem Methanolextrakt der vorher mit Äther ausgezogenen Früchte von *Ammi visnaga* (*151*).

III. Konstitutionsermittlung der Chromone.

1. Allgemeine Bemerkungen

(vgl. auch *156*).

Zur Konstitutionsaufklärung wird das vermutliche Chromon zunächst Reaktionen unterworfen, die für die Anwesenheit eines 2-Methylchromonskelettes beweisend sind.

Abbau mit Alkali. Die wichtigste unter ihnen stellt die Spaltung mit siedender 1- bis 10-proz. Lauge (oder Sodalösung) dar, die am Beispiel des Eugenitins bzw. seines Methyläthers aufgezeigt sei.

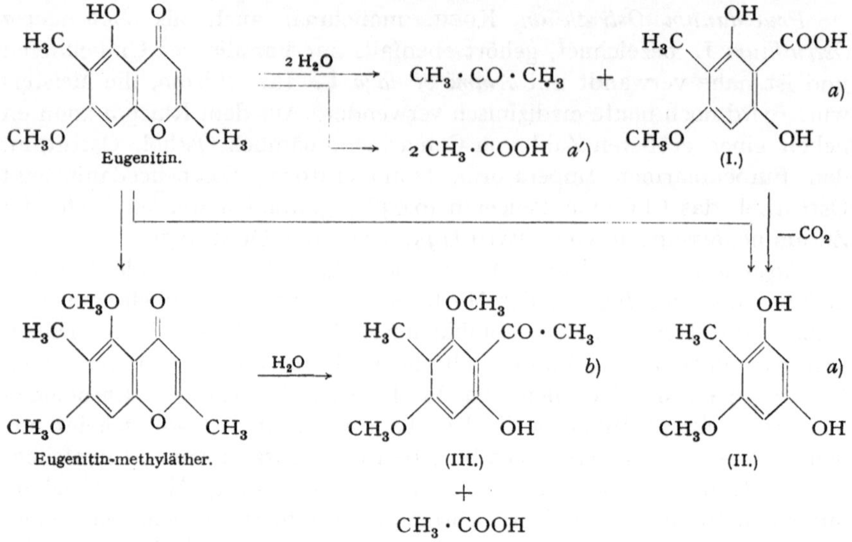

$$\underset{\text{Eugenitin.}}{\text{(HO, O, }H_3C\text{, }CH_3O\text{, }O\text{, }CH_3\text{ chromone)}} \xrightarrow{2\,H_2O}$$

$$CH_3 \cdot CO \cdot CH_3 \;+\; \underset{\text{(I.)}}{\text{(}H_3C\text{, }OH\text{, }COOH\text{, }CH_3O\text{, }OH\text{)}} \quad a)$$

$$\cdots\!\longrightarrow\; 2\,CH_3 \cdot COOH \quad a')$$

$$\Big| -CO_2$$

$$\underset{\text{Eugenitin-methyläther.}}{\text{(}CH_3O\text{, }O\text{, }H_3C\text{, }CH_3O\text{, }O\text{, }CH_3\text{)}} \xrightarrow{H_2O} \underset{\text{(III.)}}{\text{(}H_3C\text{, }OCH_3\text{, }CO\cdot CH_3\text{, }CH_3O\text{, }OH\text{)}} \quad b) \qquad \underset{\text{(II.)}}{\text{(}H_3C\text{, }OH\text{, }CH_3O\text{, }OH\text{)}} \quad a)$$

$$+$$
$$CH_3 \cdot COOH$$

Unter Aufsprengung des γ-Pyronringes entsteht aus Eugenitin ein Phenol (II) mit vier Kohlenstoffatomen weniger als das Ausgangsmaterial und Aceton und Kohlensäure, sowie nach a' wenige Prozente Essigsäure. Die intermediär auftretende Carbonsäure (I) läßt sich meistens nicht oder nur in kleinen Mengen fassen (vgl. S. 134).

Neben der Reaktion a tritt bei der Alkalieinwirkung aber noch eine weitere Reaktion b auf, die unter Aufsprengung des γ-Pyronringes zwischen $C_{(2)}$ und $C_{(3)}$ zu einem o-Oxyacetophenon-derivat (III) und Essigsäure führt. Sehr häufig verlaufen die beiden Reaktionen a und b nebeneinanderher. Das Verhältnis a/b hängt unter anderem von der Substitution am Benzolring und der Konzentration des verwendeten Alkalis ab. Freie Hydroxylgruppen begünstigen, wie aus dem Beispiel des Paares Eugenitin und Eugenitin-methyläther hervorgeht, a, während b durch starkes Alkali bevorzugt ist. Die nach a und b zu erwartenden Reaktionsprodukte lassen sich ohne Schwierigkeit isolieren; ihre Identifizierung liefert einen sicheren Beweis für das Vorliegen eines 2-Methylchromongerüstes. Das o-Oxyacetophenon-derivat (III) läßt sich in einfacher Weise wieder in das ursprüngliche Chromon zurückverwandeln (siehe S. 141).

Die *2-ständige Pyron-Methylgruppe* zeichnet sich durch erhöhte Reaktionsfähigkeit aus: mit aromatischen Aldehyden, wie Anisaldehyd, Piperonal, reagiert sie bei Gegenwart von $RO^{\ominus}$ zu gelb gefärbten Styrylchromonen (*54, 20, 137*); 2-Äthyl- oder 2-Propylgruppen zeigen diese Reaktion nicht ($Ar =$ Aryl):

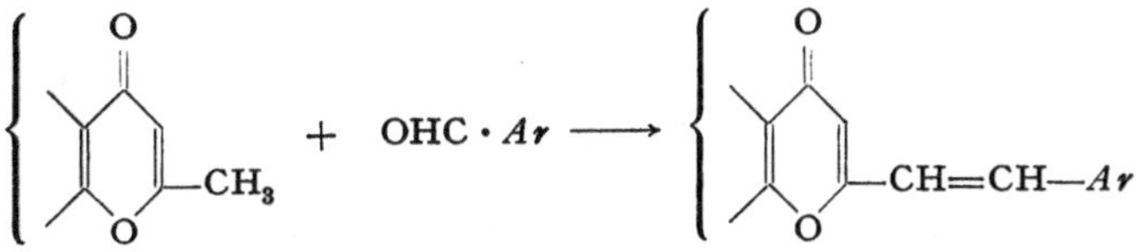

$$\left\{ \begin{array}{c} O \\ \diagdown \diagup -CH_3 \\ O \end{array} \right. + \ OHC \cdot Ar \ \longrightarrow \ \left\{ \begin{array}{c} O \\ \diagdown \diagup -CH=CH-Ar \\ O \end{array} \right.$$

Eine Ausnahme scheinen 2-Methyl-5,7-dioxychromon und seine Äther zu machen (*52*), die unter Standardbedingungen nicht mit aromatischen Aldehyden reagieren. An der 2-Methylchromonstruktur dieser Stoffe kann aber nicht gezweifelt werden (*128*). Die *2-ständige Chromon-Methylgruppe* ist auch die Ursache für eine rote Farbreaktion mit starkem Alkali (*137, 140, 14*). Mit Äthyl- oder Propylgruppen tritt die Reaktion nicht oder viel schwächer ein; sie wird verhindert durch phenolische Hydroxyl- oder Acetoxylgruppen, während Methoxylgruppen nicht stören.

Von weiteren nützlichen Farbreaktionen sei die meist violette bis blaue *Ferrichlorid-Reaktion* (*14, 10*) erwähnt, die von Chromonen mit 5-ständiger OH-Gruppe gegeben wird. Bei 5-Oxy-6,7-furano-chromonen tritt eine grüne Färbung auf. Die Hydroxylgruppe am $C_{(7)}$ bewirkt keine Farbreaktion. Chromone mit zwei freien, in Stellung 5 und 7 stehenden Hydroxylgruppen lassen beim Versetzen ihrer alkalischen Lösung mit Wasserstoffsuperoxyd eine meist blaue, wenig beständige Farbreaktion erkennen. Die Reaktion unterbleibt bei 7-Oxychromonen sowie 5-Oxy-7-alkoxychromonen (*14*). Auf die meist gelborange gefärbten Oxoniumsalze von Chromonen mit starken Mineralsäuren (Schwefelsäure, Salpetersäure, Perchlorsäure) ist schon früher hingewiesen worden. Auf die *Ultraviolett-Absorptionen* natürlicher Chromone als wichtiges Hilfsmittel zur Strukturaufklärung wird später eingegangen (S. 137).

2-Methylchromone geben in alkalischer Lösung mit Hydrazin bzw. Hydroxylamin Pyrazol- bzw. Isoxazolabkömmlinge (*135*).

2. Spezieller Teil.

Chromone aus Eugenia caryophyllata.

Die Chromone aus dieser Pflanze, die von SCHMID, MEIJER und BOLLETER näher untersucht wurden, zeichnen sich durch ihren relativ einfachen Bau aus.

Eugenin (*72, 73, 128*), $C_{10}H_8O_3$, (IV), Schmelzp. 119—120°, stellt das einfachste natürliche Chromon dar. Bei der Entmethylierung entstand das bekannte 2-Methyl-5,7-dioxychromon (V) und bei Verkochen mit Alkali verschiedener Konzentration Aceton und 2,6-Dioxy-4-methoxy-acetophenon (VI). Die 5-ständige, chelierte OH-Gruppe verhält sich kryptophenolisch: sie läßt sich mit Diazomethan und Dimethylsulfat-Soda bei 20° nicht methylieren.

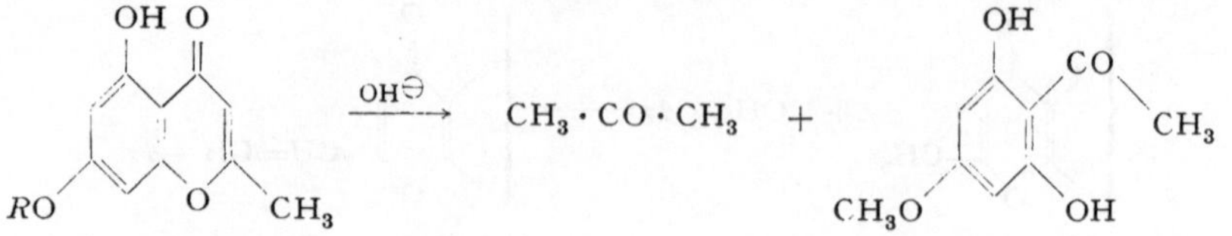

OH O $\quad\quad$ OH⊖ $\longrightarrow$ $\quad$ CH$_3$ · CO · CH$_3$ $\quad+$ $\quad$ OH $\quad$ CO $\quad$ CH$_3$

RO $\quad$ O $\quad$ CH$_3$ $\qquad\qquad\qquad\qquad\qquad\qquad$ CH$_3$O $\quad$ OH

(IV.) $R = $ CH$_3$. Eugenin.
(V.) $R = $ H. 2-Methyl-5,7-dioxy-
$\quad$ chromon.

(VI.) 2,6-Dioxy-4-methoxy-acetophenon.

Eugenitin (*116*), C$_{11}$H$_9$O$_3$·OCH$_3$, (VII), Schmelzp. 162°, *Isoeugenitin* (*119*), C$_{11}$H$_9$O$_3$ · OCH$_3$, (XI), Schmelzp. 148° und *Isoeugenitol* (*117*), C$_{11}$H$_{10}$O$_4$, (X), Schmelzp. 236°. Eugenitin ließ sich in ein Monoacetat und durch energische Methylierung unter Bedingungen, die eine Aufsprengung des γ-Pyronringes erschwerten, in den kristallisierten Dimethyläther (IX) überführen. Alkalispaltung von (VII) lieferte C-Methylphloroglucin-β-methyläther (II) und Aceton, während aus (IX) das 2-Oxy-4,6-dimethoxy-5-methylacetophenon (III) entstand. Die 7-Stellung der Methoxylgruppe im Eugenitin folgt aus seiner Unlöslichkeit in verdünnter Lauge seiner Ferrichlorid-Reaktion und der Trägheit gegenüber Diazomethan. Eine unverschlossene 7-ständige Hydroxylgruppe würde bekanntlich Lauge-

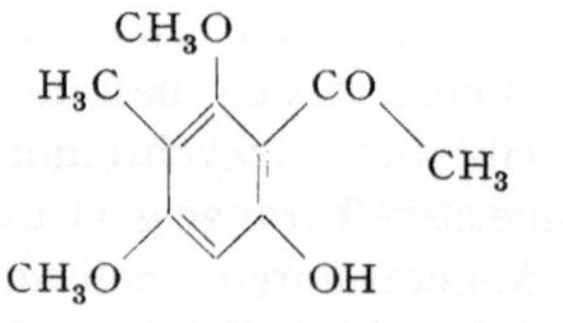

CH$_3$O $\quad$ H$_3$C $\quad$ CO $\quad$ CH$_3$ $\qquad\longleftarrow\qquad$ CH$_3$O $\quad$ O $\quad$ H$_3$C

CH$_3$O $\quad$ OH $\qquad\qquad\qquad\qquad\qquad$ CH$_3$O $\quad$ O $\quad$ CH$_3$

(III.) 2-Oxy-4,6-dimethoxy-5-methyl-
$\quad$ acetophenon.

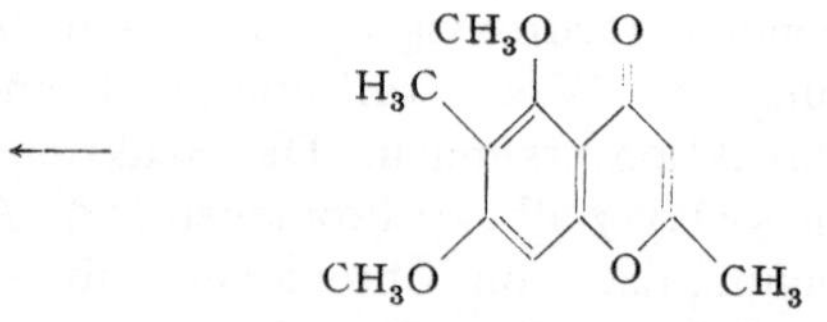

(IX.)

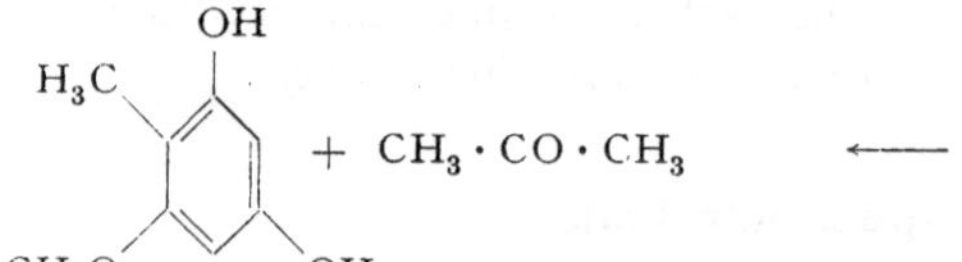

OH $\quad$ H$_3$C $\quad$ + CH$_3$ · CO · CH$_3$ $\qquad\longleftarrow$ $\quad$ OH O $\quad$ H$_3$C

CH$_3$O $\quad$ OH $\qquad\qquad\qquad\qquad\qquad$ RO $\quad$ O $\quad$ CH$_3$

(II.) C-Methyl-phloroglucin-
$\quad\beta$-methyläther.

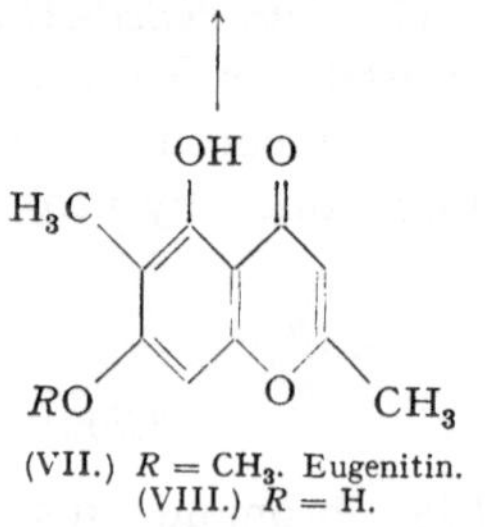

(VII.) $R = $ CH$_3$. Eugenitin.
(VIII.) $R = $ H.

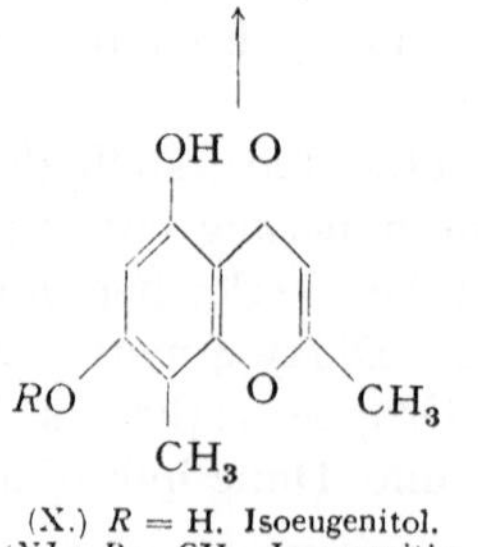

OH O $\qquad\qquad\qquad\qquad\qquad\qquad$ OH $\quad$ H$_3$C $\quad$ CO

RO $\quad$ O $\quad$ CH$_3$ $\qquad\longleftarrow\qquad$ CH$_2$
$\quad$ CH$_3$ $\qquad\qquad\qquad\qquad\qquad$ HO $\quad$ OH $\quad$ CO—CH$_3$

(X.) $R = $ H. Isoeugenitol.
(XI.) $R = $ CH$_3$. Isoeugenitin.

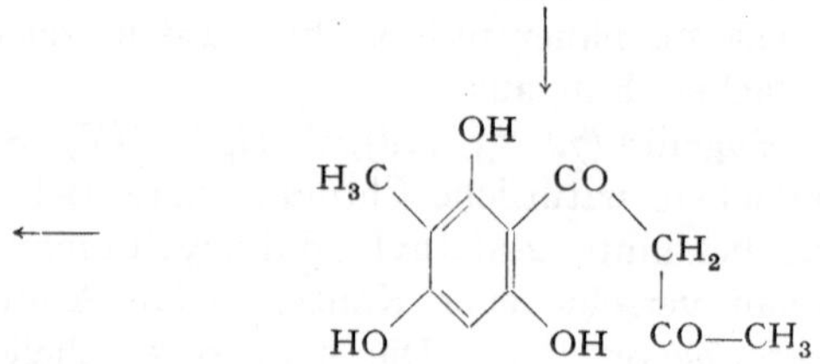

(XII.)

löslichkeit bewirken. Sie würde sich mit Diazomethan glatt methylieren lassen, würde aber keine Eisen-Reaktion geben.

Als man Eugenitin mit Jodwasserstoff erhitzte, entstand nicht das (VII) zugrunde liegende Phenol (VIII), sondern ein Isomeres (X), sehr wahrscheinlich über das Diketon (XII) (oder dessen Enol); (X) gab nämlich mit Diazomethan einen von (VII) verschiedenen Monomethyläther (XI), der die gleichen Alkalispaltstücke gab wie (VII). Da sich der Dimethyläther von (X) als verschieden von (IX) erwies, kann die Isomerie von (VII) und (XI) nur durch eine verschiedene Stellung der am Benzolkern haftenden Methylgruppe bedingt sein, die in (X) und (XI) am $C_{(8)}$-Atom haften muß. (XI) und (X) wurden später auch aus *Eugenia caryophyllata* isoliert und als *Isoeugenitin* bzw. *Isoeugenitol* bezeichnet. Aus der Lösung in Äther läßt sich Isoeugenitol (X) mit Sodalösung, Isoeugenitin (XI) langsam mit 5proz. Lauge ausziehen, während Eugenitin (VII) mit seiner stark abgeschirmten 5-ständigen phenolischen Hydroxylgruppe im Äther zurückbleibt.

Bei Di- und Trioxychromonen (ohne 3-ständige Hydroxylgruppe) ist die 5,6- bzw. 5,6,7-Anordnung gegenüber *Jodwasserstoffsäure* stabiler als die 5,8- bzw. 5,7,8-Gruppierung. So gaben die Methyläther von 2-Methyl- oder 2,3-Dimethyl-5,7,8-dioxychromon beim langen Kochen mit Jodwasserstoffsäure 2-Methyl- oder 2,3-Dimethyl-5,6,7-trioxychromon (*16, 79*) (vgl. dazu auch *158, 157, 53, 113, 32a*). Die Verhältnisse liegen hier umgekehrt wie bei den 6- bzw. 8-alkylierten Chromonen.

Weitere durch Jodwasserstoffsäure bewirkte Umlagerungen sind beim Peucenin, Khellin und Visnagin beschrieben.

Chromon aus Peucedanum Ostruthium.

Peucenin, $C_{16}H_{18}O_4$, (XIII), Schmelzp. 212°. Die Verbindung, deren Aufklärung man SPÄTH und EITER (*147*) verdankt, enthält zwei phenolische Hydroxylgruppen und ist löslich in verdünnten Alkalien. Mit Diazomethan entstand der lauge-unlösliche Monomethyläther (XIV). Ferner ließ sich eine leicht hydrierbare Doppelbindung nachweisen. Dihydropeucenin (XV) (nicht aber Peucenin) gab bei der Oxydation Isohexylsäure; anderseits bildete sich aus (XIII) mit Ozon Aceton, so daß im Peucenin der Seitenrest $(CH_3)_2C=CH—CH_2—$ vorliegen muß. Bei der alkalischen Spaltung von (XIII) traten Essigsäure und das bekannte 2,2-Dimethyl-5,7-dioxychroman (XVI), bei derselben Reaktion mit (XV) hingegen Isoamylphloroglucin (XVII) auf. Im Peucenin liegt daher ein 2-Methyl-5,7-dioxychroman mit einer Isoamylenkette in Stellung 6 oder 8 vor.

Unter der Einwirkung von Eisessig-Schwefelsäure entstanden aus Peucenin (XIII) zwei Isomere, die keine leicht hydrierbare Doppelbindung mehr enthielten und mit Alkali (XVI) gaben. Das eine, Allo-

9*

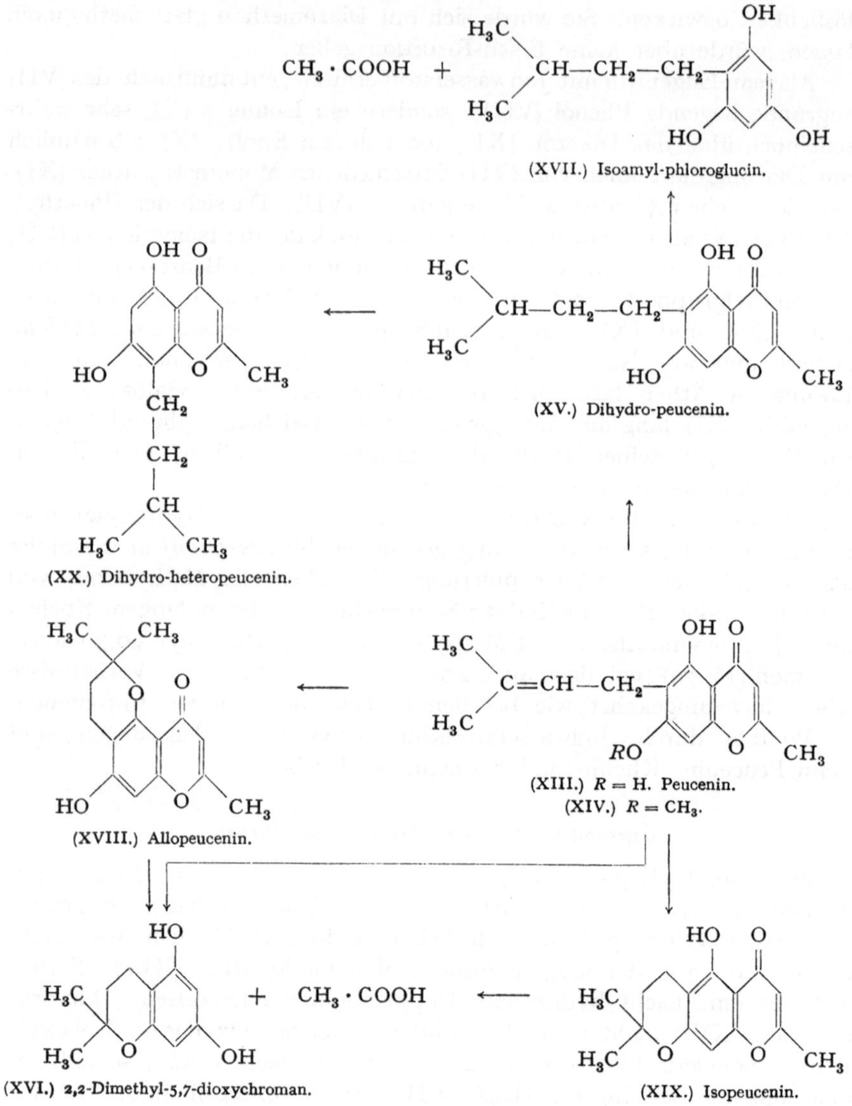

$$CH_3 \cdot COOH \; + \qquad \underset{H_3C}{\overset{H_3C}{>}}CH-CH_2-CH_2-$$

(XVII.) Isoamyl-phloroglucin.

(XX.) Dihydro-heteropeucenin.

(XV.) Dihydro-peucenin.

(XVIII.) Allopeucenin.

(XIII.) $R =$ H. Peucenin.
(XIV.) $R =$ CH$_3$.

(XVI.) 2,2-Dimethyl-5,7-dioxychroman.

(XIX.) Isopeucenin.

peucenin (XVIII), ist lauge-löslich (freie 7-ständige OH-Gruppe!), das andere, Isopeucenin (XIX), ist lauge-unlöslich. Unter der Voraussetzung, daß bei diesen für *o*-Oxy-isoamylenkörper charakteristischen Cyclisierungen (*148*) keine Umlagerung des γ-Pyronringes stattgefunden hat, beweisen die obigen Resultate die Struktur (XIII) für Peucenin. Auf Grund ihrer U. V.-Spektren sind aber (XIII), (XIV) und (XV) eindeutig den am C-Atom 6 alkylierten 5,7-Dioxychromonen zuzuordnen (*14*).

Wie Eugenitin, so erfährt auch (XVI) mit Jodwasserstoffsäure unter intermediärer Öffnung des γ-Pyronringes eine Umlagerung zum Dihydroheteropeucenin (XX) (*14*).

Chromone aus Ammi visnaga.

Khellin, $C_{12}H_8O_3 \cdot (OCH_3)_2$, (XXI), Schmelzp. 154—155°. FANTL und SALEM (*31*) fanden, daß Khellin die oben angeführte Summenformel besitzt und daß es mit Bariumhydroxyd zum Khellinon $C_{10}H_8O_2(OCH_3)_2$ (XXII) abgebaut wird. (XXII) wurde als aromatischer *o*-Oxyaldehyd und Khellin als Cumarin aufgefaßt. Die richtige Strukturformel (XXI) für Khellin wurde zuerst von SAMAAN veröffentlicht, sie wurde aber erst durch Arbeiten von SPÄTH und GRUBER (*149*) auf eine sichere experimentelle Grundlage gestellt.

Beim Erhitzen von (XXI) mit Alkalien entstanden Essigsäure und Khellinon (XXII), das als *o*-Oxyketon erkannt wurde. Dieser Befund machte für Khellin eine 2-Methylchromon-struktur wahrscheinlich, die durch Rückverwandlung von Khellinon (XXII) in Khellin (XXI) über

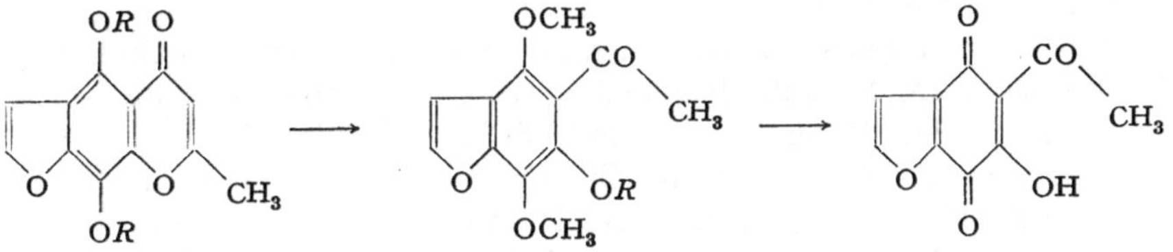

XXI.) $R = CH_3$. Khellin.
(XXI a.) $R = H$.
5,8-Bisnor-khellin.

(XXII.) $R = CH_3$. Khellinon.
(XXIV.) $R = C_2H_5$.

(XXIII.)

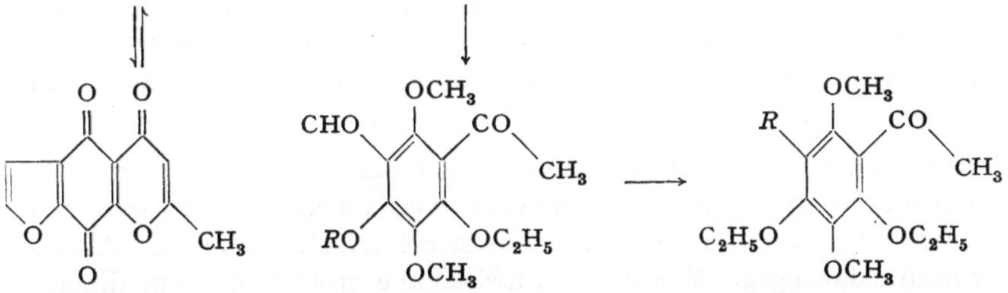

(XXX.)

(XXV.) $R = H$.
(XXVI.) $R = C_2H_5$.

(XXVII.) $R = COOH$.
(XXVIII.) $R = H$.
2,4-Diäthoxy-3,6-dimethoxy-acetophenon.

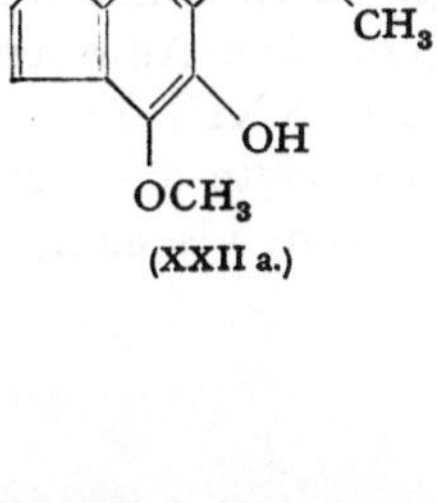

(XXII a.)

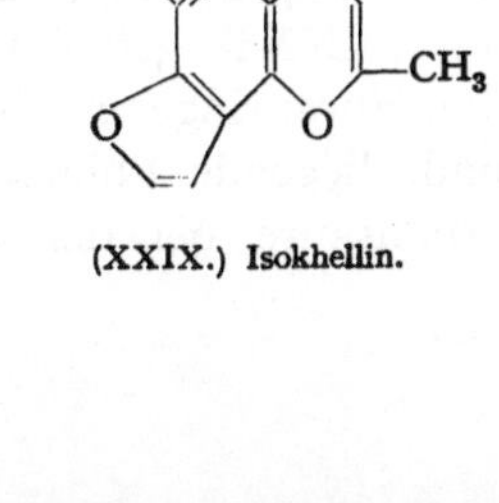

(XXIX.) Isokhellin.

das 3-Acetylchromon (*149*) oder sicherer über das Diketon (*34*, S. 141) streng bewiesen wurde.

Khellinon (XXI) lieferte mit Salpetersäure einen methoxyl-freien gelben Körper $C_{10}H_6O_5$ (XXIII), den man als *p*-Chinon auffassen konnte. Anderseits entstand aus (XXI) mit alkalischem Wasserstoffsuperoxyd 2,3-Furandicarbonsäure, so daß für Khellinon nur eine Auswahl zwischen den Strukturformeln (XXII) und (XXIIa) zu treffen war.

Die Entscheidung zugunsten von (XXII) ließ sich auf dem durch die Formeln (XXIV) bis (XXVIII) gekennzeichneten Wege erzielen: Nach Schutz der OH-Gruppe in (XXII) durch Äthylierung (XXIV) (um sie von den beiden nativen Methoxylgruppen unterscheiden zu können) wurde ozonisiert. Den resultierenden Aldehyd (XXV) haben Späth und Gruber nach Äthylierung der 5-ständigen Hydroxylgruppe zur Carbonsäure (XXVII) oxydiert und anschließend decarboxyliert. Das Abbauprodukt ließ sich als Semicarbazon mit synthetischem 2,4-Diäthoxy-3,6-dimethoxy-acetophenon (XXVIII) identifizieren. Da aus (XXIIa) 2,5-Diäthoxy-3,6-dimethoxy-acetophenon hätte entstehen müssen, steht für Khellin die Konstitutionsformel (XXI) fest.

Versuche, Khellin (XXI) mit Jodwasserstoffsäure zu entmethylieren, führten zu einer Substanz, die einen zu (XXI) isomeren Dimethyläther der Formel (XXIX) gab (Isokhellin) (*1—1b*). Entmethylierung ohne Umlagerung des Pyronringes zum 5,8-Bisnorkhellin (XXIa) gelang entweder durch Erhitzen von Khellin (XXI) mit Magnesiumjodid (*138*) oder durch Reduktion des aus (XXI) mit Salpetersäure gewonnenen Chinons (XXX) mittels Schwefeldioxyd (*82*). (Partielle Entmethylierung: S. 136.)

Kürzlich wurde auch gefunden (*134*), daß Khellin (XXI) bei milder Behandlung mit alkalischem Wasserstoffsuperoxyd 4,7-Dimethoxy-6-oxy-benzofuran-5-carbonsäure gibt. Entsprechendes gilt auch für Visnagin (XXXI) und Khellol (XLIV, S. 137).

Visnagin, $C_{12}H_8O_3 \cdot OCH_3$, (XXXI), Schmelzp. 109—111°. Auch die Strukturaufklärung dieses Chromons verdanken wir Späth und Gruber (*150*). Der zur Konstitutionsformel (XXXI) führende Abbauweg mußte dabei gegenüber dem beim Khellin erprobten etwas modifiziert werden. Visnagin ließ sich als 2-Methylchromon mit Alkali in Essigsäure und in das kristallisierte Visnaginon $C_9H_8O_3 \cdot OCH_3$ (XXXII) spalten, das seinerseits nach Kostanecki wieder in (XXXI, S. 135) zurückverwandelt werden konnte. Daneben entstand in kleinen Mengen das Salicylsäurederivat (XXXIII) (*11*), das in größerer Menge durch Wasserstoffperoxyd-Oxydation von (XXXI) gebildet wird (*134*). Die Anwesenheit eines unsubstituierten Furanringes in (XXXI) und (XXXII) sowie das beiden Stoffen zugrunde liegende Phloroglucin-gerüst zeigte das Auftreten von 2,3-Furandicarbonsäure bei der Oxydation von Visnagin, sowie der

Isolierung von Phloroglucin bei der milden Kalischmelze von Visnaginon. Für die beiden Verbindungen stehen drei Formeln zur Auswahl, von denen sich (XXXI) für Visnagin bzw. (XXXII) für Visnaginon, wie aus den untenstehenden Reaktionen hervorgeht, als richtig erwiesen:

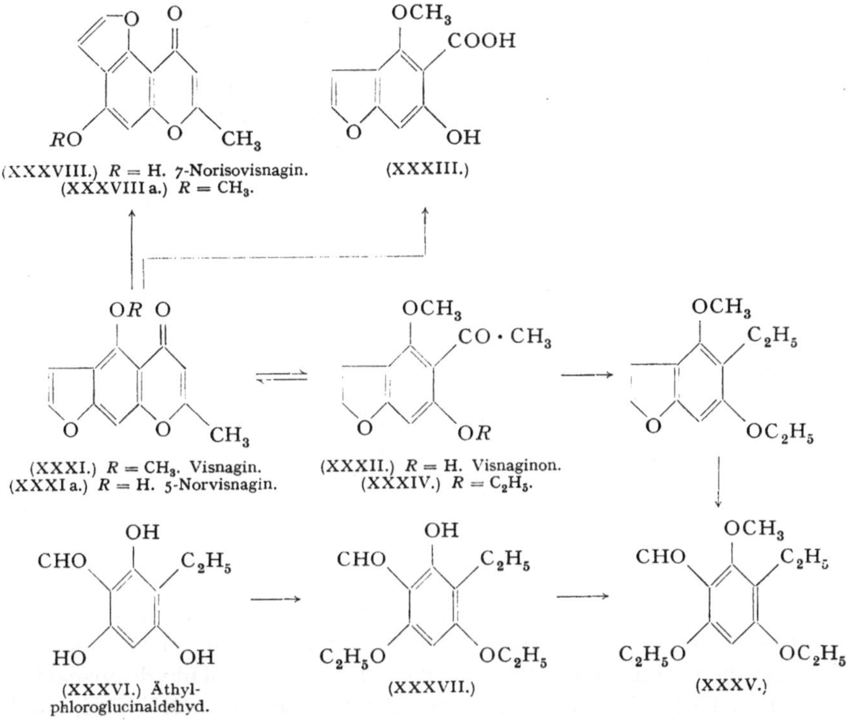

(XXXVIII.) $R = $ H. 7-Norisovisnagin.
(XXXVIII a.) $R = CH_3$.

(XXXIII.)

(XXXI.) $R = CH_3$. Visnagin.
(XXXI a.) $R = $ H. 5-Norvisnagin.

(XXXII.) $R = $ H. Visnaginon.
(XXXIV.) $R = C_2H_5$.

(XXXVI.) Äthyl-phloroglucinaldehyd.

(XXXVII.)

(XXXV.)

Das Abbauprodukt (XXXV) wurde synthetisch aus Äthyl-phloroglucin-aldehyd (XXXVI) über den Diäthyläther (XXXVII) gewonnen. Die Stellung der Äthoxylgruppen in (XXXVII) ist zwar sehr wahrscheinlich, aber nicht bewiesen. Bei anderer Lage der Äthoxylreste würde Visnagin ein angulär-anneliertes System darstellen, eine Möglichkeit, die durch die gelungene Überführung von Visnagin in Khellin [SCHÖNBERG und BADRAN (133)] dahinfällt.

In bemerkenswerter Weise ließ sich Visnagin schon durch Kochen mit verdünnter Salzsäure in das 5-Norvisnagin (XXXI a), Schmelzp. 156—158°, verwandeln (133). Methylierung führte zu (XXXI) zurück. Beim Behandeln mit Jodwasserstoffsäure hingegen bildete sich das 7-Noriso-visnagin (XXXVIII), Schmelzp. 318°, dessen Methyläther (XXXVIIIa) sich als identisch mit synthetisch gewonnenem Isovisnagin (S. 149) erwies (19). Im Gegensatz zu Eugenitin, Dihydropeucenin und Khellin lagerte sich hier nicht der γ-Pyron-, sondern der Furanring um.

Visamminol, $C_{15}H_{16}O_5$, (XXXIX), Schmelzp. 160°, $[\alpha]_D = +93°$, wurde zuerst von Smith, Pucci und Bywater (*143*) in reinem Zustand isoliert; seine Strukturaufklärung führten Bencze und Schmid aus (*10*). Der Naturstoff, der zwei $CH_3(C)$- und zwei Hydroxylgruppen besitzt, gab bei der Alkalischmelze Phloroglucin und mit Chromtrioxyd 0,6 Mol. Aceton, welches der Gruppierung $(CH_3)_2C(OH)$— entstammt. Bei der Wasserabspaltung entstand unter gleichzeitiger Allylverschiebung das optisch inaktive furanoide Abbauprodukt (XL), das mit Lauge zu (XLI) und Aceton abgebaut wurde. Ozonolyse von (XL) führte zu Essigsäure und Isobuttersäure, Ozonolyse von (XLI) zu Isobuttersäure und Phloroglucinaldehyd.

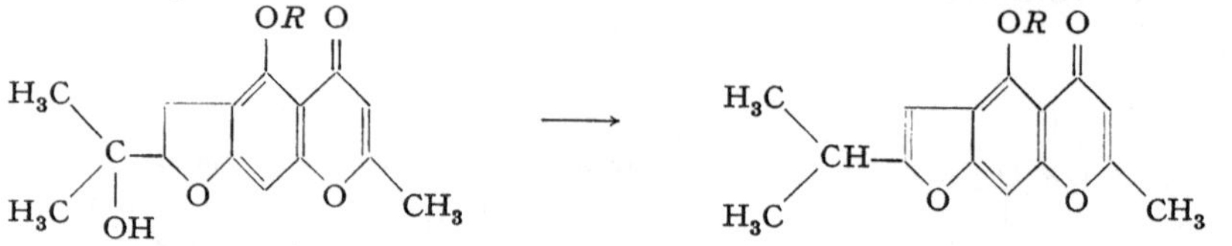

(XXXIX.) $R =$ H. Visamminol.
(XXXIXa.) $R =$ CH$_3$.

(XL.) $R =$ H.
(XLa.) $R =$ CH$_3$.

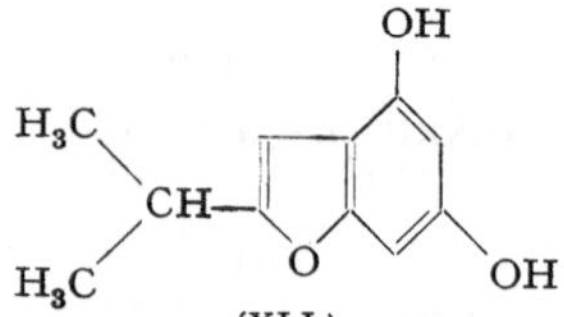

(XLI.)

Das U. V.-Spektrum von (XXXIXa) weist größte Ähnlichkeit mit demjenigen von Dihydrovisnagin (XCV, S. 150) auf, unterscheidet sich aber deutlich von den Spektren des allo- und iso-Dihydrovisnagins (XCVII, XCVI, S. 150). Auch die Absorption von (XXXIX) und Eugenitin (VII, S. 130) und ihre entsprechenden Acetate sind sehr ähnlich. Ferner sind die Spektren von (XL) und (XLa) bis auf eine kleine, durch die Isopropylgruppe bewirkte Rotverschiebung identisch mit denjenigen von 5-Norvisnagin (XXXIa) und Visnagin (XXXI). Es folgt daraus für Visamminol die lineare Struktur (XXXIX).

Das gelbe **Khellinol**, $C_{13}H_{10}O_5$, (XLII), Schmelzp. 203°, kommt in sehr kleiner Menge in *Ammi visnaga* vor und stellt das erste gefärbte natürliche Chromon dar [Bencze und Schmid (*11*)]. Es läßt sich durch partielle

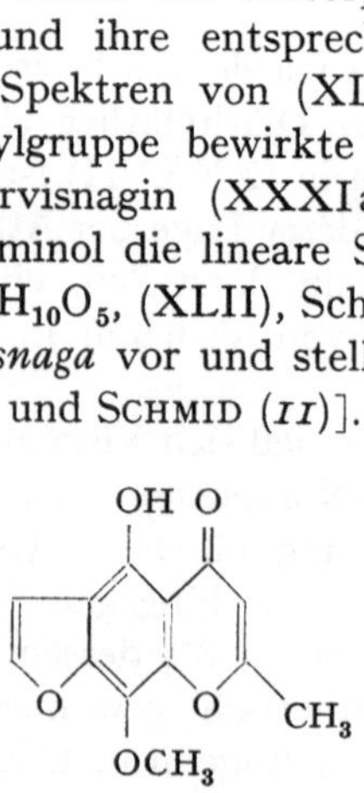

(XLII.) Khellinol.

Entmethylierung von Khellin mittels Salzsäure (78), Bromwasserstoffsäure oder Anilinhydrochlorid (132) gewinnen. Die Anwesenheit der freien 5-ständigen OH-Gruppe folgt aus seiner grünen Ferrichlorid-Reaktion, der Unlöslichkeit in verdünnter Lauge und dem U. V.-Spektrum bei Gegenwart von Aluminiumchlorid.

Khellolglucosid, $C_{19}H_{20}O_{10}$ + 2 H_2O, (XLIII), Schmelzp. 142—144°; Schmelzp., wasserfrei 174—176°, $[\alpha]_D = -33°$ (Pyridin). Die heute akzeptierte Formel für Khellolglucosid basiert auf Experimenten von FANTL und SALEM (31) sowie SPÄTH und GRUBER (151). Saure Hydrolyse des Naturstoffes gab **Khellol** (XLIV) und D-Glucose; alkalische Hydrolyse lieferte Visnaginon (XXXII) und β-Glucosido-glykolsäure, die ihrerseits mit Säuren D-Glucose abspaltete. Visnaginon (XXXII) entstand auch bei der alkalischen Spaltung von Khellol (XLIV), das sich seinerseits durch Reduktion der Oxymethylgruppe in Visnagin überführen ließ (36). Entmethylierung mittels Magnesiumjodid führt zum 5-Norkhellol (XLIVa) (132).

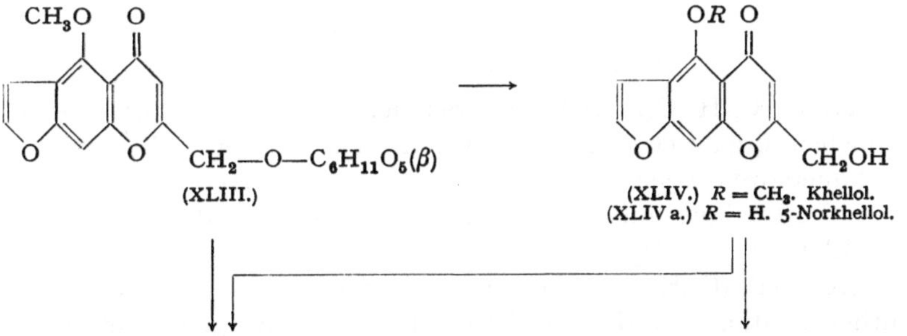

CH$_3$O O → OR O

(XLIII.) CH$_2$—O—C$_6$H$_{11}$O$_5$(β)

(XLIV.) R = CH$_3$. Khellol.
(XLIVa.) R = H. 5-Norkhellol. CH$_2$OH

(XXXII.) Visnaginon (S. 147). (XXXI.) Visnagin (S. 147).

Konstitution und U. V.-Absorptionsspektren von γ-Benzopyronen.

Das einfache 2-Methylchromon zeigt das durch den γ-Pyronring bedingte Maximum bei 296 mμ (log ε = 3,85), sowie wenig ausgeprägte selektive Absorption in der Umgebung von 250 mμ (log ε = 3,8—3,9) (38). Einführung einer Hydroxylgruppe in Stellung 7 bewirkt eine Erhöhung der langwelligen Bande ($\lambda_{max.}$ 295 mμ; log ε = 4,0), sowie das Auftreten einer Bande bei 248 mμ (log ε = 4,3). Eine ähnliche Absorption besitzt das Eugenin, das einfachste vom Phloroglucin sich ableitende Chromon (Hauptbande 290 mμ, log ε = 2,9); selektive Absorption im Gebiete von 230—260 mμ (log ε = 4,2) (14). Die Hydroxylgruppe am $C_{(7)}$ kann frei oder veräthert vorliegen. Verätherung der 5-ständigen OH-Gruppe verursacht eine hypsochrome Verschiebung von 13—19 mμ. Die Einführung eines Alkylrestes in Stellung 6, der keine mit dem Chromongerüst konjugierte Doppelbindung besitzt, bewirkt eine Verschiebung der Extremwerte um 3—5 mμ. Substitution am $C_{(8)}$ verursacht hingegen das Auftreten eines neuen Maximums bei 330 mμ (log $\varepsilon = 3,2$), das in den Spektren von Eugenin und 6-substituierten 7-Oxychromonen nur als Inflexion angedeutet ist; gleichzeitig wird das Maximum bei 290—300 mμ abgeschwächt und die Bandengruppe bei 250—260 mμ verstärkt (*14*). Ähnlich wirkt sich auch die Angliederung eines Dihydrofuranringes in 6,7 bzw. 7,8 aus (Dihydrovisnagin, Dihydrokhellin bzw. 2-Methyl-5-methoxy-4',5'-dihydro-furano-2',3' : 7,8-chromon) (*24, 10*). In den entsprechenden Furanabkömmlingen erscheint die langwellige Bande je nach der Struktur und der Anzahl von kernständigen Methoxylgruppen (o bis 2) um 14—46 mμ nach Rot verschoben (*23, 110, 10*). Die Spektren von 5-Oxychromonen erfahren durch Zusatz von Aluminiumchlorid eine charakteristische, analytisch wertvolle, bathochrome Verschiebung (*11*).

Bezüglich der Absorptionsspektren von einfachen synthetischen Chromonen sowie von Naphthopyronen vgl. (*131, 57, 27, 128*).

Die *Infrarot-Absorption* von Chromonen ist noch nicht untersucht worden. Es ist anzunehmen, daß die bei Flavonen ermittelten charakteristischen Carbonyl- und Hydroxylfrequenzen (*55*) auch für Chromone gelten.

Chromon aus Eleutherine bulbosa.

Eleutherinol, $C_{15}H_{12}O_4$, (XLV) (*27*), Schmelzp. 310° (Zers.), ist in den gebräuchlichen organischen Lösungsmitteln sehr schwer löslich. Von den vier Sauerstoffatomen liegen zwei in phenolischen Hydroxylgruppen vor, da die Verbindung mit Diazomethan in Cellosolve in den Dimethyläther (XLVI) ($C_{17}H_{16}O_4$) und mit Essigsäureanhydrid in das Diacetat (XLVII) ($C_{19}H_{16}O_6$) übergeführt wurde.

Eleutherinol-dimethyläther (XLVI) lieferte ein gelb gefärbtes Piperonyliden-Kondensationsprodukt (XLVIII), wodurch sich eine der beiden in (XLV) nachgewiesenen C-Methylgruppen als die zweiständige reaktionsfähige Chromon-methylgruppe zu erkennen gab. Beim Verkochen des Dimethyläthers (XLVI) mit 5 proz. Kalilauge bildeten sich Aceton, Essigsäure sowie die zwei Phenole (XLIX) ($C_{13}H_{13}O_3 \cdot COCH_3$) und (L) ($C_{13}H_{14}O_3$). Diese Umsetzungen beweisen die Anwesenheit eines 2-Methylchromon-Systems.

Auf Grund der U. V.-Spektren handelt es sich bei (XLIX) und (XL) um Naphthalinderivate, die noch unbekannt waren. (XLIX) wurde daher mit Kaliumpermanganat oxydiert und lieferte 3,5-Dimethoxy-phthalsäureanhydrid (LI), während (L) bei der Oxydation mit Bleitetraacetat zu zwei chinoiden Körpern führte, nämlich zu (LII), $C_{11}H_6O_2(OCH_3)_2$ und zu einer für die Analyse nicht ausreichenden Menge (LIII) vom Schmelzp. 206°; letztere zeigt Oxychinon-eigenschaften. Da mit diesen Umsetzungen das Ausgangsmaterial vollständig erschöpft und die bisher beschriebenen Verbindungen unbekannt waren, konnte die weitere Konstitutionsaufklärung nur auf synthetischem Weg erfolgen.

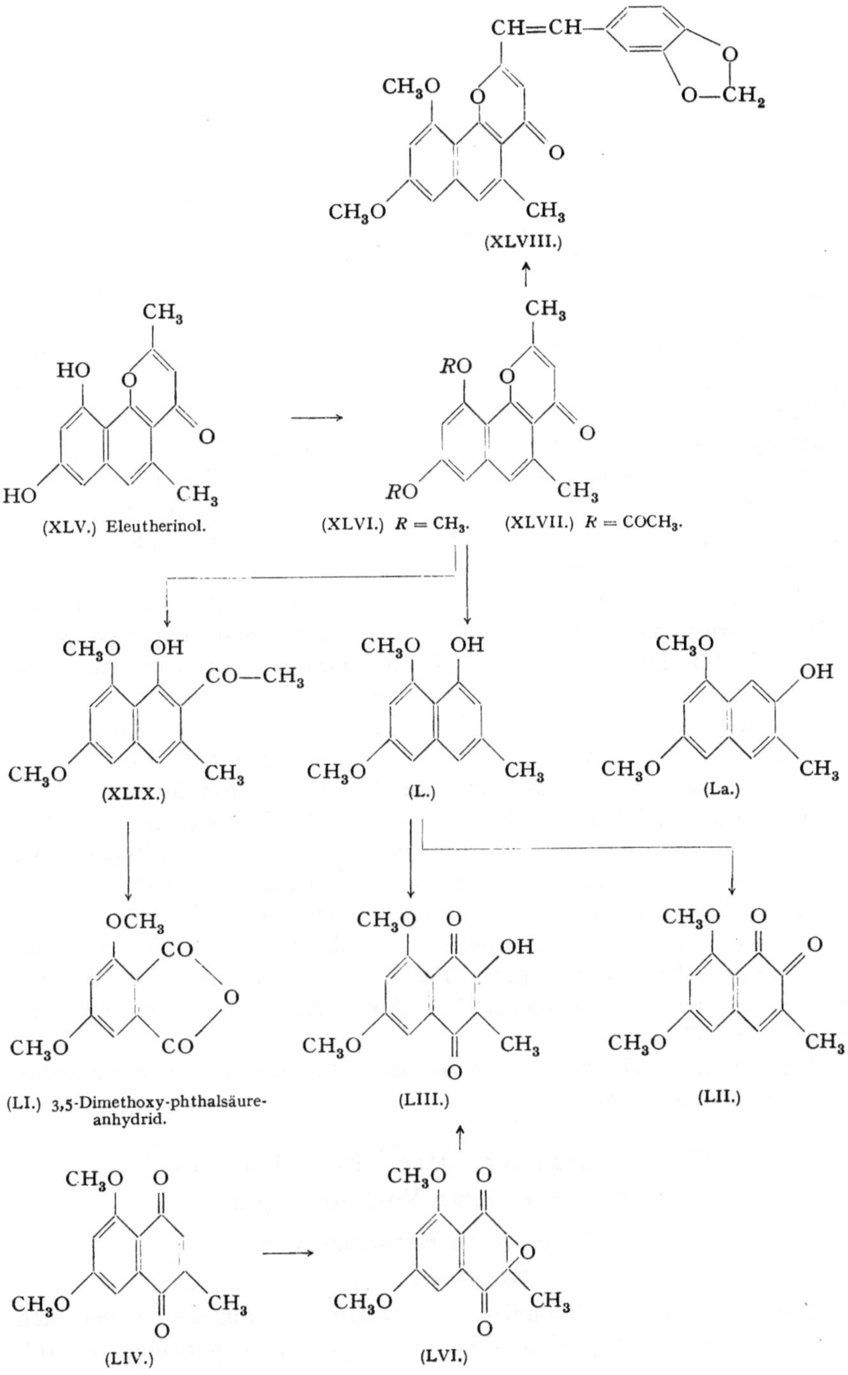

(XLVIII.)

(XLV.) Eleutherinol.

(XLVI.) $R = CH_3$. (XLVII.) $R = COCH_3$.

(XLIX.)

(L.)

(La.)

(LI.) 3,5-Dimethoxy-phthalsäure-anhydrid.

(LIII.)

(LII.)

(LIV.)

(LVI.)

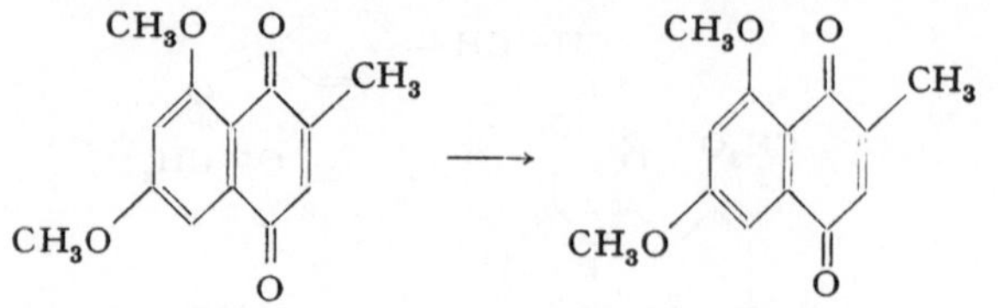

CH$_3$O   O
                CH$_3$

CH$_3$O

         O
      (LV.)

$\longrightarrow$

CH$_3$O   O
                CH$_3$

CH$_3$O

         O
      (LVII.)

Das Abbauchinon (LII) ($\lambda_{max.}$ = 408 mμ, log ε = 3,9) erwies sich verschieden von den beiden zu Vergleichszwecken hergestellten 1,4-Naphthochinonen (LIV) und (LV) ($\lambda_{max.}$ 408—410 mμ; log ε = 3,6) (*120*). Demzufolge muß (LII) ein 1,2-Naphthochinon darstellen. Modellversuche über die Bleitetraacetat-Oxydation einfacher Naphthole machten wahrscheinlich, daß es sich beim Naphthol, aus dem die beiden chinoiden Abbauprodukte (LII) und (LIII) gewonnen worden waren, um einen 2-Oxy-3-methyl-naphthalin-Abkömmling handelt. Dieses Naphthol lieferte nämlich hierbei 3-Methyl-1,2-naphthochinon und 2-Oxy-3-methyl-1,4-naphthochinon, während aus 1-Oxy-2-methyl- oder 1-Oxy-3-methyl-naphthalinen nur die entsprechenden 1,4-Naphthochinone isoliert werden konnten. Ausgehend von (LIV) hat man schließlich über das Epoxyd (LVI) das Abbauchinon (LIII) erhalten und in eindeutiger Weise identifiziert. (LIII) erwies sich als deutlich verschieden vom isomeren Oxychinon (LVII). Das o-Chinon (LII) kann dann nur mehr die Formel (LII) besitzen.

Von den beiden für das Naphthol (L) einzig in Frage kommenden Strukturformeln (L) und (L a) wurde zuerst beim Fehlen anderer Unterscheidungsmöglichkeiten auf Grund der oben angeführten Modellversuche (L a) vorgezogen. Es zeigt sich aber, daß das U. V.-Spektrum von Eleutherinol-diacetat (XLVII), in dem die auxochromen Wirkungen der beiden phenolischen Hydroxylgruppen des Eleutherinols nahezu aufgehoben sind, weitgehend mit dem Spektrum von 2-Methyl-[naphtho-1′,2′ : 6,5-pyron-(4)] übereinstimmt, sich aber sehr deutlich von der Kurve des 2,3′-Dimethyl-[naphtho-1′2′ : 5,6-pyron(4)] unterscheidet. Eleutherinol ist daher 2,3′-Dimethyl-6′,8′-dioxy-[naphtho-1′,2′-6,5-pyron(4)] [Formel (XLV), S. 140].

Eleutherinol stellt den ersten in der Natur angetroffenen Naphtho-pyron-Abkömmling dar.

IV. Synthese der natürlichen Chromone und verwandter Verbindungen.

1. Allgemeine Bemerkungen
(vgl. auch *156*).

Zur Synthese von komplizierter gebauten Chromonen eignen sich zwei verschiedene Methoden. Bei der einen (*A*), allgemeiner anwendbaren, wird zuerst der substituierte o-Oxyacetophenon-Abkömmling aufgebaut, der dann in einfacher Weise in das Chromon übergeführt wird. Die Synthese des substituierten o-Oxyacetophenons stellt häufig den schwierigsten Teil einer Chromonsynthese dar. Beim anderen Verfahren (B) werden die fehlenden Substituenten in ein bestehendes einfaches Chromon eingebaut.

Verfahren A. Die Umwandlung des o-Oxyacetophenons *a* in das 2-Methylchromon läßt sich seinerseits auf zwei Wegen erreichen: Beim ersten von KOSTANECKI (*64*) angegebenen und von WITTIG (*161, 162*) näher untersuchten Weg wird Verbindung *a* mit überschüssigem, wasserfreiem Natriumacetat und Essigsäureanhydrid längere Zeit erhitzt. Neben dem als Hauptprodukt anfallenden 3-Acetylchromon *b* bilden sich auch das 2-Methylchromon *c* und das 4-Methylcumarin *d*. Diese Synthese ist daher analytisch nicht beweisend. Verbindung *b* wird mit heißer Sodalösung (*52*) oder kalter Äthylatlösung (*49*) zu *c* entacetyliert. (Beispiele für diesen Synthesegang: *149, 45, 33.*)

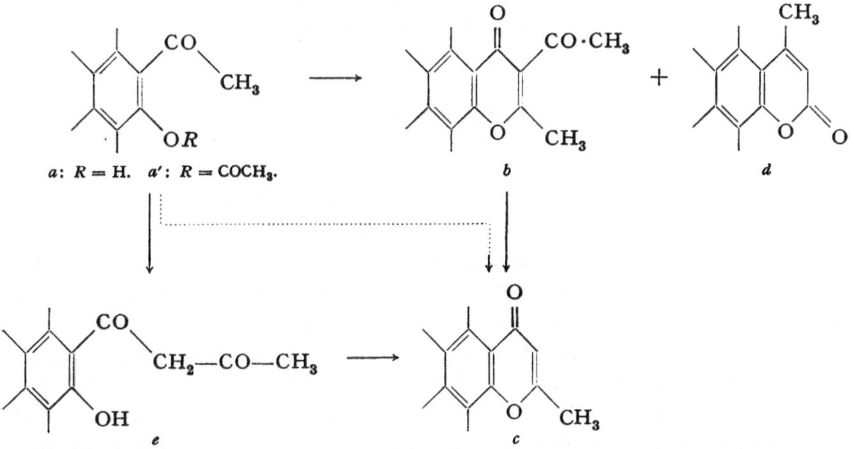

a: R = H.   a': R = COCH₃.   b   d

e   c

Eindeutig und einfacher ist der von KOSTANECKI beschriebene zweite Weg (*13, 161, 77*), bei dem *a* mit Essigester (oder einem anderen Ester) mit Natrium oder besser Natriumhydrid (*34, 130*) nach CLAISEN zum Diketon *e* kondensiert wird, das durch kurzes Erhitzen mit Säure glatt den Chromon-ringschluß (*c*) eingeht. Das Diketon *e* läßt sich aus acylierten o-Oxyacetophenonen (*a'*) auch durch intramolekulare CLAISEN-Kondensation mittels Natrium, Natriumhydrid oder Kalilauge in Pyridin bereiten (BAKER-VENKATARAMAN-Umlagerung) (*8, 69, 26, 38*). Beispiele für den zweiten Weg: Synthese von Khellin, Visnagin und seinen Verwandten.

Nach den beiden unter *A* erwähnten Verfahren lassen sich auch Polyoxy-chromone herstellen. Die CLAISEN-Kondensation mit Poly-

acetoxyacetophenon führt aber häufig zu 3-Acetylchromonen (b), die
O-Acetate lassen sich mit Salzsäure 1 : 1 ohne Ringöffnung verseifen (45);
freie Polyoxy-acetophenone sind wegen der Bildung unlöslicher Alkali-
salze nicht brauchbar. Zur Maskierung von Hydroxylgruppen in Aceto-
phenonen eignen sich die durch katalytisch erregten Wasserstoff leicht
spaltbaren Benzyläther oder in manchen Fällen die mittels verdünnter
Säure hydrolisierbaren Tetrahydro-pyranyläther (37). Auch Methyläther sind brauchbar, doch muß die Entmethylierung mit Magnesiumjodid vorgenommen werden (138, 132). Mit Jodwasserstoffsäure treten,
wie früher gezeigt wurde, häufig Umlagerungen ein.

Verfahren B. Nach diesem Verfahren sind Khellin (81, 82) und
Peucenin (14) gewonnen worden. Die einfachen Chromone sind auf
verschiedene Weise leicht zugänglich (156). Häufig werden diese Chromone
nach der Simonis-Reaktion gewonnen, die im Erhitzen eines Phenols
mit einem β-Ketoester und Phosphorpentoxyd besteht. Diese Reaktion
gibt aber an Stelle der Chromone oft Cumarine. Wie neuerdings gefunden
wurde (74), unterbleibt die Cumarinbildung, wenn das Phosphorpentoxyd
weggelassen wird und entstehender Alkohol und Wasser laufend durch
Destillation entfernt werden (75).

2. Spezieller Teil.

Chromone aus Eugenia caryophyllata.

Eugenin (IV, S. 130) entstand bei der Methylierung des bekannten
2-Methyl-5,7-dioxy-chromons mittels Diazomethan (73). Isoeugenitol
(X, S. 130) wurde aus 3-Methyl-phloroacetophenon-4,6-dimethyläther
über das Diketon 2-Oxy-3-methyl-4,6-dimethoxy-benzoylaceton und
nachfolgende Entmethylierung des Dimethoxychromons mittels Jodwasserstoff gewonnen; mit Diazomethan wurde daraus Isoeugenitin (XI,
S. 130) bereitet (118). Eugenitin (VII, S. 130) bildete sich aus C-Methyl-
phloroglucin nach der Chromonsynthese von Kostanecki nur in sehr
geringer Ausbeute (118). Besser gewann man diese Substanz aus 2-Methyl-
5,7-dioxychromon mit Methyljodid und Natriummethylat, wobei neben
Verätherung des 7-ständigen Phenolhydroxyls gleichzeitig C-Methylierung
in Stellung 6 eintrat (159). Die 5-ständige, chelierte Hydroxylgruppe
wird erst unter energischen Bedingungen methyliert. Der Naturstoff
ist kürzlich auch aus Visnagin (XXXI, S. 135) über das 2-Methyl-5,7-
dioxy-6-formylchromon erhalten worden (134).

Peucenin.

Peucenin (XIII) ließ sich durch C-Alkylierung des Mononatrium-
salzes von 2-Methyl-5,7-dioxychromon mit γ,γ-Dimethyl-allylbromid in
Benzol in etwa 4 proz. Ausbeute bereiten (14). In größerer Menge trat
bei dieser Reaktion das ringgeschlossene Iso-heteropeucenin (LVIII) auf;

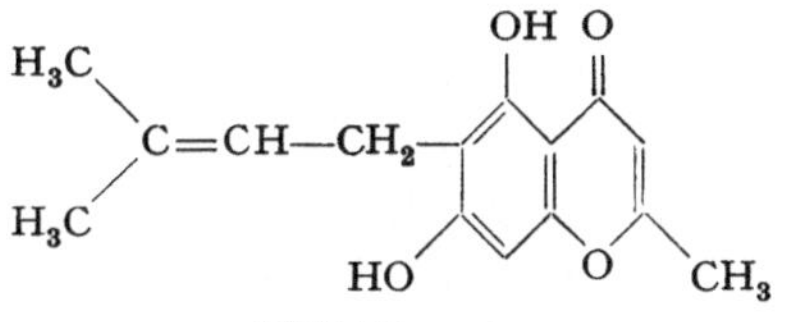

(XIII.) Peucenin.

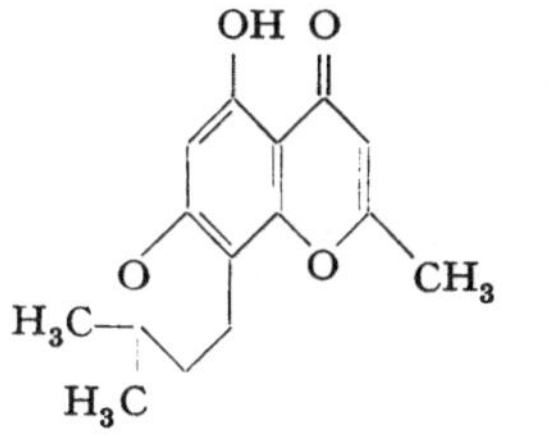

(LVIII.) Iso-heteropeucenin.

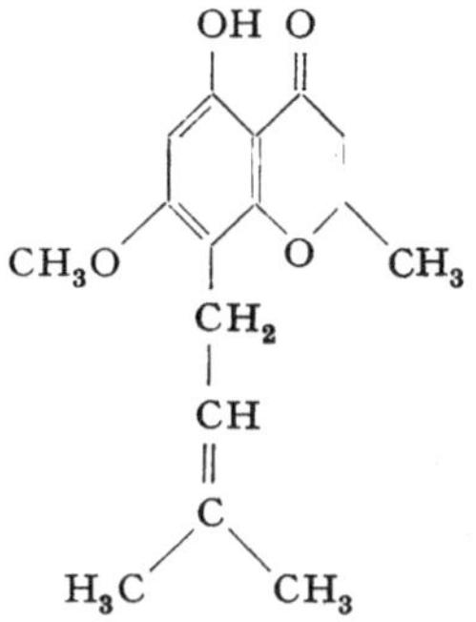

(LIX.) Heteropeucenin-methyläther.

aus den methylierten Mutterlaugen ließ sich ferner der Methyläther des Heteropeucenins (LIX) abtrennen. Substitution in Stellung 8 ist vor der in 6 begünstigt.

Khellin und Khellinon.

Das Interesse für Totalsynthese des Khellins ist offenbar durch seine pharmakologische Aktivität stimuliert worden. Von den bisher bekannt gewordenen Bildungsweisen schlagen drei, die zuerst behandelt werden sollen, ausgehend vom 2,5-Dimethoxy-resorcin, den Weg über Khellinon ein; die vierte geht vom 2-Methyl-5,7-dioxychromon aus.

CLARKE und ROBERTSON (20) führten in 2,5-Dimethoxy-resorcin eine Aldehydgruppe nach GATTERMANN ein. Im resultierenden Aldehyd (LX) wurde die reaktionsfähigere Hydroxylgruppe zuerst benzyliert (vgl. aber 33) und dann die andere mit Bromessigester zum 2-Formyl-3,6-dimethoxy-4-benzyloxy-phenoxy-essigester (LXI) umgesetzt. Cyclisierung mit Natriumäthylat (oder besser mit Magnesiummethylat) (9) führte zum Furan, das nach katalytischer Entbenzylierung den 4,7-Dimethoxy-6-oxy-cumaron-2-carbonsäureester (LXII) lieferte. Dieser Stoff gab nach C-Acetylierung (FRIEDEL-CRAFTS), Verseifung und Decarboxylierung das Khellinon (XXII), welches über das Diketon in Khellin (XXI, S. 133) umgewandelt wurde. Die Carbäthoxygruppe in (LXII) verhinderte die Absättigung der Furan-doppelbindung während der Hydrogenolyse des Benzyläthers und blockierte die reaktionsfähige Stelle im Furanring bei der FRIEDEL-CRAFTSschen Reaktion.

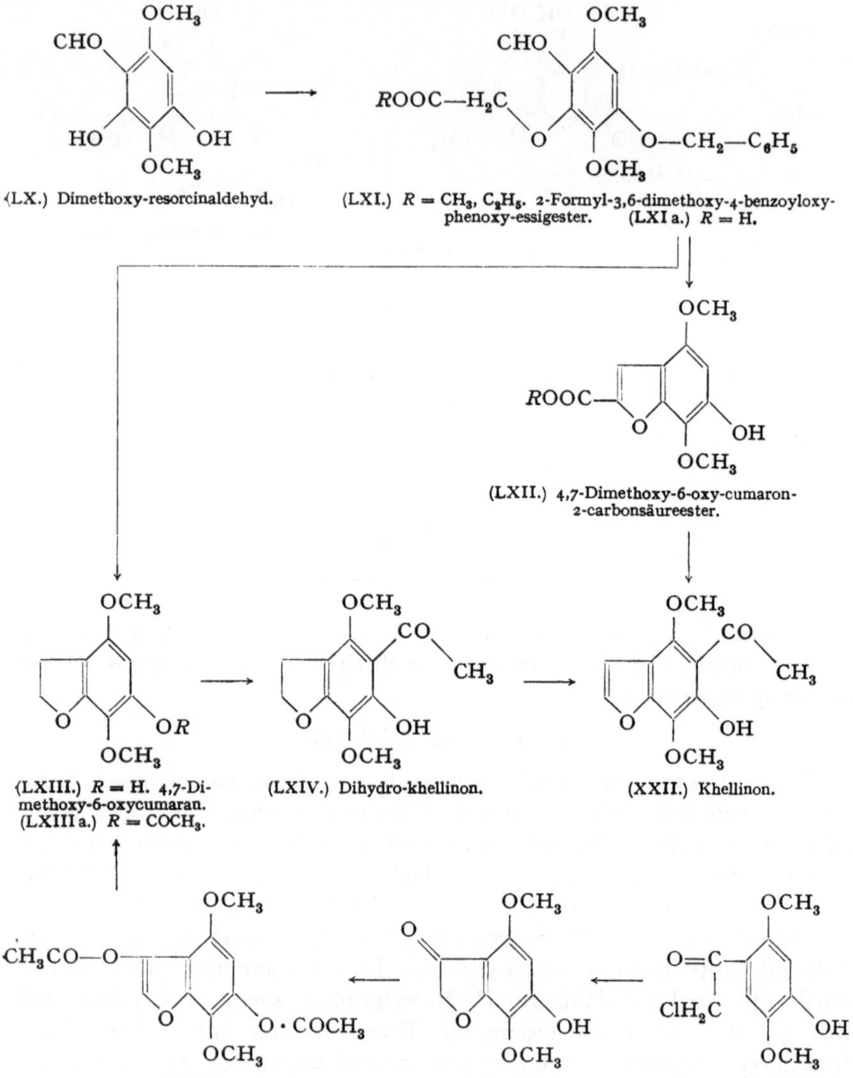

(LX.) Dimethoxy-resorcinaldehyd.

(LXI.) $R = CH_3$, C_2H_5. 2-Formyl-3,6-dimethoxy-4-benzoyloxy-phenoxy-essigester.   (LXI a.) $R = H$.

(LXII.) 4,7-Dimethoxy-6-oxy-cumaron-2-carbonsäureester.

(LXIII.) $R = H$. 4,7-Dimethoxy-6-oxycumaran. (LXIII a.) $R = COCH_3$.

(LXIV.) Dihydro-khellinon.

(XXII.) Khellinon.

(LXVII.)

(LXVI.)

(LXV.) 2,5-Dimethoxy-4,6-dioxy-ω-chloraceto-phenon.

BAXTER, RAMAGE und TIMSON (9) wählten für ihre Khellinsynthese, unabhängig von CLARKE und ROBERTSON, zunächst dieselbe Reaktions-folge. Es gelang ihnen aber nicht, in (LXII) nach GATTERMANN oder HOESCH eine C-Acetylgruppe einzuführen (vgl. die ähnlichen Verhältnisse in der Cumarinreihe). In reaktionsfähigerer Form liegt der Benzolkern im Cumaran vor. (LXI) wurde deshalb mit Natriumacetat-Essigsäure-anhydrid unter gleichzeitiger Decarboxylierung ringgeschlossen und hierauf bei erhöhtem Druck zum 4,7-Dimethoxy-6-oxycumaran (LXIII) hydriert. Letztes reagierte mit Acetonitril nach HOESCH zum Dihydro-khellinon (LXIV), welches sich durch Dehydrierung mittels Pd-Norit oder N-Bromsuccinimid (*39*) in Khellinon (XXII) überführen ließ. Aus (LXIV) wurde auch Dihydro-khellin hergestellt.

Zum 4,7-Dimethoxy-6-oxycumaran (LXIII) gelangten GARDNER, WENIS und LEE (*33*) sowie GEISSMANN und HALSALL (*39*) an Hand der Reaktionsfolge: (LXV) → (LXVI) → (LXVII) → (LXIIIa) → (LXIII). Das durch Einwirkung von Natriumacetat auf das aus 2,5-Dimethoxy-resorcin gewonnene 2,5-Dimethoxy-4,6-dioxy-ω-chloracetophenon (LXV) hergestellte Cumaranon (LXVI) wurde über das Enolacetat (LXVII) zum Cumaran (LXIIIa) reduziert. Direkte Hydrierung von (LXVI) führte nicht zum Ziel.

Im besten Falle beträgt die Ausbeute an Dihydro-khellinon (LXIV), ausgehend von 2,5-Dimethoxy-resorcin, etwa 10%. Da die Herstellung des Ausgangsmaterials, die zwar verbessert werden konnte (*39*), sowie seine Überführung in Dihydro-khellinon noch zahlreiche weitere Reaktions-stufen umfaßt, besitzen alle Khellinsynthesen vorderhand keine industrielle Bedeutung. Dasselbe trifft auch für die nachfolgende Khellinsynthese von MURTI und SESHADRI zu (*81, 82*):

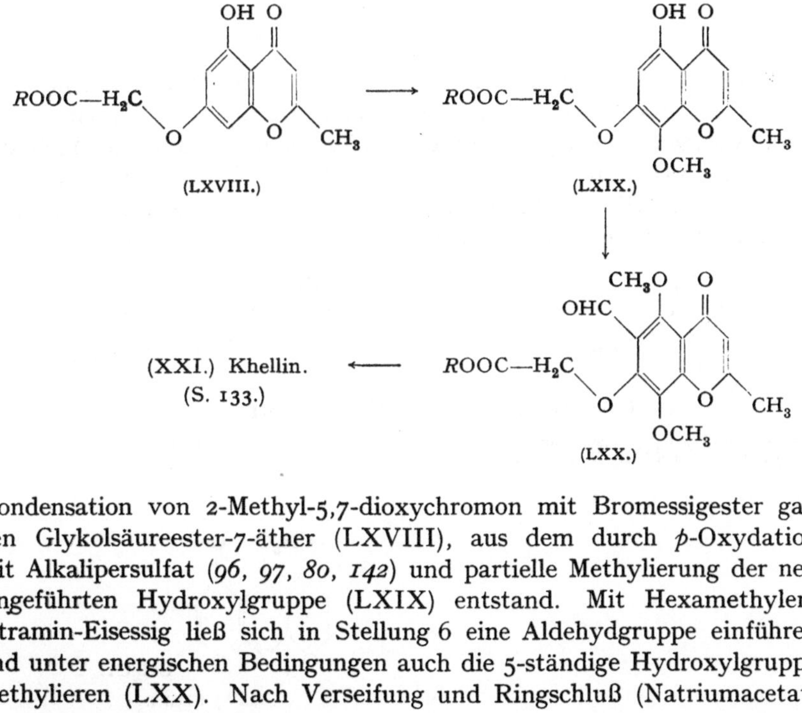

(LXVIII.) (LXIX.)

(XXI.) Khellin. ← (LXX.)
(S. 133.)

Kondensation von 2-Methyl-5,7-dioxychromon mit Bromessigester gab den Glykolsäureester-7-äther (LXVIII), aus dem durch *p*-Oxydation mit Alkalipersulfat (*96, 97, 80, 142*) und partielle Methylierung der neu eingeführten Hydroxylgruppe (LXIX) entstand. Mit Hexamethylen-tetramin-Eisessig ließ sich in Stellung 6 eine Aldehydgruppe einführen und unter energischen Bedingungen auch die 5-ständige Hydroxylgruppe methylieren (LXX). Nach Verseifung und Ringschluß (Natriumacetat-

Essigsäureanhydrid) entstand schließlich unter gleichzeitiger Decarboxylierung Khellin (XXI).

Da bei der Gewinnung von Khellin aus *Ammi visnaga* das pharmakologisch wenig aktive Visnagin (XXXI) als Nebenprodukt anfällt, fehlte es nicht an Versuchen, dieses in Khellin umzuwandeln. Schönberg und Badran (*133*) entmethylierten (XXXI) mit Salzsäure, führten in Stellung 8 eine Nitrogruppe ein und oxydierten das daraus gewonnene *p*-Aminophenol zum Chinon (XXX) (vgl. *82*). Dieses gab schließlich bei der Reduktion mit SO_2 und nachfolgender Methylierung das Khellin (3% Ausbeute, bezogen auf Visnagin). Auch durch Kernoxydation mittels Persulfat ließ sich entmethyliertes Visnagin in Khellin überführen (*78*) (Ausbeute 9%).

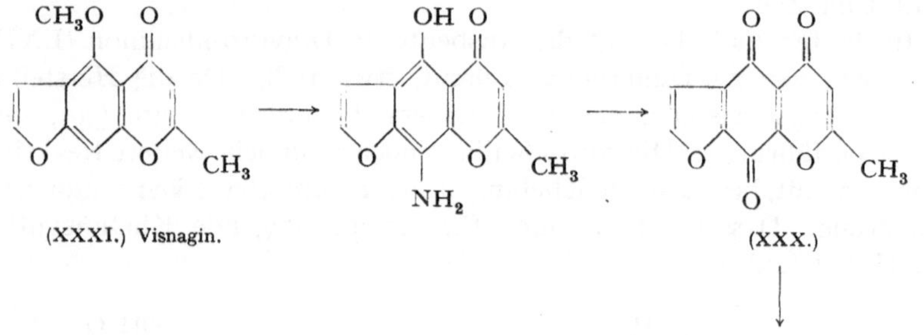

(XXXI.) Visnagin. (XXX.)

(XXI.) Khellin (S. 133).

Khellol (XLIV, s. unten), das Aglukon des natürlichen Khellolglucosids, ist von Geissman und Bolger (*38*) künstlich aufgebaut worden. Dazu wurde Visnaginon-benzyloxyacetat (LXXI) oder das analoge Acetoxyacetat (LXXII) mit Natriumhydrid in Pyridin in das entsprechende Diketon (LXXIII) umgelagert (vgl. S. 141), welches ohne Isolierung, mit Salzsäure cyclisiert und entbenzyliert bzw. entacetyliert wurde.

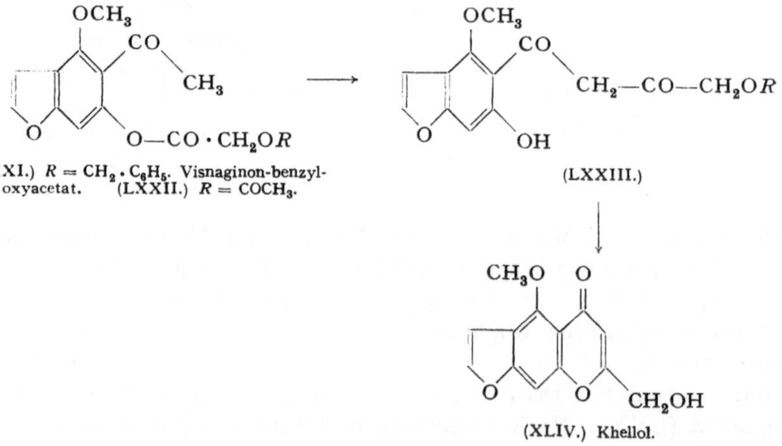

(LXXI.) $R = CH_2 \cdot C_6H_5$. Visnaginon-benzyl-oxyacetat.   (LXXII.) $R = COCH_3$.

(LXXIII.)

(XLIV.) Khellol.

Visnagin, Visnaginon und verwandte Verbindungen.

Auch für Visnaginon (XXXII) bzw. Visnagin (XXXI, S. 146) liegen mehrere, zum Teil einander recht ähnliche Synthesen vor.

GRUBER und HORVÁTH (46, 47) gingen bei ihrer Visnagin-synthese von Phloracetophenon-carbonsäureester (LXXIV) aus. Dieser gab nach dem Umsatz mit Chloracetonitril (HOESCH) und Cyclisierung mit Natriumacetat den 4,6-Dioxy-5-acetyl-cumaranon-carbonsäure-(7)-ester (LXXV). Erschöpfende Acetylierung lieferte (LXXVI), aus dem nach Absättigung der Enol-Doppelbindung 1 Mol. Essigsäure zu (LXXVII) abgespalten wurde. Nach Entfernung der Acylgruppen und schonender Methylierung mittels Diazomethan ließ sich der Monomethyl-äther (LXXVIII) gewinnen, der beim Verseifen und Ansäuern unter Kohlendioxyd-abspaltung in Visnaginon (XXXII) überging. Ausgehend von (LXXV) ließ sich Visnaginon noch auf einem anderen Wege bereiten: Eliminierung der Carbäthoxygruppe und partielle Methylierung gab 4-Methoxy-5-acetyl-6-oxy-cumaranon-(3) (LXXIX), das nach der oben geschilderten Cumaranon-Cumaron-Umwandlung zum Visnaginon führte.

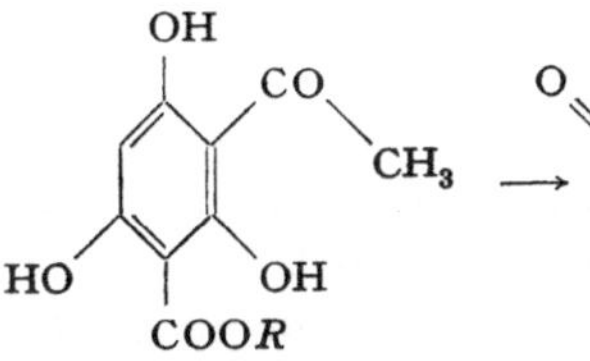

(LXXIV.) Phloracetophenon-
carbonsäureester.

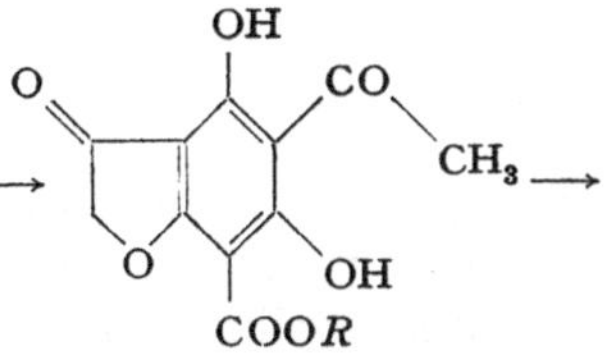

(LXXV.) 4,6-Dioxy-5-acetyl-
cumaranon-carbonsäure-(7)-ester.

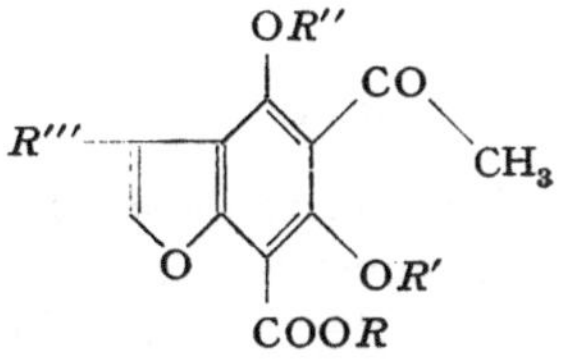

(LXXVI.) R' = R'' = COCH₃;
R''' = O·COCH₃.
(LXXVII.) R' = R'' = COCH₃;
R''' = H.
(LXXVIII.) R' = R''' = H;
R'' = CH₃.

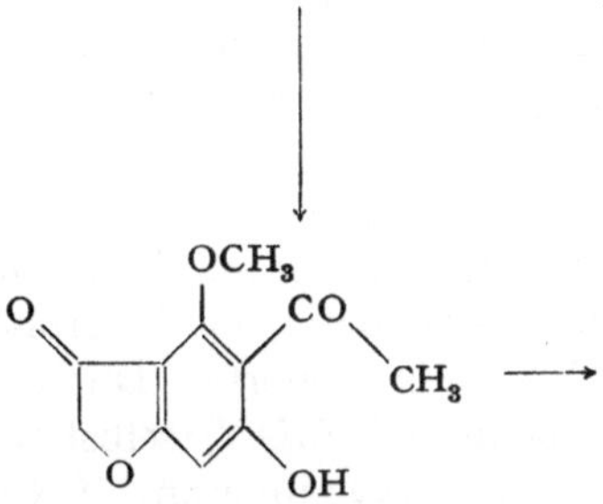

(LXXIX.) 4-Methoxy-5-acetyl-
6-oxy-cumaranon-(3).

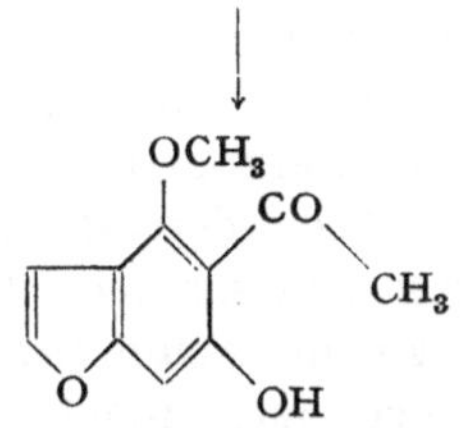

(XXXII.) Visnaginon.

DAVIES und NORRIS (24) sowie GEISSMANN und HINREINER (40) konnten Visnaginon (XXXII) vom 4,6-Dioxycumaran (LXXX) ausgehend bereiten. (LXXX) wurde aus 3,4,6-Triacetoxycumaron durch energische Hydrierung und Verseifung hergestellt. Bei der HOESCH-Reaktion gab es zur Hauptsache 4,6-Dioxy-5-acetylcumaran (LXXXI), neben wenig des Isomeren (LXXXII). Monomethylierung von (LXXXI)

mittels Methyljodid lieferte ein Gemisch der zwei möglichen Monomethyl-
äther, die sich durch Chromatographie an alkalischem Aluminiumoxyd
trennen ließen. Der fester haftende, saurere Äther stellte Dihydro-
visnaginon (LXXXIII) dar; durch Dehydrierung mittels Pd-Tierkohle
oder N-Bromsuccinimid entstand daraus Visnaginon (XXXII).

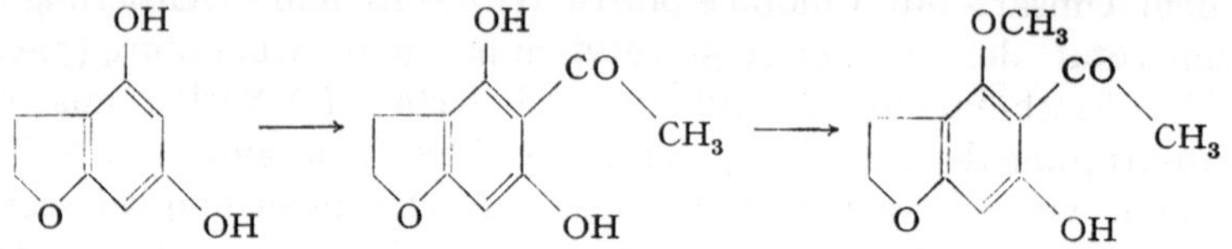

(LXXX.) 4,6-Dioxycumaran.   (LXXXI.) 4,6-Dioxy-5-acetyl-   (LXXXIII.) Dihydro-visnaginon.
                                      cumaran.

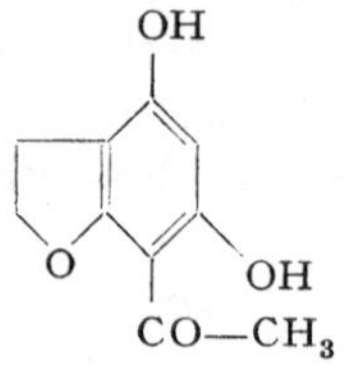

(XXXII.) Visnaginon.
(S. 147.)

(LXXXII.)

Eine prinzipielle Schwierigkeit bei Synthesen von Abkömmlingen
des symmetrisch gebauten Phloroglucins besteht bekanntlich darin, die
Seitenketten in einer bestimmten Stellung an das Phloroglucinskelett
anzuheften. Es nimmt daher nicht Wunder, daß andere, zum Teil schon
früher ausgeführte Versuche zur Visnagin-synthese nicht zum Ziel führten.
Es entstand nämlich sein Isomeres, *Isovisnagin* (LXXXIV), für das
man, wie für das *Allovisnagin* (XCIV), ein natürliches Auftreten voraus-
sehen kann.

So führte die Anwendung der Hoesch-Reaktion mit Acetonitril
oder der Friedel-Craftsschen Reaktion mit Acetylchlorid auf (LXXXV)
zum 7-Acetylderivat (LXXXVI), das nach Hydrolyse und Decarboxylierung Isovisnaginon (LXXXVII) lieferte. Durch Chromonringschluß
gelangte man zum Isovisnagin (LXXXIV) selbst (*19*).

Auch aus 4-Methoxy-6-acetoxy-cumaran (LXXXVIII), das aus
Phloroglucinmethyläther über 3,6-Diacetoxy-4-methoxycumaron bereitet
worden war, bildete sich mit Acetonitril nach Hoesch oder besser durch
Friessche Umlagerung nur Dihydro-isovisnaginon (LXXXIX), das
weiter zum Isovisnaginon (LXXXVII) dehydriert werden konnte (*45, 19*).
Schließlich entstand auch aus Phloracetophenon und Chloracetonitril
fast ausschließlich das 7-Acetylcumaranon-(3) (XC) (*51, 50*); aus dem
daraus hergestellten 4-Methoxy-6-acetoxy-7-acetylcumaron (XCI) gewann man nämlich nach Kostanecki Isovisnagin (LXXXIV) (*49, 45*).

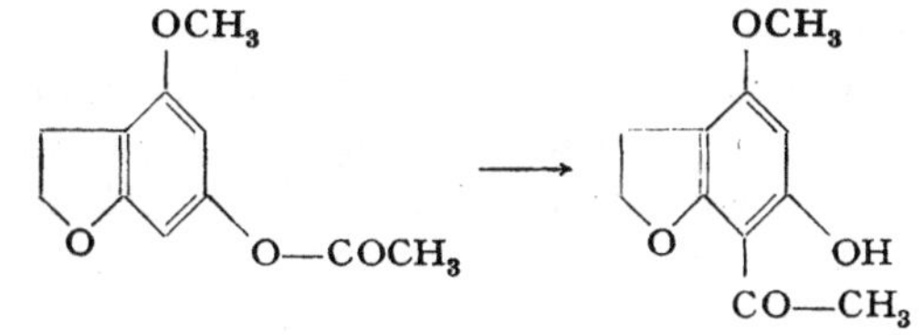

(LXXXVIII.) 4-Methoxy-6-acetoxy-cumaran.

(LXXXIX.) Dihydro-isovisnaginon.

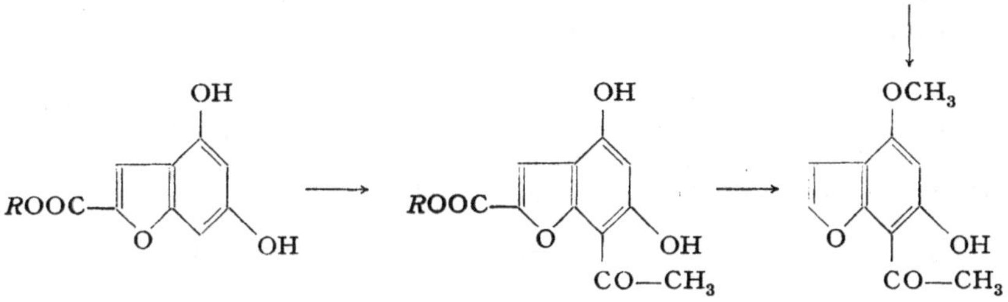

(LXXXV.)

(LXXXVI.)

(LXXXVII.) Isovisnaginon.

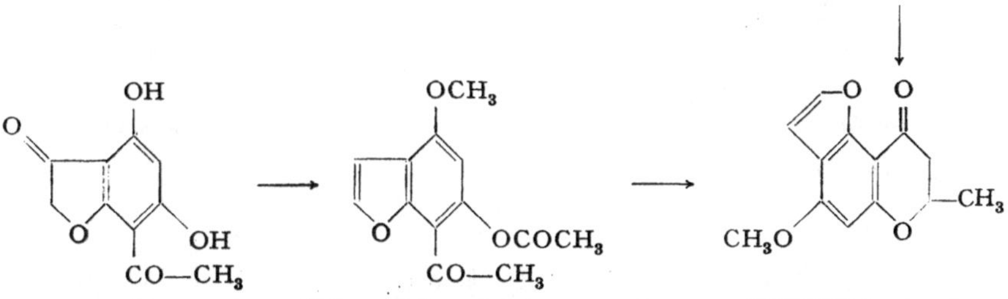

(XC.) 7-Acetyl-cumaranon.

(XCI.) 4-Methoxy-6-acetoxy-7-acetyl-cumaron.

(LXXXIV.) Isovisnagin.

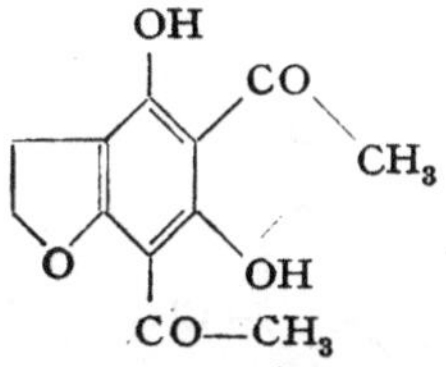

(XCII.) 4,6-Dioxy-5,7-diacetyl-cumaran.

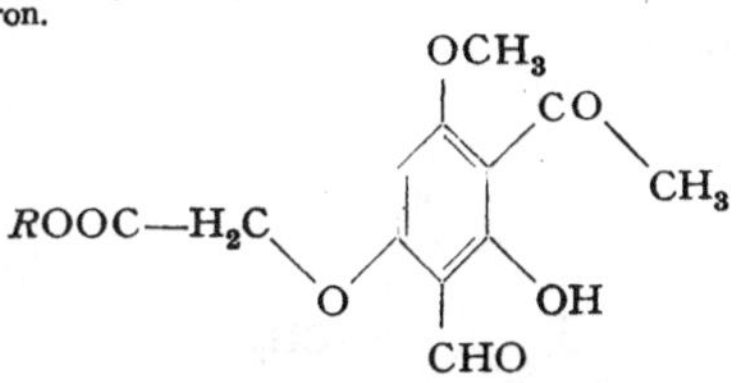

(XCIII.)

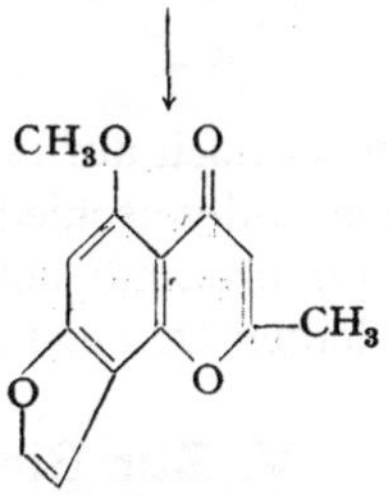

(XCIV.) Allovisnagin.

Dihydro-isovisnaginon (LXXXIX) wurde auch aus (LXXXII) (siehe oben) durch Monomethylierung bereitet (24). (LXXXII) selbst entstand

ferner durch Lauge-spaltung von 4,6-Dioxy-5,7-diacetylcumaran (XCII),
das durch doppelte Friessche Umlagerung des 4,6-Diacetoxycumarans
gewonnen worden war (*24*). Ausgehend von 2,4-Dioxy-6-methoxy-
acetophenon wurde über (XCIII) auch das zweite anguläre Visnagin-
isomere, das Allovisnagin (XCIV), aufgebaut (*94*).

Im Zusammenhang mit den Synthesen des Khellins und Visnagins sind auf
ähnlichen Wegen noch *eine Reihe weiterer Furochromone* hergestellt worden, die,
soweit sie als Naturprodukte auftreten könnten, nachstehend aufgeführt sind:
Dihydro-visnagin (XCV) (*24*); Dihydro-isovisnagin (XCVI) (*19, 24*); Dihydro-
allovisnagin (XCVII) (*24*); Dihydro-khellin (XCVIII) (*9*); 2-Methylfurano-
(3′ : 2′-6 : 7)-chromon (Khevisin) (XCIX) (*23*); 2-Methyl-4′,5′-dihydrofurano-
(3′ : 2′-6 : 7)-chromon (C) (*23, 48*); 2-Methyl-8-oxy-(8-methoxy)-4′,5′-dihydro-
furano-(3′ : 2′-6 : 7)-chromon (Ca) (*22*); 2-Methyl-2′-isopropenylfurano-(3′ : 2′-6 : 7)-
chromon (aus dem Naturprodukt Euparin) (Cb) (*61*).

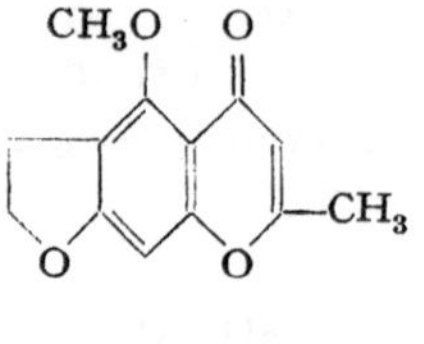

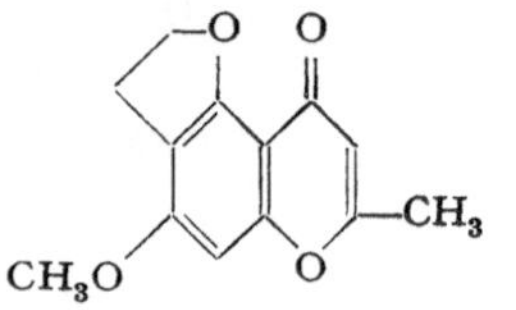

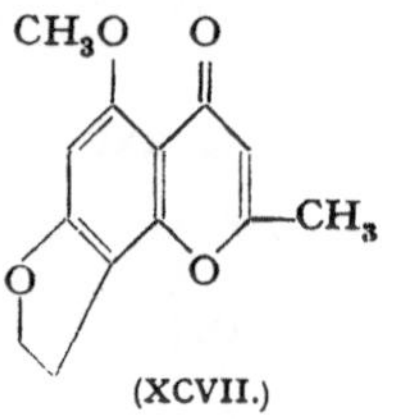

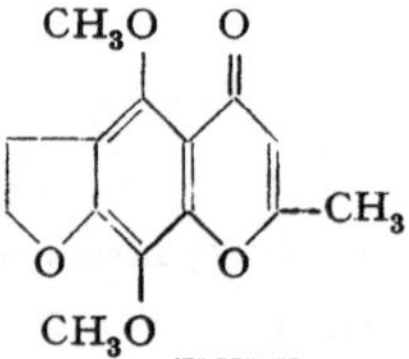

Schließlich sei noch auf Arbeiten hingewiesen, welche die Herstellung
von Homologen und verschiedenen Abkömmlingen natürlicher Chromone,
darunter der pflanzen-möglichen Chalkone und Benzalcumaranone, zum
Gegenstande haben (*1, 1a, 1b, 137, 138, 134*).

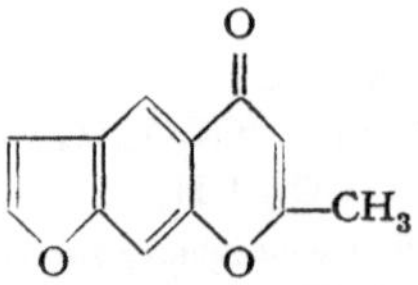

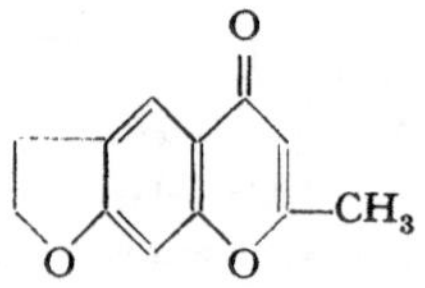

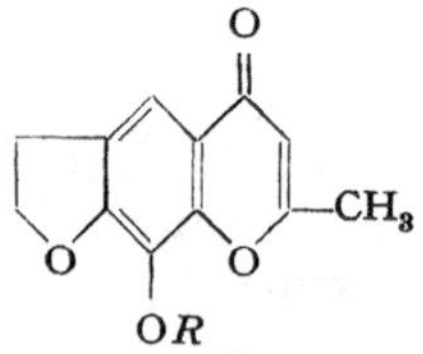

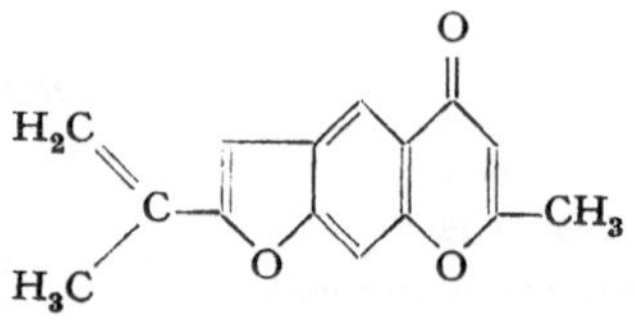

V. Zur Biogenese der Chromone.

Wenn man vom Eleutherinol, als dem bisher einzig dastehenden
Naphthopyron absieht, so scheint es, daß sich die Chromone in bio-
genetischer Hinsicht am besten zwischen Flavonen und Cumarinen ein-

reihen lassen. Die Erforschung des Pflanzenreiches auf Chromone hin ist allerdings noch sehr wenig fortgeschritten.

Die bisher isolierten Chromone stammen aus zwei Umbelliferen, die bekanntlich reich an Cumarinen sind und aus einer Myrtaceae, die mit den Umbelliferen verwandt ist (vgl. S. 126). Die in Chromonen nachgewiesenen Seitenketten, wie der Furanring und die γ,γ-Dimethylallylgruppe, sind charakteristisch für Cumarine (*146*, *25*), aber nicht für Flavone. Im (+)-Visamminol liegt ein 5'-tert.-Oxy-isopropyl-4',5'-dihydrofuranring vor, der sich auch in den Cumarinen (+)-Marmesin (*18*) (Rutaceae) und (—)-Nodakenetin (*152*, *153*) (Umbelliferae) findet. Alle bisher isolierten Chromone leiten sich vom Phloroglucin bzw. Oxyphloroglucin ab; Phloroglucin-Abkömmlinge findet man sowohl unter den Cumarinen als auch unter den Flavonen. Auf eine nahe Verwandtschaft zwischen Chromonen und Flavonen weisen der γ-Pyronring und die kernmethylierten Chromone aus *Eugenia caryophyllata* hin, wie aus einer Gegenüberstellung dieser Chromone mit gewissen flavanoiden Inhaltsstoffen aus Kiefernkernhölzern (*67*, *68*, *28*) hervorgeht.

Methylsubstituierte Cumarine sind bis heute nicht in der Natur angetroffen worden.

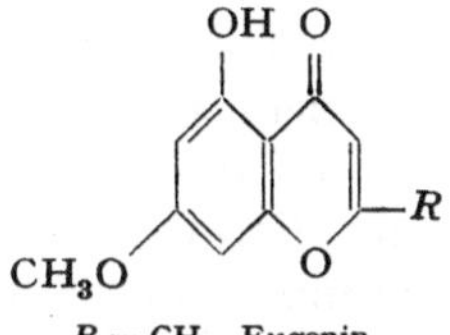

$R = CH_3$. Eugenin.
$R = C_6H_5$. Tectochrysin.

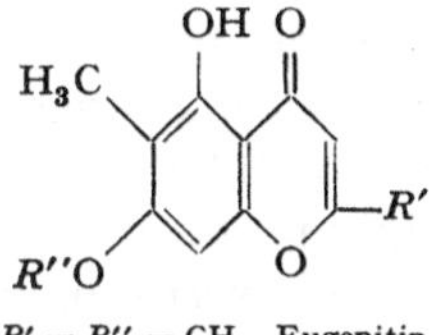

$R' = R'' = CH_3$. Eugenitin.
$R' = C_6H_5$; $R'' = H$. Strobochrysin.

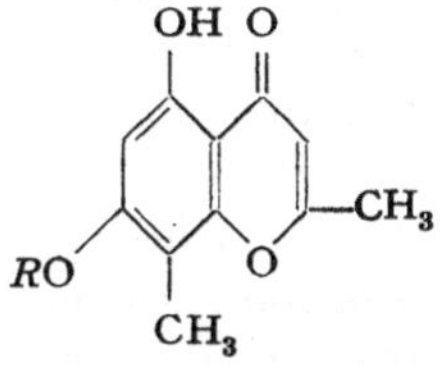

$R = H$. Isoeugenitol. $R = CH_3$. Isoeugenitin.

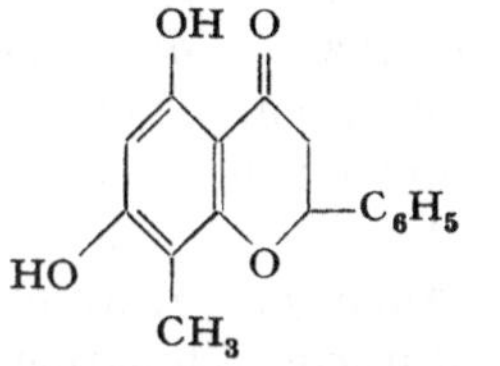

Strobopinin oder Cryptostrobin.

Ein 2,3-Dihydrochromon-gerüst kommt auch in den Rotenoiden vor; diese Stoffe werden in der Pflanze sehr wahrscheinlich über Isoflavone aufgebaut, so daß sie hier nicht weiter berücksichtigt werden müssen. Zusammenfassendes betr. Rotenoide: FEINSTEIN und JACOBSON (*31 a*).

Das gleichzeitige Vorkommen von Eugenitin bzw. Isoeugenitin und Isoeugenitol in wilden Gewürznelken läßt vermuten, daß das 3-Methyl-2,4,6-trioxy-benzoylaceton als gemeinsamer Vorläufer auftritt. Die Seitenkette kann sowohl mit der 2-ständigen Hydroxylgruppe (zu Isoeugenitin und Isoeugenitol) oder mit der 6-ständigen (zu Eugenitin) cyclisieren. Die Annahme eines solchen Vorläufers erfährt eine starke

Stütze durch die Isolierung des 2,4,6-Trimethoxy-benzoylacetons (Eugenon) (*72, 126*) und seines Cyclisierungsproduktes, des Eugenins, aus *Eugenia caryophyllata*. Eugenon ist der erste natürliche Benzoyl-aceton-Abkömmling. Das seltene Auftreten dieser Stoffe ist nicht unerwartet, da sie leicht gespalten werden oder bei Anwesenheit einer *o*-ständigen Hydroxylgruppe Cyclisierung zu einem γ-Pyron erleiden.

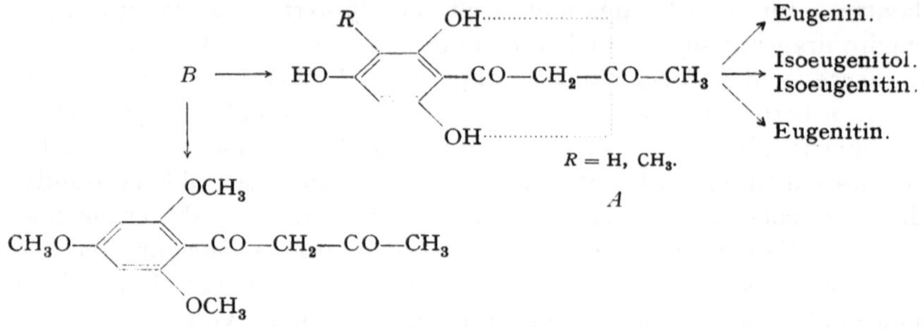

Eugenon.

Infolge der starken Cyclisierungstendenz eines *o*-Oxy-benzoylaceton-Körpers ist anzunehmen, daß *A* kein direkter Vorläufer für Eugenon darstellt (*42*). *A* und Eugenon müssen vielmehr einen gemeinsamen aromatischen Vorläufer mit keiner oder einer kürzeren Seitenkette (*B*) haben, der in un- oder partiell methylierter Form in *A*, oder nach Methylierung aller kernständigen Hydroxylgruppen mit Hilfe derselben Reaktion oder Reaktionen, in Eugenon übergeht. Über die Natur des Zwischenproduktes *B* lassen sich nur Vermutungen anstellen. Würde *B* Phloroglucin (bzw. Methylphloroglucin) oder dessen biologisches Äquivalent darstellen, so könnte man sich die Bildung von *A* und Eugenon durch die zellmögliche Kondensation mit Acetessigsäure vorstellen. Bei dieser Reaktion wären aber auch 4-Methylcumarine zu erwarten (chemisches Analogon: Simonis-Pechmann-Reaktion). Gegenwärtig sind etwa 70 natürliche Cumarine bekannt, darunter aber kein einziges 4-Methylcumarin [vgl. (*115*)], so daß diese Hypothese als hinfällig erscheint. Dasselbe Argument gilt, wenn *B* ein Phloracetophenon- (bzw. Methylphloracetophenon-) Abkömmling wäre, denn in solchen Stoffen könnte die Carbonylgruppe mit Essigsäure (oder Acetaldehyd, gefolgt von Oxydation) zu 4-Methylcumarinen, und die aktive Methylgruppe

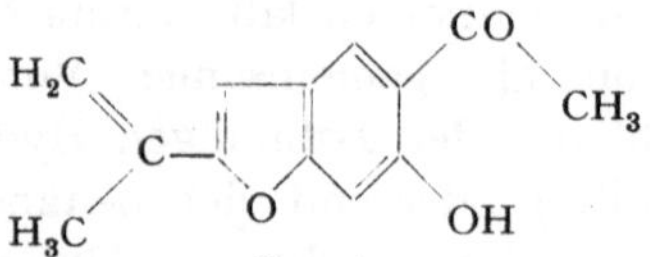

Euparin.

in gleicher Weise zu 2-Methylchromonen reagieren. Manche natürliche o-Oxyacetophenone, wie etwa das Euparin (*60, 61*) (aus *Eupatorium purpureum*), scheinen daher keine Zwischenprodukte, sondern eher Abbauprodukte von Zwischenprodukten oder von Chromonen zu sein.

REICHEL (*99, 98*) faßt hingegen auf Grund von Experimenten unter sogen. physiologischen Bedingungen o-Oxyacetophenone als direkte Zwischenprodukte der Oxychalkon- bzw. Flavanonsynthese auf.

Am plausibelsten scheint es, für B einen aromatischen oder potentiell aromatischen Körper mit einer C_3-Seitenkette anzunehmen, der mit den für die Biogenese der Flavanoide und der Cumarine postulierten substituierten oder unsubstituierten C_6—C_3-Strukturen, die praktisch ausnahmslos in *p*-Stellung zur Seitenkette eine Hydroxylgruppe tragen, sehr nahe verwandt ist. Verlängerung der C_3-Kette führt zu A und Eugenon. Die nach dem Schema $C_6 + C_3$ (Hexose + Triose) sich ergebenden Entstehungsmöglichkeiten und den strukturellen Bau solcher C_6—C_3-Vorläufer haben kürzlich GEISSMAN und HINREINER (*35, 41*) in ausführlicher und anregender Weise diskutiert, so daß sich hier eine weitere Besprechung erübrigt. Die von C_6—C_1-Strukturen sich ableitenden Naturstoffe werden danach als Abbauprodukte von C_6—C_3-Einheiten aufgefaßt. Anderseits weisen die weite pflanzliche Verbreitung von C_6—C_1-Verbindungen und die Beobachtung (*12*), daß Sedopeptulose (*D*-Altroheptulose) unter den bei der Photosynthese zuerst gebildeten Zuckern vorkommt, auf eine größere biogenetische Bedeutung der C_6—C_1-Körper hin. Die Möglichkeit ist nicht von der Hand zu weisen, daß sie auch zum Aufbau der C_6—C_3-Vorläufer dienten.

Einen direkten Hinweis auf die Bedeutung von C_6—C_3-Einheiten für die Biosynthese von Chromonen liefert vielleicht die Beobachtung von MEIJER (*72*), nach der die essentiellen Öle aus wild wachsenden Gewürznelken Chromone und Eugenon, aber kein Eugenol enthalten, während das entsprechende Öl aus Nelken kultivierter Pflanzen zum größten Teil aus Eugenol besteht. Auch die Stellung der Chromone zwischen Flavonen und Cumarinen würde so verständlich.

Kürzlich ist eine Hypothese diskutiert worden [BIRCH und DONOVAN (*12a*)], nach der die Chromone durch Verknüpfung von **Acetat**resten aufgebaut werden sollen.

Trotz ihrer spekulativen Natur mag diesen Ausführungen ein gewisser heuristischer Wert zukommen. So wird man erwarten dürfen, daß kernhydroxylierte Chromone, die in Zukunft noch aus dem Pflanzenreich isoliert werden mögen, sich ausnahmslos vom 7-Oxychromon ableiten. Als Substituenten wird man den Furanring und Prenylreste —CH_2—CH=$C(CH_3)_2$ (an C oder O gebunden) als solche oder in abgewandelten Formen (Isopropylfuranring, Dimethylchromanring und verschieden oxydierte Stufen usw.) vorfinden.

VI. Pharmakologie und therapeutische Anwendung einiger Chromone.

Vgl. hierzu auch die ausführliche Zusammenfassung von Huttrer und Dale (56); ferner (65) und (114).*

Das wichtigste Ausgangsmaterial zur Gewinnung von Chromonen, die Droge *Ammi visnaga*, besitzt eine in Ägypten seit altersher bekannte und benützte Wirkung auf die glatte Muskulatur der Uretern, Gallengänge, Gallenblase und Bronchien. Neuerdings wurde auch ihre coronarerweiternde und ihre die Herzkontraktionen verbessernde Wirkung bekannt. Ein großer Teil dieser Eigenschaften ist zweifellos auf ihren Gehalt an Khellin, Visnagin, Khellolglucosid und Visnagan zurückzuführen.

Khellin verursacht beim Hund in höheren Dosen einen kurzandauernden Blutdruckabfall, der durch Atropin nicht verhindert wird. Dieser Druckabfall ist, da eine sympatholytische Wirkung fehlt, als Folge einer direkten Einwirkung von Khellin auf die Gefäßmuskulatur aufzufassen (66). In ähnlicher Weise scheint Khellin im Herz-Lungenpräparat die Coronardurchblutung zu beeinflussen (2, 2a). 500—700 μg/ml Blut verstärken den Coronardurchfluß bedeutend, wobei die Wirkung länger als nach Amylnitritgabe anhält und etwa 4—6mal stärker ist als diejenige des Aminophyllins.

Auch am isolierten Kaninchenherz ist bei der Perfusion mit Khellinlösung eine Steigerung des mittleren Coronarflusses von 30—60% erzielt worden (32, 63). Beim Ganztier (Hund) haben intravenös verabreichte Dosen von 2,5 mg/kg eine 20—90 Minuten dauernde Verstärkung des Coronarflusses um 20—23% zur Folge. Die nur geringe gleichzeitige Blutdrucksenkung zeigt, daß die peripheren Blutgefäße anscheinend weniger empfindlich sind als die Coronararterien (7).

Demgegenüber kommen Samaan und Mitarbeiter (105, 112) zum Schluß, daß Khellin weniger auf den coronaren Durchfluß einwirkt als vielmehr die Herzkontraktionen beeinflußt. Dosen von 2—4 mg/kg (intravenös) führen hauptsächlich in der Systole zu einer Verlangsamung und zu einer Verkleinerung des Schlagvolumens. Der Druckabfall ist nach Atropingabe oder Vagusdurchtrennung geringer. Bemerkenswert ist auch die Feststellung, daß Khellin beim Hund das durch Medikamente künstlich erzeugte Kammerflimmern verhindern kann (17).

Weiter wurde gezeigt, daß bei Khellingaben das Darmvolumen zuerst leicht abnimmt, dann später infolge Erlahmen der Muskulatur zunimmt. Die erschlaffende Wirkung auf die Harnblase, die Gallengänge und Gallenblase ist etwa gleich kräftig wie die von Papaverin, aber deutlich stärker

---

* Der Verfasser dankt Herrn Dr. P. Waser (Zürich) für die Hilfe bei der Abfassung dieses Abschnittes bestens.

als diejenige von Eupaverin. Die stärkste antispastische Wirkung wird auf die Uretern ausgeübt. Die Bronchialmuskulatur erschlafft gleichfalls weitgehend, wobei Khellin auch Histamin antagonisiert (2). Meerschweinchen können so gegen lethale Dosen von Histamin-Aerosol geschützt werden.

Visnagin ist in bezug auf die Beeinflussung der Kranzarterien dem Khellin ähnlich; die spasmolytische Wirkung auf die glatten Organmuskeln ist aber nur halb so groß (2).

Khellol-glucosid. Nach SAMAAN (*106, 109, 112*) soll dieses Chromonglucosid in Dosen von 1,0—1,5 mg/kg beim Hund eine verbesserte Kontraktion des Herzmuskels und eine Erhöhung des Schlagvolumens bewirken, besonders nach Durchtrennung der Vagi. Der mit Bariumchlorid-Infusionen spastisch verminderte Coronarfluß wird durch das Chromon vergrößert. Der Blutdruck erfährt dementsprechend eine Steigerung, wahrscheinlich auch infolge einer direkten Wirkung auf das Vasomotorenzentrum. Bei isolierten Kaninchenherzen wird ebenfalls eine verbesserte Coronardurchblutung festgestellt. Andere Autoren (*2, 4, 7*) haben auf Grund von Experimenten mit isolierten Kaninchenherzen nach LANGENDORFF und isolierten Coronar-Arterienstücken die coronardilatierende Wirkung von Khellol-glucosid in Abrede gestellt. Während die stimulierende Herzwirkung feststeht, scheint die Frage einer Coronarwirksamkeit von Khellol-glucosid noch weiterer Abklärung zu bedürfen.

Das Atemzentrum wird durch Khellol-glucosid deprimiert und der Blutdruck durch die entstehende Asphyxie erhöht. Große Dosen lähmen die Atmung vor dem Herzen.

Visamminol besitzt etwa die gleich vasodilatorische Wirkung wie Khellin (*143*). *Khellinol* hingegen ist ohne Wirkung (*132*). Die Chromone aus *Eugenia caryophyllata* sind bisher noch nicht biologisch untersucht worden.

Das kristallisierte *Visnagan* aus *Ammi visnaga*, das, wie schon früher erwähnt, kein Chromon darstellt, besitzt beim Test am isolierten Kaninchenherzen eine 8mal so große coronar-dilatierende Aktivität wie Khellin (*143*). [Biologische Testierung von amorphem Visnagan (*107, 108, 15*).] Schließlich sei noch erwähnt, daß in 2-Stellung durch basische Reste substituierte Chromone im Maustest gute analgetische Wirkungen erkennen lassen (*129*).

Klinische Verwendung. Von den Chromonen ist bisher, seiner pharmakologischen Wirkungen entsprechend, nur *Khellin* therapeutisch verwendet worden. Es wirkt erschlaffend bei Spasmen der Gallenwege, der Uretern und der Gallen- und Harnblase. Bei *Angina pectoris* ist mit 0,1 g Khellin (intramuskulär oder per os appliziert) ein günstiger Effekt beobachtet worden. Diese Dosis wird täglich bis 3mal gegeben.

Die Besserung der anginösen Zustände ist in über der Hälfte der Fälle anhaltend (6), wie sich auch durch Elektrokardiogramme dokumentieren ließ. Die günstige Wirkung bei der Behandlung von Coronarsklerosen wird von anderen Autoren vermißt (141). Der akute Bronchial-Asthma-anfall kann mit einer einmaligen Dosis von 0,2—0,3 g Khellin (intra-muskulär) behandelt werden (3, 100). Auch chronische Fälle, die z. B. gegenüber Adrenalin oder Aminophyllin refraktär sind, sollen sich für eine Khellinbehandlung eignen. Die Wirkung der Droge hält bis zu 24 Stunden an. Nebenwirkungen fehlen und die Blutgerinnung bleibt unverändert. Auch bei Keuchhusten wurde Khellin klinisch erprobt (62).

VII. Zusammenhänge zwischen Konstitution und pharmakologischer Wirksamkeit von Chromonen.

In den letzten Jahren sind eine Reihe von Arbeiten erschienen (2, 7, 137, 138, 132, 131, 129, 130, 58, 59, 159), die sich mit der Verknüpfung von coronardilatorischer bzw. spasmolytischer Aktivität und dem chemischen Bau von Chromonen befassen. Trotz gewisser Unstimmig-keiten, die zum Teil auf die Verschiedenheit der angewandten Test-methoden zurückzuführen sind, lassen sich doch schon heute aus den publizierten Daten gewisse Zusammenhänge herauslesen.

So ist sicherlich die chemisch reaktionsfreudige 2-ständige Methyl-gruppe von großer Bedeutung für die spasmolytische Aktivität. Das unsubstituierte Chromon ist unwirksam, während 2-Methylchromon etwa dieselbe coronardilatorische Wirksamkeit wie Khellin besitzt. 2-Nor-khellin und 2-Norvisnagin sind schwächer wirksam als Khellin bzw. Visnagin. Verlängerung der Alkylgruppe (zu Äthyl) führt zu Wirkungs-abfall [vgl. aber (58, 59)]. In ähnlicher Weise wie die 2-ständige macht sich auch die 3-ständige Methylgruppe bemerkbar, während 2,3-dialkylierte oder 2-alkylierte-3-acylierte Chromone ohne Aktivität sind. Dies erinnert an die Bedeutung der Methylgruppe bei 2-Methylnaphthochinon und anderen physiologisch aktiven Verbindungen (95). Bei einfachen Chromonen ohne Furanring führt der Austausch des 2-ständigen H-Atoms gegen die Oxymethylgruppe, die Carboxylgruppe, gegen einen Acyl-rest oder gegen basisch substituierte Gruppen (β-Aminoäthyl, N-Methyl-piperidyl u. a. m.), zu Präparaten, die schwächer wirksam sind als Khellin. Dies trifft auch für Khellol und Khellol-glucosid zu. Unter den basisch substituierten 2-Chromonen finden sich solche, wie 2-(β-Dimethylaminoäthyl)-chromon, 2-(3'- bzw. 4'-N-Methyl-piperidyl)-chro-mon, die sich im Mäusetest durch eine bedeutend stärkere *analgetische* Wirkung als Dimethyl-aminoantipyrin auszeichnen (130).

Als aktive Präparate erweisen sich hingegen die Ester von Chromon-2-carbonsäuren, unter denen der *n*-Butyl- und der Tetrahydro-furfuryl-ester Khellin stark übertreffen (131). Auch die in Stellung 2 durch schwach basische Reste substituierten Chromone sind gut wirksam, vor allem das 2-(4'-Pyridyl)-5-methoxychromon (*129*).

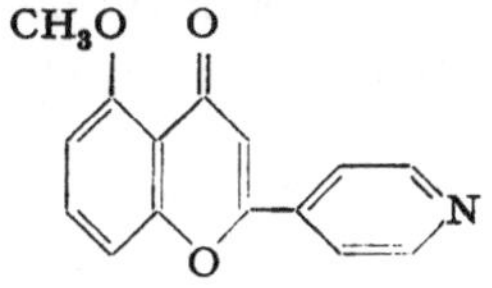

2-(4'-Pyridyl)-5-methoxychromon.

Dieser Stoff wurde nach Verfütterung an Kaninchen im Harn als 2-(4'-Pyridyl)-5-oxychromon ausgeschieden; wie schon früher (vgl. S. 135) erwähnt wurde, läßt sich die 5-ständige Methoxylgruppe auch chemisch sehr leicht spalten. Spasmolytisch stärker wirksam als Khellin ist auch das 2-(2'-Furyl)-chromon, während die Ansichten hinsichtlich der Aktivität des Flavons auseinandergehen (*137, 59*).

Der Austausch der Sauerstoffatome im γ-Pyronkern gegen Schwefelatome scheint die Wirksamkeit nur wenig zu beeinflussen. Ohne Effekt soll auch die Hydrierung der Pyron-Doppelbindung sein (*59, 131*).

Die Einführung von Methylgruppen in den Benzolkern führt zu wenig- bis unwirksamen Präparaten, während Alkoxylgruppen das Gegenteil bewirken. 2-Methyl-7-isopropoxy-chromon ist aktiver als 2-Methyl-chromon und Khellin stärker wirksam als Visnagin. Werden im Khellin die beiden in 5,8 befindlichen Methoxylgruppen gegen höhere Alkoxyreste ersetzt, so nimmt der antispastische Effekt in der folgenden Reihenfolge ab (*138*): $OCH_3 > OC_2H_5 > OC_3H_7$ (*n*) $> O \cdot CH_2{-}CH{=}CH_2$. Bei Chromonen ohne Furanring können die Verhältnisse umgekehrt liegen (*137, 59*). Auch die Stellung der Alkoxylgruppen spielt eine Rolle. Kernständige Hydroxylgruppen scheinen unwirksame Stoffe zu geben (*138, 132*; vgl. aber das aktive Visamminol).

Spasmolytisch wenig wirksam sind ferner *o*-Oxyacetophenon- und *o*-Oxybenzoylaceton-abkömmlinge (*2, 160*), auch wenn sie sich von wirksamen Chromonen ableiten.

Neben den Chromonen zeigen auch Cumarine (*59*) und Visnagan spasmenlösende Eigenschaften. Die Furocumarine aus *Ammi majus* L. sollen sich zur Behandlung der Leukodermie eignen (*29*).

Anhang.

VIII. Weitere Inhaltsstoffe aus *Eleutherine bulbosa* und einige damit verwandte Verbindungen.

Bei der näheren Untersuchung der Knollen von javanischer *Eleutherine bulbosa* stellte sich heraus, daß diese Pflanze reich an Naphthalinderivaten ist, die sonst von der Natur nur sehr selten synthetisiert werden.

Aus dem Ätherextrakt der Knollen ließen sich neben dem Naphthopyron Eleutherinol, das schon früher besprochen wurde (S. 138), drei
weitere Inhaltsstoffe, Eleutherol, Eleutherin und Isoeleutherin, gewinnen.
Auf die beiden letztgenannten, Naphthochinone darstellende Verbindungen soll später eingegangen werden.

Eleutherol, $C_{14}H_{12}O_4$, (CI), Schmelzp. 202—203°, $[\alpha]_D = + 90°$
(Chloroform); ist farblos und zeigt am Tageslicht blaue Fluoreszenz.
Seine Strukturaufklärung führten Schmid, Ebnöther und Meijer
aus (*127, 123*). Von den vier O-Atomen liegen zwei in einem Laktonring, der sich beim Ansäuern der alkalischen Lösung sofort wieder schließt,
eines als Methoxylgruppe und das vierte als kryptophenolische Hydroxylgruppe vor. Heiße verdünnte Lauge spaltete den Stoff nach der
Gleichung:

$$C_{14}H_{12}O_4 + H_2O = C_{12}H_{10}O_4 + CH_3CHO \qquad (1)$$

in optisch inaktive Eleutherolsäure (CII) und Acetaldehyd. Die Carbonsäure gab mit Ozon 3-Methoxyphthalsäure. Ihr Methyläther (CIII)
und ihr Äthyläther (CIV) nahmen bei der katalytischen Reduktion
3 Mol. Wasserstoff unter Bildung von Desmethoxy-tetrahydro-eleutherol-
säure-methyläther (CV) bzw. Desmethoxy-tetrahydro-eleutherolsäure-
äthyläther (CVI) auf. Es wird also bei der Hydrierung derjenige Ring

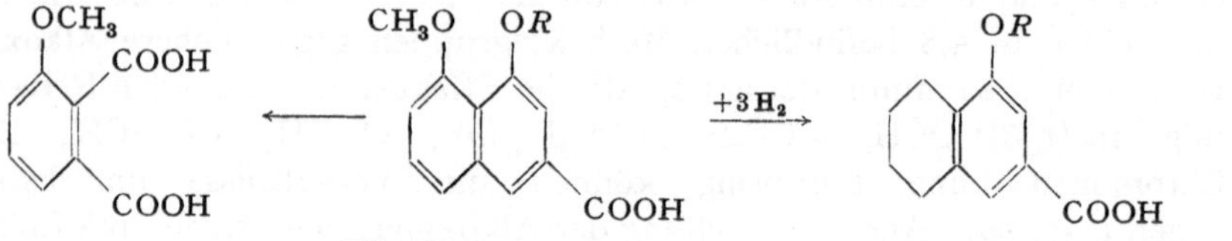

(CII.) $R = $ H. Eleutherolsäure.
(CIII.) $R = CH_3$.
(CIV.) $R = C_2H_5$.

(CV.) $R = CH_3$. Desmethoxy-
tetrahydro-eleutherolsäure-
methyläther. (CVI.) $R = C_2H_5$.

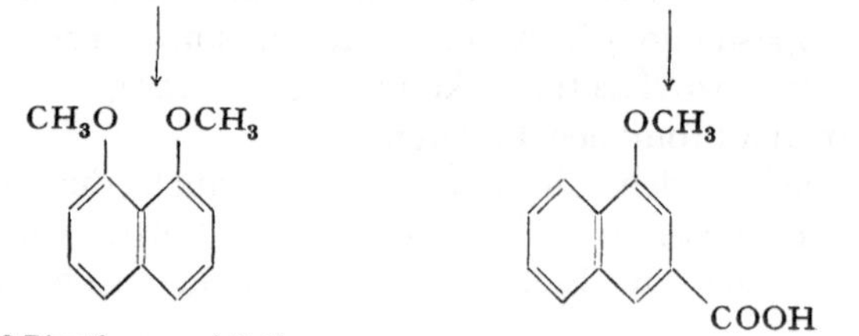

1,8-Dimethoxy-naphthalin.

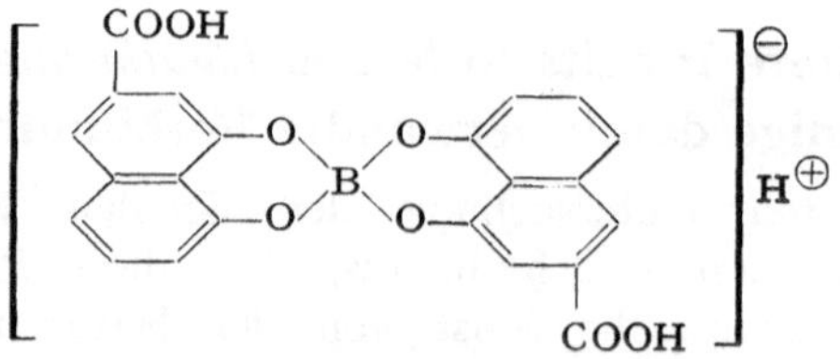

(CVII.) Borsäurekomplex der Nor-eleutherolsäure.

unter gleichzeitiger Hydrogenolyse der Methoxylgruppe reduziert, aus dem die 3-Methoxy-phthalsäure stammte. Da die Dehydrierung von (CV) zur bekannten 1-Methoxy-3-naphthoesäure führte, muß Eleutherolsäure die 1-Oxy-5- oder 8-methoxy-3-naphthoesäure darstellen. Die Lokalisierung der Methoxylgruppe folgte aus der Identifizierung des Decarboxylierungsproduktes von (CIII) als 1,8-Dimethoxynaphthalin. Die entmethylierte Eleutherolsäure gab ferner — wie 1,8-Dioxynaphthalin — einen sauren Borsäurekomplex (CVII).

Aus Gleichung (1) folgt, daß Eleutherol ein Kondensationsprodukt aus Eleutherolsäure und Acetaldehyd darstellt. Da Eleutherol selbst bei der Ozonolyse 3-Methoxyphthalsäure gab und sich durch Zinkstaubdestillation in 2-Methylnaphthalin und über sein Kupplungsprodukt mit diazotiertem Anilin in ein Derivat des 1,4-Naphthochinons überführen ließ, wie aus dem Spektrum folgte, haftet der Acetaldehyd-Rest am $C_{(2)}$. Für Eleutherol folgt dann im Hinblick auf den in ihm nachgewiesenen Laktonring und die $CH_3(C)$-Gruppe die Naphthalid-Formel CI. Die charakteristische Laugespaltung läßt sich an Hand des untenstehenden Schemas leicht verstehen, wenn man sich die Aldolnatur des Zwischenproduktes (CVIII) vor Augen hält. Eleutherolmethyläther (CI a) ist demgemäß Alkalien gegenüber stabil.

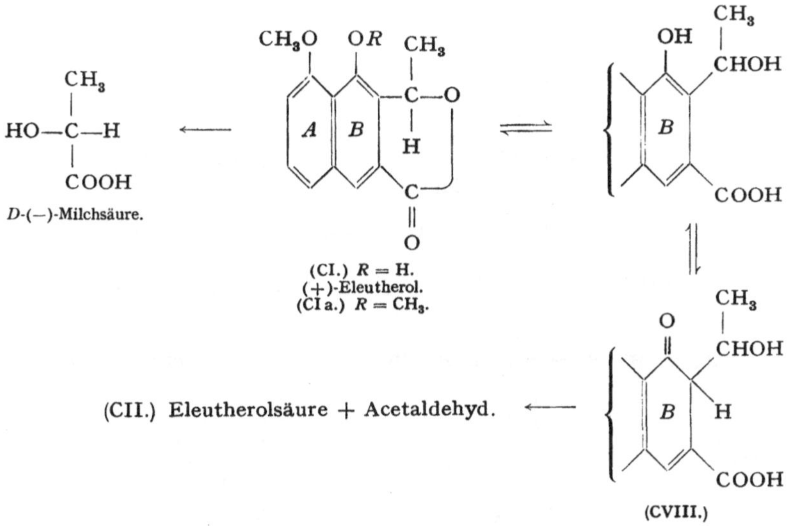

(CI.) $R = H.$
(+)-Eleutherol.
(CI a.) $R = CH_3$.

D-(—)-Milchsäure.

(CII.) Eleutherolsäure + Acetaldehyd.

(CVIII.)

Durch Ozonisation ließ sich (+)-Eleutherol ferner zur D-(—)-Milchsäure abbauen (*123, 122*), so daß dem Naturstoff die FISCHERsche Projektionsformel (CI) zukommt.

Die Synthese von (±)-Eleutherol ist kürzlich SCHMID, EBNÖTHER und HABER gelungen (*124*).

Mit Eleutherol nahe verwandt sind einige Inhaltsstoffe aus der Rinde von *Rhamnus japonica* Maxim., nämlich die beiden Glykoside α- und β-Sorinin und ihre entsprechenden Aglykone α- und β-Sorigenin. Die von Nikuni für diese Stoffe abgeleiteten Strukturformeln scheinen in einigen Punkten noch nicht völlig gesichert zu sein.

α-**Sorinin**, $C_{24}H_{28}O_{14}$, (CIX), Schmelzp. 159° (Zers.), (*84—91*), färbt sich im Licht violett. Hydrolyse mit verdünnter Säure bzw. Wasser lieferte *D*-Glucose und *D*-Xylose bzw. das Disaccharid Primverose (6-Xylosido-glucose) und das Aglykon α-*Sorigenin* (CX), $C_{13}H_{10}O_5$, Schmelzp. 227—229°. Auch α-Sorigenin (CX) verfärbt sich am Licht; es wurde als solches auch in der Rinde von *Rhamnus japonica* aufgefunden. Das Genin enthält zwei Hydroxylgruppen (die sich durch Diazomethan mit verschiedener Geschwindigkeit methylieren lassen), eine Methoxygruppe und einen Laktonring. Zinkstaubdestillation führte zu einem auf Grund des Spektrums als 2,3-Dimethylnaphthalin angesehenen

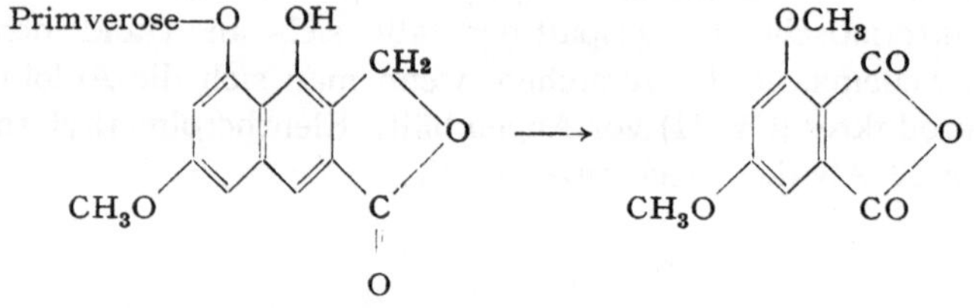

(CIX.) α-Sorinin.

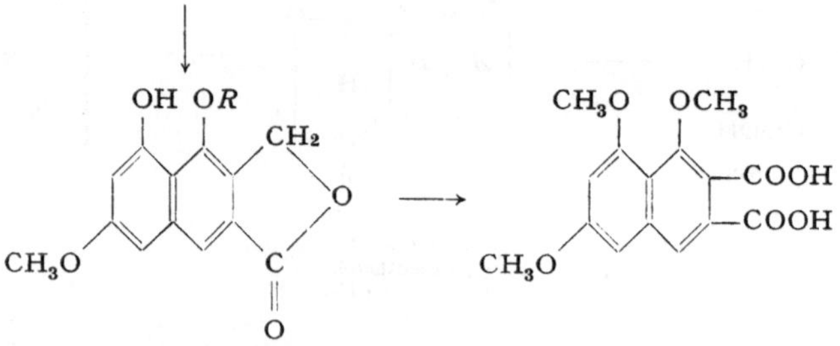

(CX.) R = H. α-Sorigenin. (CXI.) R = CH₃.

(CXII.)

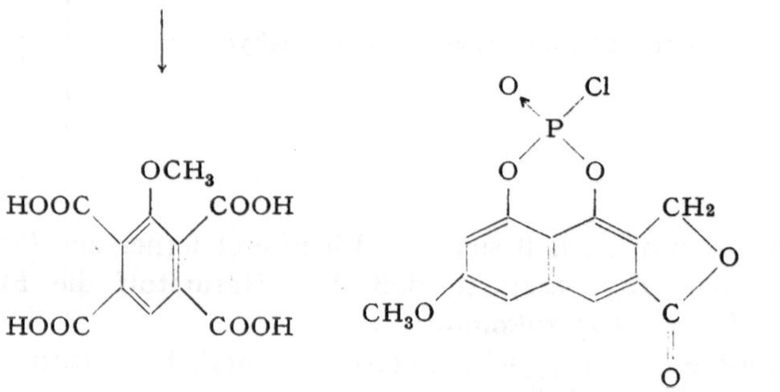

(CXIII.)

Kohlenwasserstoff, während sein Dimethyläther mit Permanganat eine Dicarbonsäure $C_{15}H_{14}O_7$ (CXII) entstehen ließ, von der sich ein Anhydrid ableitet. α-Sorigenin (CX) läßt sich daher als Lakton der 2-Oxymethyl-3-naphthoesäure auffassen. Der α-Sorigenin-monomethyläther (CXI), der auch durch Hydrolyse von α-Sorinin-methyläther bereitet wurde, gab mit Kaliumpermanganat eine methoxylhaltige Tetracarbonsäure $C_{11}H_8O_9$, die auf Grund ihres U. V.-Spektrums als Methoxypyromellithsäure angesprochen wurde. Permanganat führte α-Sorinin (CIX) in eine Oxy-methoxy-benzoldicarbonsäure über, aus der weiter mit Diazomethan 3,5-Dimethoxyphthalsäure entstand. Da α-Sorigenin (CX) ähnliche Farbreaktionen wie 1,8-Dioxynaphthalin zeigte und mit Phosphoroxychlorid einen cyclischen Phosphorsäure-chloridester (CXIII) gab, nehmen die beiden Hydroxylgruppen in (CX) die *peri*-Stellung ein. Die leichte Methylierung von α-Sorinin (CIX) machte ferner wahrscheinlich, daß seine Hydroxylgruppe nicht in *o*-Stellung zum Laktoncarboxyl steht. Für α-Sorinin bzw. α-Sorigenin lassen sich daher die Formeln (CIX) und (CX) schreiben.

β**-Sorinin** (*84, 92*) ließ sich aus den Mutterlaugen von α-Sorinin als braunes Pulver gewinnen; es gab bei der Hydrolyse neben Primverose β-*Sorigenin*, $C_{12}H_8O_5$, (CXIIIa), Schmelzp. 237—240°. Vom α-Sorigenin unterscheidet es sich durch das Fehlen der Methoxylgruppe; im übrigen gab es dieselben Reaktionen wie α-Sorigenin und ließ sich namentlich in die (CXII) entsprechende Dicarbonsäure und Methoxy-pyromellithsäure überführen. NIKUNI stellte für β-Sorigenin daher die nachfolgende Strukturformel (CXIIIa) auf. Im Glykosid haftet der Zucker am 8-ständigen Hydroxyl.

Versuche zur Synthese von α- oder β-Sorigenin haben noch nicht zum Erfolg geführt (*93*), doch gelang die Herstellung des Laktons der 1-Oxy-2-oxymethyl-3-naphthoesäure (*163*).

(CXIIIa.) β-Sorigenin.

Während Naphthochinone und hydrierte *Naphthalinabkömmlinge im Pflanzenreich* weit verbreitet auftreten, sind richtige Naphthalinderivate sehr selten isoliert worden. Naphthalin wurde in Nelkenstiel-ölen (*144*), in einer Storaxrinde (*144*), in den Blattknospen von *Betula lenta* (Birkenknospenöl) (*155*) und in *Evernia prunastri* (Eichenmoos) (*154*) nachgewiesen. In den ätherischen Ölen von *Juniperus Oxycedrus* kommt ein bisher nicht aufgeklärtes Dimethylnaphthalin vor (*76*). Aus den grünen Schalen der Walnuß *(Juglans regia)* ist vor kurzer Zeit das 1,4,5-Trioxy-naphthalin-5-β-D-glucosid (CXIV), ein Derivat des α-Hydrojuglons, isoliert worden (*21, 103*).

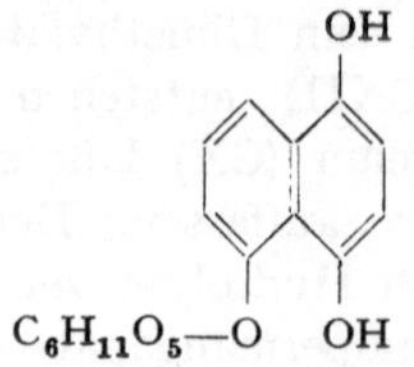

$$OH$$

$$C_6H_{11}O_5—O \quad OH$$

(CXIV.) 1,4,5-Trioxy-naphthalin-5-β-D-glucosid.

Eleutherine-chinone und Fusarubin.

Neben Eleutherinol und Eleutherol sind in den Knollen von *Eleutherine bulbosa* zwei Naphthochinone, Eleutherin und Isoeleutherin, enthalten, die sich als Derivate des bisher unbekannten 6,7-Benzo-isochromans durch interessante chemische Reaktionen auszeichnen. Mit diesen Chinonen verwandt ist Fusarubin aus dem Pilz *Fusarium solani*.

Eleutherin, $C_{15}H_{13}O_3 \cdot OCH_3$, (CXV) (*125*), kristallisiert in gelborangen Stäbchen vom Schmelzp. 175°; $[\alpha]_D = + 346°$ (Chloroform). Sein chemisches Verhalten und die große Ähnlichkeit seines U. V.-Spektrums mit demjenigen von Juglon-methyläther machte das Vorliegen eines 1,4-Naphthochinon-gerüstes wahrscheinlich. Diese Beziehung ließ sich durch Oxydation von (CXV) zur 3-Methoxyphthalsäure, welche den Ring *A* repräsentiert (und Acetaldehyd) erhärten. Die unter Standardbedingungen ausgeführte reduzierende Acetylierung und Methylierung führte überraschenderweise zu den farblosen Dihydromonoacetyl- bzw. Monomethyleleutherinen (CXVI bzw. CXVII). Erst unter energischen Verhältnissen ließ sich auch die zweite Hydroxylgruppe acetylieren bzw. methylieren (CXVIII bzw. CXIX). Bei Versuchen, durch oxydativen Abbau von Dihydroeleutherin-dimethyläther (CXIX) zu 1,4-Dimethoxypyromellithsäure (CXX) die Struktur des zweiten chinoiden Ringes (*B*) zu beweisen, erhielt man nur 3-Methoxyphthalsäure, da (CXIX) zuerst oxydativ zu Eleutherin entmethyliert wurde. Erst nach der Zerstörung des im Dihydro-eleutherin-dimethyläther (CXIX) vorliegenden potentiellen 1,4-Naphthochinonsystems ließ sich die Pyromellithsäure (CXX) fassen und mit einem aus Durochinon bereiteten Vergleichspräparat identifizieren. Zu diesem Zwecke hat man Dihydro-eleutherin-monoacetat (CXVI) katalytisch hydriert und das resultierende Tetrahydro-desmethoxydihydro-eleutherin-monoacetat (CXXI) verseifend zu (CXXII) methyliert, welches dann oxydiert wurde.

Da Eleutherin (CXV) keine C≡C-Bindung und keine nicht-chinoide Carbonylgruppe besitzt, muß der verbleibende $C_5H_{10}O$-Rest in einem an die C-Atome 2 und 3 des 8-Methoxy-1,4-naphthochinon-gerüstes angeschlossenen O-Heteroring vorliegen. Aus demselben Ring (*C*) stammt auch der Acetaldehyd. Die Struktur dieses Ringes folgte aus dem näheren Studium des bei der CLEMMENSEN-Reaktion von Eleutherin in neuartiger

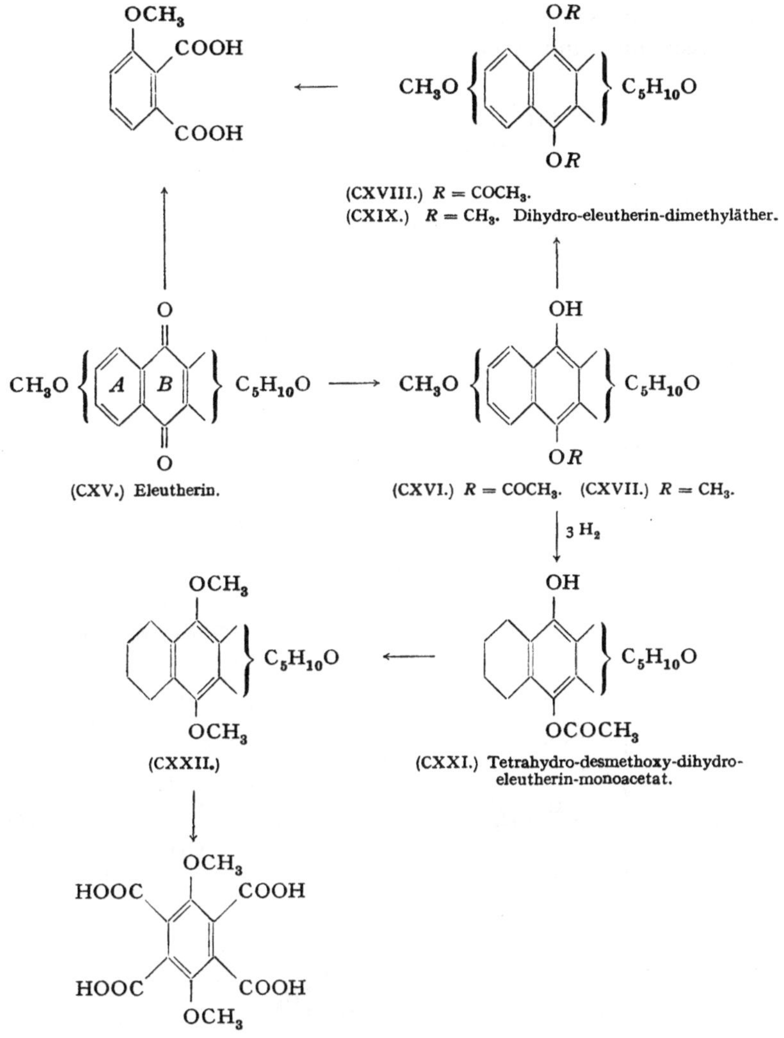

OCH₃
$$\text{OCH}_3$$

$$\begin{array}{c}\text{OCH}_3\\ \text{COOH}\\ \\ \text{COOH}\end{array} \quad \longleftarrow \quad \text{CH}_3\text{O}\left\{\begin{array}{c}\text{O}R\\ \\ \\ \text{O}R\end{array}\right\}\text{C}_5\text{H}_{10}\text{O}$$

(CXVIII.) $R = \text{COCH}_3$.
(CXIX.) $R = \text{CH}_3$. Dihydro-eleutherin-dimethyläther.

$$\text{CH}_3\text{O}\left\{\begin{array}{cc}\text{O}\\ A & B\\ \text{O}\end{array}\right\}\text{C}_5\text{H}_{10}\text{O} \quad \longrightarrow \quad \text{CH}_3\text{O}\left\{\begin{array}{c}\text{OH}\\ \\ \text{O}R\end{array}\right\}\text{C}_5\text{H}_{10}\text{O}$$

(CXV.) Eleutherin.   (CXVI.) $R = \text{COCH}_3$. (CXVII.) $R = \text{CH}_3$.

$3\,\text{H}_2$

$$\left\{\begin{array}{c}\text{OCH}_3\\ \\ \text{OCH}_3\end{array}\right\}\text{C}_5\text{H}_{10}\text{O} \quad \longleftarrow \quad \left\{\begin{array}{c}\text{OH}\\ \\ \text{OCOCH}_3\end{array}\right\}\text{C}_5\text{H}_{10}\text{O}$$

(CXXII.)   (CXXI.) Tetrahydro-desmethoxy-dihydro-
eleutherin-monoacetat.

$$\begin{array}{c}\text{OCH}_3\\ \text{HOOC}\qquad\text{COOH}\\ \\ \text{HOOC}\qquad\text{COOH}\\ \text{OCH}_3\end{array}$$

(CXX.) 1,4-Dimethoxy-pyromellithsäure.

Weise gebildeten Reduktionsproduktes (CXXIII). Dieser farblose Körper unterscheidet sich vom Eleutherin durch den Mindergehalt eines O-Atomes. Die Ringe A und B sind darin noch intakt. Mit Phthalpersäure gelangte man unter Aufnahme eines Sauerstoffatoms zu einem Oxychinon (CXXIV), das sich unter CLEMMENSEN-Bedingungen wieder in (CXXIII) zurückverwandeln ließ. Das Oxychinon gab mit Chromsäure ein mit Eleutherin isomeres Ketochinon, das ψ-Eleutherin (CXXV). Während alle Eleutherin-Abkömmlinge optisch aktiv sind, ist ψ-Eleutherin optisch inaktiv. Seine Struktur folgte aus der Ozonisierung, wobei Propionsäure (aus der Äthyl-seitenkette) und (über Acetessigsäure aus dem Aceton-rest) Essigsäure und Aceton entstanden.

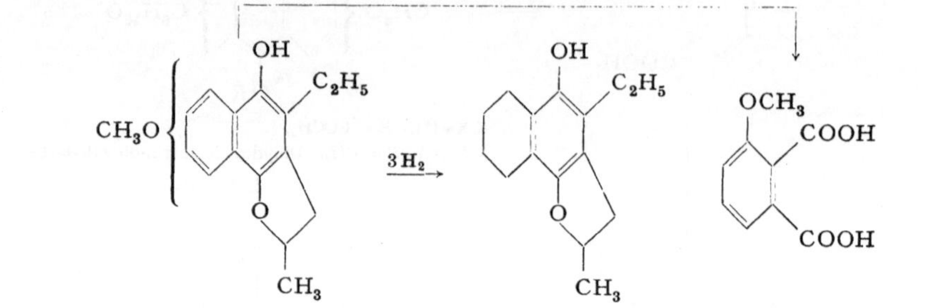

(CXXIII.) Clemmensen-Reduktionsprodukt.

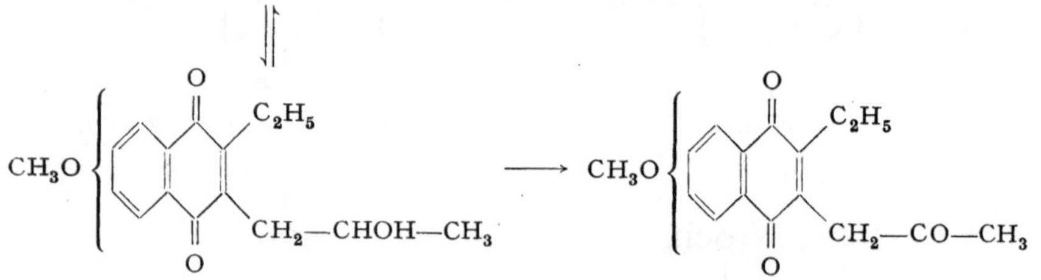

(CXXIV.)

(CXXV.) ψ-Eleutherin.

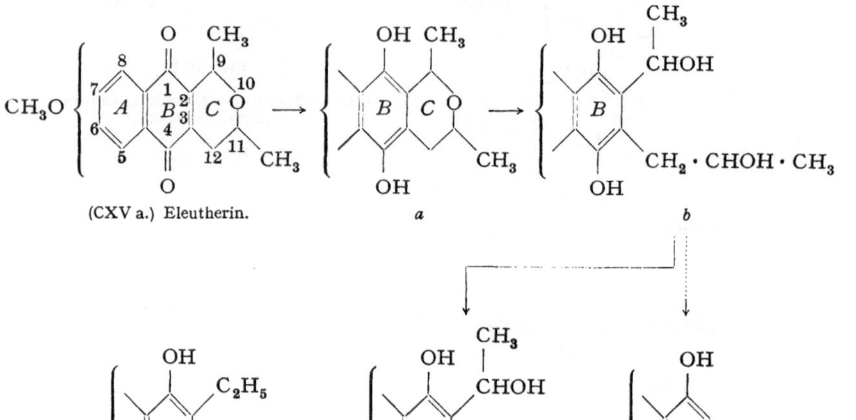

(CXV a.) Eleutherin.

a

b

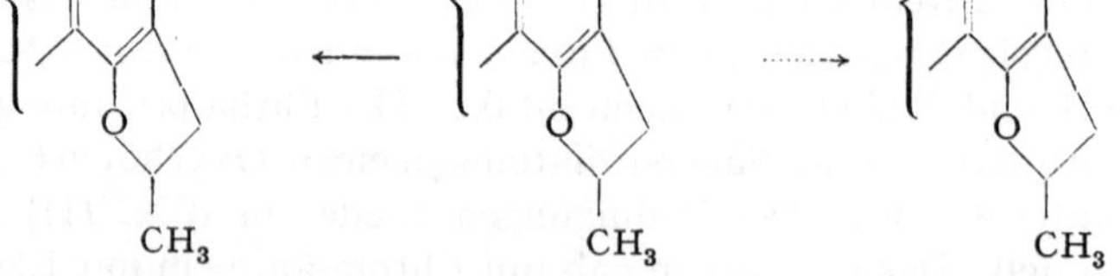

(CXXIII.)

(C.)

(CXXVI.)

Dem Eleutherin selbst ist demnach die Partialformel (CXVa), die zwei Asymmetriezentren enthält, zuzuweisen. Über die von (CXV) zum

Clemmensen-Reduktionsprodukt führende, wahrscheinliche Reaktionsfolge informieren die untenstehenden Formeln. Aldolspaltung der Zwischenprodukte *b* oder *c* bewirkt das Auftreten des in kleiner Menge isolierten äthylfreien Reduktionsproduktes (CXXVI) (vgl. die ähnlichen Verhältnisse beim Eleutherol, S. 158). Die Hydrogenolyse der $Ar \cdot CHOH \cdot CH_3$-Gruppe findet in der Reduktion von Benzylalkohol zu Toluol mit Zn(Hg)—HCl eine Parallele.

Als letztes blieb noch übrig, der Methoxylgruppe im Eleutherin (CXV) eine der beiden allein möglichen Stellungen 5 oder 8 zuzuweisen. Die Zuweisung der Stellung 8 gründet sich auf der Beobachtung, daß im Dihydro-eleutherin eine der beiden Hydroxylgruppen infolge sterischer Hinderung reaktionsträge ist, und der am Eleutherol festgestellten, stark abschirmenden Wirkung einer *peri*-ständigen Methoxylgruppe. Bei Kenntnis der Lage der reaktionsträgen OH-Gruppe ($C_{(1)}$—OH oder $C_{(4)}$—OH) im Dihydro-eleutherin-monomethyläther (CXVII) läßt sich daher auf die Lage der Methoxylgruppe im Kern *A* zurückschließen. Dihydro-eleutherin-monomethyläther gab nun bei der Clemmensen-Reaktion zur Hauptachse den Alkohol (CXXVII), eine Molekülverbindung aus 1 Mol. Dihydro-eleutherin-methyläther (CXVII) und 1 Mol. Dihydro-alloeleutherin-methyläther, und in kleiner Menge das furanoide Reaktionsprodukt (CXXIIIa), das sich von (CXXIII) nur durch entgegengesetzte Drehung unterscheidet. Ein furanoides Reduktionsprodukt mit 1-ständiger Methoxylgruppe trat nicht auf. In (CXVII) befindet sich daher die OH-Gruppe am C-Atom 1 und für Eleutherin ist mit großer Wahrscheinlichkeit die Formel (CXV) zutreffend, in der die Methoxylgruppe, wie im Eleutherol, in 8 steht.

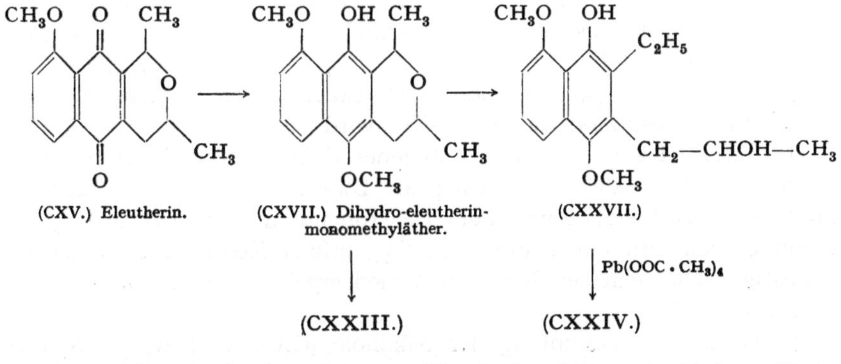

(CXV.) Eleutherin.   (CXVII.) Dihydro-eleutherin-monomethyläther.   (CXXVII.)

(CXXIII.)   (CXXIV.)

Neben Eleutherin ließ sich aus *Eleutherine bulbosa* in kleiner Menge ein weiteres Chinon, das *Iso-eleutherin*, $C_{15}H_{13}O_3 \cdot OCH_3$ (CXXVIII, S. 169), isolieren (*121*); es konnte in derselben Weise wie Eleutherin zum optisch inaktiven ψ-Eleutherin (CXXV) abgebaut werden und repräsentiert somit ein Stereoisomeres von Eleutherin. Ein weiteres

Stereoisomeres, das *Allo-eleutherin* (CXXIX, S. 169) (*121*), entstand durch partielle Racemisation von Eleutherin mittels sirupöser Phosphorsäure. Bei Zimmertemperatur stellt sich ein Gleichgewicht ein, das bei 83% Allo-eleutherin und 17% Eleutherin liegt, unabhängig davon, von welchem der beiden Chinone man ausging. Allo-eleutherin ist, wie aus den untenstehenden Daten und den I. R.-Spektren hervorgeht, der Antipode des Iso-eleutherins. Letzteres ließ sich anderseits teilweise in *Allo-iso-eleutherin* (CXXX, S. 169), den Antipoden des Eleutherins, umlagern (*121*).

Tabelle 2. Stereoisomere Eleutherine.

Eleutherin (CXV), Schmelzp. 175°, $[\alpha]_D = + 346°$ ⎫   Racemat
Allo-iso-eleutherin (CXXX), Schmelzp. 175°, $[\alpha]_D = - 342°$ ⎰   Schmelzp. 156°

Iso-eleutherin (CXXVIII), Schmelzp. 177°, $[\alpha]_D = - 46°$ ⎫   Racemat
Allo-eleutherin (CXXIX), Schmelzp. 177°, $[\alpha]_D = + 45°$ ⎰   Schmelzp. 151—152°

Dieselben Zusammenhänge zwischen Schmelzpunkten und spezifischen Drehungen findet man auch bei den Dihydroderivaten der vier Chinone.

Von Interesse sind die *konfigurativen Zusammenhänge* der vier Naphthochinone. Für das Eleutherin als dem Chinon mit der größten Rechtsdrehung sei zunächst $C_{(9)}$: (+) und $C_{(11)}$: (+)-Konfiguration definiert; Allo-iso-eleutherin besitzt dann (—) (—)-Konfiguration. Eleutherin und Allo-eleutherin gaben bei der Clemmensen-Reduktion dasselbe linksdrehende furanoide Reduktionsprodukt (CXXIII) und dasselbe rechtsdrehende Oxychinon (CXXIV). Bei Iso-eleutherin (und Allo-iso-eleutherin) liegen die Verhältnisse umgekehrt. Der Furan-ringschluß bei der Reaktion (CXV) → (CXXIII) bzw. (CXXVIII) → (CXXIIIa) verläuft optisch einheitlich. Die Oxychinone (CXXIV) bzw. (CXXIVa) ließen sich mit Phosphorsäure nicht racemisieren und gaben mit Zn(Hg)—HCl wieder dieselben furanoiden Reduktionsprodukte (CXXIII) bzw. (CXXIIIa) zurück, aus denen sie entstanden waren. Die Konfigruation der furanoiden Reduktionsprodukte bildet sich also unabhängig von der Konfiguration am $C_{(9)}$ aus und es ist daher dieses Kohlenstoffatom, welches mit Phosphorsäure racemisiert wird.

Theoretisch ist zu erwarten, daß jenes C-Atom racemisiert, welches in dem bei der Säure-umlagerung als Zwischenprodukt auftretenden Kation die positive Ladung trägt. Das Ion mit der Ladung am $C_{(9)}$ ist gegenüber dem mit der Ladung am $C_{(11)}$ infolge Resonanzstabilisierung begünstigt. Die relativen Konfigurationen ergeben sich daher wie folgt (*Tabelle 3, S. 167*).

Die sterische Verknüpfung der Alkohole (CXXIV) bzw. (CXXIVa) mit den entsprechenden Chinonen ist durch den Weg (CXV) → → (CXVII) → (CXXVII) → (CXXIV) gegeben, bei dem das Asymmetriezentrum 9 unberührt bleibt.

Aus Eleutherin entstand, wie erwähnt, bei der Reduktion das linksdrehende furanoide Reduktionsprodukt (CXXIII), aus seinem konfigurativ

Tab. 3. Konfigurations-zusammenhänge der Eleutherine.

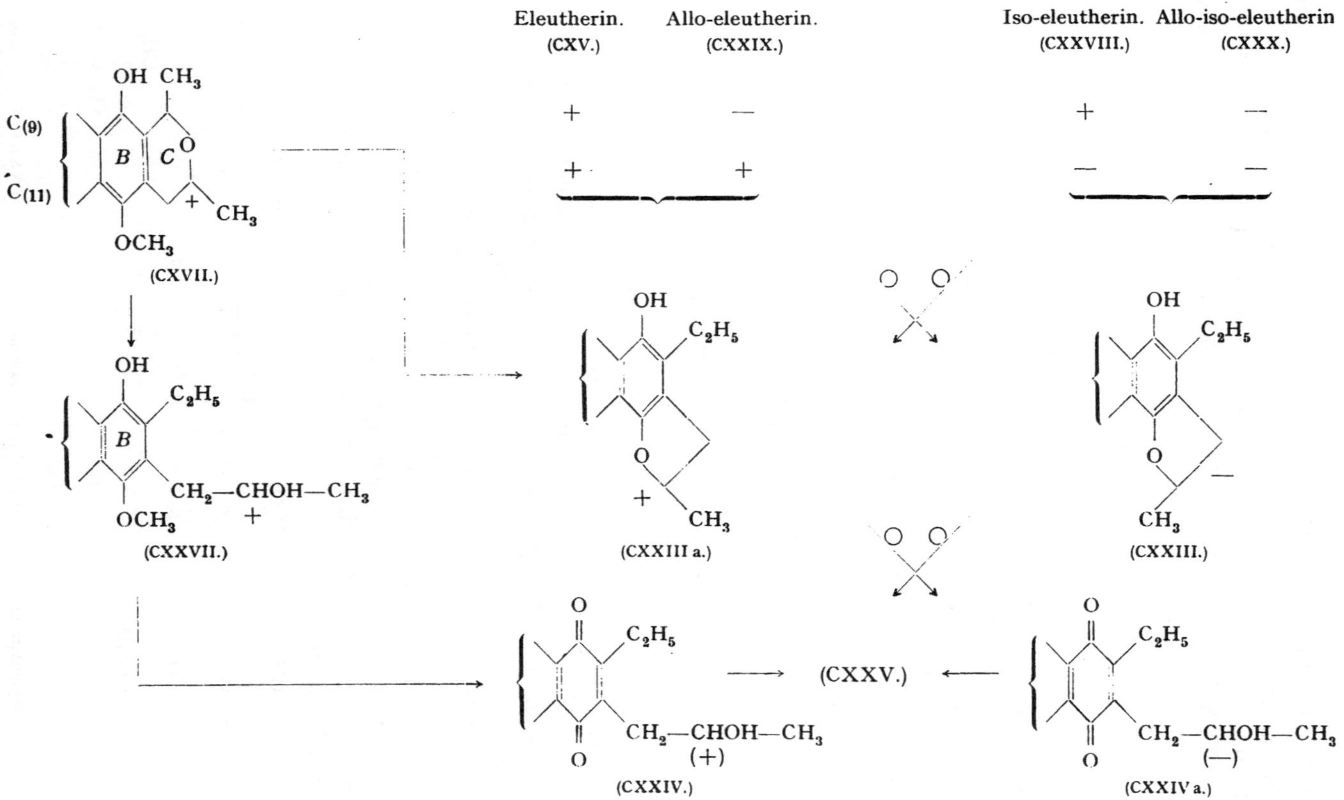

Eleutherin. (CXV.)	Allo-eleutherin. (CXXIX.)	Iso-eleutherin. (CXXVIII.)	Allo-iso-eleutherin. (CXXX.)
+	+	+	—
+	—	—	—

OH CH₃

C_2H_5

CH_3

(CXXIII.)

C_2H_5

CH_3

(CXXIIIa.)

OH

C_2H_5

CH_2—CHOH—CH₃
(—)
(CXXIVa.)

C_2H_5

CH_2—CHOH—CH₃
(+)
(CXXIV.)

(CXXV.)

OH CH₃

CH_3

B C

OCH₃

(CXXVII.)

OH

C_2H_5

B

CH_2—CHOH—CH₃

OCH₃

(CXXVII.)

$C_{(9)}$

$C_{(11)}$

identischen Dihydro-monomethyläther (CXVII) hingegen der rechtsdrehende Stoff (CXXIIIa). Das heißt nichts anderes, als daß, beruhend
auf einem verschiedenen Reaktionsmechanismus, entweder bei der
Reduktion (CXV) → (CXXIII) oder bei der Umsetzung (CXVII) →
→ (CXXIIIa) WALDENsche Umkehrung eingetreten sein muß. Zweifellos
hat die Konfigurations-Umkehr bei der ersten Reaktion, einem intramole-
kularen nucleophilen Mechanismus entsprechend, stattgefunden. Dihydro-
eleutherin-dimethyläther (CXVIII) wurde mit Zn(Hg)—HCl nur am
$C_{(9)}$ partiell racemisiert. Bei der zweiten Reaktion ist eine tautomere
Ketoform von (CXVII) vermutliches Zwischenprodukt. Bei der oxydativen
Ringöffnung der Furane zu den Oxychinonen, die ebenfalls unter WALDEN-
scher Umkehrung vor sich geht, greift das Hydroxyl-kation aus der Per-
säure vermutlich am $C_{(11)}$ an. Der Methyläther (CXVII) erwies sich
nämlich Phthalpersäure gegenüber als stabil; erst mit Bleitetraacetat
trat Oxydation zum Chinon ein.

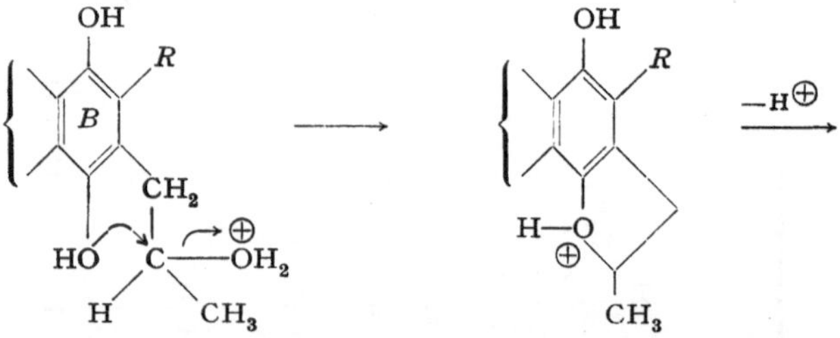

Schließlich ist es gelungen, die Konfiguration der C-Atome 9 und 11
in den Eleutherine-chinonen und ihren Abbauprodukten auf die Konfiguration der Zucker und Aminosäuren zu beziehen (*122*). Bei der
Ozonisierung des Oxychinons (CXXIV) entstand (+)-β-Oxybuttersäure.
Auf Grund der auf die gebräuchliche Projektionsformel für *D*-Glycerin-
aldehyd bezogenen Konfiguration von (+)-β-Oxybuttersäure kommt
dem (+)-Oxychinon (CXXIV) somit die untenstehende FISCHERsche
Projektionsformel zu. Die Konfiguration desselben Asymmetriezentrums
in den furanoiden Reduktionsprodukten und in den Chinonen ist damit
auch bestimmt.

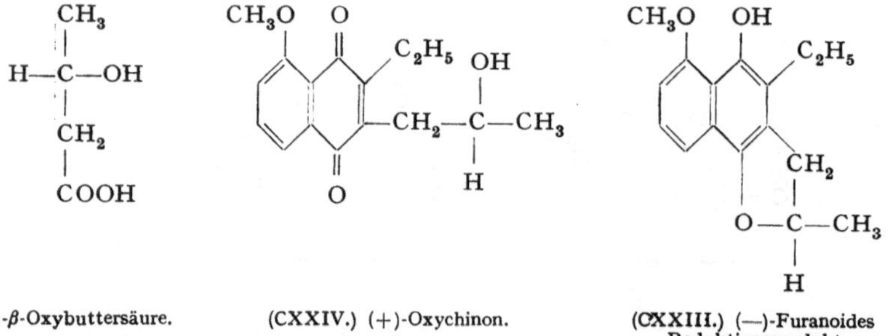

(+)-β-Oxybuttersäure.   (CXXIV.) (+)-Oxychinon.   (CXXIII.) (—)-Furanoides
                                                                                Reduktionsprodukt.

Ein optischer Vergleich der dihydrierten Eleutherine-chinone mit dem ähnlich gebauten, konfigurativ aufgeklärten (+)-Eleutherol ermöglichte die Aussage, daß im Eleutherin (CXV) und Iso-eleutherin (CXXVIII) dem C-Atom 9 dieselbe Konfiguration zukommt, wie dem Asymmetriezentrum im (+)-Eleutherol. In den Dihydro-eleutherin- und Dihydro-isoeleutherin-Derivaten besitzt das Asymmetriezentrum 9, wie im (+)-Eleutherol, einen positiven Drehungsbeitrag. Aus den unten aufgeführten Zahlen ist weiter zu entnehmen, daß die Molekularrotation von (+)-Eleutherol relativ wenig bei der Acetylierung, stark hingegen bei der Methylierung abfällt. In ganz ähnlicher Weise verhalten sich die Molekularrotationen von Dihydro-eleutherin-monomethyläther (CXVII) und Dihydro-isoeleutherin-monomethyläther (CXXXI) bei den entsprechenden Umwandlungen. Diese parallel verlaufenden Drehungsänderungen sprechen im Sinne der FREUDENBERGschen Verschiebungsregel für identische Konfigurationen, da sich hierbei der Drehungsbeitrag des Zentrums 11, wie sich auf Grund der optischen Superpositionsregel berechnen läßt, kaum ändert.

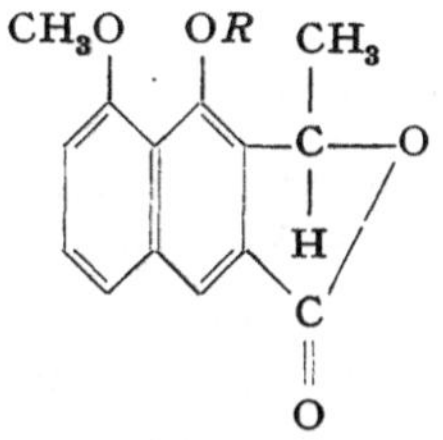

Eleutherol.

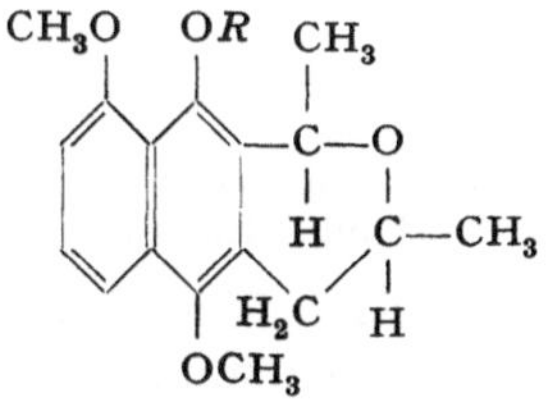

Dihydro-eleutherin-mono-
methyläther.

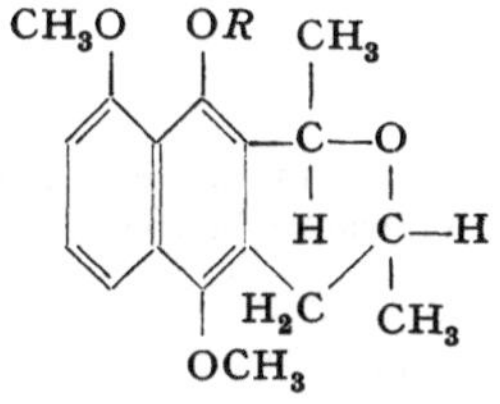

Dihydro-iso-eleutherin-mono-
methyläther.

$[M]_D$		$[M]_D$		$[M]_D$	
$R = H$:	$+ 220°$	$R = H$:	$+ 593°$	$R = H$:	$+ 174°$
$R = COCH_3$:	$+ 195°$	$R = COCH_3$:	$+ 419°$	$R = COCH_3$:	$+ 59°$
$R = CH_3$:	$+ 95°$	$R = CH_3$:	$+ 272°$	$R = CH_3$:	$- 94°$

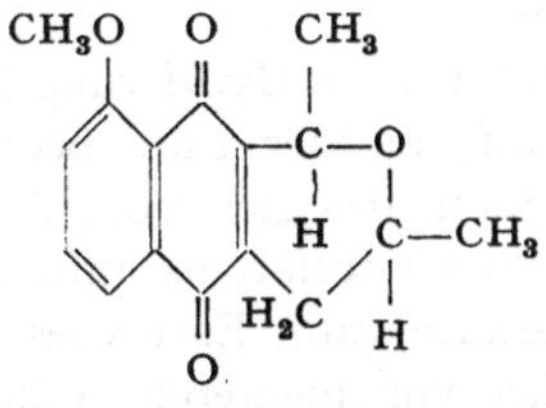

(CXV.) Eleutherin.
[Spiegelbild: Allo-iso-eleutherin (CXXX.)].

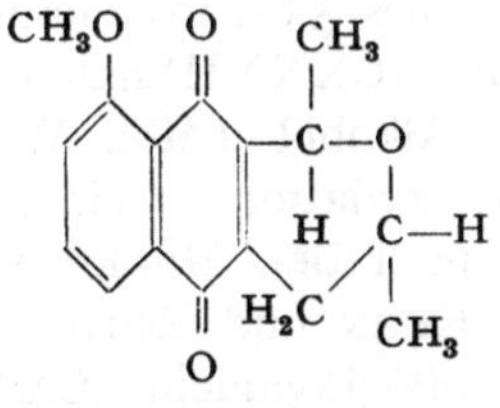

(CXXVIII.) Iso-eleutherin.
[Spiegelbild: Allo-eleutherin (CXXIX.)].

Aus den Projektionsformeln ist zu entnehmen, daß im Eleutherin und Allo-iso-eleutherin die beiden Methylgruppen im Isopyranring *cis*, im Allo- und Iso-eleutherin *trans* angeordnet sind.

Hinsichtlich der *Biogenese* der Inhaltsstoffe aus *Eleutherine bulbosa* könnte man annehmen, daß zunächst ein 1,8-Dioxynaphthalin-Vorläufer sich mit Isopren oder dessen biologischem Äquivalent kondensiert; Oxydation dieses Zwischenproduktes könnte dann die drei Naturstoffe Eleutherol, Eleutherin und Iso-eleutherin geben, die hinsichtlich des $C_{(9)}$ dieselbe Konfiguration besitzen. Der Zusammenhang zwischen Eleutherin und Eleutherinol [(XLV), S. 139] ist evident.

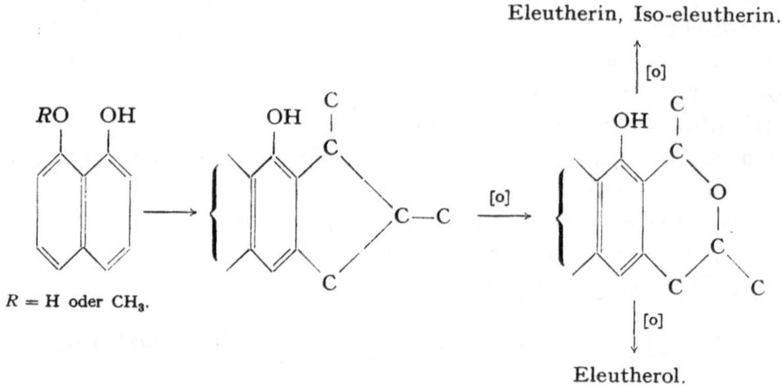

Eleutherin, Iso-eleutherin.

Eleutherol.

$R = H$ oder CH_3.

Eleutherin zeigte gegenüber *Pyococcus aureus* und *Streptococcus haemolyticus* eine schwache bakteriostatische Wirkung. Eleutherol war ohne Aktivität (*122*).

Fusarubin, $C_{15}H_{14}O_7$, (CXXXII). Einer ganz anderen Quelle, näm-lich dem Pilz *Fusarium solani*, entstammt das von Ruelius und Gauhe (*101*) untersuchte Chinon Fusarubin, das sich auch, wie schon erwähnt, vom 6,7-Benzoisochroman ableitet. Fusarubin kristallisiert in roten Prismen vom Schmelzp. 218° (Zers.). Neben einer Methoxyl- und einer C-Methylgruppe finden sich zwei phenolische OH-Gruppen vor. Ein Sauerstoffatom gehört einer sehr reaktionsfreudigen alkoholischen Hydroxylgruppe an, die sich schon mit kalter verdünnter methanolischer bzw. alkoholischer Salzsäure veräthern ließ. Fusarubin besitzt das Spektrum eines 5,8-Dioxy-naphthochinons.

Fusarubin (CXXXII) nahm bei der katalytischen Hydrierung in Essig-ester oder Alkohol 2 Mol. Wasserstoff auf; an der Luft beobachtete man Rück-oxydation zu einem roten Chinon, das sich vom Ausgangs-material durch den Mindergehalt eines Sauerstoffatoms unterscheidet und nach Kuhn und Roth 2 Mol. Essigsäure gab. Es erwies sich als identisch mit Javanicin (CXXXIII), das vor mehreren Jahren von Arnstein und Cook (*5*) neben Oxy-javanicin aus *Fusarium javanicum* gewonnen worden war. Oxy-javanicin ist sehr wahrscheinlich mit Fusarubin identisch. Bei der analogen Hydrierung von Fusarubin in Eisessig bildete sich nach der Rück-oxydation ein vom Javanicin ver-schiedenes Desoxy-fusarubin (CXXXIV). in dem die alkoholische

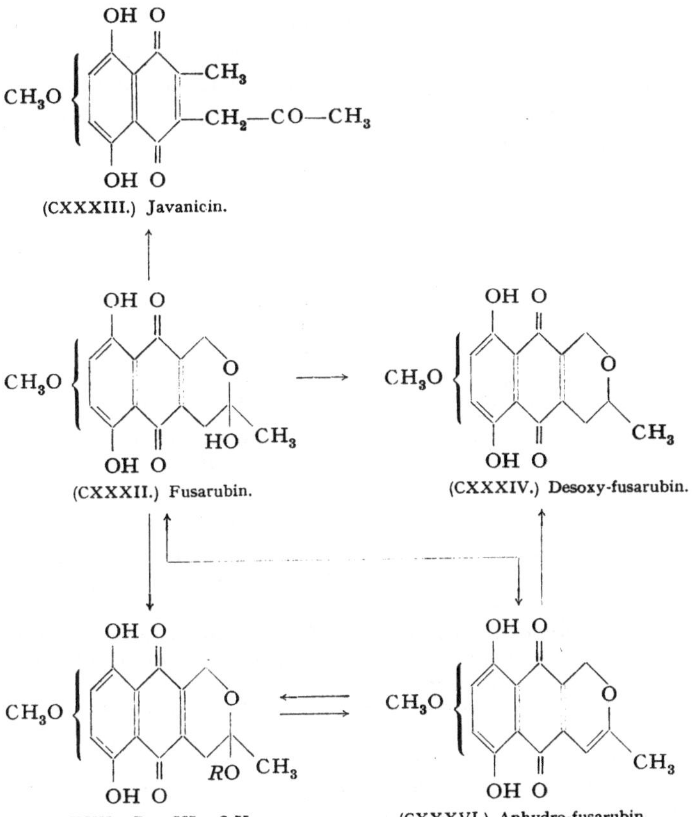

CH_3O ⎰ ⎱ —CH_3

OH O

CH_2—CO—CH_3

OH O

(CXXXIII.) Javanicin.

CH_3O ⎰ ⎱

OH O

O

HO CH_3

OH O

(CXXXII.) Fusarubin.

CH_3O ⎰ ⎱

OH O

O

CH_3

OH O

(CXXXIV.) Desoxy-fusarubin.

CH_3O ⎰ ⎱

OH O

O

RO CH_3

OH O

(CXXXV.) $R = CH_3$; C_2H_5.

CH_3O ⎰ ⎱

OH O

O

CH_3

OH O

(CXXXVI.) Anhydro-fusarubin.

Hydroxylgruppe fehlt. Diese Ergebnisse lassen für Fusarubin nur mehr die cyclische Halbketalformel (CXXXII) zu. Die glatte Ätherbildung (CXXXV), das inerte Verhalten Carbonyl-reagenzien gegenüber und die leichte Wasserabspaltung zu Anhydro-fusarubin (CXXXVI) — kurzes Erwärmen in Eisessig genügte — stehen mit dieser Formulierung in bestem Einklang. Anhydro-fusarubin ließ sich ohne Schwierigkeiten in (CXXXII), (CXXXIV) und (CXXXV) umwandeln. Infolge Mutarotation der Cyclohalbketal-gruppierung — andere Asymmetriezentren fehlen — tritt Fusarubin optisch aktiv auf.

Fusarubin zeigte eine schwache entwicklungshemmende Wirkung gegenüber Tuberkelbazillen.

RUELIUS und GAUHE (*101*) haben bemerkt, daß im Kulturfiltrat von gewissen *Fusarium solani*-Pilzen Fusarubin in reduzierter Form und an einen wasserlöslich machenden Rest *(Fusarubinogen)* gebunden vorliegt. Die wasserlösliche Leukoverbindung wurde in schwach alkalischem Medium zu einem wasserlöslichen Farbstoff oxidiert, der als

kristallines Ammoniumsalz isoliert werden konnte (*102*). Der wasser-
löslich-machende Rest ließ sich äußerst leicht abspalten und dann als
Schwefelsäure identifizieren. Bei dem „wasserlöslichen Fusarubin"
handelt es sich daher um einen phenolischen Schwefelsäure-ester des
Fusarubins der Formel (CXXXVII).

(CXXXVII.) Schwefelsäure-ester des Fusarubins.
$R' = SO_3H$; $R'' = H$, oder $R' = H$; $R'' = SO_3H$.

Vermutlich treten auch noch andere Oxychinone im Mycel oder im Kulturfiltrat von Schimmelpilzen als Schwefelsäureester auf.

Literaturverzeichnis.

1. Abu-Shady, H. A. and T. O. Soine: Experiments with Khellin. I. The Preparation of Desmethylkhellin and some of its Derivatives. J. Amer. Pharmaceut.
Assoc. **41**, 325 (1952).

1a. — — Experiments with Khellin. II. The Synthesis of 5,6-Dimethoxy-2-
methylfuro(2′,3′,7,8)chromone and Its Identity with Isokhellin. J. Amer.
Pharmaceut. Assoc. **41**, 403 (1952).

1b. — — Experiments with Khellin. III. The Formation of Desmethylisokhellin
from Khellin. J. Amer. Pharmaceut. Assoc. **41**, 429 (1952).

2. Anrep, G. V., G. S. Barsoum, and M. R. Kenawy: The Pharmacological
Action of the Crystalline Principle of *Ammi visnaga* L. J. Pharm. Pharmacol. **1**,
164 (1949).

2a. Anrep, G. V., G. S. Barsoum, M. R. Kenawy, and G. Misrahy: *Ammi visnaga*
in Treatment of Angine Syndrome. Brit. Heart J. **8**, 171 (1946).

3. — — — — Therapeutic Uses of Khellin. Method of Standardization.
Lancet **252**, 557 (1947).

4. Anrep, G. V., M. R. Kenawy, G. S. Barsoum, and I. R. Fahmy: Gazz. Fac.
Med. Cairo **14**, 1 (1947). (Originalarbeit nicht zugänglich.)

5. Arnstein, H. R. V. and A. H. Cook: Production of Antibiotics by Fungi.
III. Javanicin. An Antibacterial Pigment from *Fusarium javanicum*. J. Chem.
Soc. (London) **1947**, 1021.

6. Ayad, H.: Khellin in Angina Pectoris. Lancet **254**, 305 (1948).

7. Bagouri, M. M.: The Coronary Vasodilator Action of the Crystalline Principle
of *Ammi visnaga*. J. Pharm. Pharmacol. **1**, 177 (1949).

8. Baker, W.: Molecular Rearrangement of Some *o*-Acyloxy-acetophenones
and the Mechanism of the Production of 3-Acylchromones. J. Chem. Soc.
(London) **1933**, 1381.

9. Baxter, R. A., G. R. Ramage, and J. A. Timson: Furochromones. I. The
Synthesis of Khellin. J. Chem. Soc. (London), Supplementary Issue No. I,
1949, 30.

10. Bencze, W. und H. Schmid: Über die Konstitution des Visamminols. Experientia **10**, 12 (1954).

11. — — (unveröffentlichte Resultate).

12. BENSON, A. A., J. A. BASSHAM, M. CALVIN, A. G. HALL, H. E. HIRSCH, S. KAWAGUCHI, V. LYNCH, and N. E. TOLBERT: The Path of Carbon in Photosynthesis. XV. Ribulose and Sedoheptulose. J. Biol. Chem. **196**, 703 (1952).

12a. BIRCH, A. J. and F. W. DONOVAN: Studies in Relation to Biosynthesis. I. Some Possible Routes to Derivates of Orcinol and Phloroglucinol. Austral. J. Chem. **6**, 360 (1953).

13. BLOCH, M. und ST. V. KOSTANECKI: Über das β-Methyl-3-oxypheno-γ-pyron (β-Methyl-3-oxychromon). Ber. dtsch. chem. Ges. **33**, 471 (1900).

14. BOLLETER, A., K. EITER und H. SCHMID: Synthese des Peucenins. Helv. Chim. Acta **34**, 186 (1951).

15. CAVALLITO, C. J. and H. E. ROCKWELL: Isolation of Visnagan, the Amorphous Coronary-Dilator Principle of *Ammi visnaga*. J. Organ. Chem. (U. S. A.) **15**, 820 (1950).

16. CHAKRAVORTY, D. K., S. K. MUKERJEE, V. V. S. MURTY, and T. R. SESHADRI: Nuclear Oxydation in Flavones and Related Compounds. XXXV. Isomerization of 5,7,8-Hydroxychromones into 5,6,7-Hydroxychromones. Proc. Indian Acad. Sci., Sect. A **35**, 34 (1952).

17. CHARLIER, R. et E. PHILIPPOT: Action empêchante de la Khelline pour la fibrillation ventriculaire engendrée par l'association chloroforme-adrénaline. Arch. int. Pharmacodynamie **81**, 404 (1950).

18. CHATTERJEE, A. and S. S. MITRA: On the Constitution of the Active Principles Isolated from the Matured Bark of *Aegle marmelos*, CORREÂ. J. Amer. Chem. Soc. **71**, 606 (1949).

19. CLARKE, J. R., G. GLASER, and A. ROBERTSON: Furano-compounds. VIII. The Synthesis of *iso*Visnagin and a Partial Synthesis of Visnagin. J. Chem. Soc. (London) **1948**, 2260.

20. CLARKE, J. R. and A. ROBERTSON: Furano-compounds. IX. The Synthesis of Khellin and Related Compounds. J. Chem. Soc. (London) **1949**, 302.

21. DAGLISH, C.: The Isolation and Identification of a Hydrojuglone Glycoside Occurring in the Walnut. Biochemic. J. **47**, 452 (1950).

22. DAVIES, J. S. H. and T. DEEGAN: Furanochromones. III. The Synthesis of 8-Methoxy-2-methylfurano-(3′ : 2′-6 : 7)-chromone and its Derivatives. J. Chem. Soc. (London) **1950**, 3202.

23. DAVIES, J. S. H., P. A. McCREA, W. L. NORRIS, and G. R. RAMAGE: Furanochromones. IV. Synthesis of 2-Methylfurano-(3′ : 2′-6 : 7)-chromone and Derivatives thereof. J. Chem. Soc. (London) **1950**, 3206.

24. DAVIES, J. S. H. and W. L. NORRIS: Furanochromones. II. The Synthesis of Visnagin and Related Compounds. J. Chem. Soc. (London) **1950**, 3195.

25. DEAN, F. M.: Naturally Occurring Coumarins. Fortschr. Chem. organ. Naturstoffe **9**, 225 (1952).

26. DOYLE, B. G., F. GOGAN, J. E. GOWAN, J. KEANE and T. S. WHEELER: Transformation Reactions. I. The Mechanism of the Transformation of *o*-Aroyl-oxyacetoarones into *o*-Hydroxy-diaroylmethanes and the Synthesis of Flavones. Sci. Proc. Roy. Dublin Soc. (N. S.) **24**, 291 (1948).

27. EBNÖTHER, A., TH. M. MEIJER und H. SCHMID: Über Eleutherinol, ein natürliches Naphthopyron. (Inhaltsstoffe aus *Eleutherine bulbosa* (MILL.) URB. VI.) Helv. Chim. Acta **35**, 910 (1952).

28. ERDTMAN, H.: Über einige Inhaltsstoffe des Kernholzes der Coniferenordnung Pinales. Ihre taxonomische, physiologische und biochemische Bedeutung. Holz als Roh- und Werkstoff **11**, 245 (1953).

29. FAHMY, I. R. and H. ABU-SHADY: Isolation and Properties of Ammoidin, Ammidin and Majudin. Quart. J. Pharm. Pharmacol. **21**, 499 (1948).

30. Fahmy, I. R., N. Badran and M. F. Messeid: Photoelectric Colorimetric Method for the Estimation of Khellin. J. Pharm. Pharmacol. **1**, 529, 535 (1949).

31. Fantl, P. und S. I. Salem: Chellol-Glucosid. Biochem. Z. **226**, 166 (1930).

31a. Feinstein, L. and M. Jacobson: Insecticides Occurring in Higher Plants. Fortschr. Chem. organ. Naturstoffe **10**, 423 (1953).

32. Fellows, E. J., K. F. Killam, J. J. Toner, R. A. Dailey, and E. Macko: Comparative Pharmacology of Khellin, Visnagin and Khellolglucoside. Federat. Proc. (Amer. Soc. exp. Biol.) **9**, 271 (1950).

32a. Gallagher, K. M., A. C. Hughes, M. O'Donnel, E. M. Philbin, and T. S. Wheeler: Rearrangement in the Demethylation of 2'-Methoxyflavones. J. Chem. Soc. (London) **1953**, 3770.

33. Gardner, T. S., E. Wenis, and J. Lee: The Synthesis of Khellin Derivatives. J. Organ. Chem. (U. S. A.) **15**, 841 (1950).

34. Geissman, T. A.: The Structure of Khellin. J. Amer. Chem. Soc. **71**, 1498 (1949).

35. — The Chemistry of Flower Color Variation. J. Chem. Education **26**, 657 (1949).

36. — Chromones. IV. The Conversion of Khellol into Visnagin. Derivatives of Khellol and Visnagin. J. Amer. Chem. Soc. **73**, 3355 (1951).

37. — Chromones. V. The Preparation of 2-Methyl-7-hydroxychromone and 2-Methyl-5,8-dimethoxy-7-hydroxychromone. J. Amer. Chem. Soc. **73**, 3514 (1951).

38. Geissman, T. A. and J. W. Bolger: Chromones. VI. The Synthesis of Khellol. J. Amer. Chem. Soc. **73**, 5875 (1951).

39. Geissman, T. A. and T. G. Halsall: Chromones. III. A Total Synthesis of Khellin. J. Amer. Chem. Soc. **73**, 1280 (1951).

40. Geissman, T. A. and E. Hinreiner: Chromones. II. The Synthesis of Visnaginone. J. Amer. Chem. Soc. **73**, 782 (1951).

41. — — Theories of the Biogenesis of Flavonoid Compounds. I. Botan. Rev. **18**, 77 (1952).

42. — — Theories of the Biogenesis of Flavanoid Compounds. II. Botan. Rev. **18**, 165 (1952).

43. Grindley, D. N.: *Ammi visnaga:* Composition of the Fatty Acids Present in the Seed Fat. J. Sci. Food Agr. (England) **1**, 53 (1950).

44. Gruber, W.: Diss. Universität Wien, **1938** (Originalarbeit nicht zugänglich).

45. Gruber, W. und K. Horváth: Synthese von Isovisnagin. Monatsh. Chem. **80**, 563 (1949).

46. — — Synthese von Visnagin. (Kurze Mitt.) Monatsh. Chem. **80**, 874 (1949).

47. — — Synthese von Visnagin. II. Mitt. über Furochromone. Monatsh. Chem. **81**, 819 (1950).

48. — — Synthese von Dihydrokevisin. III. Mitt. über Furochromone. Monatsh. Chem. **81**, 828 (1950).

49. Gruber, W. und F. E. Hoyos: Resynthese des Visnagins und Synthese des Isovisnaginons. Monatsh. Chem. **78**, 417 (1948).

50. — — Über Dicarbonylderivate von Phenolen. III. Mitt. Monatsh. Chem. **80**, 303 (1949).

51. Gruber, W. und F. Traub: Über Dicarbonylderivate von Phenolen. II. Mitt. Monatsh. Chem. **77**, 414 (1947).

52. Gulati, K. C., S. R. Seth, and K. Venkataraman: Synthetical Experiments in the Chromone Group. XIII. Hydroxy-2-styrylchromones. J. Chem. Soc. (London) **1934**, 1765.

53. Hattori, S.: Über die Entmethylierung des Wogonins. Ber. dtsch. chem. Ges. 72, 1914 (1939).

54. Heilbron, I. M., H. Barnes, and R. A. Morton: Chemical Reactivity and Conjugation: The Reactivity of the 2-Methyl Group in 2 : 3-Dimethylchromone. J. Chem. Soc. (London) 123, 2559 (1923).

55. Hergert, H. L. and E. F. Kurth: The Infrared Spectra of Lignin and Related Compounds. I. Characteristic Carbonyl and Hydroxyl Frequencies of Some Flavones, Flavanones, Chalcones and Acetophenones. J. Amer. Chem. Soc. 75, 1622 (1953).

56. Huttrer, Ch. P. and E. Dale: The Chemistry and Physiological Action of Khellin and Related Products. Chem. Rev. 48, 543 (1951).

57. Jacobson, C. R., K. R. Brower, and E. D. Amstutz: Some New Coumarins and Chromones and their Ultraviolet Absorption Spectra. J. Organ. Chem. (U. S. A.) 18, 1117 (1953).

58. Jongebreur, G.: The Relation between the Chemical Constitution and the Pharmacological Activity of Some Synthetic Pyrones. Pharmac. Weekbl. 86, 661 (1951).

59. — Relation between the Chemical Constitution and the Pharmacological Action, Especially on the Coronary Vessels of the Haert, of Some Synthesized Pyrones and Khellin. Arch. int. Pharmacodynamie 90, 384 (1952).

60. Kamthong, B. and A. Robertson: Furano-compounds. III. Euparin. J. Chem. Soc. (London) 1939, 925.

61. — — Furano-compounds. V. The Synthesis of Tetrahydroeuparin and the Structure of Euparin. J. Chem. Soc. (London) 1939, 933.

62. Khalil, A. and A. Safwat: Use of Visammin in Treatment of Whooping-Cough. Amer. J. Dis. Child. 79, 42 (1950).

63. Killam, K. F. and E. J. Fellows: Pharmacology of Khellin. Federat. Proc. (Amer. Soc. exp. Biol.) 9, 291 (1950).

64. Kostanecki, St. v. und A. Rozycki: Über eine Bildungsweise von Chromon-derivaten. Ber. dtsch. chem. Ges. 34, 102 (1901).

65. Lesser, M. A.: Khellin. Drug and Cosmetic Ind. 67, 480, 556 (1950).

66. Lian, C. et R. Charlier: Etude expérimentale et clinique de la Khelline. Acta Cardiol. 5, 373 (1950).

67. Linstedt, G.: Constituents of Pine Heartwood. XXVI. A General Discussion. Acta Chem. Scand. 5, 129 (1951).

68. Linstedt, G. and A. Misiorny: Constituents of Pine Heartwood. XXIX. A Synthesis of Strobochrysin Dimethyl Ether (5,7-Dimethoxy-6-methylflavone). Acta Chem. Scand. 6, 1212 (1952).

69. Mahal, H. S. and K. Venkataraman: Synthetical Experiments in the Chromone Group. XIV. The Action of Sodamide on 1-Acyloxy-2-aceto-naphthones. J. Chem. Soc. (London) 1934, 1767.

70. Malosse, Th.: Sur l'*Ammi visnaga*. Thèse, Montpellier, 1881. Amer. J. Pharm. 53, 639 (1881).

71. Matzke, O.: Diss. Universität Wien, 1945 (Originalarbeit nicht zugänglich).

72. Meijer, Th. M.: Eugenine and Eugenone, New Compounds from Wild-Growing Clove Varieties. (*Eugenia Caryophyllata* Thungb.) Rec. trav. chim. Pays-Bas 65, 843 (1946).

73. Meijer, Th. M. und H. Schmid: Über die Konstitution des Eugenins. Helv. Chim. Acta 31, 1603 (1948).

74. Mentzer, C., D. Molho et P. Vercier: Sur un nouveau mode de condensation d'esters β-cétoniques et de phénols en chromones. C. R. hebd. Séances Acad. Sci. 232, 1488 (1951).

75. Mentzer, C. et P. Vercier: Sur la condensation du méthylmalonate d'éthyle avec les phénols. C. R. hebd. Séances Acad. Sci. **232**, 1674 (1951).

76. Mousseron, M., R. Granger et M. Ronoayroux: Sur la constitution de l'essence et de l'huile pyrogénée de *Juniperus oxycedrus* L. C. R. hebd. Séances Acad. Sci. **208**, 1411 (1939).

77. Mozingo, R.: 2-Ethylchromone. Org. Synth. **21**, 42 (1941).

78. Mukerjee, S. K. and T. R. Seshadri: Nuclear Oxidation in Flavones and Related Compounds. XXXVIII. A Transformation of Visnagin to Khellin. Proc. Indian Acad. Sci., Sect. A **35**, 323 (1952).

79. Mukerjee, S. K., T. R. Seshadri, and S. Varadarajan: Nuclear Oxidation in Flavones and Related Compounds. XXXVII. Isomerization of 2,3-Dimethyl-5,7,8-trihydroxychromone. Proc. Indian Acad. Sci., Sect. A **35**, 82 (1952).

80. Murti, V. V. S. and T. R. Seshadri: Nuclear Oxidation in Flavones and Related Compounds. XI. New Synthesis of Nobiletin. Proc. Indian Acad. Sci., Sect. A **27**, 217 (1948).

81. — — Synthesis of Khellin. J. Sci. Ind. Research (India), Sect. B **8**, No. 6, 112 (1949) [Chem. Abstr. **44**, 1501 (1950)].

82. — — Nuclear Oxidation in Flavones and Related Compounds. XXIII. A Synthesis of Khellin. Proc. Indian Acad. Sci., Sect. A **30**, 107 (1949).

83. Mustapha, I. C. R.: Sur le principe actif de l'*Ammi visnaga*. C. R. hebd. Séances Acad. Sci. **89**, 442 (1879).

84. Nikuni, Z.: A New Glycoside from the Bark of *Rhamnus japonica*. Bull. Agr. Chem. Soc. Japan **14**, 25 (1938) [Chem. Zbl. **1938** II, 77].

85. — The Distribution of α-Sorinin and α-Sorigenin. Bull. Agr. Chem. Soc. Japan **15**, 15 (1939) [Chem. Zbl. **1939** I, 4344].

86. — On the Constitution of α-Sorigenin. Bull. Agr. Chem. Soc. Japan **15**, 43 (1939) [Chem. Zbl. **1939** II, 1292].

87. — Studies on the Components of the Bark of *Rhamnus japonica*. V. Bull. Agr. Chem. Soc. Japan **17**, 92 (1941) [Chem. Abstr. **36**, 4814 (1942)].

88. — Composition of *Rhamnus japonica* Bark. Rept. Japan. Assoc. Advancement Sci. **17**, 82 (1942) [Chem. Abstr. **44**, 3477 (1950)].

89. — The Components of the Bark of *Rhamnus japonica*. VI. The Chemical Structures of α-Sorinin and α-Sorigenin. Bull. Agr. Chem. Soc. Japan **18**, 41 (1942) [Chem. Abstr. **45**, 5140 (1951)].

90. — The Components of the Bark of *Rhamnus japonica*. VII. The Chemical Character of the Lacton Ring of α-Sorinin and its Derivatives. Bull. Agr. Chem. Soc. Japan **18**, 59 (1942) [Chem. Abstr. **45**, 5141 (1951)].

91. Nikuni, Z. and H. Hayashi: Studies on the Components of the Bark of *Rhamnus japonica*. IV. Determination of the Nucleus of α-Sorigenin. Bull. Agr. Chem. Soc. Japan **15**, 158 (1939) [Chem. Zbl. **1940** I, 1936].

92. Nikuni, Z. and H. Hitsumoto: Studies on the Components of the Bark of *Rhamnus japonica*. VIII. The Chemical Structure of β-Sorinin. J. Agr. Chem. Soc. Japan **20**, 283 (1944) [Chem. Abstr. **45**, 5141 (1951)].

93. Nikuni, Z., K. Yagi, and Y. Yagyu: Studies on the Components of the Bark of *Rhamnus japonica*. IX. Attempts to Synthesize „Sorigenin". Mem. Inst. Sci. Ind. Research (Japan) **6**, 84 (1948).

94. Phillipps, G. H., A. Robertson, and W. B. Whalley: Furano-compounds. XII. Some Analogues of Khellin and Visnagin. J. Chem. Soc. (London) **1952**, 4951.

95. Platt, B. C. and Th. M. Sharp: N[1]-Sulphanilamides Derived from Aminoquinoxalines and Aminomethylquinoxalines. J. Chem. Soc. (London) **1948**, 2129.

96. RAJAGOPALAN, S. and T. R. SESHADRI: Nuclear Oxidation in the Flavones and Related Compounds. XIV. Constitution of Quercetagitrin. Proc. Indian Acad. Sci., Sect. A **28**, 31 (1948).

97. RAO, K. Ve., K. Vi. RAO, and T. R. SESHADRI: Nuclear Oxidation in the Flavone Series. II. Synthesis of Norwogonin and Isowogonin. Proc. Indian Acad. Sci., Sect. A **25**, 427 (1947).

98. REICHEL, L.: Bildung der Oxychalkone und Oxyflavanone bei Gegenwart von Aminosäuren. Naturwiss. **32**, 215 (1944).

99. REICHEL, L. und W. BURKART: Über Bildungs- und Umwandlungsbedingungen des o-Oxychalkons und des Flavanons. Chemie und Biochemie der Pflanzenstoffe. VI. Ber. dtsch. chem. Ges. **74**, 1802 (1941).

100. ROSEMAN, R. H., A. P. FISHMAN, S. R. KAPLAN, H. G. LEVIN, and L. N. KATZ: Observations on the Clinical Use of Khellin. J. Amer. Med. Assoc. **143**, 160 (1950).

101. RUELIUS, H. W. und A. GAUHE: Über Fusarubin, einen Naphthochinonfarbstoff aus Fusarien. Liebigs Ann. Chem. **569**, 38 (1950).

102. — — Über einen Oxynaphthochinon-schwefelsäure-ester aus Fusarien. Liebigs Ann. Chem. **570**, 121 (1950).

103. — — Isolierung und Konstitution eines Hydrojuglon-glucosides aus den grünen Schalen der Walnuß. Liebigs Ann. Chem. **571**, 69 (1951).

104. SAMAAN, K.: The Isolation and Properties of Visammin, Visammidin, Visnaginin, Visnagidin, Khellinin, Khellidin and Visnagan. Quart. J. Pharm. Pharmacol. **4**, 14 (1931).

105. — The Pharmacological Action of Visammin. Quart. J. Pharm. Pharmacol. **5**, 6 (1932).

106. — The Pharmacological Action of Khellinin. Quart. J. Pharm. Pharmacol. **5**, 183 (1932).

107. — *Ammi visnaga:* A Study of Certain Constituents. Quart. J. Pharm. Pharmacol. **6**, 13 (1933).

108. — Visnagan as a Coronary Dilator. Quart. J. Pharm. Pharmacol. **18**, 82 (1945).

109. — Khellinin as a Coronary Dilator. Quart. J. Pharm. Pharmacol. **19**, 135 (1946).

110. SAMAAN, K., A. M. HOSSEIN, and M. S. EL RIDI: A Spectrophotometric Method for Identification and Assay of Visammin and Khellinin. Quart. J. Pharm. Pharmacol. **20**, 502 (1947).

111. — — — The Absorption Spectra of Visammin and of Khellinin as Indications of the Chemical Structures of the Two Compounds. Quart. J. Pharm. Pharmacol. **20**, 504 (1947).

112. SAMAAN, K., A. M. HOSSEIN, and I. FAHIM: The Response of the Heart to Visammin and to Khellinin. J. Pharm. Pharmacol. **1**, 538 (1949).

113. SASTRI, V. D. N. and T. R. SESHADRI: Synthesis of 5:7:8-Trihydroxyflavones and Their Derivatives. Proc. Indian Acad. Sci., Sect. A **24**, 243 (1946).

114. SCHINDLER, H.: Über den echten Amnei, *Ammi visnaga* (L.) LAM., eine Khellinhaltige, spasmolytisch wirksame mediterrane Droge. Pharmazie 8, 176 (1953).

115. SCHMID, H.: Über die Methyläther des 4-Methyl-5,7-dioxycumarins. (Eine Bemerkung zur Konstitution des Eugenins.) Helv. Chim. Acta **30**, 1661 (1947).

116. — Über die Inhaltstoffe von *Eugenia caryophyllata* (L.) THUNBG. III. Isolierung und Konstitution des Eugenitins. Helv. Chim. Acta **32**, 813 (1949).

117. SCHMID, H. und A. BOLLETER: Über die Inhaltstoffe von *Eugenia caryophyllata* (L.) THUNBG. IV. Isolierung des Isoeugenitols. Helv. Chim. Acta **32**, 1358 (1949).

118. — — Synthese des Isoeugenitols und verwandter Verbindungen. Helv. Chim. Acta **33**, 917 (1950).

119. Schmid, H. und A. Bolleter: Über die Inhaltstoffe von *Eugenia caryophyllata* (L.) Thunbg. V. Isolierung des Isoeugenitins. Helv. Chim. Acta 33, 1770 (1950).
120. Schmid, H. und M. Burger: Synthese von 3-Methyl-6,8-dimethoxy-1,4-naphthochinon (oder 2-Methyl-5,7-dimethoxy-1,4-naphthochinon) und 2-Methyl-6,8-dimethoxy-1,4-naphthochinon (oder 3-Methyl-5,7-dimethoxy-1,4-naphthochinon). Helv. Chim. Acta 35, 928 (1952).
121. Schmid, H. und A. Ebnöther: Isolierung und Konstitution des Isoeleutherins. Allo- und Alloiso-eleutherin. (Inhaltstoffe aus *Eleutherine bulbosa*. IV.) Helv. Chim. Acta 34, 561 (1951).
122. — — Über die Konfiguration der Eleutherine-Chinone. (Inhaltstoffe aus *Eleutherine bulbosa* (Mill.) Urb. V.) Helv. Chim. Acta 34, 1041 (1951).
123. Schmid, H., A. Ebnöther und M. Burger: Zur Konstitution des Eleutherols. (Inhaltstoffe aus *Eleutherine bulbosa* (Mill.) Urb. II.) Helv. Chim. Acta 33, 609 (1950).
124. Schmid, H., A. Ebnöther und R. G. Haber: Unveröffentlichte Versuche.
125. Schmid, H., A. Ebnöther und Th. M. Meijer: Über die Konstitution des Eleutherins. (Inhaltstoffe aus *Eleutherine bulbosa* (Mill.) Urb. III.) Helv. Chim. Acta 33, 1751 (1950).
126. Schmid, H. und Th. M. Meijer: Über die Konstitution des Eugenons. Helv. Chim. Acta 31, 748 (1948).
127. Schmid, H., Th. M. Meijer und A. Ebnöther: Über die Konstitution des Eleutherols. (Inhaltstoffe aus *Eleutherine bulbosa* (Mill.) Urb. I.) Helv. Chim. Acta 33, 595 (1950).
128. Schmid, H. und H. Seiler: Über einige synthetische Naphthopyrone. Helv. Chim. Acta 35, 1990 (1952).
129. Schmutz, J., R. Hirt, F. Künzle, E. Eichenberger und H. Lauener: Synthese von basisch substituierten Chromonen. Helv. Chim. Acta 36, 620 (1953).
130. Schmutz, J., R. Hirt und H. Lauener: Synthese von 2-Carbonyl-chromonen. Helv. Chim. Acta 35, 1168 (1952).
131. Schmutz, J., H. Lauener, R. Hirt und M. Sanz: Chromonderivate: U. V.-Absorptionsspektren; coronardilatorische Wirkung. Helv. Chim. Acta 34, 767 (1951).
132. Schönberg, A. and G. Aziz: Furochromones and Coumarins. VI. Demethylation of Xanthotoxin, Khellin and Khellol with Aniline Hydrochloride and Magnesium Iodide. J. Amer. Chem. Soc. 75, 3265 (1953).
133. Schönberg, A. and N. Badran: Khellin from Visnagin. J. Amer. Chem. Soc. 73, 2960 (1951).
134. Schönberg, A., N. Badran and N. A. Starkowsky: Furo-chromones and Coumarins. VII. Degradation of Visnagin, Khellin and Related Substances. Experiments with Chromic Acid and Hydrogen Peroxide, and a Synthesis of Eugenitin. J. Amer. Chem. Soc. 75, 4992 (1953).
135. Schönberg, A. and M. M. Sidky: Furochromones and Coumarins. VIII. Action of Hydrazin Hydrate and Hydroxylamine on Khellin, Khellol and Visnagin. J. Amer. Chem. Soc. 75, 5128 (1953).
136. Schönberg, A. and A. Sina: Xanthotoxin from the Fruits of *Ammi majus* L. Nature (London) 161, 481 (1948).
137. — — Khellin and Allied Compounds. J. Amer. Chem. Soc. 72, 1611 (1950).
138. — — On Visnagin and Khellin and Related Compounds. A Simple Synthesis of Chromone. J. Amer. Chem. Soc. 72, 3396 (1950).
139. — — Experiments with Xanthotoxin and Imperatorin Obtained from the Fruits of *Ammi majus* L. J. Amer. Chem. Soc. 72, 4826 (1950).
140. — — Colour Tests. II. A Characteristic Test for 2-Methyl-4-pyrones. J. Chem. Soc. (London) 1950, 3344.

141. Scott, R. C., A. Iglauer, R. S. Green, J. W. Kaufman, B. Berman, and J. McGutte: Studies on the Effect of Oral and Parenteral Administration of Visammin (Khellin) in Patients with Angina Pectoris. Circulation **3**, 80 (1951).

142. Seshadri, T. R.: Nuclear Oxidation in the Flavones and Related Compounds. XIII. A Discussion of the Results. Proc. Indian Acad. Sci., Sect. A **28**, 1 (1948).

143. Smith, E., L. A. Pucci, and W. G. Bywater: Crystalline Visnagan. Science (New York) **115**, 520 (1952).

144. Soden, H. v. und W. Rojahn: Über das Vorkommen von Naphthalin in ätherischen Ölen. Pharmaz. Ztg. **47**, 779 (1902) [Chem. Zbl. **1902** II, 1117].

145. Späth, E.: Die natürlichen Cumarine und ihre Wirkung auf Fische. Monatsh. Chem. **69**, 75 (1936).

146. — Die natürlichen Cumarine. Ber. dtsch. chem. Ges. **70** A, 83 (1937).

147. Späth, E. und K. Eiter: Über die Konstitution des Peucenins. IV. Mitt. über natürliche Chromone. Ber. dtsch. chem. Ges. **74**, 1851 (1941).

148. Späth, E., K. Eiter und Th. Meinhard: Chroman- und Cumaron-Ringschlüsse bei einigen natürlichen Cumarinen. LIX. Mitt. über natürliche Cumarine. Ber. dtsch. chem. Ges. **75**, 1623 (1942).

149. Späth, E. und W. Gruber: Die Konstitution des Khellins (aus *Ammi visnaga*). I. Mitt. über natürliche Chromone. Ber. dtsch. chem. Ges. **71**, 106 (1938).

150. — — Die Konstitution des Visnagins (aus *Ammi visnaga*). II. Mitt. über natürliche Chromone. Ber. dtsch. chem. Ges. **74**, 1492 (1941).

151. — — Die Konstitution des Chellol-glucosids aus *Ammi visnaga* Lam. III. Mitt. über natürliche Chromone. Ber. dtsch. chem. Ges. **74**, 1549 (1941).

152. Späth, E. und P. Kainrath: Die Konstitution des Nodakenins aus *Peucedanum decursivum* Maxim. XX. Mitt. über natürliche Cumarine. Ber. dtsch. chem. Ges. **69**, 2062 (1936).

153. Späth, E. und E. Tyray: Über die Konstitution des Nodakenins aus *Peucedanum decursivum* Maxim. L. Mitt. über natürliche Cumarine. Ber. dtsch. chem. Ges. **72**, 2089 (1939).

154. Stoll, M. und W. Scherrer: Über das Extraktöl des Eichenmooses. Congr. chim. ind. Paris **17**, I, 205 (1937) [Chem. Zbl. **1938** II, 2512].

155. Treibs, W.: Beweis der Identität der Betulenolsäure mit der Homocaryophyllensäure. (II. Mitt. über Betulenole.) Ber. dtsch. chem. Ges. **71**, 612 (1938).

156. Wawzonek, St.: Chromones, Flavones and Isoflavones. In: R. C. Elderfield, Heterocyclic Compounds **2**, p. 229. New York: J. Wiley. 1951.

157. Wessely, F. und K. Kallab: Über eine Umlagerung in der Flavonreihe. Monatsh. Chem. **60**, 26 (1932).

158. Wessely, F. und G. H. Moser: Synthese und Konstitution des Skutellareins. Monatsh. Chem. **56**, 97 (1930).

159. Whalley, W. B.: A New Synthesis of Eugenitin. J. Amer. Chem. Soc. **74**, 5795 (1952).

160. Wiley, P. F.: Chromones and Related Compounds as Bronchodilators. J. Amer. Chem. Soc. **74**, 4329 (1952).

161. Wittig, G.: Über einfache Chromon- und Cumarin-Synthesen. Ber. dtsch. chem. Ges. **57**, 88 (1924).

162. Wittig, G., Fr. Bangert und H. E. Richter: Zur Erschließung der Benzo-γ-pyrone. Liebigs Ann. Chem. **446**, 155 (1926).

163. Yagi, K.: Synthesis of Sorigenin. Mem. Inst. Sci. Ind. Research (Japan) **8**, 200 (1951).

(Eingelaufen am 2. Dezember 1953.)

The Configuration of Polypeptide Chains in Proteins.

By **Linus Pauling** and **Robert B. Corey**, Pasadena, California.

With 41 Figures.

Contents.

Introduction.

The properties of proteins are determined not only by the sequence of amino-acid residues in the polypeptide chains, but also by the configuration of the chains—the way in which the chains are coiled or folded. It is probable that denaturation, the loss of some of the specific properties of a native protein, may in many cases be the result simply of a change in configuration of the polypeptide chains, without any change whatever in the sequence of amino-acid residues.

During the past few years great progress has been made in the attack on the determination of the sequence of amino-acid residues in the polypeptide chains of proteins, through the work of Sanger and his collaborators (*109, 110*) and of other investigators. There has also been significant progress in the attack on the problem of the configuration of polypeptide chains, largely through the application of the X-ray diffraction technique. These two aspects of the problem of the structure of proteins are to a considerable extent independent of one another; some of the stable configurations of polypeptide chains seem to permit an essentially arbitrary sequence of residues of nearly all of the amino acids, without significant steric hindrance. The side-chains of the residues may then be embroidered upon a polypeptide fabric of well-defined configuration without necessarily any significant change in the configuration; the properties of the resulting proteins would, of course, be determined not only by the basic fabric of the polypeptide chains, but also by the pattern of the side-chains. In general, however, it is to be expected that the stable configurations of polypeptide chains in proteins are themselves determined by the nature and sequence of the side-chains. In particular, proline and hydroxyproline residues, which prevent hydrogen-bond formation by the nitrogen atoms of the corresponding amide groups, may be expected to have a great effect in altering or determining the polypeptide chain configuration [Pauling, Corey, and Lavine (*95*)].

Early work on the investigation of proteins and other natural products by the X-ray diffraction method has been discussed in the review by Kratky and Mark (*57*) published in this Series sixteen years ago. Much of the recent progress has been the result of precise X-ray diffraction studies of crystalline amino acids, peptides, and other simple substances related to proteins, which were discussed three years ago in this Series by Corey (*31*). A brief survey of X-ray theory and techniques was also given in this article. In the following pages we shall discuss first the developments of the past three years in the determination of the structure of simple substances related to proteins, and the significance of the results with respect to the dimensions of the amide group in the polypeptide chain. The helical configurations and layer configurations of hydrogen-bonded polypeptide chains that have been formulated by use of the principles obtained from the study of simple substances are then described. The last Sections of our paper present a summary of the evidence for the occurrence of some of these configurations in fibrous proteins and globular proteins, with brief mention also of synthetic polypeptides.

I. The Dimensions of the Amide Group.

1. The Amino Acids.

Structures of crystalline amino acids that have been determined before 1950 have been described in the review article by COREY (31). The first of these structures to be determined was that of glycine, by ALBRECHT and COREY (1). The structure of DL-alanine, $CH_3 \cdot CH(NH_2) \cdot$ $\cdot COOH$, was then determined by LÉVY and COREY (59) in 1941, and refined somewhat by DONOHUE (42) in 1949. A precise determination of the structure of L-threonine (threo-α-amino-β-hydroxy-n-butyric acid), $CH_3 \cdot CHOH \cdot CH(NH_2) \cdot COOH$, was then made by SHOEMAKER, DONOHUE, SCHOMAKER, and COREY (113), with use of three-dimensional Fourier series and new computational techniques. This structure and those of glycine and alanine are described in detail in the earlier review (31).

Since then a detailed structure determination of L-hydroxyproline has been published by DONOHUE and TRUEBLOOD (45, 46), following an independent preliminary investigation by ZUSSMAN (126, 127). A similarly detailed study of DL-serine has also been made, by SHOEMAKER, BARIEAU, DONOHUE, and LU (112). Less precise structure determinations have been made of L-glutamine, by COCHRAN and PENFOLD (28), DL-glutamic acid hydrochloride, by DAWSON (40), and DL-methionine and DL-norleucine, by MATHIESON (66, 67).

The general nature of these structures, except for L-hydroxyproline, is not significantly different from that of those discussed in the earlier review (31). Some of the results of the investigation are contained in *Tables 1, 2, and 3* (pp. 198, 201).

a) The Crystal Structure of L-Hydroxyproline.

The determination of the structure of hydroxyproline, 4-hydroxy-
$$\overline{CH_2}$$
pyrrolidine-2-carboxylic acid, $HO \cdot CH \cdot CH_2 \cdot NH \cdot CH \cdot COOH$, is important for several reasons: it established unambiguously the relative configurations of the carbon atom carrying the hydroxyl group and the α-carbon atom, thus relating the configurations of the sugars and the amino acids [NEUBERGER (71)], and it provides structural information required for arriving at the fold or bend which a proline or hydroxyproline residue might be expected to effect in a polypeptide chain

[PAULING (78)]. The analysis of crystals of *L*-hydroxyproline (45) was based upon the relative intensities of about 750 X-ray reflections. A successful trial structure was derived from an interpretation of the complete three-dimensional Patterson function. This structure was refined (46) first by use of two-dimensional Fourier projections of electron density and least squares treatments, and finally by three successive three-dimensional least squares treatments and the calculation of two three-dimensional electron density distributions. Small maxima were observed

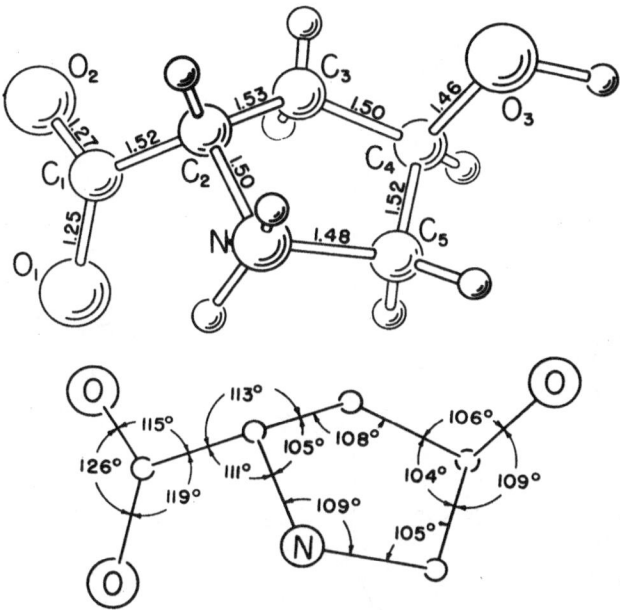

Fig. 1. A drawing of the molecule of *L*-hydroxyproline showing interatomic distances (in Ångströms) and bond angles found by X-ray analysis. [From: Acta Crystallogr. **5**, 419 (1952).]

in the electron density plot at positions close to where hydrogen atoms might be expected, but they were not sufficiently well defined to be of use in establishing the positions of the hydrogen atoms.

The molecular dimensions of hydroxyproline are shown in *Figure 1*. As in the molecules of the other amino acids, the average C—C distance is significantly smaller than the value 1.5445 Å. found in diamond [LONSDALE (60)]. It is interesting to note that the bond lengths appear to alternate in the same manner as that already observed in threonine (31, 113). The five-membered pyrrolidine ring is appreciably puckered; $C_{(4)}$, the carbon atom carrying the hydroxyl group, is about 0.4 Å. from the plane defined (within 0.03 Å.) by the other four atoms of the ring, and is on the opposite side of this plane from the carboxyl group. The α-carbon atom and the atoms of the carboxyl group lie in a plane, as expected on theoretical grounds and as found in other carboxylic acids.

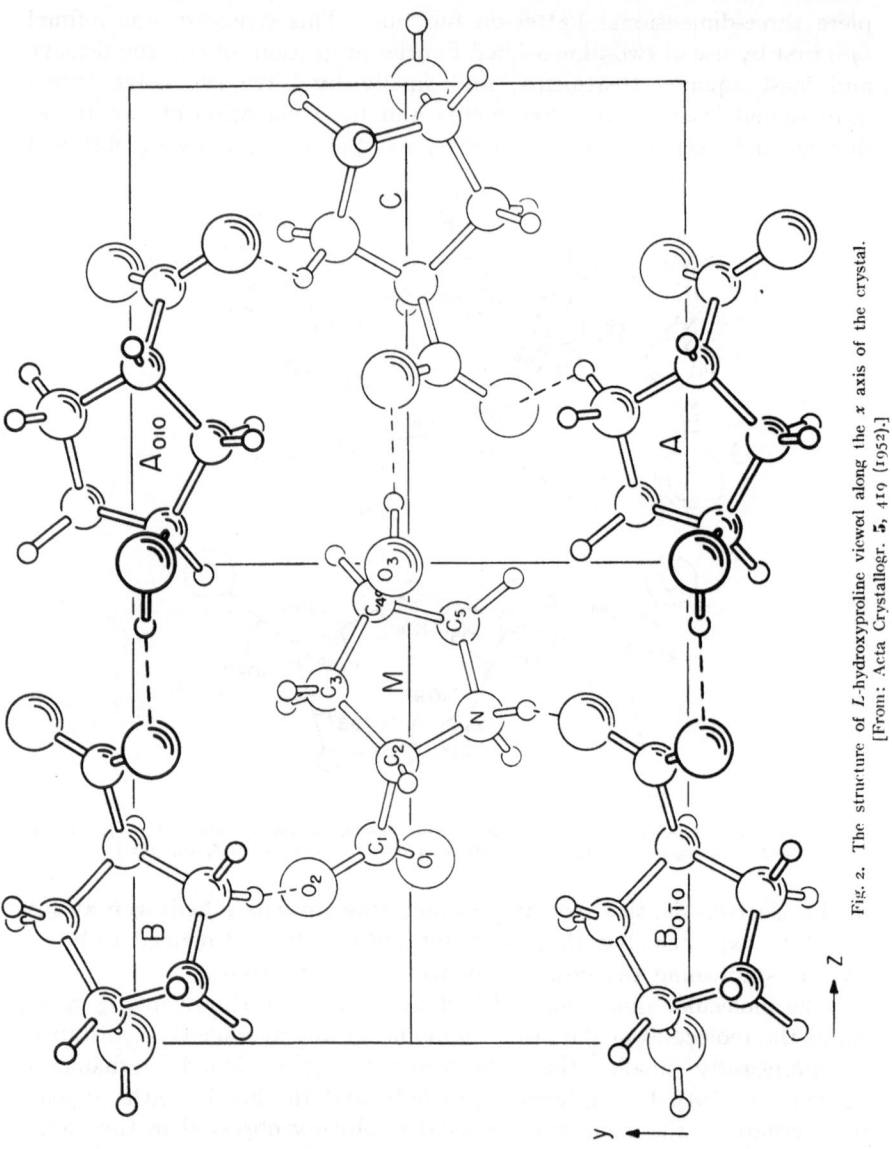

Fig. 2. The structure of L-hydroxyproline viewed along the x axis of the crystal. [From: Acta Crystallogr. **5**, 419 (1952).]

Figures 2 and *3* show the arrangement of molecules of hydroxyproline in the crystal as viewed along the *x* and *y* axes respectively. The way in which the molecules are tied together by hydrogen bonds is analogous in some respects to that previously found in crystals of *L*-threonine: in two dimensions they are connected by hydrogen bonds and electrostatic

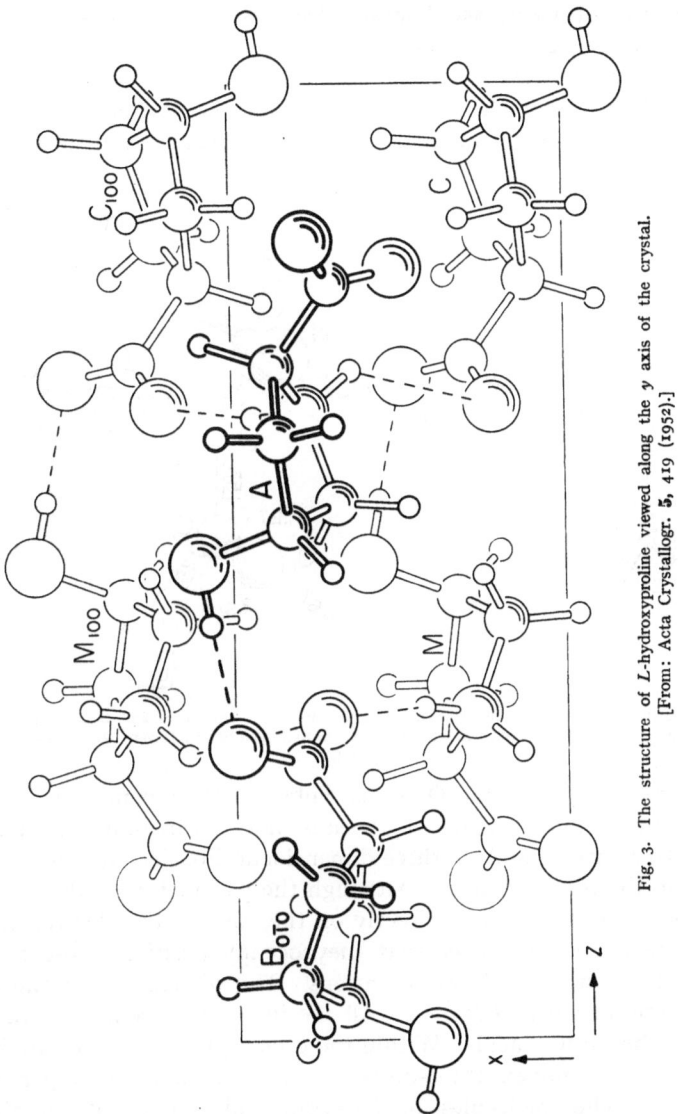

Fig. 3. The structure of *L*-hydroxyproline viewed along the *y* axis of the crystal. [From: Acta Crystallogr. **5**, 419 (1952).]

forces between the amino nitrogen atom and oxygen atoms of the carboxyl groups, and in the third dimension by bonds between hydroxyl groups and carboxyl oxygen atoms.

b) The Crystal Structure of DL-Serine.

The determination (*112*) of the crystal structure of *DL*-serine (α-amino-β-hydroxypropionic acid), $HO \cdot CH_2 \cdot CH(NH_2) \cdot COOH$, was based on the intensities of about 850 X-ray reflections. As in the analysis of its analog threonine, a satisfactory trial structure could be obtained only through the interpretation of the complete three-dimensional Patterson function. Because of the high resolution of the Patterson function it was possible to use it for making the first improvements in the positional

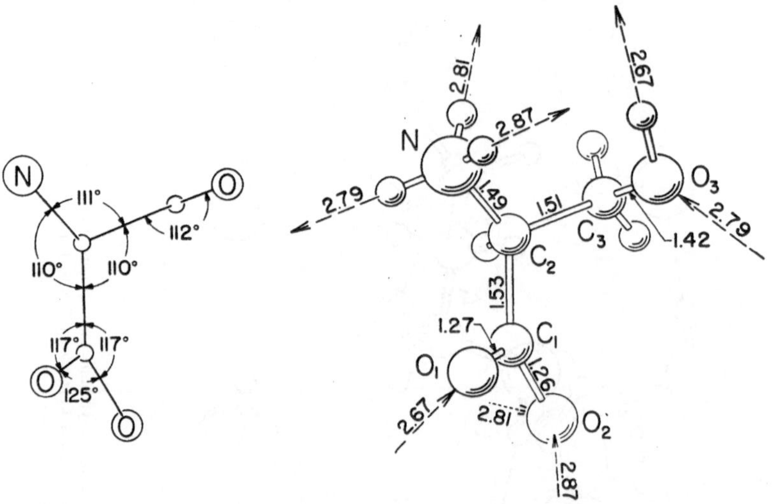

Fig. 4. A drawing of the molecule of serine showing interatomic distances and bond angles found by X-ray analysis. Dashed arrows represent the directions of hydrogen bonds. [From: Acta Crystallogr. **6**, 241 (1953).]

parameters of the C, N, and O atoms. Subsequent refinement was carried out in the customary manner by means of two three-dimensional least squares treatments and a three-dimensional Fourier synthesis based directly on the intensity data. Although the parameters of the hydrogen atoms could not be assigned directly on the basis of the electron density plots, there is good evidence that they occupy positions close to those expected from structural considerations. The dimensions of the serine molecule are shown in *Figure 4*. All are in good agreement with those found in other amino acids. Within the crystal, the molecules are bound together by a complex three-dimensional network of hydrogen bonds *(Figure 5)*. The molecules of *DL*-serine, like those of *DL*-alanine, apparently possess a size and shape that permit them to solve the problems of crystallization very neatly. As in crystals of *DL*-alanine (*31, 59*), the positively charged NH_3^+ group forms three strong hydrogen bonds tetrahedrally arranged about the nitrogen atom. Two of these bonds

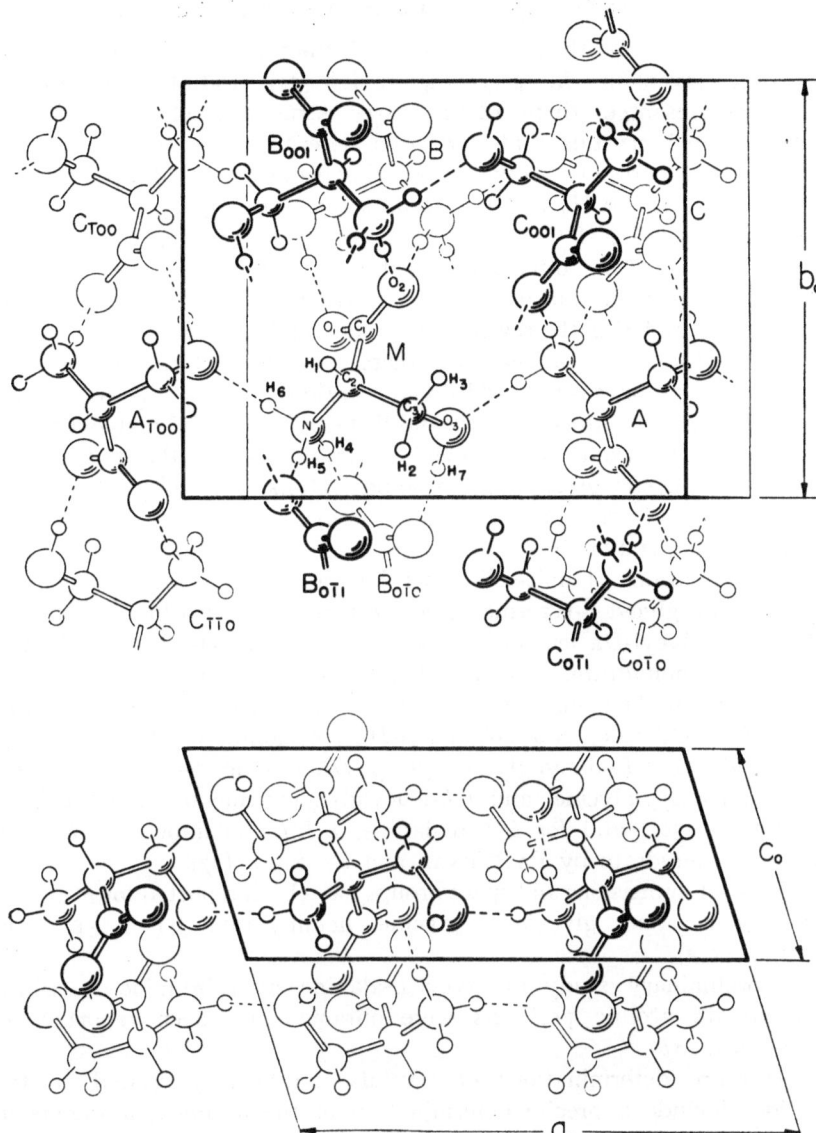

Fig. 5. Two views of the structure of *DL*-serine: above, perpendicular to the *ab* plane; and, below, along the *b* axis of the crystal. [From: Acta Crystallogr. **6**, 241 (1953).]

are formed to carboxyl oxygen atoms and the third to the hydroxyl oxygen atom, all in different molecules. All intermolecular contacts are normal, and the packing throughout the crystal is good.

2. Simple Peptides and Related Substances.

A summary of the early X-ray diffraction studies of peptides has been given in the preceding review (*31*). Detailed descriptions are also presented there of the three structures of peptides and closely related crystals that had been determined by 1950. These include the structures of 2,5-diketopiperazine [Corey (*30*)]; β-glycylglycine, $NH_2 \cdot CH_2 \cdot CO \cdot NH \cdot CH_2 \cdot COOH$ [Hughes and Moore (*52*)]; and N-acetyl-glycine, $CH_3 \cdot CO \cdot NH \cdot CH_2 \cdot COOH$ [Carpenter and Donohue (*26*)]. Inferences about the interatomic distances and bond angles in poly-peptide chains as obtained from these structures were summarized by Corey (*31*), and slightly revised by Corey and Donohue (*32*).

During the last four years complete structure determinations have been reported for several additional peptide crystals. These include α-glycylglycine [Hughes, Biswas, and Wilson (*51*)]; N,N'-diglycylcystine [Yakel and Hughes (*123*)]; glycyl-*L*-asparagine [Katz, Pasternak, and Corey (*53*)]; cysteylglycine-sodium iodide [Dyer (*47*)], glycyl-*L*-tyrosine hydrochloride [Smits and Wiebenga (*116*)], and glycyl-*L*-tryptophan dihydrate [Pasternak (*75*)].

Determinations of unit-cell dimensions and space groups have been reported for glycyl-*L*-tyrosine, glycyl-*L*-tyrosine hydrochloride, glycyl-*DL*-serine, glycyl-*DL*-leucine, glycyl-*L*-alanine, glycyl-*L*-alanine hydro-chloride monohydrate, glycyl-*L*-alanine hydrobromide monohydrate, glycyl-*L*-tryptophan dihydrate, glycyl-*L*-valine hydrobromide, glycyl-*L*-valine hydrochloride, and *DL*-alanyl-*DL*-methionine by Tranter (*117, 118, 118a*); for *DL*-alanyl-glycine, glycyl-*DL*-alanine, *L*-leucylglycine di-hydrate, *L*-leucylglycine, and *DL*-leucylglycylglycine by Leonard and Pasternak (*58*); and for *D,D*-dialanine, *L,L,L*-trialanine, and *D,D*-di-alanine hemihydrate by Pasternak and Leonard (*77*).

Unit-cell dimensions and space groups have been reported for α-glycyl-glycylglycine and glycylglycylglycine hemihydrate by Yakel and Hughes (*122*).

A preliminary study of glycyl-*L*-alanine hydrobromide in which approximate atomic positions were obtained has been reported by Tranter (*118b*).

Structure determinations of crystals significantly related to the peptides include a precise determination of the atomic parameters in urea [Vaughan and Donohue (*120*)], and the determinations of the structure of urea oxalate [Schuch, Merritt, and Sturdivant (*111*)], oxamide [Romers (*106*); Sly (*115*)], and parabanic acid [Blum and Davies (*18*)].

a) The Crystal Structure of α-Glycylglycine.

The dimensions of the simplest linear peptide, glycylglycine, $NH_2 \cdot CH_2 \cdot CO \cdot NH \cdot CH_2 \cdot COOH$, were first determined by Hughes

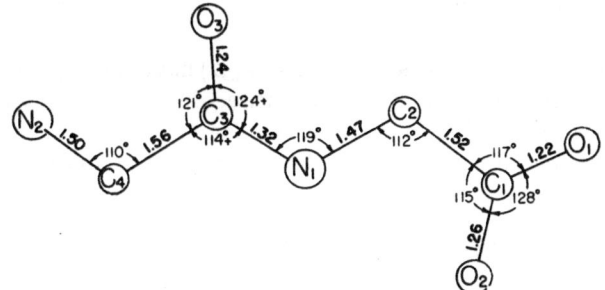

Fig. 6. A drawing showing the dimensions of a molecule of glycylglycine as found by X-ray analysis of the α crystalline form.

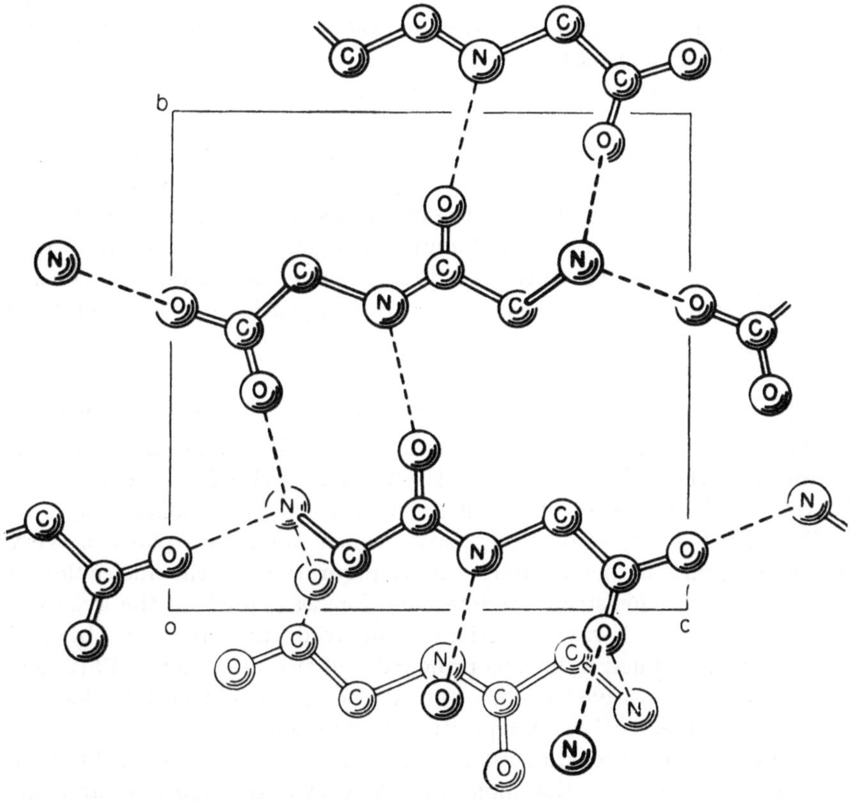

Fig. 7. The structure of crystals of α-glycylglycine projected along the a axis onto the (100) face of the crystal. Two molecules of one layer are shown, together with some atoms of adjacent molecules and one molecule of an underlying layer.

and MOORE (52) in a two-dimensional analysis of the β form, one of the three crystalline modifications originally described by BERNAL (15).

This structure of β-glycylglycine is described in the earlier review in this Series (31).

Recently HUGHES, BISWAS, and WILSON (51) have carried out a thorough analysis of α-glycylglycine based on three-dimensional intensity data for more than 2000 X-ray reflections. Half-cell Patterson projection diagrams and special theoretical methods were used to derive the approximate structure, which has been refined by three three-dimensional least squares treatments of the intensity data and by the calculation of two half-cell Fourier projection plots of electron density. Although the refinement of the atomic positions is not quite completed, the dimensions of the molecule have already been determined with a precision much greater than that attained in the analysis of β-glycylglycine; subsequent revisions of the interatomic distances and bond angles shown in *Figures 6* and *7* (p. 189) and listed in *Table 3* (p. 201) should not involve changes greater than 0.01 Å. or 1°, respectively.

The dimensions of the glycylglycine molecule as it exists in crystals of α-glycylglycine are shown in Figure 6. As in β-glycylglycine, the molecule is a dipolar ion and is essentially planar except for the terminal amino nitrogen atom, which lies 0.73 Å. below the molecular plane. Figure 7 shows the way in which the molecules are arranged in the crystal of α-glycylglycine. The terminal amino nitrogen atom forms a hydrogen bond to a carboxyl atom of each of three different molecules. The peptide nitrogen atom forms a hydrogen bond to the peptide keto oxygen atom of an adjacent molecule.

b) The Crystal Structure of N,N'-Diglycyl-L-cystine Dihydrate.

The determination of the structure of N,N'-diglycyl-*L*-cystine dihydrate, $[NH_2 \cdot CH_2 \cdot CO \cdot NH \cdot CH \cdot (COOH) \cdot CH_2 \cdot S—]_2 \cdot 2 H_2O$, by YAKEL and HUGHES (123) has provided data concerning the —CH_2—S—S—CH_2— disulfide link, which doubtless plays a very prominent part in determining the configuration of the molecules of many proteins. This structure determination was based on the observed intensities of 850 X-ray reflections. The trial structure was obtained and gradually refined by means principally of two-dimensional Patterson diagrams and successive two-dimensional Fourier plots of electron density. The final refinement of the positional parameters of the atoms was completed by two three-dimensional least squares treatments of the data.

The dimensions of the molecule are shown in *Figure 8*, and the arrangement of the molecules in the crystal in *Figure 9*. The interatomic distances and bond angles are in general agreement with those found in other peptides. It is to be noted that in this structure the amide group is not quite planar, but is bent in a manner corresponding to a rotation of 6° about the C—N peptide bond. This rotation would correspond to a strain energy of about 0.4 kcal/mole if it is assumed that the two structures I and II (p. 200) contribute in the ratio 60 : 40. The lengths of the C—O and C—N bonds (1.21 and 1.35 Å. respectively) when compared with those characteristic of planar structures (1.24 and 1.32 Å.)

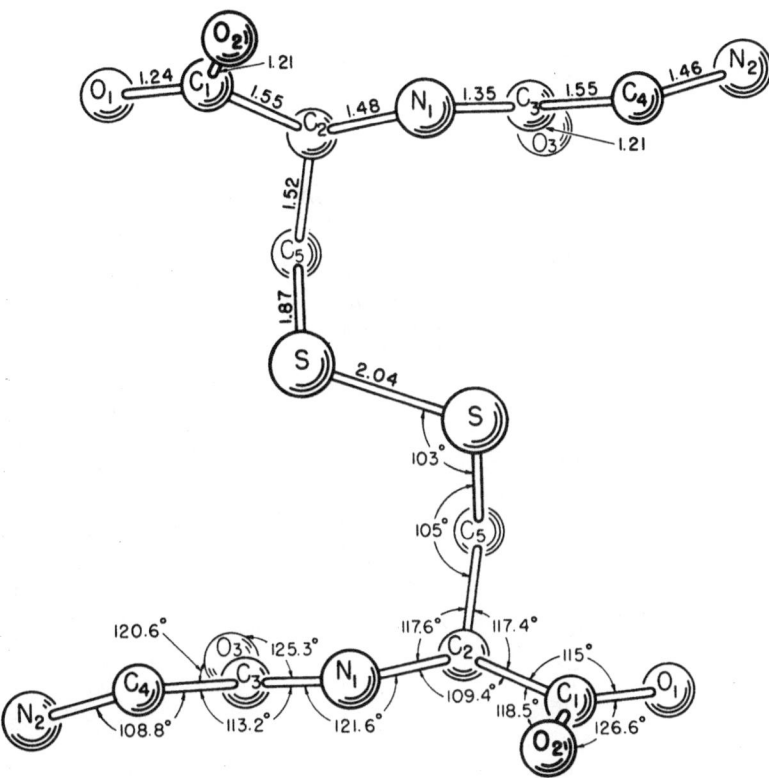

Fig. 8. A drawing of the molecule of N,N'-diglycyl-L-cystine showing the interatomic distances and bond angles found by X-ray crystal analysis. The two halves of the molecule are identical; a twofold rotation axis passes through the center of the S-S bond perpendicular to the plane of the paper.

indicate less resonance in the amide group, so that the strain is probably less than this amount.

In contrast to the planar configuration of α-glycylglycine, the planes of the carboxyl and amide groups in the glycylglycyl chain of diglycyl-cystine form a dihedral angle of 19° ± 2°. The C—S—S—C dihedral angle is 101° ± 2°, very near the value found for the S—S—S—S dihedral angle in S_8 [WARREN and BURWELL (121)], and discussed theoretically by PAULING (80). The two glycylglycine arms of the molecule are separated about 6.3 Å. by the —CH_2—S—S—CH_2— bridge. Internal ionization

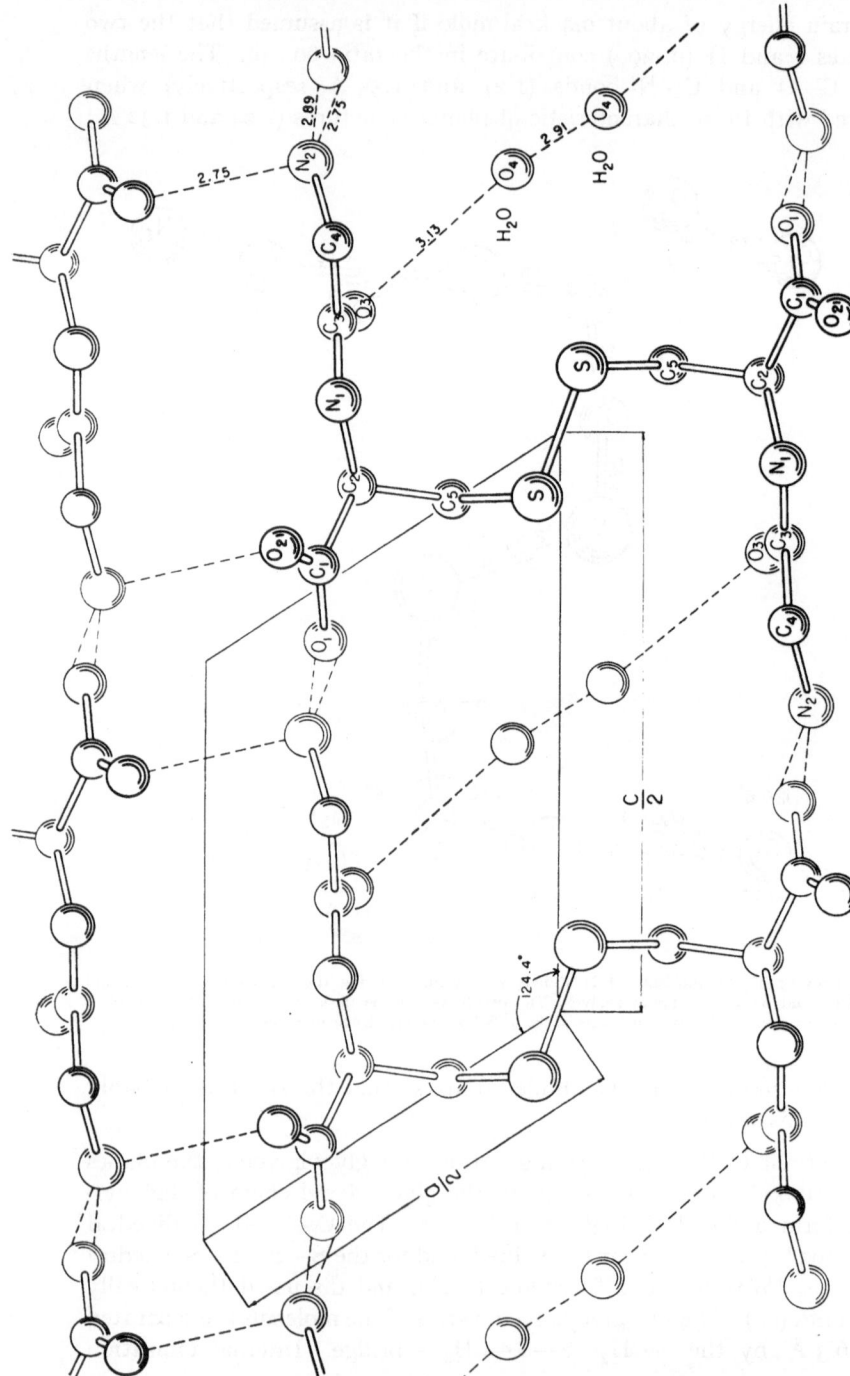

Fig. 9. A drawing of the structure of N,N'-diglycyl-L-cystine dihydrate viewed along the b axis of the crystal. The dashed lines represent hydrogen bonds. The nitrogen atom N_2 forms one hydrogen bond (2.89 Å) to oxygen atom O_1 of the adjacent molecule and one hydrogen bond (2.75 Å) to the corresponding O-atom in the molecule directly below.

of the molecule is clearly indicated by the dispɔsition of the hydrogen bonds around the terminal amino nitrogen atom.

As in other peptide and amino acid structures, the formation of hydrogen bonds is the dominant factor in controlling the molecular packing in the crystal. In the direction of the *b* axis the molecules are held together by strong hydrogen bonds between terminal amino groups and carboxyl groups, and by weaker bonds involving the peptide amide group and water molecules. Strong hydrogen bonds between terminal amide groups and carboxyl groups also firmly connect the molecules in the *a* and *c* directions.

c) The Crystal Structure of Glycyl-L-asparagine.

Additional data concerning the dimensions of the peptide bond and also information about the dimensions and configuration of the amino

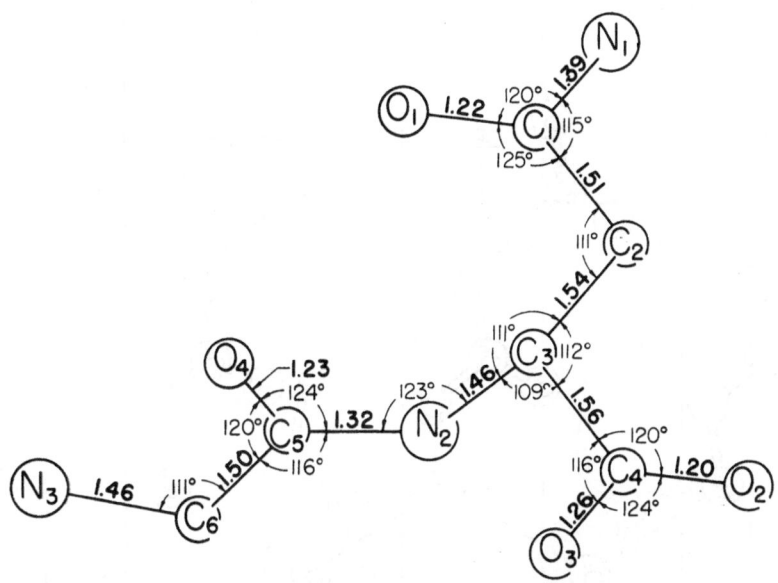

Fig. 10. A drawing of the molecule of glycyl-*L*-asparagine showing bond lengths and bond angles found by X-ray crystal analysis. [From: Acta Crystallogr. (in press).]

acid asparagine are provided by the determination by PASTERNAK, KATZ, and COREY (76) of the crystal structure of glycyl-*L*-asparagine, $NH_2 \cdot CH_2 \cdot CO \cdot NH \cdot CH(CH_2 \cdot CO \cdot NH_2) \cdot COOH$. This determination was based upon the intensities of about 1000 X-ray reflections. The complete three-dimensional Patterson vector diagram was necessary for arriving at a satisfactory trial structure. The positional parameters of the atoms were refined, first by means of three successive two-

dimensional Fourier projections and two three-dimensional least squares
calculations, and finally by a three-dimensional Fourier synthesis and
another three-dimensional least squares treatment. Although in the

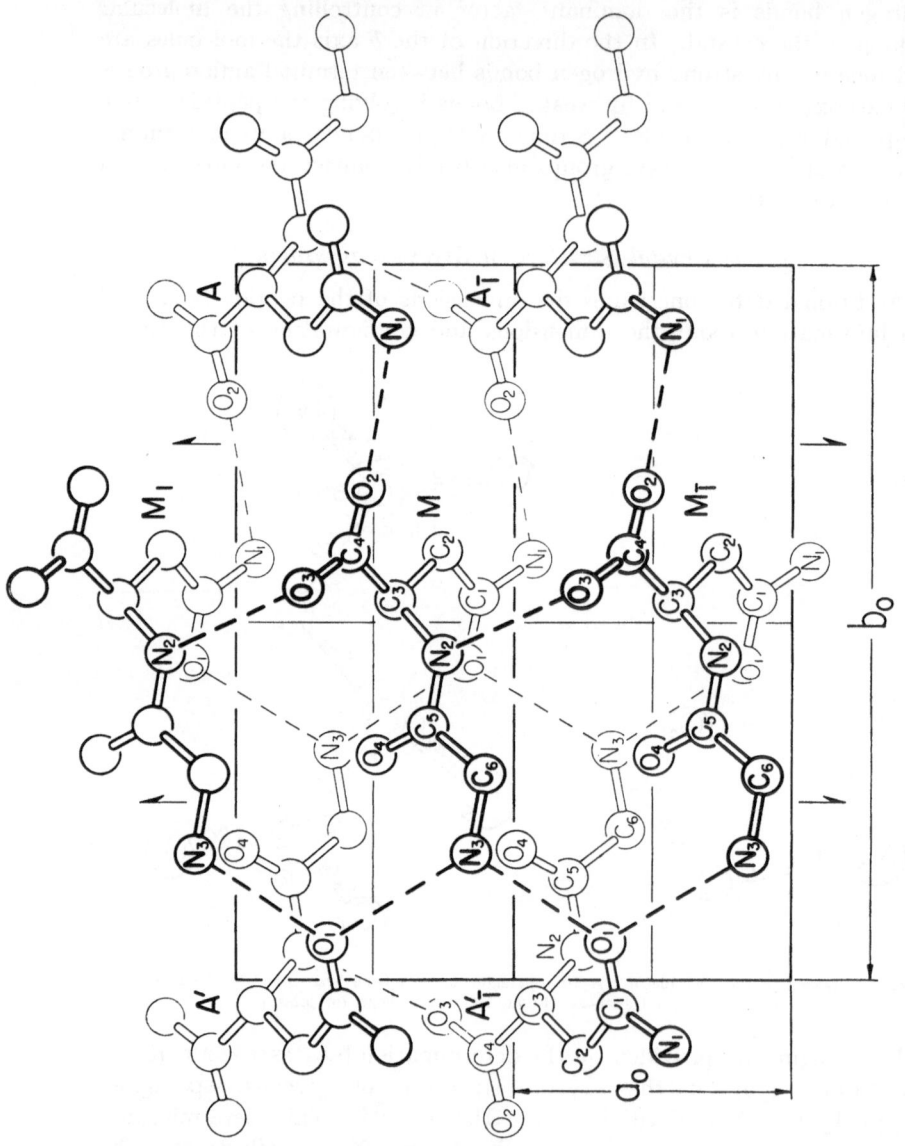

Fig. 11. A drawing of a portion of the structure of glycyl-L-asparagine viewed along the c axis of the crystal. In the direction of the a axis identical molecules stack neatly on top of one another and are held together by hydrogen bonds (dashed lines) between their peptide amide nitrogen atoms and carboxyl oxygen atoms. Other N—H···O hydrogen bonds link these molecules to those in adjacent stacks at the right and left. [From: Acta Crystallogr. (in press).]

last Fourier plot of electron density maxima appeared approximately
where hydrogen atoms would be expected, definite positions could not
be assigned to the hydrogen atoms on the basis of the X-ray data.

The dimensions of the molecule of glycyl-*L*-asparagine are shown in *Figure 10*. The peptide amide group is planar within the accuracy of the determination, and its dimensions are in good agreement with those found in other linear peptides and related compounds. There is good evidence that the terminal amino group of the peptide chain is in the form of a charged NH_3^+ group. The atoms $O_{(1)}$, $C_{(1)}$, $N_{(1)}$, and $C_{(2)}$ of the terminal amide group are also coplanar, as would be expected. The planes of the terminal amide group and the carboxyl group nearly coincide, the angle between them being only 3°. The carbon atoms of these groups form an extended chain in which the C—C bond lengths probably do not deviate significantly from the average 1.54 Å. The planar succinamic acid part of the molecule and the planar peptide group approximately perpendicular (85°) to it are easily recognized in the Figures. The shape of the molecule and the formation of hydrogen bonds determine the rather complex packing of the molecules in the crystal. In the direction of the *a* axis adjacent molecules pack neatly together, as shown by the central column of molecules in *Figure 11*, each one being held to the next by a hydrogen bond between its peptide amide nitrogen atom and a carboxyl oxygen atom of its neighbor below it in the crystal. These molecular stacks are formed into two-dimensional sheets through hydrogen bonds which connect terminal amino nitrogen atoms with terminal amide oxygen atoms and terminal amide nitrogen atoms with carboxyl oxygen atoms. In the third dimension the crystal is built up by the packing together of layers similar to the one just described. These layers (not shown in the Figures) are held together by hydrogen bonds between carboxyl oxygen atoms and terminal amino and amide nitrogen atoms. This structure seems to be unique in the fact that the carbonyl oxygen atom of the peptide group forms no hydrogen bonds at all; it makes good van der Waals contacts with the atoms which surround it, the only nitrogen atom which is near it being in a position unfavorable for the formation of a hydrogen bond.

d) The Crystal Structure of Glycyl-L-tryptophan Dihydrate.

The molecular dimensions of the amino acid tryptophan have not yet been determined by X-ray crystal analysis. These data, and additional information about the structure of peptides, are to be obtained by means of a thorough analysis of crystals of glycyl-*L*-tryptophan dihydrate.

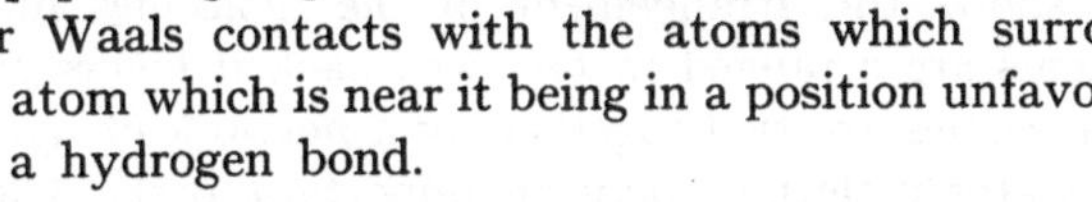

$$\text{—CH}_2\text{—C—NH—CO—CH}_2\text{—NH}_3^+ \cdot 2H_2O$$
$$| \atop \text{COO}^-$$

Glycyl-*L*-tryptophan dihydrate.

This structure determination [PASTERNAK (75)] is based on intensity data from about 1600 X-ray reflections. A satisfactory trial structure was derived from an analysis of the complete three-dimensional Patterson vector diagram, and was refined by successive Fourier projections of electron density. A projection of one molecule along the *b* axis is shown in *Figure 12*, in which only the nitrogen and oxygen atoms are labelled, the others being carbon atoms. The atoms of the indole ring system and the carbon atoms attached to it are nearly, if not perfectly, coplanar. The bonds connecting the ring nitrogen atom to its two adjacent carbon atoms appear to be very short (*ca.* 1.30 Å.). The configuration of the peptide group is the same as that found in other peptides: the five atoms

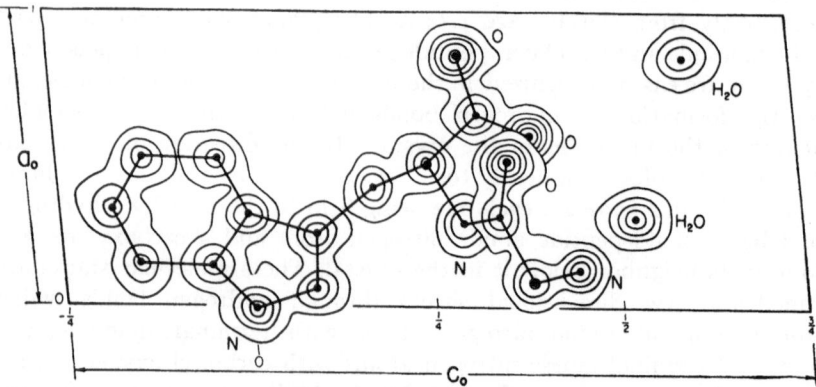

Fig. 12. A Fourier projection of the electron density in one-half of the unit cell of glycyl-*L*-tryptophan dihydrate on the (010) plane of the crystal. The positions of the C, N, and O atoms and of water molecules are clearly shown. Atoms which are not labeled are carbon atoms.

associated with it are coplanar and the C—N peptide bond is short, near 1.30 Å.

Figure 13 shows the arrangement of the molecules in the crystal. The ring systems are arranged in two well-packed layers and the polar ends of the molecules are tied together by a network of hydrogen bonds in which the water molecules play an important part. The exhaustive three-dimensional analysis of this crystal is not yet completed.

3. The Configuration of the Amide Group.

a) Dimensions of the Amide Group.

The configuration of the amide group as inferred by Corey and Pauling (*33*) from the experimental results for simple substances is shown in *Figure 14* (p. 199). This configuration involves only minor refinements from that described in the previous review (*31*).

The values of bond lengths as derived from values found in crystals of amino acids, peptides, and related substances are given in *Table 1*. The changes from the previous values (*31*) are the increase of the value for C'—O from 1.23 Å. to 1.24 Å., and a maximum change of 6° in the angles between the coplanar bonds around the nitrogen atom.

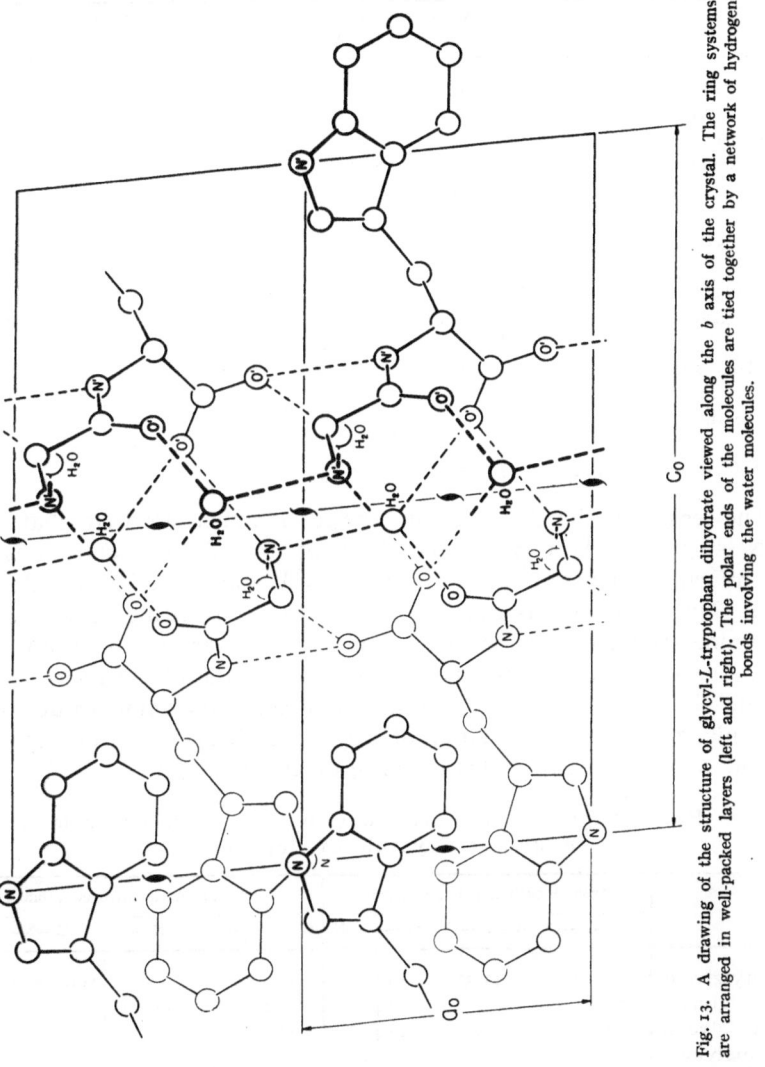

Fig. 13. A drawing of the structure of glycyl-L-tryptophan dihydrate viewed along the *b* axis of the crystal. The ring systems are arranged in well-packed layers (left and right). The polar ends of the molecules are tied together by a network of hydrogen bonds involving the water molecules.

The experimental values for the length of the αC—N single bond are in the neighborhood of 1.49 Å. for amino-acid crystals, and 1.47 Å. for the peptides; the latter value has been selected as the more significant for the amide group in polypeptide chains. The αC—C' single bond, with the average reported value 1.53 Å., seems to be significantly shorter than the C—C distance found in diamond, 1.5445 Å. (60).

Table 1. Bond Lengths of the Amide Group as Derived from Those Found in Three-Dimensional Analyses of Crystals of Amino-Acids, Peptides, and Related Compounds.

	N—αC (Å.)	C'—αC (Å.)	C'—O (Å.)	C'—N (Å.)	Reference
DL-alanine	1.50	1.54	—	—	(42)
L-threonine..............	1.49	1.52	—	—	(113)
DL-serine	1.49	1.53	—	—	(112)
L-hydroxyproline..........	1.50	1.52	—	—	(46)
N-acetylglycine	1.45	1.50	1.24	1.32	(26)
		1.51			
β-glycylglycine*	1.48	1.53	1.23	1.29	(52)
		1.53			
α-glycylglycine...........	1.47	1.56	1.24	1.32	(51)
N,N'-diglycylcystine	1.48	1.56	1.21	1.35	(124)
		1.52			
glycyl-L-asparagine	1.46	1.50	1.23	1.32	(76)
urea	—	—	1.26	1.34	(120)
urea oxalate	—	—	1.26	1.34	(111)
				1.35	
Selected value	1.47	1.53	1.24	1.32	—

The length of the C'—N bond, 1.32 Å., is much less than that of a C—N single bond, 1.47 Å., and that of the C'—O bond, 1.24 Å., is significantly greater than the value 1.215 Å. corresponding to a double bond. These dimensions can be ascribed to a resonating structure for the amide group, with about 40% double-bond character for the C'—N bond and about 60% double-bond character for the C'—O bond.

The N—H distance can be assumed to have the same value as in ammonia, 1.00 Å. The formation of a hydrogen bond would be expected to lead to an increase in this distance of about 0.03 Å.

Table 2. Bond Angles of the Amide Group as Derived from Those Found in Crystals of Peptides.

	Around carbonyl carbon atom			Around amide nitrogen atom		
	αC—C'—O	N—C'—O	αC—C'—N	C'—N—αC	C—N—H	αC—N—H
N-acetylglycine	121.0°	121.3°	117.7°	119.6°	—	100° ± 10°**
β-glycylglycine	121	125	114	122	—	—
α-glycylglycine	121.1	124.2	114.4	119.3	—	—
N,N'-diglycyl-cystine.....	120.6	125.3	113.2	121.6	—	—
glycyl-L-asparagine .	120	124	116	123	—	—
Selected values.	121	125	114	123	123	114

* Two-dimensional analysis.
** Polarized infra-red study (72).

The bond angles of the amide group as derived from those found in crystals of peptides are given in *Table 2*. The selected values differ by an average of 3° from those given in the previous review (*31*). The earlier values were based entirely upon the determined structures of N-acetyl-glycine and β-glycylglycine, which were not in good agreement with one another. The careful structure determinations of peptides that have been carried out since 1950 have given results in good general agreement with those for β-glycylglycine, and the average deviation of the observed values from the selected values is less than 1°. The values of the angles C'—N—H and αC—N—H have been selected largely from theoretical considerations. The only experimental value is that due to NEWMAN and BADGER (*72*), from the measurement of the polarization of the absorption band of N-acetylglycine at 3340 cm^{-1}, corresponding to the N—H stretching frequency. The selected value, 114°, for the angle C—N—H is not far from the upper limit given by these authors.

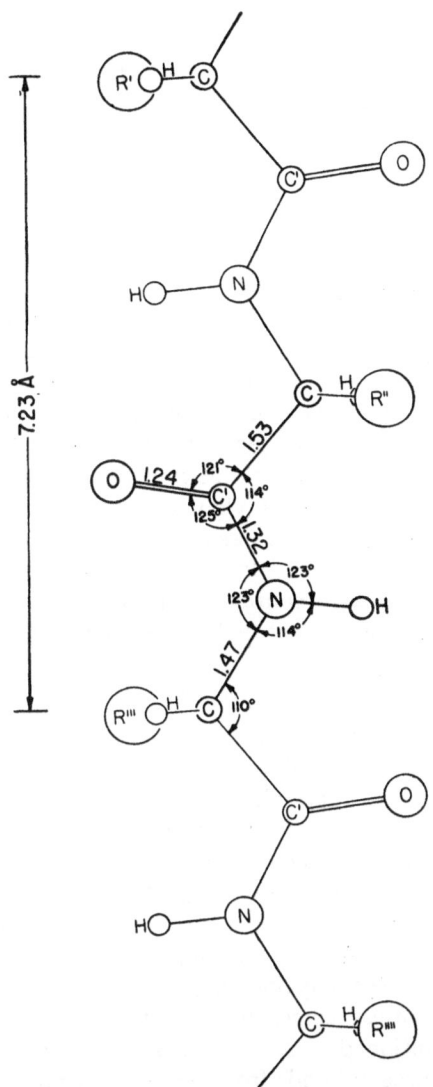

Fig. 14. A drawing of a fully extended polypeptide chain showing the dimensions of the amide group derived from X-ray analyses of crystals. [From: Proc. Roy. Soc. (London) **B 141**, 10 (1953).]

b) Effects of Resonance.

As a result of resonance between the two structures (I) and (II) in the ratio 60 : 40, as indicated by the interatomic distances, the C'—N bond assumes a significant amount of double-bond character, which would stabilize a planar configuration of the amide group. Nearly all of the structure determinations of amides have verified the planarity of the amide group to within 0.03 Å. In α-glycylglycine and N,N'-di-glycylcystine the carbonyl carbon atom and the three atoms bonded to it are within 0.03 Å. of a common plane, but the position of the α-carbon

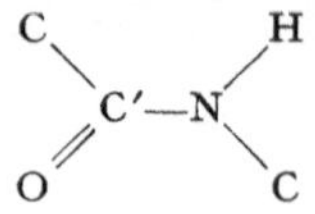

(I.) 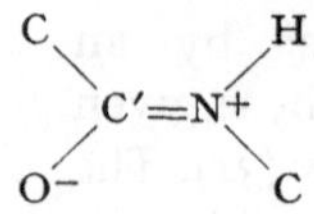   (II.)

atom bonded to the nitrogen atom corresponds to a rotation of 5 to 6° around the C'—N peptide bond. The amide group was found to be planar to within the rather wide limits of error of the determination of the structure in sodium and potassium benzyl penicillins [Crowfoot, Bunn, Rogers-Low, and Turner-Jones (39)]. A small, but perhaps real deviation from planarity has been found in a recent more accurate re-investigation of potassium benzyl penicillin [Pitt (100)]. The reported structure of cysteylglycine-sodium iodide (47) places the α-carbon atom in such a position as to correspond to a rotation of more than 40° around the C'—N bond from the planar configuration; it seems likely that this deviation in configuration of the amide group from that found in other substances is not real, but is the result of an error in the atomic parameters.

c) Properties of N—H····O Hydrogen Bonds.

The hydrogen bond is an important structural feature of all sub-stances related to proteins. The hydrogen atoms attached to nitrogen atoms in crystals of amino acids, peptides, and related substances are almost without exception involved in the formation of N—H····O hydrogen bonds. The observed dimensions of these bonds are given in Table 3 [Donohue (43); Corey and Pauling (33)]. Almost all of the N—H····O distances are within 0.12 Å. of 2.79 Å. A few exceptionally large values (3.10 Å. in L-threonine, 3.17 Å. in L-hydroxyproline, 3.03 Å. in N-acetylglycine, 3.07 Å. in β-glycylglycine, 3.03 Å. in glycylasparagine) doubtless correspond to strained hydrogen bonds. Donohue (43) has mentioned that there is greater than average deviation of these longer hydrogen bonds from the tetrahedral angle with a C'—N bond; moreover, in some of the crystals it can be seen that shortening the long hydrogen bond to a normal value would lead to strain with respect to other structural features.

Almost without exception the oxygen atom involved in an N—H····O hydrogen bond lies close to the apparent N—H direction. Presumably the greatest stability occurs when the line N—H····O is straight.

Although it might be anticipated from consideration of the partial covalent character of the H····O interaction [Pauling (79)] that the

Table 3. Dimensions of N—H····O Hydrogen Bonds Found
in Crystals of Amino Acids and Peptides.

Crystal	N—H····O (Å.)	Angle C—N····O	
		Angle found	Deviation from 110°
DL-alanine	2.80	105°	5°
	2.84	103	7
	2.88	116	6
L-threonine	2.80	98	12
	2.90	116	6
	3.10	132	22
DL-serine	2.79	99	11
	2.81	98	12
	2.87	121	11
L-hydroxyproline........	2.69	102, 113	8, 3
	3.17	81, 133	29, 23
N-acetylglycine	3.03	132	22
β-glycylglycine*	2.68	100	10
	2.80	115	5
	2.81	99	11
	3.07	131	21
α-glycylglycine..........	2.67	118	8
	2.77	114	4
	2.67	88	22
	2.75	117	7
N,N′-diglycylcystine	2.75	84.5	25.5
	2.89	129.8	19.8
	2.75	111.8	1.8
glycyl-L-asparagine	2.88	108.9	1.1
	2.93	111.3	1.3
	2.88	112.5	2.5
	2.86	116.8	6.8
	3.03	132.8	22.8
	2.75	117.9	7.9
Most probable value......	2.79 ± 0.12		

favored orientation of the N—H····O hydrogen bond about an oxygen
atom of an amide group would be in the plane of the amide group and
at an angle of about 125° with the C′—O bond, there is little evidence
that this orientation is at all favored with respect to others, and at the
present time it may be assumed that the oxygen atom may form a hydrogen
bond in any direction.

The planar amide group has been found to have the *trans* configuration
in all simple substances containing the amide group except cyclic compounds such as cyanuric acid, parabanic acid, and diketopiperazine, in
which the *cis* configuration is required by the structure of the molecules
themselves. MIZUSHIMA and co-workers (70) have obtained dipole-moment

---

* Two-dimensional analysis.

evidence that indicates that the *trans* configuration is significantly more stable than the *cis* configuration for this group. From infrared absorption studies of amides and N-substituted amides Badger and Rubalcava (9) concluded that the *trans* configuration may well be stabilized by more than 2 kcal mole^{-1}.

d) Estimations of Stabilization and Strain Energies.

The simple structures that have been determined do not in general conform exactly to the structural principles described above; small deviations in the length and straightness of hydrogen bonds and planarity of the amide group, in particular, are found to be present. The study of the structures suggests that these deviations result from the geometrical impossibility of finding a structure for a particular substance that conforms exactly to the structural principles. The stable structure for the crystal is the geometrically possible structure that minimizes the free energy of the substance; as an approximation we may say that it minimizes the strain energy resulting from deviation from an ideal structure corresponding exactly to the best configuration for each structural feature.

Estimates of the magnitude of the strain energy have been made by Pauling and Corey (85, 81, 89). The results of these estimates may be presented by a statement of the amount of deviation from the favored values of the structural parameters that cause instability amounting to 0.1 kcal mole^{-1}. The reliability of the estimates is in general not high; but it seems likely that the strain energy is not in error by as great a factor as 2.

The known values of the force constants of bonds show that the stretching or compression of the single bonds to the α-carbon atom by 0.02 Å. and of the conjugated bonds C'—O and C'—N by 0.01 Å. corresponds to a strain energy of 0.1 kcal mole^{-1}. The same strain energy corresponds to a deviation of 3° in bond angles from the favored values.

The resonance energy stabilizing the planar amide group is about 30 kcal mole^{-1}. A distortion of the planar amide group by rotating the planes of the two ends of the group to form the dihedral angle ∂ gives rise to the strain energy 0.1 kcal mole^{-1} when ∂ has a value of about 3°.

The best estimate that has been made of the strain energy of the N—H····O hydrogen bonds is based on a value calculated for the O—H····O hydrogen bond, which has length 2.76 Å., and presumably is similar in its properties to the N—H····O bond. The compressibility of ice, 12×10^{-6} cm^2 kg^{-1}, corresponds to the strain energy 0.1 kcal mole^{-1} for stretching or compression of the hydrogen bond by 0.09 Å. It must, of course, be recognized that similar changes in the length of the hydrogen bond might also result from change in the electric charges or other characteristics of the atoms involved; in particular, the formation of a hydrogen bond by the oxygen atoms of an amide group increases the hydrogen-bond-forming power of the hydrogen atom of the same amide group.

A rough estimate of the strain energy of bending the hydrogen bond has been made, according to which a bend of the angle N—H----O by 6°, from the value 180°, leads to the strain energy 0.1 kcal mole⁻¹.

An inconclusive discussion has been given of the favored orientations around the single bonds connecting the α-carbon atom with the amide group (89). Experimental evidence indicates that some orientations are favored relative to others, and that the difference in energy of favored and unfavored orientations is of the order of magnitude of 1 kcal mole⁻¹. The favored orientations are those in which one of the three other single bonds formed by the α-carbon atom lies in the plane of the amide group.

The energy of stabilization of a structure through electronic van der Waals interaction is great enough to require that stable structures be essentially close-packed. A rough value for the van der Waals energy can be obtained by application of the LONDON equation,

$$E = -\sum 38 \, \frac{R_A \, R_B}{r_{AB}{}^6} \text{ kcal mole}^{-1}.$$

Here the summation is to be taken over pairs of atoms, A and B, in the structure. The symbol r_{AB} is the distance between the two atoms A and B, in Ångströms, and R_A and R_B are the mole refractions of the atoms or groups A and B. The equation is obtained from approximate second-order perturbation theory with use of the average value 14 ev for excitation energy of the atoms or groups. The mole refraction R of the amide group including the α-carbon atom can be taken as 13 cm³.

The uncertainty in application of the equation results mainly from the neglect of van der Waals repulsion energy; this equation leads to a greater and greater stabilization energy as the atoms get closer and closer together, whereas in fact when the interatomic distance is equal to the sum of the van der Waals radii there is a significant repulsion energy, which cancels out a large fraction of the van der Waals attraction energy. The equation has been applied to the α helix and the γ helix, described in the following Section. With use of the accepted values of van der Waals radii for atoms, the α helix can be described as essentially close-packed, with no hole along the axis of the helix, whereas there is a hole, approximately 2 Å. in diameter, along the axis of the γ helix. Application of the LONDON equation shows that the instability resulting from the presence of this hole amounts to 4 kcal mole⁻¹ per residue, in comparison with the α helix (92).

II. Helical Configurations of Polypeptide Chains.

It was not until 1950 that any precisely described configurations for polypeptide chains in proteins were proposed. In that year BRAGG, KENDREW, and PERUTZ (*23*) published atomic coordinates for more than a dozen configurations, and discussed them in connection with their X-ray studies of hemoglobin and myoglobin. Several of their structures are helical structures. The atomic coordinates conform reasonably well with the structural principles formulated above, except that of planarity of the amide group; many of them involve significantly large deviations from planarity.

At the same time PAULING, COREY, and BRANSON were carrying out an investigation of possible helical configurations of the polypeptide chain that satisfy reasonably well all of the assumed structural principles, including the planarity of the amide group and the formation of N—H⋯O hydrogen bonds by all of the N—H and C—O groups in the polypeptide chains (*82, 94*). These investigators discovered two helical structures in which all of the amino-acid residues are equivalent to one another, except for differences in the nature of the side-chains, and all of the structural principles are well satisfied except (for one structure) the principle of closest packing. These two structures, which they named the α helix and the γ helix, are described in the following paragraphs, together with some other helical structures that have been discussed during recent years. The α helix seems to be the most widely occurring configuration of polypeptide chains in proteins.

1. The α Helix.

A diagrammatic representation of the α helix is given in *Figure 15*. The end of the polypeptide chain in each segment at the left in this Figure is to be continued in the next segment, beginning at the right; that is, this diagram is to be considered as folded onto the surface of a cylinder, with vertical axis. A drawing of the structure is shown in *Figure 16*, and a projection onto a plane perpendicular to the helical axis in *Figure 17*.

The α helix has about 3.6 residues per turn, and the smallest ring of atoms containing covalent bonds and one hydrogen bond involves 13 atoms. This ring is obtained by following along the polypeptide chain from, say, the nitrogen atom of an amide group until the carbon atom of the third more distant amide group is reached, which lies nearly directly above it, after one turn of the helix, and then tracing down along the hydrogen bond C—O H—N. BRAGG, KENDREW, and PERUTZ (*23*) have introduced a nomenclature to describe simple helical structures, in which the number of residues per turn is written, with the

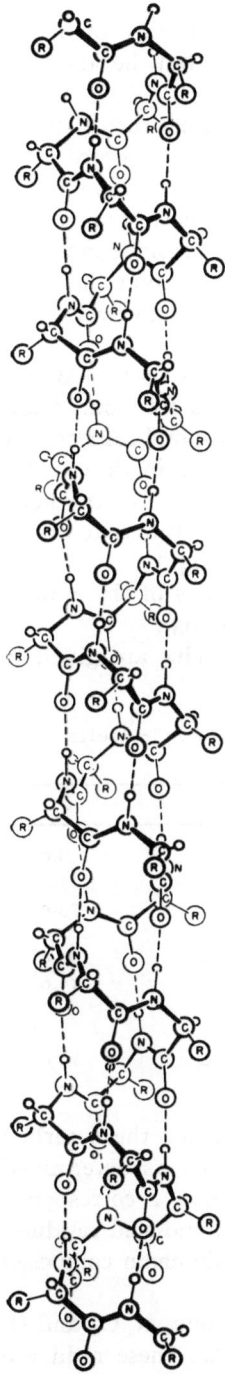

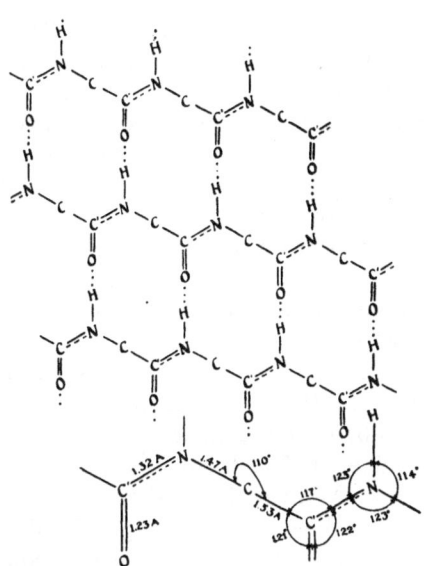

Fig. 15 (*above*). A diagrammatic representation of the α helix.
[From: Proc. Nat. Acad. Sci. (U. S. A.) **37**, 235 (1951).]

Fig. 16 (*left*). A drawing of the α helix.

number of atoms in a hydrogen-bonded ring
as subscript. With this nomenclature the α
helix is given the symbol 3.6$_{13}$.

It is interesting that BRAGG, KENDREW, and
PERUTZ did not discuss any helical structures
except those with an integral number of amide
groups per residue. A discussion of helical structures of this sort, without atomic coordinates,
however, had been published earlier by HUGGINS

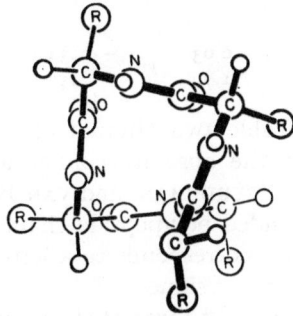

Fig. 17. A projection of the α helix on a plane perpendicular to
the helical axis.

(50); Huggins also mentioned the possibility of the existence of helical structures with nonintegral amide groups per turn, but he did not describe any such structure in detail.

The α helix was first predicted to have 3.69 residues per turn, a fiber-axis length per turn (pitch of the helix) of 5.44 Å., and, accordingly, a length per residue of 1.47 Å. along the fiber axis. Later it was pointed out (83) that the fiber-axis length per residue 1.47 Å. corresponds to a hydrogen-bond distance 2.75 Å., and that the length per residue increases by 0.01 Å. for every 0.03 Å. increase in the hydrogen-bond distance. The favored value 2.79 Å. corresponds accordingly to 1.48 Å. for the fiber-axis length per residue.

The number of residues per turn is determined largely by the C'—αC—N bond angle at the α-carbon atom: it varies from 3.60 for bond angle 108.9° to 3.67 for bond angle 110.8°. These ratios correspond respectively to 18 residues in five turns of the helix, and 11 residues in three turns of the helix. With hydrogen-bond length 2.79 Å., the pitch of the helix with 3.60 residues per turn is predicted to be 5.33 Å., and that of the helix with 3.67 residues per turn is predicted to be 5.44 Å.

Details of the trigonometric calculation of dimensions of the α helix have been published by Low and Grenville-Wells (63).

Atomic coordinates for the 18-residue 5-turn α helix are given in *Table 4*.

Table 4. Atomic Coordinates for the 18-Residue 5-Turn α Helix x, y, z, ϱ in Å.

Atom	x	y	z	ϱ	θ
C.........	0.00	0.00	0.00	2.29	0.0°
N	1.16	0.00	0.89	1.59	27.8°
C'	2.42	0.00	0.44	1.61	73.8°
O.........	2.69	0.00	— 0.76	1.74	82.0°
C*........	3.52	0.00	1.50	2.29	100.0°
βC_1	— 1.33	0.20	0.76	3.34	— 17.6°
or					
βC_2	— 0.03	— 1.34	— 0.76	3.34	17.6°
Axis	1.76	1.47	—	0.00	—

In this Table two alternative positions are listed for the β-carbon atom. With the absolute configuration of amino acids as determined by Bijvoet, Peerdeman, and van Bommel (17), position *1* corresponds to a right-handed polypeptide chain composed of L-amino-acid residues, and position *2* corresponds to a left-handed polypeptide chain composed of L-amino-acid residues.

The distances of three of the atoms of the amide group, N, C', and O, from the axis of the helix are 1.57, 1.61, and 1.76 Å. These radii are only about 0.1 Å. greater than the accepted van der Waals radii of the atoms, and accordingly the α helix can be considered to be essentially close-packed, without any cavity along its axis.

A value for the diameter of a polypeptide chain of a protein with the α-helical configuration can be calculated from the properties of proteins. The average residue weight in most proteins is about 105, and the density is about 1.33 g cm^{-3}. These values correspond to a volume of 131 Å.^{3} per residue. We may take the axial length per residue as 1.48, which leads to 89 Å.^{2} for the cross-sectional area, and to a diameter of 10.6 Å. We may accordingly predict that the polypeptide chain of proteins with the configuration of the α helix has the approximate diameter 10.5 Å. The van der Waals radius of the α helix may vary from a minimum of about 4.5 Å., in the neighborhood of glycine residues, to a maximum considerably greater than 5.3 Å., depending upon the configuration of long side-chains of the larger amino-acid residues.

2. The γ Helix and Other Helixes.

The γ helix, shown in *Figure 18*, has about 5.2 residues per turn, and a fiber-axis length per residue of about 0.95 Å. The γ helix can be given the symbol 5.2_{17}. It is a satisfactory structure with respect to all structural features except the van der Waals stabilization through close packing of atoms. The distances of the atoms N, C′, and O of the amide group from the axis of the helix are 2.67, 2.66, and 2.65 Å., respectively. These values are on the average 1.0 Å. greater than for the α helix. There is accordingly a hole along the axis of the γ helix, with diameter approximately 2 Å. This diameter is less than the van der Waals diameter of the oxygen atom; it is accordingly

Fig. 18. A drawing of the γ helix. [From: Proc. Nat. Acad. Sci. (U.S.A.) **37**, 205 (1951).]

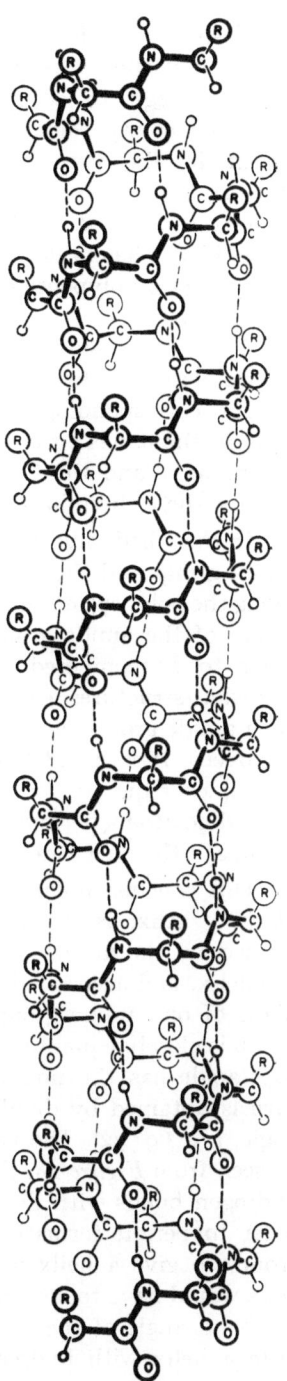

impossible for the cavity to be filled by water molecules, or by other molecules (helium atoms and hydrogen molecules alone are small enough to occupy such a cavity). Accordingly, we conclude that the cavity remains empty. Application of the London equation, given in the preceding Section (p. 203), to the γ helix and the α helix has led to the conclusion that there is a great instability of the γ helix because of the large deviation from close packing of the atoms. No evidence has been obtained for the existence of the γ helix in proteins or synthetic polypeptides, and it seems likely that the structure is unsatisfactory.

The α helix and the γ helix differ in an interesting qualitative respect. If we describe the positive direction of a polypeptide chain as the direction corresponding to the sequence NH · CO · CHR . . . , then we see that in the α helix the N—H group is directed in the positive direction of the axis, and the C=O group is directed negatively, whereas in the γ helix these directions are reversed.

In the α helix hydrogen bonds are formed between the NH group of an amide group and the CO group of the amide group third removed from it along the polypeptide chain; in the γ helix between an NH group and the CO group of the amide group fifth removed from it along the chain; this leads to 13-membered and 17-membered rings, respectively. We may ask what structures would be formed if hydrogen bonds occurred between each amide group and the second removed along the chain, or the fourth removed. It is found that the amount of deviation from the structural principles required for the corresponding helixes of the γ series, with 14 and 20 atoms per hydrogen-bonded ring, respectively, is so great as to eliminate the structures from consideration.

This conclusion can easily be reached by consideration of the nature of the γ helix, as shown in Figure 18, p. 207. It is seen that the plane formed by the atoms N—αC—C′ is nearly perpendicular to the axis of the helix, and accordingly that the dihedral angle formed by the vertical plane of one amide group and the vertical plane of an adjacent amide group is closely equal to the N—αC—C′ bond angle of the α-carbon atom. This angle has the normal value 109° 28′; the number of residues per turn is obtained by dividing 360° by the supplement of the tetrahedral angle, i. e. 69° 32′. The value obtained is 5.18 residues per turn. As can be seen from *Figure 18*, this places each amide group in position to form hydrogen bonds with the fifth amide group from it along the chain. In order to permit formation of a hydrogen bond with the fourth amide group, to give a helix with a 14-membered hydrogen-bonded ring, the helix would have to have about 4.2 residues per turn, which would require the bond angle of the α-carbon atom to be about 94°; and similarly to form a helix with hydrogen bonds to the sixth amide group removed the bond angle of the α-carbon atom would have to be about 122°. The large strain energy associated with these bond angles eliminates these structures.

Two other structures of the α type are 2.2_{10} and 4.4_{16}. The first of these, which is topologically similar to structures discussed by HUGGINS (50) and BRAGG, KENDREW, and PERUTZ (23), has been treated in detail by DONOHUE (44), and the second has been discussed by LOW and BAYBUTT (62) and LOW and GRENVILLE-WELLS (63). Each of them involves only rather small deviations from the optimum structural features given in the preceding section. It seems not unlikely that the 2.2_{10} helix is twisted too tightly, and that the atoms of the amide groups are forced into van der Waals contacts with one another that are too small, resulting in instability through van der Waals repulsion. The 4.4_{16} helix has an axial cavity about 1 Å. in diameter, and some strain in the α-carbon bond angle and the hydrogen bonds; the total amount of strain has been estimated to be 1 kcal per mole. A discussion of all factors affecting the stability of these helixes has been published by DONOHUE (44).

III. Sheets of Polypeptide Chains.

The idea that some fibrous proteins contain sheets of nearly completely extended polypeptide chains has a long history. In 1923 it was suggested by BRILL (25), on the basis of a study of the X-ray diagrams of silk, that silk contains long polypeptide chains, and in 1928 MEYER and MARK (68) pointed out that the fiber-axis repeat of silk, 7.0 Å., corresponds to the length of two residues in a nearly fully extended polypeptide chain, as calculated with the use of the approximately known values of inter-atomic distances and bond angles. MEYER and MARK also suggested that the peptide chains were strongly attracted to one another by forces between the CO and NH groups of adjacent chains. Then in 1931 it was discovered by ASTBURY and STREET (6) that hair, wool, and other fibers of the keratin class can be stretched into an extended form, which they named "β keratin", and they suggested that the polypeptide chains in β keratin are in the nearly completely extended configuration. The recognition of the interaction of the carbonyl and imino groups as involving the formation of N—H----O hydrogen bonds was made in 1936 by MIRSKY and PAULING (69) and HUGGINS (49). It was not until 1951 that it was recognized that the hydrogen-bonded sheet in which all of the amide groups are coplanar and the polypeptide chains are completely extended, represented in *Figure 19* (p. 210), is rendered impossible for all proteins or polypeptides except polyglycine by van der Waals repulsion of the side-chains, as shown in the Figure [PAULING and COREY (89)].

The two structurally most satisfactory sheet configurations of nearly completely extended polypeptide chains are the antiparallel-chain pleated

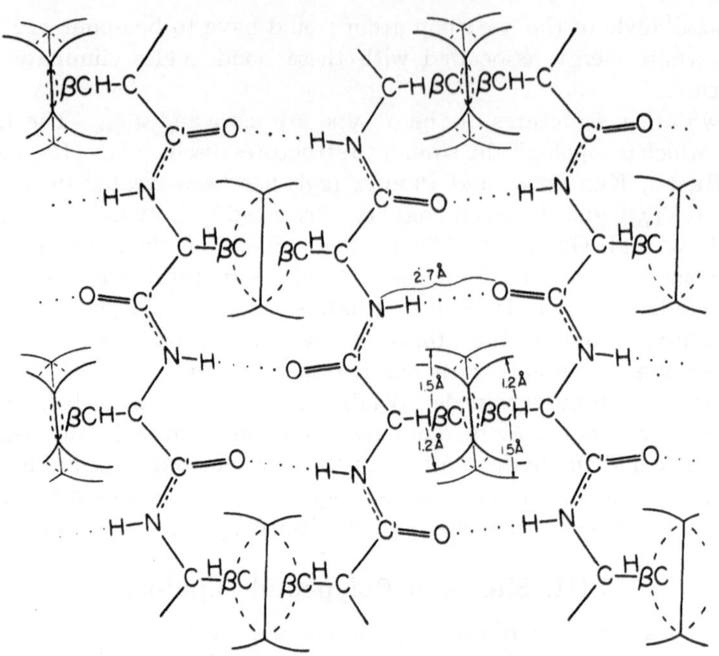

Fig. 19. A diagrammatic representation of a hydrogen-bonded structure of fully-extended polypeptide chains with alternate chains oppositely oriented, showing steric hindrance between β-carbon atoms of adjacent chains. [From: Proc. Nat. Acad. Sci. (U. S. A.) **37**, 729 (1951).]

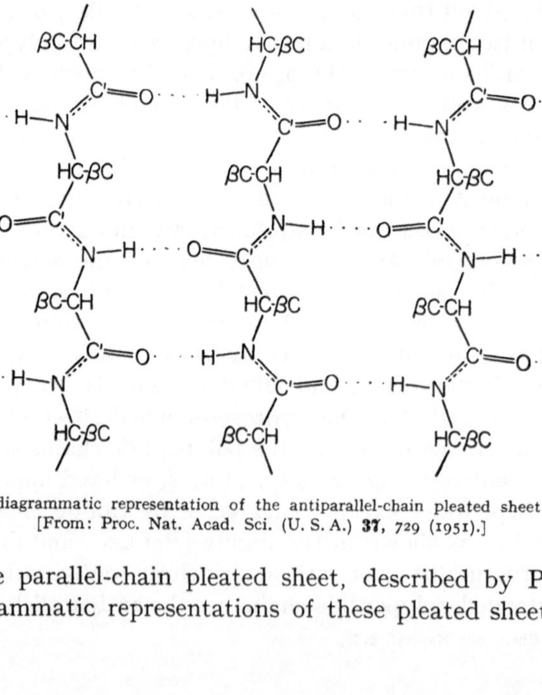

Fig. 20. A diagrammatic representation of the antiparallel-chain pleated sheet structure. [From: Proc. Nat. Acad. Sci. (U. S. A.) **37**, 729 (1951).]

sheet and the parallel-chain pleated sheet, described by PAULING and COREY. Diagrammatic representations of these pleated sheets are given

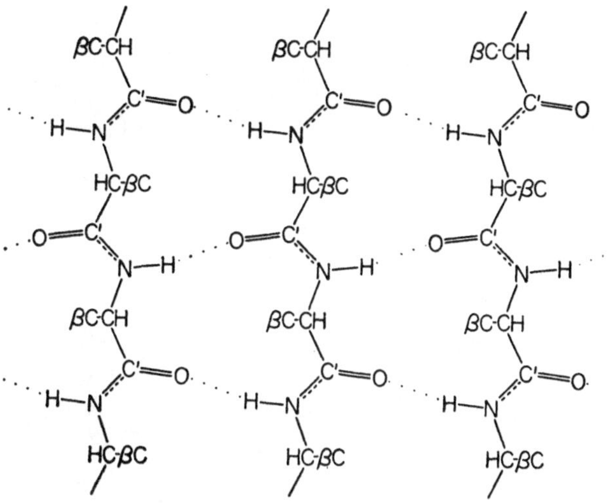

Fig. 21. A diagrammatic representation of the parallel-chain pleated sheet structure.
[From: Proc. Nat. Acad. Sci. (U. S A.) **37**, 729 (1951).]

Table 5. Atomic Coordinates for Two Pleated Sheets.

A. Antiparallel-Chain Pleated Sheet.

$a_0 = 9.50$ Å., $b_0 = 7.00$ Å., $c_0 = 1.00$ Å. (assumed).
Four atoms in x, y, z; $\bar{x}$, $\frac{1}{2} + y$, $\bar{z}$; $\frac{1}{2} - x$, $\bar{y}$, z;
$\frac{1}{2} + x$, $\frac{1}{2} - y$, $\bar{z}$.

Atom	x	y	z
C.......	0.034	− 0.005	− 0.70
N	− 0.030	0.173	− 0.20
C′	0.051	0.320	0.21
O.......	0.180	0.326	0.22
βC......	0.024	− 0.005	− 2.24

B. Parallel-Chain Pleated Sheet.

$a_0 = 4.85$ Å., $b_0 = 6.50$ Å., $c_0 = 1.00$ Å. (assumed).
Two atoms in x, y, z; $\bar{x}$, $\frac{1}{2} + y$, $\bar{z}$.

Atom	x	y	z
C.......	0.012	0.000	− 0.98
N	− 0.066	0.186	− 0.26
C′	0.118	0.315	0.28
O.......	0.371	0.295	0.25
βC......	− 0.093	0.014	− 2.45

in *Figures 20* and *21*, and drawings of the structures are given in *Figures 22* and *23* (p. 212). The atomic coordinates (91) are presented in *Table 5*.

In these pleated-sheet configurations the polypeptide chains are bent in such a way that the trace of the plane of each amide group and

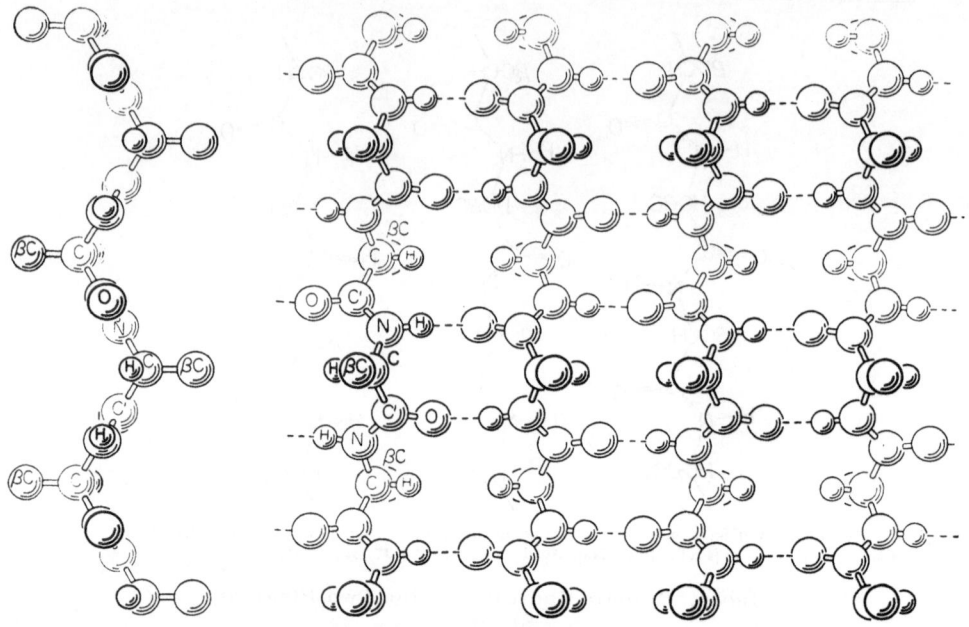

Fig. 22. A drawing representing the antiparallel-chain pleated sheet structure.
[From: Proc. Nat. Acad. Sci. (U. S. A.) **37**, 729 (1951).]

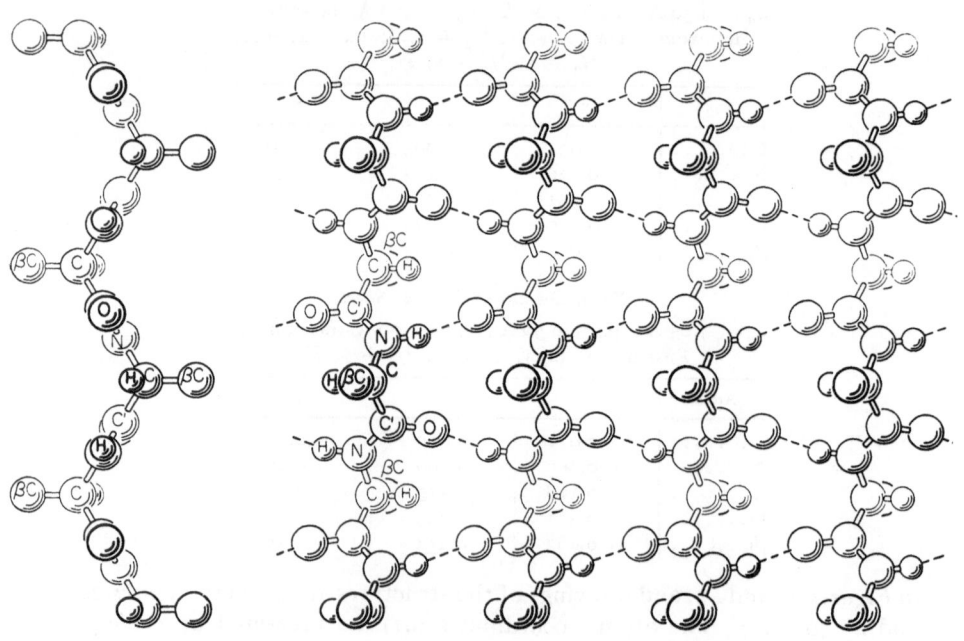

Fig. 23. A drawing representing the parallel-chain pleated sheet structure.
[From: Proc. Nat. Acad. Sci. (U. S. A.) **37**, 729 (1951).]

the median plane of the sheet is horizontal; that is, perpendicular to the direction of the axis of the polypeptide chain. The deformation results in an orientation of the α-carbon atom that provides room for the side-chains, without steric hindrance; the direction of the bond αC—βC is nearly perpendicular to the plane of the sheet, in each case. The amount of pleating of the sheet that achieves the minimum strain in the hydrogen bonds is somewhat less for the antiparallel-chain pleated sheet than for the parallel-chain pleated sheet, leading to a predicted fiber axis length of 7.00 Å. for the antiparallel-chain pleated sheet and 6.50 Å. for the parallel-chain pleated sheet.

Figures 22 and *23* are drawn in such a way as to represent pleated sheets formed of *D*-amino acid residues; in order to represent proteins, the mirror images of these structures must be taken.

In addition to these two pleated sheets, a polar pleated sheet of *trans* amide groups has been described, which is, however, satisfactory only for polypeptide chains with alternate *D* and *L* residues, or with glycine residues alternating with *D* or *L* residues of other amino acids (*85*).

Two other structures, the antiparallel-chain rippled sheet and the parallel-chain rippled sheet, are obtained by reflecting alternate chains in *Figures 22* and *23* in the plane of the paper; they are represented topologically by the same diagrams, *Figures 20* and *21*, as the corresponding pleated sheets. The rippled-sheet structures might be satisfactory for an equimolal mixture of *D* polypeptides and *L* polypeptides, or for polyglycine. The predicted values of the fiber axis identity distance are the same as for the corresponding pleated sheets (*91*). Two pleated-sheet configurations of polypeptide chains involving both *cis* and *trans* amide groups have also been formulated (*90*).

IV. The Structure of Fibrous Proteins.

The interpretation of X-ray diffraction photographs of proteins and of other information bearing on the structure of these substances is not an easy task. Since the first X-ray photographs of proteins were made, over thirty years ago, by HERZOG and JANCKE (*48*), hundreds of papers have been published in this field, and scores of protein structures have been proposed. The path to the present point, which we hope is somewhat near the goal, has been a long and tortuous one.

The following discussion of the structure of proteins is far from complete. In it little mention is made of proposed protein structures other than those that have been precisely described, by the assignment of atomic coordinates. No effort has been made to present the large body of incomplete structural information, such as the structural classification of proteins with use of their X-ray diagrams, and the identity distances, dimensions of unit cells, and other structural information that can be obtained from the X-ray diagrams when a more complete analysis cannot be made. We have attempted instead to give in a few pages the principal arguments that have led to the assignment of certain well-defined configurations to the atoms of the polypeptide chains (except for side-chain atoms) in some proteins, together with some indication of the reliability of the assignments and the nature of the problems that must still be solved.

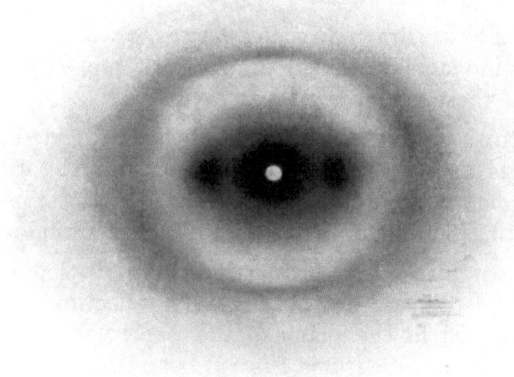

Fig. 24. An X-ray diffraction photograph of horse hair, a typical α-keratin diagram (fiber axis vertical).

Herzog and Jancke (48) discovered three structural types of fibrous proteins, characterized by their X-ray diagrams. The diagrams obtained by passing an incident beam of monochromatic X-radiation through the fibrous protein oriented with the fiber axis perpendicular to the X-ray beam (fiber axis along the vertical, in the figures) are shown as *Figures 24, 25,* and *26,* of horse hair, silk, and kangaroo-tail tendon, respectively. The valuable investigations of Astbury and his collaborators have led to the discovery that very many fibrous proteins in their natural state give diagrams similar to that of *Figure 24;* Astbury has named this diagram the "α-keratin diagram", and the corresponding structure is called the "α-keratin structure". Astbury and Woods (8) discovered that when hair or another α-keratin protein is stretched, after treatment with steam or alkali, it may undergo great extension, of about 100 per cent, with a great change in the X-ray diagram. The new diagram, called the "β-keratin diagram", and corresponding to the "β-keratin structure", is somewhat similar to that of silk, shown in *Figure 25,* although not so well defined. Gelatin and collagen are the principal representatives of the collagen structure, which gives the characteristic X-ray diagram shown in *Figure 26*.

It is our opinion that at the present time the basic polypeptide-chain configurations are known for the α-keratin proteins and for silk and the β-keratin proteins, but not for collagen and gelatin.

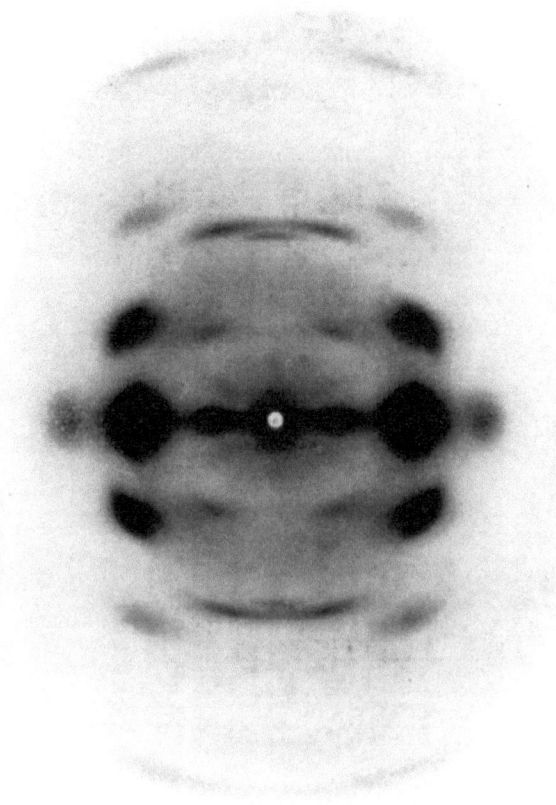

Fig. 25. An X-ray photograph of silk fibroin (fiber axis vertical).

1. The Structure of the α-Keratin Proteins.

Proteins that give X-ray diagrams like that in *Figure 24* have been grouped by ASTBURY under the name the keratin-myosin-epidermin-fibrinogen class. Hair, wool, horn, and fingernail produce nearly identical X-ray diagrams. The diagram of porcupine quill is characterized by much greater detail, but is recognizable as an α-keratin diagram (*64*). HERZOG

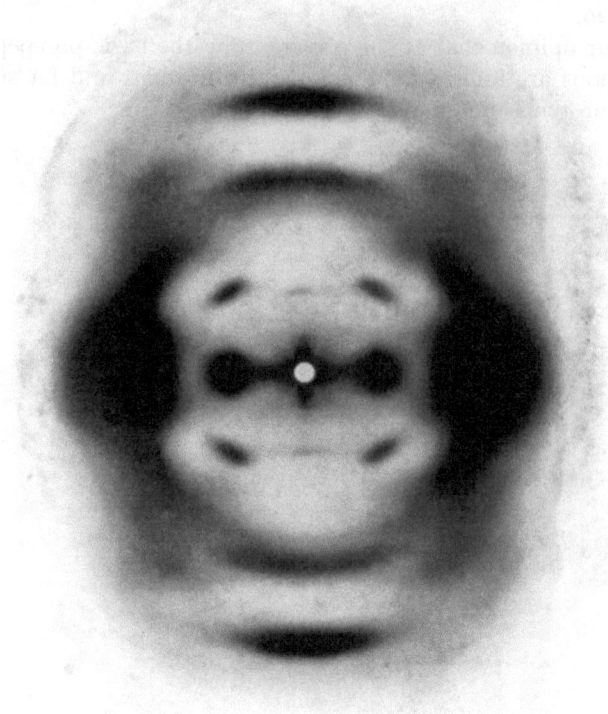

Fig. 26. An X-ray photograph of kangaroo-tail tendon, a highly oriented form of collagen (fiber axis vertical).

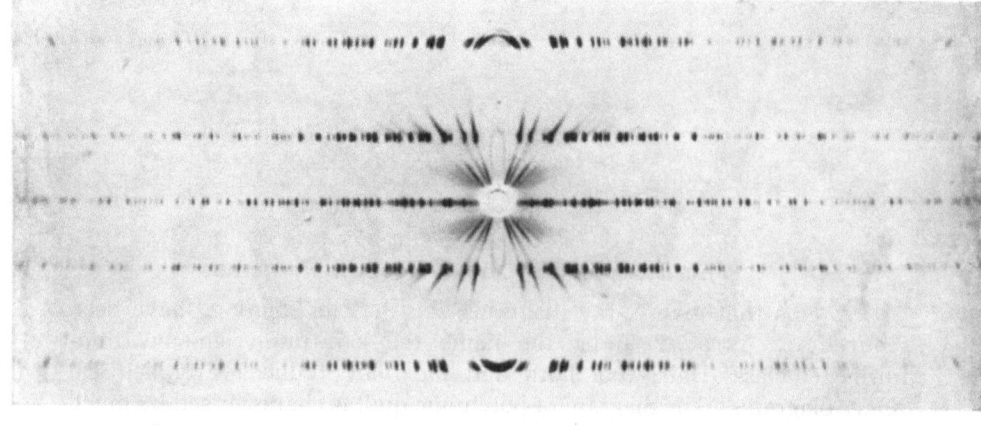

Fig. 27. An X-ray diffraction photograph of a crystal of glycylasparagine rotated around its *a* axis.

and JANCKE (48) obtained essentially α-keratin photographs from muscle, and the principal proteins of muscle (actomyosin, myosin) have been found to produce this diagram (19). Epidermis (41, 10), its principal protein epidermin (107, 107a), fibrinogen and fibrin (10), and bacterial flagella (7) are in this class.

An idea of the difficulty of discovering the structure of a fibrous protein with use of the X-ray diagrams can be obtained by the comparison of the diagram for hair, *Figure 24*, and an X-ray photograph of a crystal of a simple substance, such as glycylasparagine, *Figure 27*. Hundreds of spots, representing X-ray diffraction maxima, are to be seen on the latter photograph; a complete series of photographs for a crystal of this sort may provide values of the intensities of as many as one thousand non-equivalent X-ray reflections. In the glycylasparagine molecule there are thirteen atoms other than hydrogen, and the job of the X-ray crystallographer is to determine the values of three parameters, the x, y, and z coordinates relative to the origin of the unit cell, for each of the atoms. The one-thousand experimental intensity values provide enough information to permit the accurate evaluation of these thirty-nine parameters. A protein such as keratin is far more complicated than the glycylasparagine crystal, and its X-ray diffraction pattern provides a much smaller amount of information than is provided by the X-ray photographs of the simple substance.

a) Some Interpretations of the X-ray Pattern.

It may be illuminating to discuss some of the erroneous interpretations that were given to the X-ray diagram of α-keratin during the period 1930 to 1950. These interpretations were based upon the principal features of the X-ray diagram [the strong meridional reflection, above and below the central image in *Figure 24* (p. 214), and the two strong equatorial reflections], the mechanical properties, especially the extensibility, and other properties, such as the infrared absorption spectrum.

The BRAGG spacing of the strong meridional reflection is 5.15 Å. The similarity of the X-ray diagram of stretched hair to that of silk indicated a similar configuration of the polypeptide chains in stretched hair and in silk, and this configuration had been described as essentially that of a fully extended or nearly extended polypeptide chain by MEYER and MARK in 1928 (see the following Section). ASTBURY and STREET (6) accordingly suggested that in the α form the polypeptide chains of hair and other proteins of this class are folded, as shown in *Figure 28*. In this proposed structure for α-keratin there are three amino-acid residues in the length 5.15 Å. along the fiber axis. The fiber-axis length of three residues in β-keratin is about 10.0 Å., so that the α β transition would, according to this proposal, be accompanied by an extension of about 100 per cent.

Another structure, represented in *Figure 29*, was formulated by Huggins (*50*), and has been discussed by Zahn (*125*), Simanouti and Mizushima (*114*), Ambrose and Hanby (*3*), and Robinson and Ambrose (*105*). This structure has two residues in the length 5.15 Å., and permits only 30 per cent extensibility to the β configuration. The structure described below, composed of α helixes, and

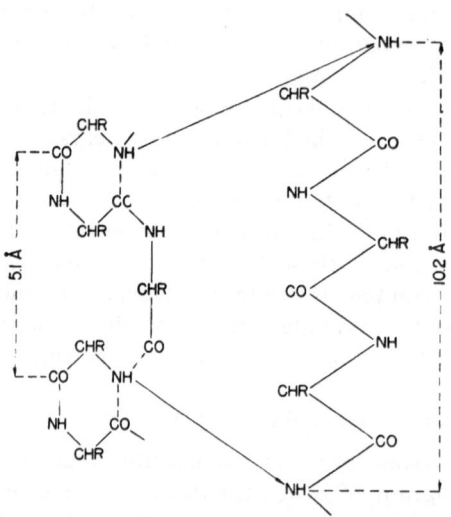

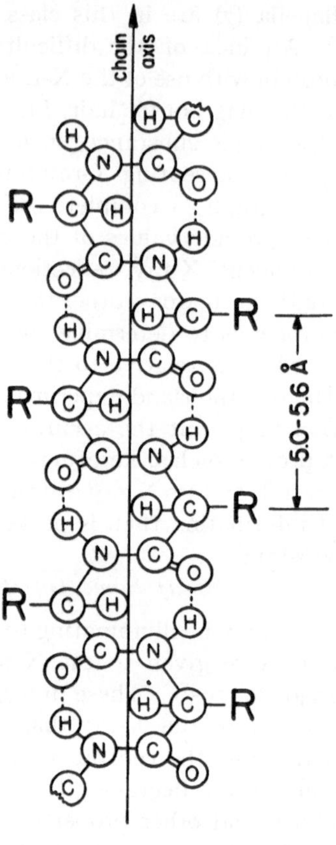

Fig. 28 (*above*). A diagrammatic representation of a method of folding of the polypeptide chains in hair and other proteins of the α-keratin class which was proposed in 1931.

Fig. 29 (*right*). An arrangement of the polypeptide chain which was proposed in 1942 for proteins of the α-keratin class.

now considered to be the correct one, places about 3.5 residues of each polypeptide chain in the axial length 5.15 Å., and the explanation of the occurrence of the strong 5.15 Å. meridional reflection on the X-ray photographs is a rather complicated one, which probably could not have been discovered from analysis of the X-ray diagrams alone, but required the prediction of the α helix from structural principles for its discovery. With 3.5 residues in the axial length 5.15 Å., this structure leads to a predicted extensibility of 126 per cent, in excellent agreement with the experimental value 127 per cent reported by Rudall (*108*) for myosin, epidermis, and fibrin.

When the α helix and the γ helix were formulated, on the basis of structural principles, the question arose as to whether one or the other of these helical structures, rather than the ribbon-like structures of *Figures 28* and *29*, might not represent the polypeptide chain in α-keratin.

The γ helix could be eliminated at once by the consideration of its diameter. The average residue weight of α-keratin proteins given by amino-acid analyses is about 110, and the density of the proteins is about

Fig. 30. A hexagonal packing of α helixes.

1.30 g./cm.³ From these values the volume occupied by one residue is calculated to be 141 Å.³. The axial length per residue predicted for the γ helix is 0.98 Å.; hence the average cross-sectional area of the molecule is 144 Å.². If the assumption is made that the molecule with the configuration of the γ helix has the form of a circular cylinder, as is indicated by the configuration, then the diameter of the cylinder is calculated to be 13.5 Å. It would be expected that the cylindrical molecules would pack together in hexagonal packing, as shown in *Figure 30*, and that the principal equatorial reflection would be the reflection (11·0), with interatomic distance equal to the distance between adjacent layers of molecules. The principal equatorial reflection, seen as the dark spots to the right and left of the central image in *Figure 24* (p. 214), is observed to have Bragg spacing about 9.8 Å.—the blackening extends over the region from 9 Å. to 11 Å. The corresponding distance between centers of the cylindrical molecules is larger by the factor $2/\sqrt{3}$; that is, it is about 11.3 Å. It is clearly impossible that the molecules with the configuration of the γ helix, and diameter about 13.5 Å., be present in α-keratin.

b) The Occurrence of the α Helix.

The corresponding calculation for the α helix, with axial length 1.48 Å. per residue, gives 11.0 Å. as the diameter of the molecule. This is close enough to the value indicated by the principal equatorial reflection to justify further consideration of the α helix as the configuration of the α-keratin proteins.

An arrangement of α helixes in hexagonal packing, as shown in *Figure 30*, does not, however, provide any explanation of the occurrence on the X-ray diagrams of the strong 5.15 Å. meridional reflection. The only strong meridional reflections that would be expected are those corresponding to 1.48 Å., equal to the axial length per residue, or to integral multiples of this spacing, in case that side-chains of different kinds were repeated in the regular way. It could be predicted that off-meridional reflections would occur on layer lines corresponding to a spacing equal to the pitch of the helix, predicted to be about 5.4 Å., and an effort was made to account for the 5.15 Å. reflection as resulting from the overlapping of off-meridional reflections from poorly oriented specimens (*86*).

A. *Synthetic polypeptides.* Support of the idea that the polypeptide chains in the α-keratin proteins have the configuration of the α helix was provided by the results of investigations of fibrous synthetic polypeptides. X-ray photographs of poly-γ-methyl-*L*-glutamate (*Figure 31*), poly-γ-benzyl-*L*-glutamate, and other synthetic polypeptides published by Bamford, Hanby, and Happey (*13*) are similar in a general way to the α-keratin photographs, but the 5.15 Å. meridional reflection is not present—there are instead some strong off-meridional reflections on layer lines corresponding to spacing about 5.4 Å. It was pointed out by Pauling and Corey (*84*) that the X-ray data for poly-γ-methyl-*L*-glutamate could be indexed with a hexagonal unit, corresponding to hexagonal packing of cylindrical molecules of the size expected for the α helix of this polypeptide, and with identity distance 27.5 Å. along the fiber axis. The strongest non-equatorial reflections lie on the fifth layer line, with effective spacing 5.5 Å., corresponding to the pitch of the

α helix, and weaker reflections lie on other layer lines. The identity distance represents five turns of the helix, which has 18 residues in five turns, or 3.60 residues per turn, which lies within the predicted limits 3.60 to 3.67. Moreover, measurements with polarized infrared radiation were reported by AMBROSE and ELLIOTT (2) to show that the C=O and N—H groups are oriented nearly parallel to the fiber axis, as required by the α helix. The same orientation of these groups was shown by the infrared measurements to occur in the α-keratin proteins also.

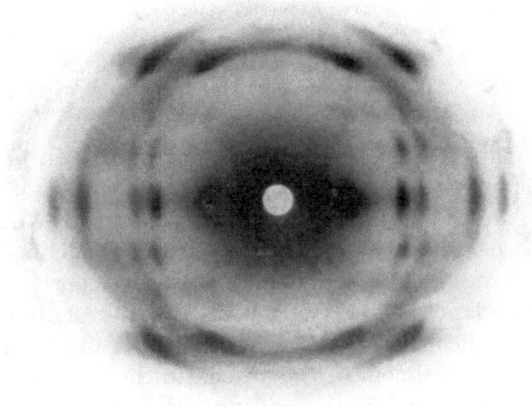

Fig. 31. An X-ray diffraction photograph of a highly oriented specimen of poly-γ-methyl-L-glutamate. [From: C. H. BAMFORD et al., Proc. Roy Soc. (London) **B 141**, 49 (1953).]

Very strong additional evidence for the occurrence of the α helix in synthetic polypeptides has been obtained through the investigations of COCHRAN, CRICK, and VAND (27), who derived a theoretical expression for the intensities of X-ray scattering by a helical molecule. The theory leads to the prediction of the relative intensities of the different layer lines in the X-ray photograph of a crystal containing helical molecules in parallel orientation, and the theoretical expression for an 18-residue 5-turn helix was found to be in good agreement with the observed pattern of 28 layer lines for poly-γ-methyl-L-glutamate (11). Another modification of poly-γ-methyl-L-glutamate was also discovered by BAMFORD and collaborators (12) which has 29 residues in 8 turns of the helix, corresponding to 3.625 residues per turn, and for this modification also there is good agreement between the layerline intensities and the values predicted by the theory of COCHRAN, CRICK, and VAND. A modification of poly-L-alanine with 47 residues in 13 turns (3.615 residues per turn) has also been reported recently (12 a).

B. The α-Keratin Proteins. The most convincing additional evidence for the presence of the α helix in the α-keratin proteins was provided through the efforts of Perutz (98). Perutz pointed out that the presence or absence of a reasonably strong meridional reflection with spacing about 1.48 Å., the axial length of the α-helix per residue, could be taken as strong evidence of the presence or absence of α helixes. The description of the α helix given by Perutz (97) is illuminating:

"If the α helix is compared to a spiral staircase with the residues as steps, then the height of each step is 1.5 Å. and the height of each turn 5.4 Å., making 3.6 steps per turn. It takes 18 steps or 5 turns until a step is found exactly in a vertical line above the starting point. Hence the true 'repeat' pattern is 18 × 1.5 = 27 Å. . . . . The 3.6-residue helix, if present in all proteins and polypeptides of the α type, should give rise to a reflection at 1.5 Å. spacing from planes perpendicular to the fiber axis, corresponding to the axial repeat of residues along the chain."

Although α helixes might be present and shifted in position relative to one another in such a way as to cause the intensity of this reflection to be zero, the likelihood of such a structure is not great. No such reflection had in 1950 been reported for hair, horn, fingernail, or any synthetic polypeptide. A meridional reflection had been reported for porcupine quill and venus-clam shell-closing muscle by MacArthur (64), and the presence of the reflection had been quoted by Pauling and Corey (86) in support of the assignment of the α helix to these proteins, but without emphasis of the importance of the evidence. Perutz found on investigation that the reflection was present on X-ray photographs of suitably oriented samples of poly-γ-benzyl-L-glutamate, hair, muscle, porcupine quill, and hemoglobin, and it has been mentioned by Perutz (97) that Astbury has observed it for epidermin, tropomyosin, fibrin, and bacterial flagella. The reflection has also been observed in poly-γ-methyl-L-glutamate and other synthetic polypeptides (12). No other well-defined configuration of the polypeptide chain that has so far been proposed would account for the presence of this reflection. *Figure 32* is an X-ray diffraction photograph of a horse hair showing this reflection at 1.5-Å. spacing.

Some additional evidence of the presence of the α helix in the α-keratin proteins has been given by the work of Riley and Arndt (103, 104), involving the application of the radial-distribution method. Pauling and Corey (88) had calculated radial-distribution curves for the α helix and the γ helix, and compared them with an experimental curve for hemoglobin. A refinement of this work was undertaken by Riley and Arndt, who investigated, in addition to a number of globular proteins, a sample of disoriented α-keratin, and found the radial-distribution curve to agree well with the theoretical curve for the α helix.

The present consensus concerning the occurrence of the α helix in proteins is well summarized by Kendrew (54) in a recent report of the

Conference on the Structure of Proteins held in Pasadena, 21–25 September, 1953:

"It cannot be said that the helix as a structural principle had entered into the fundamentals of our thinking up to the time [1 May 1952] of the Royal Society conference (5): indeed on that occasion there was strong disagreement as to the existence of helical chains. The Pasadena conference revealed that the helix has now come into its own with a vengeance ... It would appear from the discussion at this conference that the great majority of workers in the field would now agree that the α helix is the basic chain configuration present in α-polypeptides and α-forms of fibrous

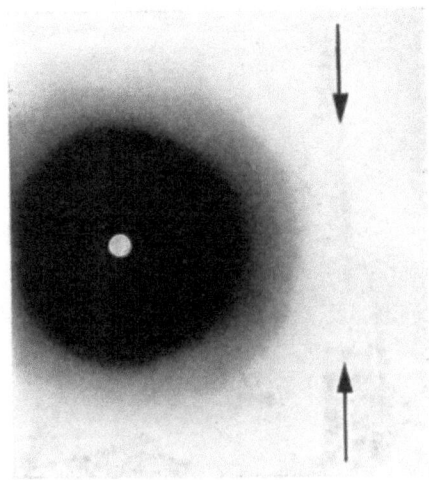

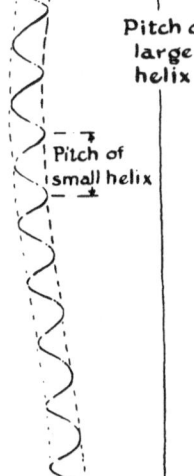

Pitch of large helix

Pitch of small helix

Fig. 32 (*above*). An X-ray diffraction photograph of a horse hair taken with the fiber axis horizontal, and making an angle of 60° with the X-ray beam. The sharp arc at the right is the reflection corresponding to 1.5 Å.

Fig. 33 (*right*). A compound helix with pitch of the large superimposed helix equal to 12.5 times the pitch of the small helix. [From: Nature **171**, 59 (1953).]

proteins; the discussion centred not so much about evidence for the existence of the α helix as about details such as the modes of aggregation of helices, methods for turning corners, and evidence for left- or right-handedness, etc., mainly in reference to fibrous proteins."

The problem of the origin of the 5.15 Å. meridional reflection was solved through the suggestion, made simultaneously by CRICK (*34, 37, 38*) and PAULING and COREY (*93*), that the α-keratin proteins contain α helixes that are twisted about one another, to form more complex structures. CRICK pointed out that it is reasonable to describe the α helix as a cylinder with protuberances and cavities on its surface, the protuberances (bumps) being the amino-acid side-chains, and the cavities (holes) the spaces between the side-chains. He suggested that in a stable protein the adjacent α helixes should be so placed and oriented relative to one another as to permit the bumps of an α helix to fit into the holes of adjacent α helixes, and pointed out that a good fit could be obtained in case that adjacent x helixes were inclined to one another at an angle of about 18°. He suggested that this inclination could be achieved by coiling the molecules about one another, the coiling involving a certain amount of strain energy of compressing and stretching the C—O H—N hydrogen bonds.

Pauling and Corey had reached a similar conclusion on a different postulatory basis. They pointed out that an α helix for a polypeptide chain involving the repetition of different

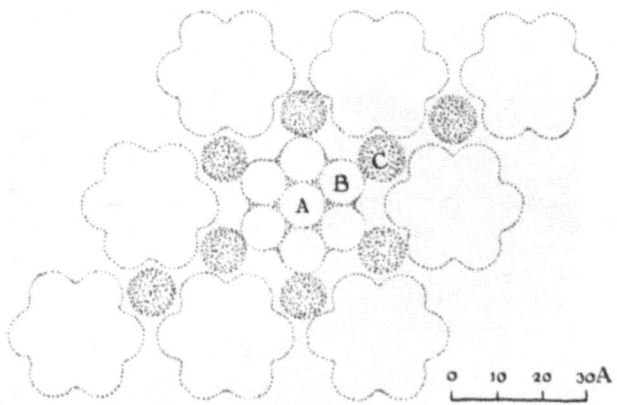

Fig. 34 (*left*). Six compound α helixes twisted around a seventh straight α helix to form a seven-strand cable. [From: Nature **171**, 59 (1953).]

Fig. 35 (*above*). A cross-sectional view of a proposed structure for the α-keratin proteins, showing the seven-strand cables AB_6 and the interstitial compound helixes C. The protein chains are not so nearly circular in cross-section as indicated in the drawing, and space would be filled more effectively than is indicated. [From: Nature **171**, 59 (1953).]

amino-acid residues according to a pattern would not be expected to have a straight axis. The changes in the nature of the side-chain groups might well cause the hydrogen-bond distance to vary by 0.1 or 0.2 Å. about its average value, 2.79 Å., either directly through the interaction of the side-chains with the carbonyl and imino groups of the amide groups, or indirectly by steric hindrance or van der Waals attraction. If the lengthening and shortening of the hydrogen bonds occurred regularly, the axis of the helix would be distorted into a large helix. For example, if in a polypeptide with a repeating unit of four amino-acid residues of different sorts two of the hydrogen bonds are longer than the other two by 0.2 Å. the α helix would describe a large helix with pitch about 66 Å., the length of 44 residues, and radius about 1.5 Å. A compound helix of this sort is shown in *Figure 33* (p. 223).

A radius of 10 Å. for the large helix would permit six molecules to twist about a seventh straight α helix, to form the seven-strand cable shown in *Figure 34*. The outer α helixes in such a seven-strand cable might have a repeating unit of seven residues, which would correspond to 97.2 per cent of two turns of the α helix, so that one turn of the large helix would occur in about 35 turns of the α helix.

A detailed structure for the α-keratin proteins was suggested by PAULING and COREY (*93*). It involves hexagonal packing of seven-strand cables of α helixes, as shown in *Figure 35*, with individual α helixes occupying the interstitial positions. The calculation given above leads to the value 11.0 Å. for the diameter of a cylindrical protein molecule with the configuration of the α helix. Accordingly a seven-strand α cable would be expected to be about 33 Å. in diameter. The equatorial reflections given by the α-keratin proteins can be interpreted in terms of a hexagonal unit with a_0 equal to about 32.4 Å.; in particular, this unit accounts for the observed rather weak reflection with spacing 28 Å., this being the distance between adjacent planes formed by rows of the seven-strand cables. The structure explains the presence of the 5.15 Å. meridional reflection as resulting from a repeating unit of seven amino-acid residues. The axial length of this repeating unit is calculated to be 10.36 Å. It might give rise to a weak first-order reflection, but should give rise to a strong second-order reflection, because of the reinforcement of the two turns of the α helix (tilted at an angle with the fiber axis) in the repeating unit of seven residues. The predicted spacing of the strong meridional reflection produced in this way is 5.18 Å., which is in satisfactory agreement with the observed spacing.

It is possible that some other, perhaps more satisfactory way of coiling the α helixes about one another can be found than the proposed method, involving seven-strand cables. There exists strong evidence that polypeptide chains with the configuration of the α helix are present in the α-keratin proteins, and that these molecules are not simply arranged parallel to one another, but are twisted about one another. There exists a certain amount of evidence supporting the detailed structure shown in *Figure 35*, p. 224, but this structure cannot as yet be said to have been shown to be the correct one.

2. The Structure of Silk and the β-Keratin Proteins.

Silk was one of the proteins investigated by HERZOG and JANCKE (*48*) in 1920. A detailed investigation of silk was reported in 1923 by BRILL (*25*), who concluded that the amino-acid residues were present in the substance as polypeptide chains, with two residues of each of four chains passing through the unit cell. In 1928 MEYER and MARK (*68*) discussed the dimensions of the unit cell in comparison with the predicted dimensions

of polypeptide chains, and concluded that the polypeptide chains in silk
fibroin are nearly completely extended in the direction of the axis of the
fiber.

The identity distance in the direction of the fiber axis of silk is easily
found from the X-ray photograph (*Figure 25,* p. 215) by measurement
of the distances between the well-defined layer lines. The values that
have been reported are near 7.0 Å.; values between 6.95 Å. and 7.2 Å.
have been found by different investigators and for silk of different kinds
(*25, 68, 55, 56, 119*). ASTBURY and STREET (*6*), in reporting the discovery
that wool can be stretched into a β form, gave the value about 6.68 Å. for
the identity distance. Similar values have been found also for other
proteins with the β-keratin structure.

After the recognition of the importance of the hydrogen bond in
proteins, it was at first assumed that silk and the β-keratin proteins
have the planar-sheet structure, shown in *Figure 19,* p. 210. As a result
of the work of COREY on the dimensions of the polypeptide chain it
was recognized that the chains are not completely extended in these
proteins, because the identity distance is less than the predicted value
7.23 Å. for completely extended chains; moreover, it was seen that
steric hindrance would prevent the assumption of the planar-sheet
structure by any polypeptide except polyglycine.

The pleated-sheet structures were then discovered. The parallel-
chain pleated sheet and the antiparallel-chain pleated sheet were found
during an investigation of the consequences of the assumption that
certain orientations about the single bonds to the α-carbon atom are
favored over other orientations (*89*). This assumption leads the value
6.68 Å. for the identity distance in the fiber-axis direction. A study (*91*)
was then made of the consequences of the postulate that it is more
important that the hydrogen bond be linear than that favored orientations
about the α-carbon atoms be assumed; this postulate leads to the
predicted value 7.00 Å. for the identity distance for the antiparallel-
chain pleated sheet, and 6.50 Å. for the parallel-chain pleated sheet.
It is possible to form pleated sheets of intermediate type, and intermediate values of the identity distance would be predicted for them.

Both silk fibroin and β keratin show strong equatorial reflections
with spacings about 4.65 Å., corresponding to the distance between
polypeptide chains in a sheet. They also show reflections with larger
spacing, corresponding to the distance between sheets; this spacing
is about 10 Å. for the β-keratin proteins. All of the features of the
X-ray diagrams are compatible with the assumption that these proteins
have one or another of the pleated-sheet structures.

The close approximation of the observed fiber-axis identity distance
for silk fibroin, about 7.00 Å., and the predicted value for the anti-

parallel-chain pleated sheet, 7.00 Å., strongly suggests that silk fibroin is to be assigned the antiparallel-chain pleated-sheet structure. An unpublished investigation (*65*) involving comparison of calculated and observed values of the intensities of X-ray reflections of silk fibroin has permitted the conclusion to be reached that *Bombyx mori* silk fibroin consists of pairs of antiparallel-chain pleated sheets in which in each polypeptide chain the glycine residues, which alternate with other residues, lie on the side of the sheet opposite the direction of the other sheet, with the remaining residues, principally alanine, having their side-chains directed into the space between the two sheets. All of the side-chains of the sheet (other than hydrogen atoms) are thus on the same side, the pairs of sheets being oriented in such a way as to include the side-chains in the region between them.

The identity distance 6.68 Å. for the β-keratin proteins favors the parallel-chain pleated sheet over the antiparallel-chain pleated sheet, but it is possible that the structure involves both parallel and anti-parallel orientations, perhaps at random. Although no verification of the pleated-sheet structure for the β-keratin proteins has yet been made through the comparison of calculated and observed intensities of X-ray photographs, it is highly probable that this structure is the correct one.

3. Collagen and Gelatin.

The most striking feature of the X-ray photographs of collagen and gelatin (*Figure 26*, p. 216) is the strong meridional reflection with spacing 2.86 Å. The equatorial reflections correspond to hexagonal packing of circular cylinders with diameter about 12 Å. From these dimensions and the density, 1.34 g./cm.³, it can be calculated that there are about three amino-acid residues per molecule in the length 2.86 Å.

The only precisely described structure that has been proposed for collagen (*87*) is one in which three nearly extended polypeptide chains, involving a sequence of two amide groups with the *cis* configuration and one with the *trans* configuration in each chain, are twisted about one another, and held to one another by lateral hydrogen bonds. There are two hydrogen bonds formed by each group of three residues; the third residue is considered to be proline or hydroxyproline, which is prevented from forming a hydrogen bond because of the replacement of hydrogen by carbon on the nitrogen atom. Although this structure is satisfactory in some respects—it accounts for some of the features of the X-ray diagram, and for the orientation of NH and CO groups indicated by the work with polarized infrared radiation—it is unsatisfactory in others, and has been abandoned.

It was concluded by Cohen and Bear (*29*) by application of the theory of Cochran, Crick, and Vand (*27*) that the gelatin molecule has a helical form, with seven units, probably involving three residues, in two turns of the helix. Recently a somewhat different conclusion has been reached independently by Bear (*14*), Randall (*101*), and Pasternak (*74*); namely, that the helical molecule has ten units in three turns (or seven turns) of the helix, the identity distance along the helical axis being 28.6 Å.

Despite the great amount of work that has been expended on the problem by several different investigators during recent years, no reasonable alternative structure to the three-chain structure originally proposed for collagen and gelatin has yet been found.

V. The Structure of Globular Proteins.

During recent years many X-ray crystallographers have been vigorously attacking the problem of the structure of globular proteins.

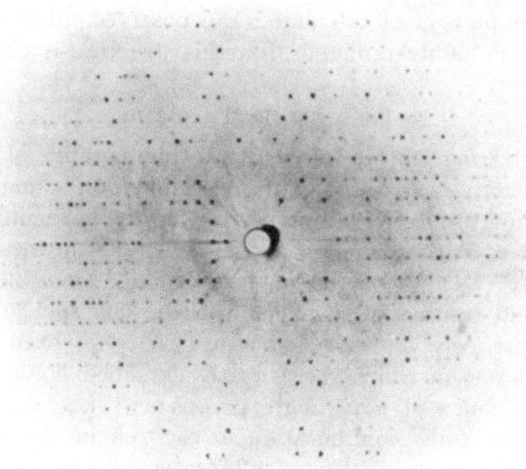

Fig. 36. An X-ray diffraction photograph of a crystal of lysozyme chloride grown at pH 4.5 and photographed in contact with its mother liquor by Palmer. The crystal was oscillated around its *c* axis through an angle of 3^0. Palmer has found that this form of lysozyme chloride is tetragonal with unit cell dimensions $a_0 = 79.1 \pm 0.1$ and $c_0 = 37.9 \pm 0.2$ Å., each unit cell containing eight molecules of the protein (*73*).

Despite the great effort that they have expended in the work, the investigators have not yet succeeded in determining the detailed structure of any one of these substances.

There is very good evidence that some globular proteins contain segments of polypeptide chains which have the form of rods about 10 Å. in diameter. These rods can be identified with some confidence as having the configuration of the α helix. Details of structure, such as the exact

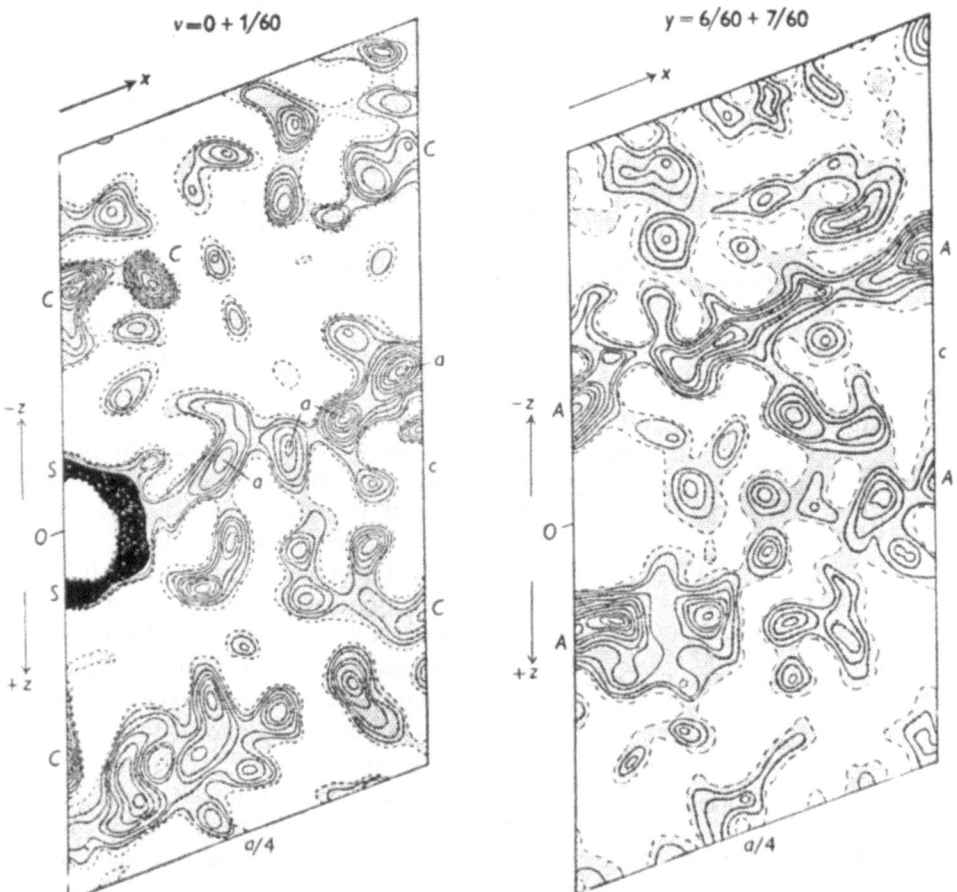

Fig. 37. Sections of the Patterson vector diagram of horse ferrihemoglobin [PERUTZ (96)], indicating rods of vector density in the direction of the x axis with coordinates near $z = 0$ (left) and $z = + \frac{1}{6}$ and $- \frac{1}{6}$ (right). [From: Proc. Roy Soc. (London) **A 195**, 474 (1949).]

location of the α helix segments relative to one another in the molecule and the way in which the polypeptide chain goes from one segment to another, remain unknown.

It was found by BERNAL and CROWFOOT (16) that it is possible to obtain excellent X-ray photographs from crystals of globular proteins. They prepared photographs of pepsin crystals which were kept in equilibrium with the saturated solution during the preparation of the photograph. About half a hundred crystalline globular proteins have now been subjected to X-ray investigation. An excellent review of the work has been published by Low (*61*).

A representative photograph, that of lysozyme hydrochloride taken by Palmer (*73*), is reproduced as *Figure 36* (p. 228). From a complete set of photographs made with this crystal the intensities of about 10,000 independent X-ray diffraction maxima can be obtained. The molecule contains about 1000 atoms (other than hydrogen atoms, which contribute very little to the X-ray diffraction pattern)—the molecular weight is about 14,000. There are accordingly about 3000 parameters to be determined in the course of the investigation of the structure of the crystal. It can be seen that the amount of experimental information is great enough to permit, in principle, the approximate location of most of the atoms in the molecule. Unfortunately, however, no straightforward method of proceeding from the X-ray data to the structure of the crystal is known, and X-ray investigators are not yet able to utilize to full advantage the experimental information that they obtain.

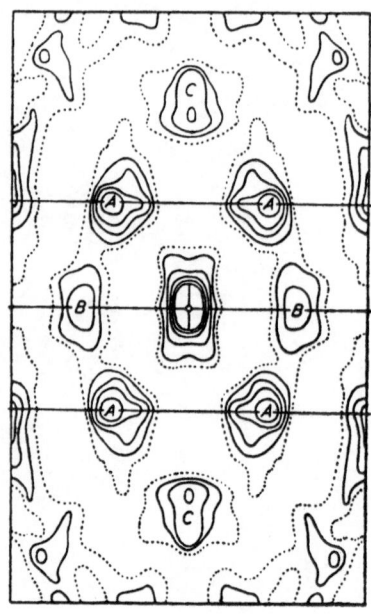

Fig. 38. A projection of the Patterson diagram of horse ferrihemoglobin along the *a* axis [Perutz (*96*)], showing a rough hexagonal array of rods of vector density. [From: Proc. Roy Soc. (London) **A 195**, 474 (1949).]

The progress that has been made on the attack on the structure of globular proteins may be illustrated by the consideration of horse ferrihemoglobin, which has been carefully studied by Perutz and his collaborators. The observed intensities for many thousands of reflections were first used in the calculation of a three-dimensional Patterson function (*20, 96*). The positions of maxima in this function, relative to the origin, represent important interatomic vectors in the crystal. Perutz concluded from consideration of the Patterson diagram that there are rods present in the crystal. Some of the evidence is indicated in *Figure 37*. In the section on the left of Figure 37, representing the sum for the sections $y = 0$ and $y = 1/60$ through the Patterson diagram, there is indication of a rather poorly defined rod along the line $z = 0$; that is, along the x axis of the crystal. There are maxima along this rod with a spacing of about 5 Å.; these are indicated by the letters *a*.

These maxima were discussed in greater detail by BRAGG, KENDREW, and PERUTZ (*23*).

The sum of the Patterson sections at $y = {}^6/_{60}$ and ${}^7/_{60}$ is shown on the right side of *Figure 37*. It is seen that there is some indication of rods along the x axis, with z coordinate $+\,{}^1/_6$ and $-\,{}^1/_6$. The relative positions of the rods are shown in *Figure 38*, which represents the projection of the Patterson diagram along the a axis. From these

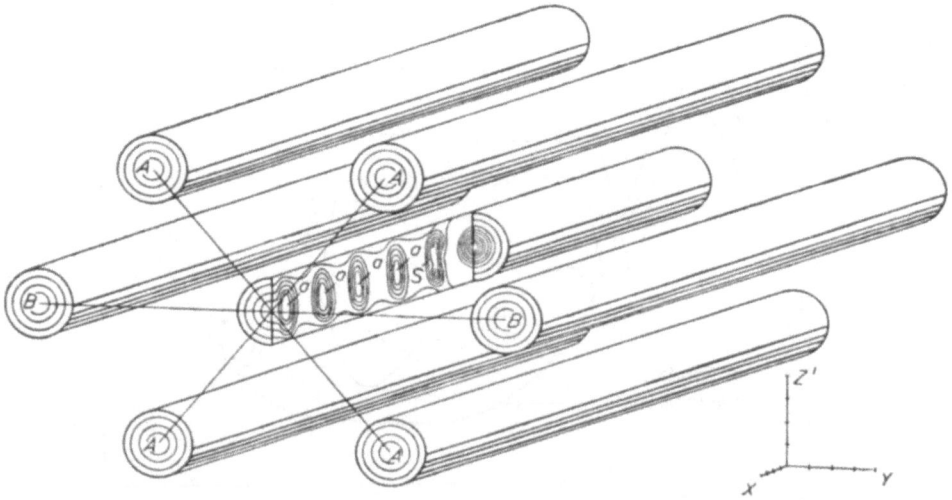

Fig. 39. An idealized picture of the rod-like features in the Patterson vector diagram of horse ferrihemoglobin [PERUTZ (*96*)]. The scale marks on the axial cross give Ångströms. [From: Proc. Roy Soc. (London) **A 195**, 474 (1949).]

data PERUTZ drew the conclusion that the structure can be represented approximately as shown in *Figure 39*. The rods are in a rough hexagonal array, and their centers are approximately 10.5 Å. apart.

Several arguments were then used by PAULING and COREY (*88*) in support of their suggestion that these rods are polypeptide chains with the configuration of the α helix. Their argument was based on three points: the diameter of the rods, the positions of the maxima a along the rods, and the radial distribution function. The diameter of the rods, as reported by PERUTZ, 10.5 Å., is close to the value calculated for the α helix. BRAGG, KENDREW, and PERUTZ (*23*) had, in fact, concluded before they had learned about the α helix that the rods in hemoglobin probably have the same configuration as the molecules in α keratin. The positions of the peaks a along the rods were reported by BRAGG, KENDREW, and PERUTZ as about 5 Å., 11.5 Å., 16.5 Å., 21.5 Å., 27 Å., 32 Å., etc.; the values for the α helix were given by PAULING and COREY (*88*) as 5.1 Å., 10.6 Å., 16.7 Å., 21.4 Å., 27.5 Å., 32.6 Å., etc., in excellent accordance with the experimental points. The radial distribution function (which is the three-dimensional Patterson diagram averaged over the angular coordinates, to become a function of the radius only) for hemoglobin has a pronounced maximum at about 5.0 Å.; this maximum is indicated in *Figure 37*, as the very heavy region surrounding the origin. The comparison of a rough experimental radial distribution function for horse ferrihemoglobin, obtained from PERUTZ's

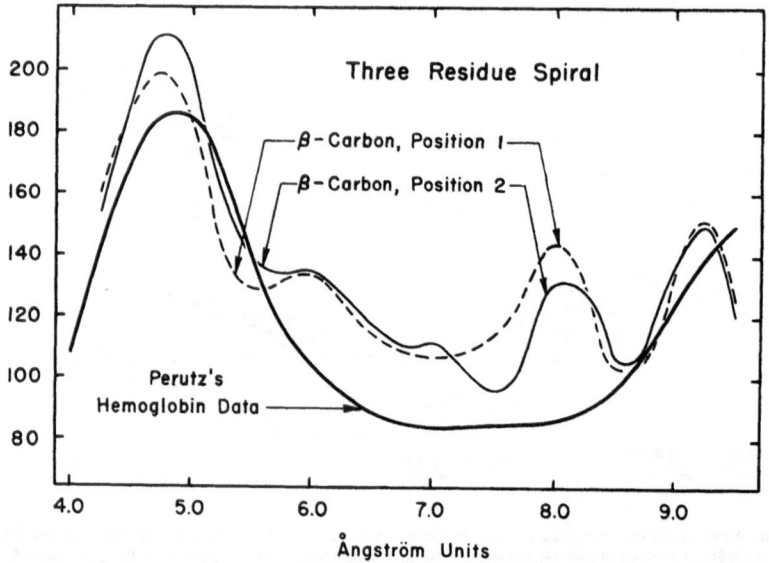

Fig. 40. A comparison of a radial distribution function calculated for the α helix (the left-handed helix described as β-carbon in position 2, and the right-handed helix described as β-carbon in position 1) with the experimental radial distribution function for horse ferrihemoglobin. [From: Proc. Nat. Acad. Sci. (U. S. A.) **37**, 282 (1951).]

data, and the theoretical functions for the α helix (the left-handed helix, described as β-carbon in position 2, and the right-handed helix, described as β-carbon in position 1) is shown in *Figure 40*, and a similar comparison for the γ helix is shown in *Figure 41*. It is seen that the comparison can be taken to eliminate the γ helix, and to provide some support for the α helix.

A discussion of the three-dimensional Patterson diagram of horse ferrihemoglobin has been shown by CRICK (35) to lead to the conclusion that the α helix segments are not more than about 20 Å. long, and that they are considerably disoriented from parallelism to the x axis in the crystal. CRICK (36) has also published a discussion of the total intensity of X-ray reflections of horse ferrihemoglobin at spacings near 10 Å., and has come to the conclusion that the observations are compatible with a structure consisting mainly of α helixes, which are not necessarily parallel to one another.

Additional support for the identification of the rods in hemoglobin with the α helix and evidence of the presence of the α helix in a number of other globular proteins have been provided by the studies of RILEY and ARNDT (*103*, *104*). These investigators had obtained experimental radial distribution curves by the Fourier inversion of the X-ray diagram of disoriented protein specimens, and had reported that all of the globular proteins studied by them seem to have the α helix as their principal

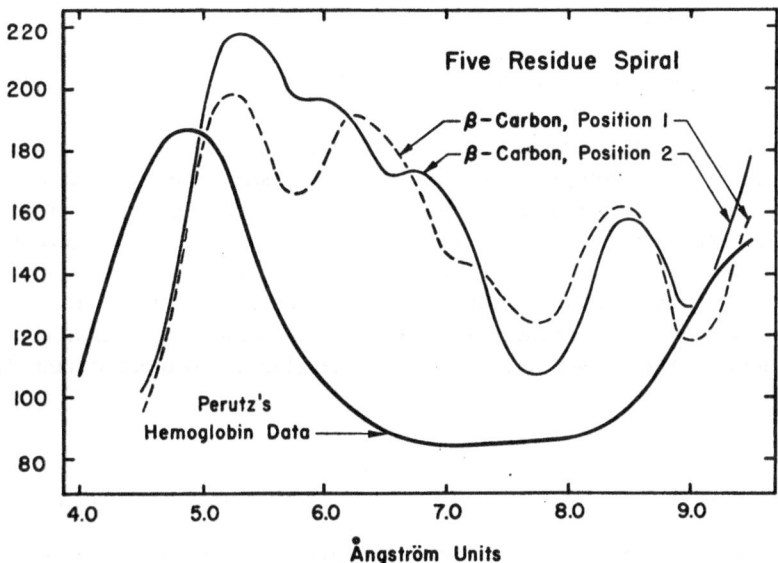

Five Residue Spiral

β−Carbon, Position 1
β−Carbon, Position 2

Perutz's
Hemoglobin Data

Ångström Units

Fig. 41. Radial distribution functions calculated for the γ helix compared with the experimental radial distribution function for horse ferrihemoglobin. [From: Proc. Nat. Acad. Sci. (U. S. A.) **37**, 282 (1951).]

structural feature. They have attempted to distinguish between the left-handed α helix and the right-handed α helix, and have reported that bovine serum albumin and some other globular proteins are probably composed mainly of left-handed α helixes, whereas hemoglobin and some others are probably composed mainly of right-handed α helixes. Their results indicate that both kinds of α helixes are present in insulin; they have identified SANGER's chain *A* with the right-handed α helix, and his chain *B* with the left-handed α helix (*4*). Globular proteins described by RILEY and ARNDT as consisting principally of α helixes include bovine serum albumin, hemoglobin, myoglobin, cytochrome c, insulin, lysozyme, ovalbumin, β-lactoglobulin, edestin, and pepsin [RILEY (*102*)].

Patterson diagrams have been computed from X-ray intensities for a number of other globular-protein crystals. For some of these proteins evidence has been obtained of the presence of rods about 10.5 Å. in diameter in roughly parallel orientation, whereas for others this evidence is lacking, perhaps because the α helixes (indicated to be present by the work of Riley and Arndt) are oriented in several directions, thus confusing the Patterson diagram. A detailed discussion of the results that have been obtained is given in the review by Low (61).

A vigorous attempt is being made by X-ray crystallographers at the present time to determine the phases of the Fourier terms in the function representing the electron distribution itself. The Patterson diagram, which can be calculated from the X-ray intensities without knowledge of the phases, provides only incomplete information about the structure, whereas the electron-distribution function, which could be calculated from the intensities if the phases of the terms were also known, gives a complete description of the structure. In particular, Bragg and Perutz (24, 21, 22, 99) have made very significant progress in the determination of the Fourier phases for hemoglobin, although they are still far from the complete solution of the problem. The results that they have already obtained are so significant that we may look forward with some confidence to seeing the complete solution of the problem of the crystal structure of some globular protein during the coming decade.

References.

1. Albrecht, G. and R. B. Corey: The Crystal Structure of Glycine. J. Amer. Chem. Soc. 61, 1087 (1939).
2. Ambrose, E. J. and A. Elliott: The Structure of Synthetic Polypeptides. II. Investigation with Polarized Infra-red Spectroscopy. Proc. Roy. Soc. (London), Ser. A 205, 47 (1951).
3. Ambrose, E. J. and W. E. Hanby: Evidence of Chain Folding in a Synthetic Polypeptide and in Keratin. Nature (London) 163, 483 (1949).
4. Arndt, U. W. and D. P. Riley: Intra-helix S—S Linked Structures for Insulin. Nature (London) 172, 245 (1953).
5. Astbury, W. T. (leader of the discussion): A Discussion on the Structure of Proteins. Proc. Roy. Soc. (London), Ser. B 141, 1 (1953).
6. Astbury, W. T. and A. Street: X-ray Studies of the Structure of Hair, Wool, and Related Fibres. I. General. Philos. Trans. Roy. Soc. (London), Ser. A 230, 75 (1931).
7. Astbury, W. T. and C. Weibull: X-ray Diffraction Study of the Structure of Bacterial Flagella. Nature (London) 163, 280 (1949).
8. Astbury, W. T. and H. J. Woods: X-ray Studies of the Structure of Hair, Wool, and Related Fibres. II. The Molecular Structure and Elastic Properties of Hair Keratin. Philos. Trans. Roy. Soc. (London), Ser. A 232, 333 (1933).
9. Badger, R. M. and H. Rubalcava: The Infrared Absorption Spectra of Amides in Solution and Their Relation to the Spectra of Polypeptides. Proc. Nat. Acad. Sci. (U. S. A.) 40, 12 (1954).
10. Bailey, K., W. T. Astbury and K. M. Rudall: Fibrinogen and Fibrin as Members of the Keratin-Myosin Group. Nature (London) 151, 716 (1943).

11. BAMFORD, C. H., L. BROWN, A. ELLIOTT, W. E. HANBY and I. F. TROTTER: Structure of Synthetic Polypeptides. Nature (London) **169**, 357 (1952).

12. — — — — — Some New Investigations on the Structure of Synthetic Polypeptides. Proc. Roy. Soc. (London), Ser. B **141**, 49 (1953).

12 a. — — — — — Alpha- und Beta-Forms of Poly-*L*-Alanine. Nature (London) **173**, 27 (1954).

13. BAMFORD, C. H., W. E. HANBY and F. HAPPEY: The Structure of Synthetic Polypeptides. I. X-ray Investigation. Proc. Roy. Soc. (London), Ser. A **205**, 30 (1951).

14. BEAR, R. S.: Conference on the Structure of Proteins, Pasadena, Calif., Sept. 1953.

15. BERNAL, J. D.: The Crystal Structure of the Natural Amino Acids and Related Compounds. Z. Kristallogr., Mineral., Petrogr. **78**, 363 (1931).

16. BERNAL, J. D. and D. CROWFOOT: X-ray Photographs of Crystalline Pepsin. Nature (London) **133**, 794 (1934).

17. BIJVOET, J. M., A. F. PEERDEMAN and A. J. VAN BOMMEL: Determination of the Absolute Configuration of Optically Active Compounds by Means of X-rays. Nature (London) **168**, 271 (1951).

18. BLUM, J. and D. R. DAVIES: The Crystal Structure of Parabanic Acid (unpublished).

19. BOEHM, G. und H. H. WEBER: Das Röntgendiagramm von gedehnten Myosinfäden. Kolloid-Z. **61**, 269 (1932).

20. BOYES-WATSON, J., E. DAVIDSON and M. F. PERUTZ: An X-ray Study of Horse Methaemoglobin. I. Proc. Roy. Soc. (London), Ser. A **191**, 83 (1947).

21. BRAGG, L.: X-ray Analysis of the Haemoglobin Molecule. Proc. Roy. Soc. (London), Ser. B **141**, 67 (1953).

22. — Application of X-ray Optics to Proteins. Conference on the Structure of Proteins, Pasadena, Calif., Sept. 1953.

23. BRAGG, L., J. C. KENDREW and M. F. PERUTZ: Polypeptide Chain Configurations in Crystalline Proteins. Proc. Roy. Soc. (London), Ser. A **203**, 321 (1950).

24. BRAGG, L. and M. F. PERUTZ: The Structure of Haemoglobin. Proc. Roy. Soc. (London), Ser. A **213**, 425 (1952).

25. BRILL, R.: Über Seidenfibroin. I. Liebigs Ann. Chem. **434**, 204 (1923).

26. CARPENTER, G. B. and J. DONOHUE: The Crystal Structure of N-Acetylglycine. J. Amer. Chem. Soc. **72**, 2315 (1950).

27. COCHRAN, W., F. H. C. CRICK and V. VAND: The Structure of Synthetic Polypeptides. I. The Transform of Atoms on a Helix. Acta Crystallogr. **5**, 581 (1952).

28. COCHRAN, W. and B. R. PENFOLD: The Crystal Structure of *L*-Glutamine. Acta Crystallogr. **5**, 644 (1952).

29. COHEN, C. and R. S. BEAR: Helical Polypeptide Chain Configuration in Collagen. J. Amer. Chem. Soc. **75**, 2783 (1953).

30. COREY, R. B.: The Crystal Structure of Diketopiperazine. J. Amer. Chem. Soc. **60**, 1598 (1938).

31. — X-ray Diffraction Studies of Crystalline Amino Acids and Peptides. Fortschr. Chem. organ. Naturstoffe **8**, 310 (1951).

32. COREY, R. B. and J. DONOHUE: Interatomic Distances and Bond Angles in the Polypeptide Chain of Proteins. J. Amer. Chem. Soc. **72**, 2899 (1950).

33. COREY, R. B. and L. PAULING: Fundamental Dimensions of Polypeptide Chains. Proc. Roy. Soc. (London), Ser. B **141**, 10 (1953).

34. CRICK, F. H. C.: Is α-Keratin a Coiled Coil? Nature (London) **170**, 882 (1952).

35. — The Height of the Vector Rods in the Three-dimensional Patterson of Haemoglobin. Acta Crystallogr. **5**, 381 (1952).

36. Crick, F. H. C.: The Strength of the 10-Å. Reflexions in Haemoglobin. Acta Crystallogr. 6, 600 (1953).
37. — Fourier Transform of a Coiled Coil. Acta Crystallogr. 6, 685 (1953).
38. — The Packing of α-Helices: Simple Coiled Coils. Acta Crystallogr. 6, 689 (1953).
39. Crowfoot, D., C. W. Bunn, B. W. Rogers-Low and A. Turner-Jones: The X-ray Crystallographic Investigation of the Structure of Penicillin. Chap. 11. The Chemistry of Penicillin. (Ed., H. T. Clarke, J. R. Johnson and R. Robinson) Princeton: Univ. Press. 1949.
40. Dawson, B.: The Crystal Structure of DL-Glutamic Acid Hydrochloride. Acta Crystallogr. 6, 81 (1953).
41. Derksen, J. C. and G. C. Heringa: Szymonowicz-Festschr., Polska Gaz. Lekarska 15, 532 (1936).
42. Donohue, J.: The Crystal Structure of DL-Alanine. II. Revision of Parameters by Three-dimensional Fourier Analysis. J. Amer. Chem. Soc. 72, 949 (1950).
43. — The Hydrogen Bond in Organic Crystals. J. Physic. Chem. 56, 502 (1952).
44. — Hydrogen Bonded Helical Configurations of the Polypeptide Chains. Proc. Nat. Acad. Sci. (U. S. A.) 39, 470 (1953).
45. Donohue, J. and K. N. Trueblood: The Crystal Structure of Hydroxy-L-Proline. I. Interpretation of the Three-dimensional Patterson Function. Acta Crystallogr. 5, 414 (1952).
46. — — The Crystal Structure of Hydroxy-L-Proline. II. Determination and Description of the Structure. Acta Crystallogr. 5, 419 (1952).
47. Dyer, H. B.: The Crystal Structure of Cysteyl-glycine Sodium Iodide. Acta Crystallogr. 4, 42 (1951).
48. Herzog, R. O. und W. Jancke: Über den physikalischen Aufbau einiger hochmolekularer organischer Verbindungen. 1. vorl. Mitt. Ber. dtsch. chem. Ges. 53, 2162 (1920); Festschrift der Kaiser Wilhelm-Ges., S. 118 (1921).
49. Huggins, M. L.: Hydrogen Bridges in Organic Compounds. J. Organ. Chem. (U. S. A.) 1, 407 (1936).
50. — The Structure of Fibrous Proteins. Chem. Rev. 32, 195 (1943).
51. Hughes, E. W., A. B. Biswas and J. N. Wilson: The Crystal Structure of α-Glycylglycine. Acta Crystallogr. (to be published).
52. Hughes, E. W. and W. J. Moore: The Crystal Structure of β-Glycylglycine. J. Amer. Chem. Soc. 71, 2618 (1949).
53. Katz, L., R. A. Pasternak and R. B. Corey: Configuration of the Peptide Link and of Asparagine in Glycyl-L-Asparagine. Nature (London) 170, 1066 (1952).
54. Kendrew, J. C.: Structure of Proteins. Nature (London) 173, 57 (1954).
55. Kratky, O.: Über Seidenfibroin. II. Z. physik. Chem., Abt. B 5, 297 (1929).
56. Kratky, O. und S. Kuriyama: Über Seidenfibroin. III. Z. physik. Chem., Abt. B 11, 363 (1930).
57. Kratky, O. und H. Mark: Anwendung physikalischer Methoden zur Erforschung von Naturstoffen: Form und Größe dispergierter Moleküle. Röntgenographie. Fortschr. Chem. organ. Naturstoffe 1, 255 (1938).
58. Leonard, J. E. and R. A. Pasternak: The Unit-cell Dimensions and the Space Groups of Some Simple Peptides of Glycine, Alanine, and Leucine. Acta Crystallogr. 5, 150 (1952).
59. Lévy, H. A. and R. B. Corey: The Crystal Structure of DL-Alanine. J. Amer. Chem. Soc. 63, 2095 (1941).
60. Lonsdale, K.: Divergent-beam X-ray Photography of Crystals. Trans. Roy. Soc. (London), Ser. A 240, 244 (1947).

61. Low, B. W.: In: The Proteins, Vol. 1, chapter 4, p. 235. (Ed., H. Neurath and K. Bailey) New York: Acad. Press. 1953.

62. Low, B. W. and R. B. Baybutt: The Pi Helix—A Hydrogen Bonded Configuration of the Polypeptide Chain. J. Amer. Chem. Soc. **74**, 5806 (1952).

63. Low, B. W. and H. J. Grenville-Wells: Generalized Mathematical Relationships for Polypeptide Chain Helices. The Coordinates of the π Helix. Proc. Nat. Acad. Sci. (U. S. A.) **39**, 785 (1953).

64. MacArthur, I.: Structure of α-Keratin. Nature (London) **152**, 38 (1943).

65. Marsh, R. E., L. Pauling and R. B. Corey: The Structure of Silk Fibroin (to be published).

66. Mathieson, A. McL.: The Crystal Structure of the Dimorphs of DL-Methionine. Acta Crystallogr. **5**, 332 (1952).

67. — Polymorphism of DL-Norleucine. Acta Crystallogr. **6**, 399 (1953).

68. Meyer, K. H. und H. Mark: Über den Aufbau des Seiden-Fibroins. Ber. dtsch. chem. Ges. **61**, 1932 (1928).

69. Mirsky, A. E. and L. Pauling: On the Structure of Native, Denatured, and Coagulated Proteins. Proc. Nat. Acad. Sci. (U. S. A.) **22**, 439 (1936).

70. Mizushima, S., T. Simanouti, S. Nagakura, K. Kuratani, M. Tsuboi, H. Baba and O. Fujioka: The Molecular Structure of N-Methylacetamide. J. Amer. Chem. Soc. **72**, 3490 (1950).

71. Neuberger, A.: Stereochemistry of Amino Acids. Adv. Protein Chem. **4**, 297 (1948), especially p. 321.

72. Newman, R. and R. M. Badger: The Infrared Spectra of N-Acetylglycine and Diketopiperazine Polarized Radiation at 25° and at — 185° C. J. Chem. Physics **19**, 1147 (1951).

73. Palmer, K. J.: Lysozyme Chloride. I. X-ray Diffraction Study of Lysozyme Chloride. Abstr. of Papers, 1st Intern. Union Crystallogr., p. 17 (1948); cf. Structure Reports for 1947–1948, **11**, 729.

74. Pasternak, R. A.: Conference on the Structure of Proteins, Pasadena, Calif., Sept. 1953.

75. — The Crystal Structure of Glycyl-L-Tryptophan Dihydrate (unpublished).

76. Pasternak, R. A., L. Katz and R. B. Corey: The Crystal Structure of Glycyl-L-Asparagine. Acta Crystallogr. **7**, 225 (1954).

77. Pasternak, R. A. and J. E. Leonard: The Unit-cell Dimensions and the Space Groups of Some Alanyl Peptides. Acta Crystallogr. **5**, 152 (1952).

78. Pauling, L.: A Theory of the Structure and Process of Formation of Antibodies. J. Amer. Chem. Soc. **62**, 2643 (1940).

79. — Discussion, Colloques Intern. Centre Nat. Rech. Sci. XVIII. La liaison chimique. Paris, avril 1948, p. 155.

80. — On the Stability of the S_8 Molecule and the Structure of Fibrous Sulfur. Proc. Nat. Acad. Sci. (U. S. A.) **35**, 495 (1949).

81. — Les protéines. Rapports et discussions. 9e Conseil de Chimie, Inst. Intern. Chimie Solvay. (Ed., R. Stoops) Bruxelles. 1953, p. 63.

82. Pauling, L. and R. B. Corey: Two Hydrogen-bonded Spiral Configurations of the Polypeptide Chain. J. Amer. Chem. Soc. **72**, 5349 (1950).

83. — — Atomic Coordinates and Structure Factors for Two Helical Configurations of Polypeptide Chains. Proc. Nat. Acad. Sci. (U. S. A.) **37**, 235 (1951).

84. — — The Structure of Synthetic Polypeptides. Proc. Nat. Acad. Sci. (U. S. A.) **37**, 241 (1951).

85. — — The Pleated Sheet; A New Layer Configuration of Polypeptide Chains. Proc. Nat. Acad. Sci. (U. S. A.) **37**, 251 (1951).

86. — — The Structure of Hair, Muscle, and Related Proteins. Proc. Nat. Acad. Sci. (U. S. A.) **37**, 261 (1951).

87. Pauling, L. and R. B. Corey: The Structure of Fibrous Proteins of the Collagen-Gelatin Group. Proc. Nat. Acad. Sci. (U. S. A.) **37**, 272 (1951).
88. — — The Polypeptide-Chain Configuration in Hemoglobin and Other Globular Proteins. Proc. Nat. Acad. Sci. (U. S. A.) **37**, 282 (1951).
89. — — Configurations of Polypeptide Chains with Favored Orientations Around Single Bonds: Two New Pleated Sheets. Proc. Nat. Acad. Sci. (U. S. A.) **37**, 729 (1951).
90. — — Two Pleated Sheet Configurations of Polypeptide Chains Involving Both Cis and Trans Amide Groups. Proc. Nat. Acad. Sci. (U. S. A.) **39**, 247 (1953).
91. — — Two Rippled Sheet Configurations of Polypeptide Chains, and a Note about the Pleated Sheets. Proc. Nat. Acad. Sci. (U. S. A.) **39**, 253 (1953).
92. — — Stable Configurations of Polypeptide Chains. Proc. Roy. Soc. (London), Ser. B **141**, 21 (1953).
93. — — Compound Helical Configurations of Polypeptide Chains: Structure of Proteins of the α Keratin Type. Nature (London) **171**, 59 (1953).
94. Pauling, L., R. B. Corey and H. R. Branson: The Structure of Proteins: Two Hydrogen-bonded Helical Configurations of the Polypeptide Chain. Proc. Nat. Acad. Sci. (U. S. A.) **37**, 205 (1951).
95. Pauling, L., R. B. Corey and L. R. Lavine: (unpublished).
96. Perutz, M. F.: An X-ray Study of Horse Methaemoglobin. II. Proc. Roy. Soc. (London), Ser. A **195**, 474 (1949).
97. — Crystallography. Structure of Proteins and Related Compounds. Annu. Rep. Progr. Chem. **48**, 361 (1951).
98. — New X-ray Evidence on the Configuration of Polypeptide Chains. Nature (London) **167**, 1053 (1951).
99. — A Fourier Projection of Hemoglobin. Conference on the Structure of Proteins, Pasadena, Calif., Sept. 1953.
100. Pitt, G. J.: A Refinement of the Crystal Structure of Potassium Benzyl-penicillin. Acta Crystallogr. **5**, 770 (1952).
101. Randall, J. T.: Conference on the Structure of Proteins, Pasadena, Calif., Sept. 1953.
102. Riley, D. P.: Conference on the Structure of Proteins, Pasadena, Calif., Sept. 1953.
103. Riley, D. P. and U. W. Arndt: New Type of X-ray Evidence on the Molecular Structure of Globular Proteins. Nature (London) **169**, 138 (1952).
104. — — X-ray Scattering by Some Native and Denatured Proteins in the Solid State. Proc. Roy. Soc. (London), Ser. B **141**, 93 (1953).
105. Robinson, C. and E. J. Ambrose: Atomic Models. 2. The Use of Atomic Models in Investigating Stable Configurations of Protein Chains. Trans. Faraday Soc. **48**, 854 (1952).
106. Romers, C.: The Structure of Oxamide. Acta Crystallogr. **6**, 429 (1953).
107. Rudall, K. M.: Fibrous Proteins. Sympos. Soc. Dyers Colourists, England, 1946, p. 15.
107 a. — The Proteins of the Mammalian Epidermis. Adv. Protein Chem. **7**, 253 (1952).
108. — Elastic Properties and α,β-Transformation of Fibrous Proteins. Proc. Roy. Soc. (London), Ser. B **141**, 39 (1953).
109. Sanger, F. and E. O. P. Thompson: The Amino-acid Sequence in the Glycyl Chain of Insulin. 2. The Investigation of Peptides from Enzymic Hydrolysates. Biochemic. J. **53**, 366 (1953).
110. Sanger, F. and H. Tuppy: The Amino-acid Sequence in the Phenylalanyl Chain of Insulin. 2. The Investigation of Peptides from Enzymic Hydrolysates. Biochemic. J. **49**, 481 (1951).

111. SCHUCH, A. F., L. L. MERRITT, Jr. and J. H. STURDIVANT: The Crystal Structure of Urea Oxalate (unpublished).

112. SHOEMAKER, D. P., R. E. BARIEAU, J. DONOHUE and C. S. LU: The Crystal Structure of *DL*-Serine. Acta Crystallogr. **6**, 241 (1953).

113. SHOEMAKER, D. P., J. DONOHUE, V. SCHOMAKER and R. B. COREY: The Crystal Structure of L_s-Threonine. J. Amer. Chem. Soc. **72**, 2328 (1950).

114. SIMANOUTI, T. and S. MIZUSHIMA: Intramolecular Rotation and the Structure of High Polymers. I. The Structure of Polypeptide Chain. Bull. Chem. Soc. Japan **21**, 1 (1948).

115. SLY, W. G. and J. H. STURDIVANT: The Crystal Structure of Oxamide (unpublished).

116. SMITS, D. W. and E. H. WIEBENGA: The Crystal Structure of Glycyl-*L*-Tyrosine Hydrochloride. Acta Crystallogr. **6**, 531 (1953).

117. TRANTER, T. C.: Unit-cell Dimensions and Space Groups of Synthetic Peptides. I. Glycyl-*L*-Tyrosine, Glycyl-*L*-Tyrosine Hydrochloride, Glycyl-*DL*-Serine, and Glycyl-*DL*-Leucine. Acta Crystallogr. **5**, 843 (1952).

118. — Unit-cell Dimensions and Space Groups of Synthetic Peptides. II. Glycyl-*L*-Alanine, Glycyl-*L*-Alanine Hydrochloride, Glycyl-*L*-Alanine Hydrobromide, and Glycyl-*L*-Tryptophane. Acta Crystallogr. **6**, 805 (1953).

118 a. — Unit-cell Dimensions and Space Groups of Synthetic Peptides. III. Glycyl-*L*-valine Hydrobromide, Glycyl-*L*-valine Hydrochloride, and *DL*-Alanyl-*DL*-methionine. Acta Crystallogr. **7**, 134 (1954).

118 b. — Crystal Structure of Glycyl-*L*-alanine Hydrobromide. Nature (London) **173**, 221 (1954).

119. TROGUS, C. und K. HESS: Zur Kenntnis der natürlichen Seiden und ihres Verhaltens gegen Säuren und Basen. Biochem. Z. **260**, 376 (1933).

120. VAUGHAN, P. and J. DONOHUE: The Structure of Urea. Interatomic Distances and Resonance in Urea and Related Compounds. Acta Crystallogr. **5**, 530 (1952).

121. WARREN, B. E. and J. T. BURWELL: The Structure of Rhombic Sulfur. J. Chem. Physics **3**, 6 (1935).

122. YAKEL, H. L., Jr. and E. W. HUGHES: The Unit-cell Dimensions and Space Groups of Two Modifications of Crystalline Glycylglycylglycine. Acta Crystallogr. **5**, 847 (1952).

123. — — The Structure of N,N'-Diglycyl-*L*-cystine Dihydrate. J. Amer. Chem. Soc. **74**, 6302 (1952).

124. — — The Crystal Structure of N,N'-Diglycyl-*L*-cystine Dihydrate. Acta Crystallogr. **7**, 291 (1954).

125. ZAHN, H.: Über die Struktur des α-Keratins. Z. Naturforsch. **2** b, 104 (1947).

126. ZUSSMAN, J.: The Structure of Hydroxyproline. Acta Crystallogr. **4**, 72 (1951).

127. — The Structure of Hydroxyproline. Acta Crystallogr. **4**, 493 (1951).

(Received, December 1, 1953.)

Column Chromatography in the Study
of the Structure of Peptides and Proteins.

By **W. A. Schroeder**, Pasadena, California.

With 12 Figures.

Contents.

Introduction.

Only ten years ago, SYNGE (*196*) ended a review on the significance of the partial hydrolytic products of proteins with this statement:

"To conclude, it seems that the main obstacle to progress in the study of protein structure by methods of organic chemistry is inadequacy of technique rather than any theoretical difficulty. It is likely that the development of new methods of work in this field will lead us to a much clearer understanding of the proteins."

In the intervening years, there has been a tremendous upsurge in the development of methods for the study of proteins and the diligent application of these methods by many investigators is clearly evidenced by the published papers on this subject.

Not the least important of these procedures is the chromatographic method which is receiving ever-increasing use. In 1944, CONSDEN, GORDON, and MARTIN (*23*) devised the modern methods of paper chromatography which in their almost innumerable major and minor modifications have been applied to a variety of problems related to protein structure. At about this same time, numerous attempts were being made to devise methods for separating amino acids by column chromatography on a variety of adsorbents [reviewed by MARTIN and SYNGE (*114*)]. Although paper chromatography was successful from the beginning in separating almost all of the common amino acids by two dimensional chromatography on a single sheet, an equally complete resolution by column chromatographic methods in 1945 required a complicated series of chromatograms on a variety of ill-defined adsorbents: indeed, no one apparently has ever tried a complete analysis of amino acids in a protein hydrolysate by such methods. More recently, however, the ready separation of complicated mixtures of amino acids on columns of starch or ion exchange resin has been achieved by MOORE and STEIN and these excellent methods are receiving application not only in the quantitative determination of amino acids but also in the fractionation of peptides and in the purification of proteins.

In the present review, we shall be concerned in the main with column chromatography as it applies to the study of the structure of proteins and shall touch upon paper chromatography only when it is the sole chromatographic procedure which has been used in conjunction with useful or potentially useful chemical methods. The exclusion of paper chromatography from this review necessitates that some useful methods and interesting results be ignored. The emphasis will be placed on generally applicable procedures and hence, we do not intend a complete listing of all published chromatographic studies of amino acids, peptides, and proteins, some of which studies were devised for specific purposes only. Although we shall refer to applications of the methods, we shall not be concerned with the interpretation of the chromatographic results as they relate to the structure of proteins.

The review will consider three broad subjects: (1) The separation of amino acids and the determination of the amino acid composition of peptides and proteins; (2) The determination of amino acid sequence in proteins in terms of the identification of terminal residues and of the fractionation and identification of peptides; and (3) The separation and purification of proteins themselves.

I. The Separation of Amino Acids and the Determination of the Amino Acid Composition of Peptides and Proteins.

One of the more successful of the earlier attempts to resolve mixtures of amino acids was the procedure for the separation of acetyl amino acids by partition chromatography on silica gel as devised originally by Martin and Synge (*113*) and amplified in further work by them and their collaborators (*69*). The ability of this method to separate phenylalanine, leucine + isoleucine, valine, methionine, proline, tyrosine, and alanine in the form of their acetyl derivatives filled a need which was not met by chemical methods then in vogue. Despite this fact, the method has received little use by others than Martin, Synge and their collaborators with the exception of a very thorough study of the potentialities of the method by Tristram (*206*) who used *ad hoc* mixtures which approximated the composition of several protein hydrolysates. The decline of this method may be ascribed not only to the fact that its application was limited to relatively few amino acids but also to the fact that paper chromatography appeared soon after (*23*) and rapidly came into high favor. At the present time, this method may be considered to be of historical interest only. Separations in other ways on other adsorbents in general were even more limited than those of the acetyl amino acids and it is unlikely that they will ever be put to use especially in view of the excellent methods which are now available. These methods which employ starch or an ion exchange resin as the adsorbent will now be discussed in some detail.

1. Analytical Determination of Amino Acids by Chromatography.

a) Separation of Amino Acids on Starch.

Elsden and Synge (*53*) and Synge (*197*) first used starch for the chromatographic separation of amino acids but the method was limited to a few amino acids. MOORE and STEIN, however, have brought the separation of all of the common amino acids on starch to a high degree of perfection.

The preparation and operation of starch chromatograms has been described by MOORE and STEIN (*124, 125, 189*) with meticulous care and attention to detail, even, in fact, to the test tube brush which is recommended for use in cleaning the many test tubes required by the method. In the elaboration of the final procedure, so many variables have been explored and the effects of the variations have been so completely described that it is difficult to see how anyone who followed the description of procedure exactly could fail to·obtain satisfactory results.

For the analytical separation of the amino acids, the dimensions of the starch columns are 30 × 0.9 cm. The correct packing of the column is insured by careful control of the water content of the starch. The column is formed by pouring a slurry of starch in butanol into the chromatographic tube and then packing by applying air pressure. The type of conditioning which the column next receives is dependent upon the nature of the developers which will be used during the chromatogram. At the end of the conditioning, the sample of amino acids totaling 2 to 5 mg. is added, rinsed in, and development is begun. The rate of solvent flow through starch columns is very slow; for a 30-cm. column, it is adjusted to 2.0 to 2.5 ml. per hour per cm.^{2} of cross-sectional area. As a result the duration of the chromatogram may extend in some instances over a period of a week and for practical application the use of automatic equipment is required. STEIN and MOORE (*189*) devised a fully automatic fraction collector which permitted the effluent of the chromatograms to be divided into fractions of the order of one ml. The size of the fraction will depend upon the size of the chromatographic column and 0.5-ml. fractions were collected from the analytical columns. The effectiveness of the chromatographic separation was assessed by reacting the amino acids of these fractions quantitatively with ninhydrin. If then the quantity so determined colorimetrically in each fraction is plotted against the volume of effluent, the zones of the amino acids appear as a series of discrete peaks. Because the colorimetric reaction with ninhydrin is quantitative, the amount of each amino acid can be determined by integrating the area under each peak.

The results which may be obtained by such methods of starch chromatography are shown in *Fig. 1*. This figure shows the separation of a synthetic mixture of 17 amino acids and ammonium chloride on analytical size columns (30 × 0.9 cm.) by means of four types of developer. The most effective developer which is 1 : 2 : 1 *n*-butyl alcohol-*n*-propyl alcohol-0.1 *N*-HCl followed by 2 : 1 *n*-propyl alcohol-0.5 *N*-HCl separates proline, threonine, aspartic acid, serine, glycine, ammonia,

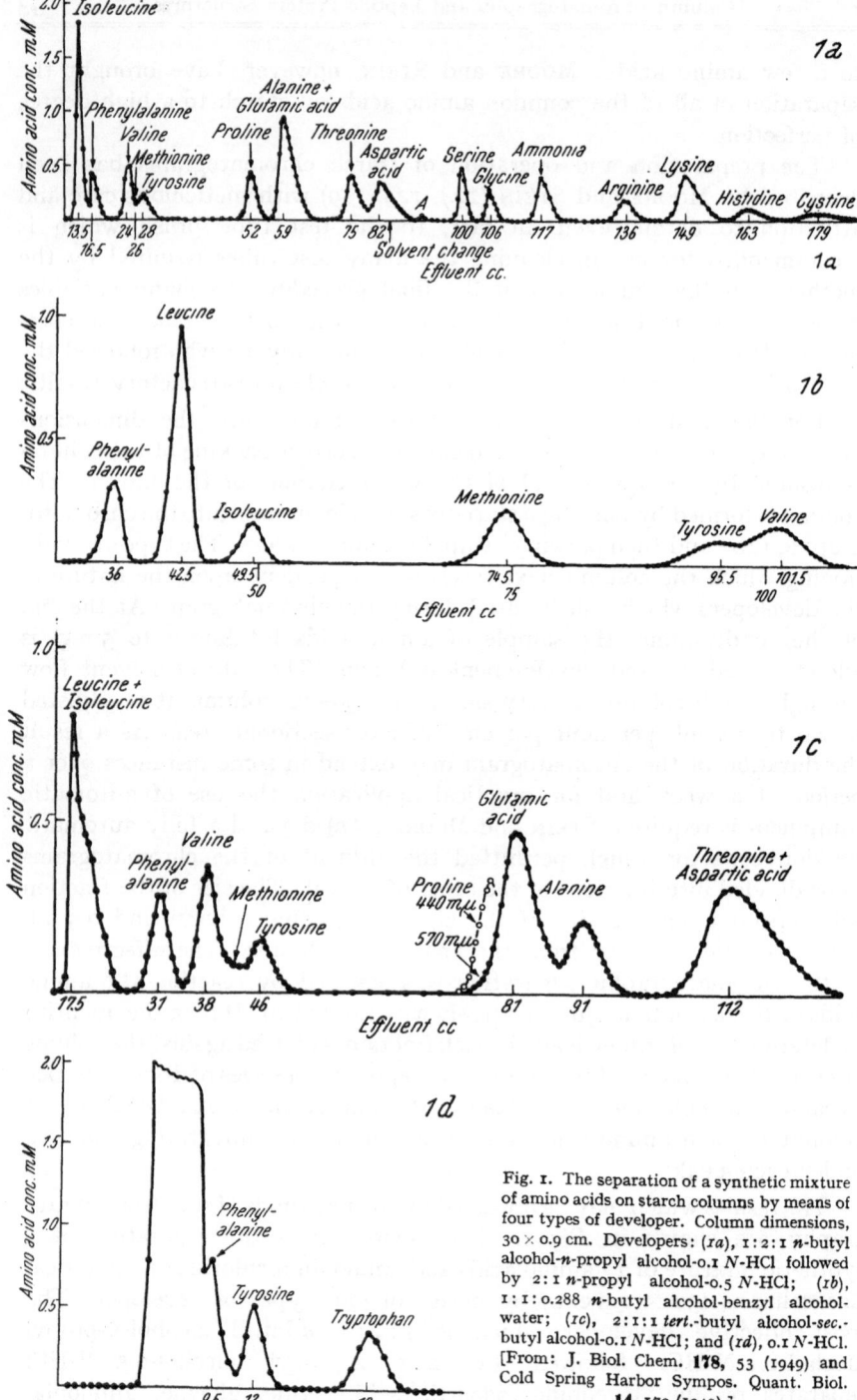

Fig. 1. The separation of a synthetic mixture of amino acids on starch columns by means of four types of developer. Column dimensions, 30 × 0.9 cm. Developers: (1a), 1:2:1 n-butyl alcohol-n-propyl alcohol-0.1 N-HCl followed by 2:1 n-propyl alcohol-0.5 N-HCl; (1b), 1:1:0.288 n-butyl alcohol-benzyl alcohol-water; (1c), 2:1:1 tert.-butyl alcohol-sec.-butyl alcohol-0.1 N-HCl; and (1d), 0.1 N-HCl. [From: J. Biol. Chem. **178**, 53 (1949) and Cold Spring Harbor Sympos. Quant. Biol. **14**, 179 (1949).]

arginine, lysine, histidine, and cystine as distinct peaks (*Fig. 1 a*) and permits the quantity of each to be calculated. On this chromatogram, however, the separation of leucine, isoleucine, phenylalanine, valine, methionine, and tyrosine is far from adequate and glutamic acid and alanine appear as a single peak. The first six may be separated by the use of 1 : 1 : 0.288 *n*-butyl alcohol-benzyl alcohol-water as the developer (*Fig. 1 b*), and glutamic acid and alanine by the use of 2 : 1 : 1 *tert.*-butyl alcohol-*sec.*-butyl alcohol-0.1 *N*-HCl (*Fig. 1 c*). The separation of tyrosine and valine (*Fig. 1 b*) is greatly improved if the chromatogram is maintained at 15° (*175*); this change in conditions does not influence the other separations of this type of chromatogram. Finally, tryptophan may be isolated by the use of 0.1 *N*-HCl as developer (*Fig. 1 d*). Thus, by means of four chromatograms which use four portions of a protein hydrolysate it is possible to fractionate all of the common amino acids and ammonia into more or less distinct peaks and to determine quantitatively the amount of each.

By extensive studies with mixtures of amino acids of known composition, MOORE and STEIN have shown that the recoveries of the individual amino acids are 100 ± 3 per cent of the amounts in the starting mixture. One may with confidence expect, therefore, that the analytical results from hydrolysates of proteins will be correct to the same degree. MOORE and STEIN have discussed the identification of the peaks from a starch chromatogram. Absolute identification, of course, will require isolation and characterization but when one is working with peptide or protein hydrolysates the possibility of error in identification is greatly minimized by the fact that the effluent volume at which a given amino acid emerges from the column is constant to ± 5 to 10 per cent and is not influenced by the presence or absence of other amino acids; furthermore, the relative volumes at which the various amino acids emerge are even more constant than the absolute volumes.

The methods of starch chromatography represent the first quantitative method by which all of the amino acids in a mixture may be determined on a micro scale with excellent accuracy and with assurance that components which are present only in small amount have not been overlooked. It may seem that the procedure is time consuming and tedious compared to other methods but it should be remembered that when four chromatograms have been completed one determination of all amino acids will have been made and that such information can be collected in less than two weeks with even a modest amount of equipment. If one considers how much time and effort must often be expended in the chemical determination of just one amino acid [Reference may be made to the excellent compilation of such methods by BLOCK and BOLLING (*8*)], it will be abundantly clear that in starch chromatography we have a more rapid and, most important, a more precise method for determining amino acids. This praise of the starch chromatographic method should not be taken to mean that the determination of the amino acid composition of a protein is now only a two-week's task: the accurate determination of the amino acid composition of a protein is still an arduous task even by this method. The initial acquisition and construction of equipment, the gaining of experience with the method, the running of replicate determinations on replicate hydrolyses of the protein are time consuming, and the thousands of colorimetric estimations with the concomitant washing of test tubes are tedious but the effort is likely to yield much valuable information.

b) Separation of Amino Acids on Ion Exchange Resins.

More recently Moore and Stein (126) have described a method for the separation and determination of the amino acids on the sulfonated polystyrene resin, Dowex-50. In the main, the general procedure differs little from that of starch chromatography. Prior to the packing of the chromatographic column, the Dowex-50 (250–500 mesh) is rather elaborately purified and then packed into the tube as a slurry in an appropriate buffer.

Two sizes of column are required: the larger which is 100 × 0.9 cm. is water jacketed in order to maintain elevated temperatures whereas the smaller which is 15 × 0.9 cm. is operated at room temperature. The developers are a series of buffers with p_H values ranging from 3.41 to 11.0 and the temperatures are varied from 25° to 75°. The rate of solvent flow through the colums is adjusted in both instances to 4 ml. per hour and 1-ml. fractions are collected. The colorimetric ninhydrin procedure may be applied to these fractions with only minor modification.

The nature of the chromatogram which is obtained on the 100-cm. columns is shown in *Fig. 2a*. The generally excellent separation of all of the common amino acids has been achieved on a single chromatogram which requires about a week of continuous operation. Despite the fact that all of the amino acids may appear as discrete peaks, the quantitative determination of the basic amino acids requires that they be determined by means of the short column (*Fig. 2b*) from which they may be estimated with the same accuracy and precision as in starch chromatography. The recovery of all of the other amino acids from the 100-cm. columns is quantitative.

c) Comparison and Discussion of Starch and Ion Exchange Methods.

The ion exchange method undoubtedly has definite advantages over the starch method in requiring the use of fewer chromatograms. In addition, biological fluids of high salt content may be chromatographed directly on the ion exchanger without the desalting which is necessary

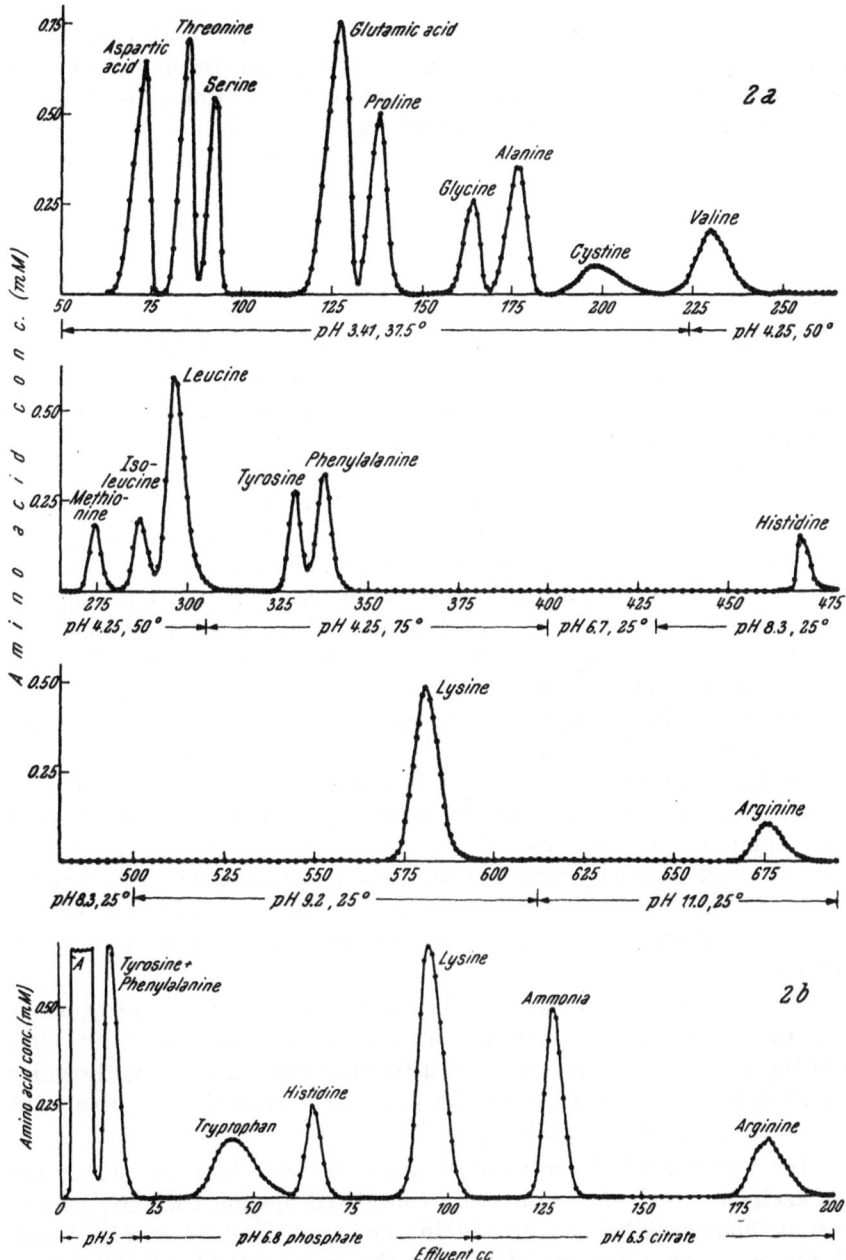

Fig. 2. The separation of a synthetic mixture of amino acids on the sulfonated polystyrene resin, Dowex-50, with buffers of the designatet pH and at the temperature shown. (2 a), Column dimensions, 100 × 0.9 cm. (2 b), Column dimensions. 15 × 0.9 cm. Temperature, 25°. [From: J. Biol. Chem. **192**, 663 (1951).]

before application to starch columns; desalting is known to produce decomposition of arginine [STEIN and MOORE (*192*)]. Furthermore, the greater capacity of the ion exchanger is of especial use in the isolation of pure compounds where starch columns have an additional disadvantage because the effluent is contaminated by carbohydrate. On the other hand, starch chromatography has the advantage of a simpler experimental set-up in that the temperature of the chromatogram need not be rigidly controlled. Changes of developer are no problem in either method but change of temperature which is required during ion exchange chromatography does introduce greater complication. Perhaps the greatest advantage of the ion exchange method lies in the fact that the columns may be used over and over indefinitely and, in addition, may be put out of service for long periods with no loss of efficiency. The reviewer has observed that a column which was out of service for 10 months behaved normally when it was returned to use.

The order with which the amino acids emerge from the two types of adsorbent is very different: compounds which are difficultly separable on the ion exchanger (for example, tyrosine and phenylalanine) part with great ease on starch and *vice versa* (alanine and glutamic acid). It may be that a combination of both methods would be advantageous. For example, the separation of leucine, isoleucine + methionine and phenylalanine + tyrosine requires very close control of conditions on Dowex-50 columns whereas no difficulty is experienced on starch. The resolution of phenylalanine and tyrosine on the ion exchanger seems to be especially difficult, and mention should be made of DUSTIN, CZAJKOWSKA, MOORE, and BIGWOOD's statement (*45*) that this pair was separated on 100-cm. columns by developing with p_H 4.25 at 75° from the beginning. It may actually be that in quantitative work greater accuracy and simplicity would be obtained by discontinuing the 100-cm. columns of Dowex-50 after the emergence of valine and then determining leucine, isoleucine, methionine, phenylalanine, and tyrosine on a starch column.

The efficiency of these adsorbents in effecting the separation of structurally similar compounds is truly amazing: SHULGIN, LIEN, GAL, and GREENBERG (*178*) have separated DL-threonine and DL-allothreonine on Dowex-50, and mention will be made in a later Section about the separation of similar peptides (p. 278).

The writer has had considerable experience with both the starch and ion exchange methods and it is his considered opinion that any user of the methods will be able to obtain the same excellent results which MOORE and STEIN did only if he pays the same meticulous attention to detail in the preparation of adsorbent and columns and in the preparation of developers. This is not meant to say that variations are not possible.

Indeed, they are but they should be made only after sufficient experience has been gained so that the possible effect of the variations can be assessed. Articles which describe the application of these methods to various problems often contain valuable descriptions of variations in method and apparatus and attention will now be called to some of these.

The chromatograms on the 100-cm. columns of Dowex-50 require about a week for completion if they are continued through the emergence of the basic amino acids, but it is possible to discontinue operation, for example, over the weekend, and to resume it without deleterious effect on the separations [Bowes (*12*)]. Moore and Stein (*126*) recommend that samples be placed on the 100-cm. columns from a volume of 1 to 2 ml. at a p$_H$ of 2.5 to 3.0. However, if the p$_H$ is between 1.5 to 2.0 it is possible to increase the volume of sample solvent at least 10-fold without greatly disturbing the separations (*169*). Mention has already been made of the importance of temperature in the separation of tyrosine and valine on starch columns.

Although caution must be exercised in changing the experimental conditions of the chromatograms if satisfactory separations are to be obtained, a much wider latitude is possible in the choice of equipment. The methods obviously require a fraction collector and since Stein and Moore's original description of their drop-counting fraction collector (*189*), no less than 23 collectors have been described although not all were designed for use with starch or ion exchange chromatography. It is possible to choose a design which has a rotating turntable by means of which the receivers are brought successively under the chromatographic column or one in which a moving arm distributes the effluent to stationary receivers. Fractions may be measured by siphoning, drop-counting, weighing, or timing. In some collectors, the turntable or moving arm moves continuously from one receiver to the next whereas in others motion is intermittent. Some are designed to receive only rather large fractions but still others may be varied over a wide range beginning with 0.5 ml. or less. Most machines collect from a single chromatogram in a single row of receivers so that receivers must be removed and replenished when a single revolution is completed; however, one moves in a spiral and a few others have devices for changing from row to row of receivers. Controls vary from elaborate electronic circuits to the very simplest mechanisms. An electronic electrolytic drop recorder such as that described by Marsh and Seibert (*111*) might prove useful in conjunction with a fraction collector. So many variants have been described that the potential constructor should by a combination of devices be able to supply his needs or he may even purchase such a machine commercially.

References to fraction collectors: Boggs, Cuendet, Dubois, and Smith (*10*); Brimley and Snow (*16*); Crook and Datta (*28*); Cuckow, Harris, and Speed (*29*);

DESREUX (38); DIMLER, VAN CLEVE, MONTGOMERY, BAIR, CASTLE, and WHITEHEAD (40); DURSO, SCHALL, and WHISTLER (44); EDELMAN and MARTIN (47); EDMAN (48); FITCH and RUSSELL (55); GILSON (65); GRANT and STITCH (70); HARRIS (80); HICKSON and WHISTLER (83); HOUGH, JONES, and WADMAN (88); JAMES, MARTIN, and RANDALL (89); LIEN, PETERSEN, and GREENBERG (104); MADER and MADER (109); PHILLIPS (136); SCHRAM and BIGWOOD (166); SCHROEDER and COREY (170); STEIN and MOORE (189); VARNER and BULEN (210); and WINGO and BROWNING (217). See also MADER and MADER (110) for accessory equipment.

A number of modifications have been made in the operation of the ninhydrin procedure for the analysis of the effluent fractions. MOORE and STEIN (124) have used a large set of matched calibrated test tubes in which the fraction could be collected, reacted with ninhydrin, diluted, and then read spectrophotometrically without removal from the test tube. SCHROEDER, KAY, and WELLS (175) have used unmatched test tubes; after the reaction had been completed and the spectrophotometric reading was to be taken, the solution was poured into a photometer tube, the optical density was read, the solution was poured out, and after the photometer tube had been drained, the same operation was carried out with the next tube without rinsing the photometer tube. The draining can be made so complete that the error is negligible.

The pipetting machines which MOORE and STEIN have used for the addition of reagents or diluents may be replaced by burets, as they suggest, or the simple semi-automatic pipet recently described by SCHRAM, DUSTIN, MOORE, and BIGWOOD (167). The automatic pipet of NYE (127) is also excellent for the addition of diluent.

For economy and availability, 1 : 1 *iso*propyl alcohol-water (175) or 1 : 1 ethyl alcohol-water (167) may be substituted for 1 : 1 *n*-propyl alcohol-water as the diluent in the ninhydrin procedure. SCHRAM *et al.* give other valuable modifications (167).

d) Application of Starch and Ion Exchange Methods to the Analysis of Peptides and Proteins.

Concurrently with the development of the starch chromatographic method, STEIN and MOORE (191) applied it to the determination of the amino acid composition of *β-lactoglobulin* and *bovine serum albumin*. In both analyses, the sum of the amino acids determined accounted for essentially 100 per cent of the nitrogen in the starting protein and more than 97 per cent of the weight. The chromatographic values were compared with those previously obtained by chemical, microbiological, and isotopic methods. Often the agreement was excellent and where disagreement existed, the chromatographic method sometimes showed the presence of a larger and sometimes the presence of a smaller amount than other methods. In view of the excellent analyses of synthetic mixtures by the starch method, it may well be that the chromatographic results are more accurate than those obtained by other methods some of which are indirect and others of which would have estimated only the *L* antipode and hence be in error if racemization had occurred during hydrolysis.

PIERCE and DU VIGNEAUD (*137, 138*) have used starch chromatography to determine the amino acid composition of two preparations of high potency *oxytocic hormone* obtained from the posterior lobe of the pituitary gland. Hydrolysates of both preparations gave analyses which showed the presence of one equivalent each of leucine, isoleucine, tyrosine, proline, glutamic acid, aspartic acid, glycine, and cystine and three equivalents of ammonia in the molecule. The analytical results which were thus obtained have now been substantiated by the total synthesis of the molecule by DU VIGNEAUD, RESSLER, SWAN, ROBERTS, KATSOYANNIS, and GORDON (*46*). The structure is as follows:*

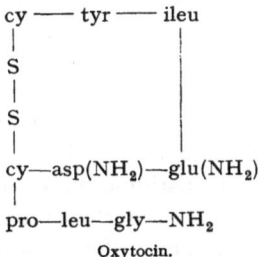

Oxytocin.

Starch chromatography has been used to compare the amino acid contents of the *hemoglobins* of normal negroes and of sickle-cell anemics [SCHROEDER, KAY, and WELLS (*175*)] in an attempt to determine whether differences of amino acid content are responsible for the different electrophoretic mobilities of the two types of hemoglobin. In the course of the analyses, 10 to 15 determinations of each of seventeen amino acids and ammonia were made by means of about 70 starch chromatograms. Statistical consideration of the results indicated the possibility of minor differences in content of several neutral amino acids but not in the content of the acidic and basic amino acids which by their presence or absence would be able to alter the electrophoretic properties of the molecules.

---

* The *abbreviations for peptides* as used here and elsewhere in this review follow the common methods now in use as summarized by SANGER (*158*). The first three (or four) letters of each name usually are used for the name of the amino acid. *ala* = = alanine, *arg* = arginine, *asp* = aspartic acid, *asp-(NH₂)* = asparagine, *cyS-Scy* = = cystine, *glu* = glutamic acid, *glu-(NH₂)* = glutamine, *gly* = glycine, *his* = = histidine, *hypro* = hydroxyproline, *ileu* = isoleucine, *leu* = leucine, *lys* = = lysine, *met* = methionine, *orn* = ornithine, *phe* = phenylalanine, *pro* = proline, *ser* = serine, *thr* = threonine, and *val* = valine. Thus, ala-gly-ileu denotes alanylglycylisoleucine whereas ala-(gly, ileu) designates that alanine contains the free amino group but that the sequence of glycine and isoleucine is undetermined.

Comparison of the amino acid composition as determined by starch chromatography with the many previous fragmentary analyses of human hemoglobin in general showed agreement with at least one previous determination. The contents of ammonia, aspartic acid, glutamic acid, and serine were estimated for the first time. Only one residue of isoleucine per 66,700 molecular weight was shown to be present; the power of the method is clearly indicated by the fact that although in most chromatograms only 10 to 15 micrograms of isoleucine were present, a very definite zone was always observed.

GLANZMANN and SIGNER (68) have determined the amino acid compositions of unfractionated *silk fibroin* and of various fractions of *silk fibroin*. They developed the chromatograms only with 2 : 1 *n*-propyl alcohol-0.5 *N*-HCl, a developer which does not separate such pairs as leucine + isoleucine, tyrosine + valine, alanine + glutamic acid and arginine + ammonia. They concluded that the least soluble fraction contained more leucine + isoleucine, aspartic acid, and arginine + ammonia and less alanine + glutamic acid and glycine than the most soluble.

The amino acid composition of sheep *adrenocorticotropic protein* ("ACTH") has been investigated by MENDENHALL (118) and compared with two independent microbiological analyses on the same sample. The three determinations agree on histidine only but in eight instances two of the three methods agree and the chromatographic data are a part of seven of these agreements. The differences in the other amino acids are not large except that the chromatographic values for methionine are low and for proline are high.

The ratio of amino acids in salmine is such that CORFIELD and ROBSON (24) found it necessary to make minor modifications in starch chromatographic methods in order to achieve satisfactory accuracy in all determinations. They find that isoleucine, alanine, valine, glycine, serine, proline, and arginine are present in the molar ratios of 1 : 1 : 3 : 4 : 7 : 6 : 50 and that the calculated minimum molecular weight is about 10,000. Their estimates of arginine and glycine agree with the values of other authors but their estimates of other amino acids tend to differ. Their quantitative procedure dispensed with the colorimetric ninhydrin method and instead used the radiochemical method of BLACKBURN and ROBSON (7) which employs radioactive copper.

CRAIG and collaborators have employed ion exchange chromatography for the initial analysis of naturally occuring peptides and proteins which had first been purified by the CRAIG method of countercurrent distribution (26). The analytical results of HAUSMANN and CRAIG (82) show that *polypeptin*, an antibiotic from cultures of *Bacillus krzemieniewski*, contains three residues of *L*-α,γ-diaminobutyric acid, two residues of *L*-leucine, one each of *L*-threonine, *D*-valine, *L*-isoleucine, and *D*-phenylalanine, and an unidentified acid. The quantitative data were determined by the ion exchange method but countercurrent distribution was used for isolation in the quantity required for analysis and the determination of configuration.

In like manner, CRAIG, HAUSMANN, and WEISIGER (27) have determined that *bacitracin A* is composed of two residues of *L*-isoleucine, two residues of aspartic acid (isolated as *DL*), and one residue each of *D*-phenylalanine, *L*-leucine, *L*-cysteine, *D*-glutamic acid, *L*-histidine, *L*-lysine, and *D*-ornithine. PORATH (*141*) isolated *bacitracin A* by carrier displacement chromatography on charcoal and has recently (*142*) shown its structure to be the following:

$$
\begin{array}{c}
\text{S}\!-\!\!\!\! \\
| \\
R\!-\!\text{lys}\!-\!\text{glu}\!-\!\text{cy}\!-\!\text{ileu}\!-\!\text{leu} \\
\quad\quad\quad\ | \quad\quad\quad\quad\ | \\
\quad\quad\ \text{asp} \quad\quad\quad\ \text{phe} \\
\quad\quad\quad\ | \quad\quad\quad\quad\ | \\
\quad\text{asp}\!-\!\text{his}\!-\!\text{orn}\!-\!\text{ileu}
\end{array}
$$

Bacitracin A.

The nature of *R* is unknown but it is not an amino acid. The ε-amino group of lysine is not free and the sulfur of cysteine is present in a heterocyclic ring of undetermined structure.

Although physical methods, physiological activity, and immunological specifity have not shown any distinguishing differences in *insulins* from various species, SANGER (*157*) obtained evidence that serine, glycine, threonine, and alanine were present in different amounts in the *A* fractions of oxidized beef, pork, and sheep insulins. LENS and EVERTZEN (99) analyzed beef and pork insulins by starch chromatography but were unable to obtain convincing proof of dissimilarity although SANGER was able to repeat his results with their samples of the two insulins. HARFENIST and CRAIG (*76a, 77*) have now placed these differences on a quantitative basis by means of ion exchange chromatography of hydrolysates of

Table 1. Differences in the Amino Acid Content of Insulins from Several Species.

Amino Acid	Type of Insulin		
	Beef	Pork	Sheep
Serine			—1*
Threonine...........................		+1	
Glycine..............................			+1
Alanine..............................		—1	
Valine		—1	
Isoleucine		+1	

* The designations +1 or —1 indicate the difference in the number of residues of a given amino acid in a given type of insulin as compared to the other two kinds.

insulins which had been purified by countercurrent distribution. The results are presented in *Table 1*, p. 253. Note that the total number of residues is the same in each insulin.

HAUSMANN (*81*) has found the usual array of amino acids in crystalline *inorganic pyrophosphatase* with the exception of cystine and hydroxy-proline and with the addition of a component which may be ethanolamine. Quantitative analysis by ion exchange methods accounts for 98 per cent of the weight and 100 per cent of the nitrogen in the protein.

In a new determination of the basic amino acids in *lysozyme*, MONIER, GENDRON, JUTISZ, and FROMAGEOT (*123*) have confirmed the original results of LEWIS, SNELL, HIRSCHMANN, and FRAENKEL-CONRAT (*100*) by microbiological methods which showed the presence of 1 residue of histidine, 6 residues of lysine, and 11 residues of arginine in the molecule of lysozyme.

SIMMONDS (*179*) has recently made the first definitive analysis of the leucine and isoleucine contents of *wool*.

The amino acid composition of T_3 *bacteriophage* as analyzed by FRASER and JERREL (*62*) has no unusual features. The host bacterium may be grown under a variety of conditions without influencing the amino acid composition of the T_3 bacteriophage which is qualitatively parallel to that of T_4 bacteriophage.

SMITH and STOCKELL (*182*) find that the amino acid composition of *carboxypeptidase* agrees with a minimum molecular weight of 34,440. All amino acids are present but the content of the sulfur amino acids is low and that of serine and threonine is high.

Crystalline *papain* has no methionine, one histidine residue, much tyrosine, and, in addition, the other common amino acids [SMITH, STOCKELL, and KIMMEL (*183*)].

In their analyses of carboxypeptidase and papain, *Smith* and co-workers noted the presence of peptides in some hydrolysates and were led to determine the effect of extended periods of hydrolysis upon the content of the various amino acids. They found that serine, threonine, aspartic acid, and lysine decreased in quantity with time apparently because of destruction; valine leucine, and isoleucine increased in quantity apparently because of release from peptides; and the other amino acids were not influenced by extended hydrolysis. The ease of hydrolysis as well as the extent of destruction differed in each protein.

The methods of ion exchange chromatography have also been applied to problems which are not directly concerned with the amino acid composition of proteins. Thus, STEIN has studied the excretion of amino acids in cystinuria (*187*) and also the amino acid constituents of normal urine (*188*) while TALLAN (*198, 199 a*) has identified the compound "X" which STEIN had found in normal urine as 3-methylhistidine.

Moore and collaborators have published several papers dealing with the determination of amino acids in *foods*. The first one [Schram, Dustin, Moore, and Bigwood (*167*)] is concerned mainly with methodology. The second paper [Dustin, Czajkowska, Moore, and Bigwood (*45*)] discusses what effect hydrolysis in the presence of large amounts of carbohydrate has on the determination of amino acids by ion exchange chromatography. When synthetic mixtures of amino acids and starch or glucose in which the amino acids constituted only 2 per cent of the mixture were hydrolyzed in large volumes of 6 N-HCl (200 ml. per 1 to 2 g. of amino acids and starch), the decomposition products of the carbohydrates did not interfere with the chromatographic determination nor was the observed recovery of an amino acid lowered by as much as 3 per cent. The methods have been applied to the determination of the amino acids in cassava flour and in human milk (*22 a*, *186 a*).

As previously mentioned, Shulgin, Lien, Gal, and Greenberg (*178*) have separated *DL*-threonine and *DL-allothreonine* on Dowex-50. Lien and Greenberg (*103*) have also used it in studies of the interconversion of amino acids *in vivo* and *in vitro*. They have chromatographed on the hydrogen form of Dowex-50 by a procedure which has been briefly described by Stein and Moore (*190*) and which used 1.5 *M*-HCl, 2.5 *M*-HCl, and then 4.0 *M*-HCl as developers. The incomplete separation of certain amino acids and the strong acids which must be used with the hydrogen form make the method much less useful when sensitive substances are to be chromatographed and have led to the use of the sodium form and the methods which have been described above.

2. Isolation of Amino Acids by Chromatography.

The analytical methods which we have been discussing have employed micro quantities of sample and the colorimetric determination of the individual amino acids. Often, however, the actual isolation of an amino acid on a macro scale may be necessary. Such needs may arise when the absolute identification of an amino acid must be made by elementary analysis and chemical characterization, when the isolation of an amino acid from natural sources constitutes the best method of preparation, or when isotopically labeled amino acids must be isolated from hydrolysates of proteins which had been formed biosynthetically *in vivo* by various organisms. Although the starch and ion exchange methods were devised on a microanalytical scale, the columns may nevertheless be increased to 8 cm. in diameter without loss of efficiency (*189*), and this fact naturally has led to their use in isolative procedures. Thus, Åqvist (*1 a*) isolated N[15]-labeled amino acids by means of starch columns and Dowex-50 columns in the hydrogen form. The chromatograms were 8 cm. in diameter and the starch had sufficient capacity for the hydrolysate of 250–300 mg.

of protein while that of the Dowex-50 was for 25–50 g. Before the ion exchange chromatography was used, an electrophoretic separation into acidic, basic, and neutral amino acids was made and certain final separations used chemical means. Considerable difficulty was met in obtaining crystalline products which were free of contaminants from the starch or resin. FRANTZ, FEIGELMAN, WERNER, and SMYTHE (61) voiced a similar complaint of contamination by carbohydrate although this disadvantage was not serious for their purposes. They used *Thiobacillus thiooxidans* for the biosynthesis of protein with C^{14} and separated the randomly labeled amino acids on 2.1-cm. starch columns by the usual developers.

If the ion exchange procedure in the sodium form is used on a preparative scale, one is immediately faced with the problem of desalting the effluent before the isolation can be made. Electrolytic desalting has been widely used but the danger of decomposition is always present [STEIN and MOORE (192)]. Desalting by ion exchange would seem to be of great advantage although no generally applicable method seems to have been devised. Reference may be made to the following papers for various applications of desalting: BRENNER and FREY (15), HIRS, MOORE, and STEIN (86), PIEZ, TOOPER, and FOSDICK (139), STEIN (188), and TALLAN and STEIN (199). If the isolation of free amino acids or peptides is not essential, desalting can also be effected through dinitrophenylation [SCHROEDER, HONNEN, and GREEN (172)].

The disadvantages of contamination or desalting are eliminated (1.) in the method of HIRS, MOORE, and STEIN which uses several ion exchange resins and volatile buffers with a sample of 2.5 g. and (2.) in the method of PARTRIDGE and collaborators which uses displacement development on ion exchangers with a sample of 10 to 280 g.

a) Method of HIRS, MOORE, and STEIN (85).

Fig. 3 presents a flow sheet of the successive chromatograms by means of which HIRS, MOORE, and STEIN (85) separate a mixture of amino acids on Dowex-50 and Amberlite IR-4 B through development with ammonium formate and ammonium acetate buffers. Although, as we have seen in Fig. 2, p. 247, the sodium form of Dowex-50 is able to resolve all amino acids on a single chromatogram, ammonium Dowex-50 is less satisfactory and several overlaps occur. As a result the more complex series of chromatograms outlined in *Fig. 3* is necessary.

The chromatographic columns were 7.5 cm. in diameter and were run at 25°. Between chromatograms, the solvent and buffer must be removed; the solvent was removed by means of a rotating Craig evaporator [CRAIG, GREGORY, and HAUSMANN (25)] and the buffer by sublimation in a flask surrounded by an electric heating mantle. The sublimations require 12 to 20 hours. After each amino acid had been resolved as a single peak on a chromatogram, the isolation was done in the same way and followed by crystallization. It should be noted that the analysis of the effluent fractions by the ninhydrin procedure requires a preliminary sublimation of the buffers because the presence of ammonia interferes with the colorimetric method. The entire procedure is carefully described in the original paper (*85*).

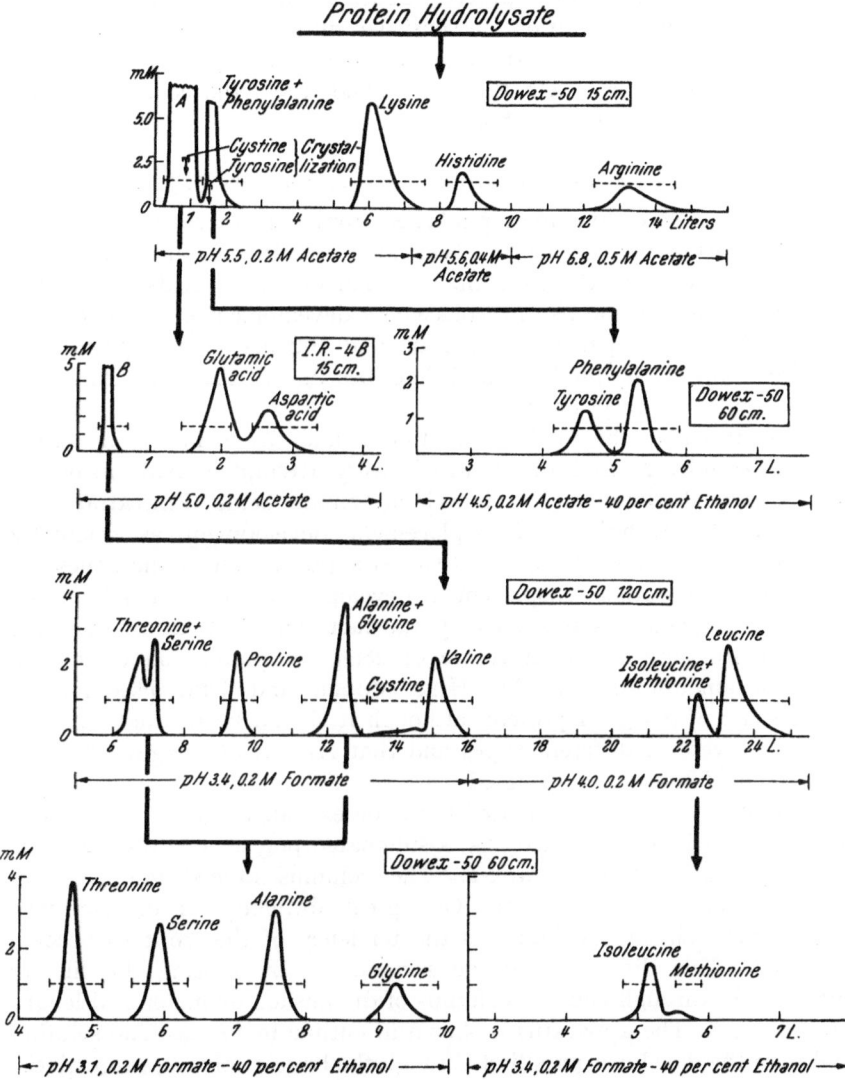

Fig. 3. Isolation of amino acids on ion exchange chromatograms through the use of volatile buffers. Columns, 7.5 cm. in diameter. Sample, acid hydrolysate of 2.5 g. of bovine serum albumin. [From: J. Biol. Chem. **195**, 669 (1952).]

When this scheme was applied to an acid hydrolysate from 2.5 g. of bovine serum albumin, all of the amino acids with the exception of methionine were isolated in an analytically pure form with an average yield of 66 per cent. Each amino acid was present as the L antipode except for cystine which was about half racemized.

The milder conditions of a procedure such as this make it more generally applicable to the separation of sensitive substances such as peptides which one would hesitate to expose to the strong acids required as developers by the hydrogen form. On the other hand, the required sublimations certainly complicate the method.

b) Method of PARTRIDGE and Collaborators (131).

In a series of eight papers published from 1949 to 1952, PARTRIDGE in collaboration with BRIMLEY, WESTALL, and PEPPER has traced the development of a method for the isolation of amino acids on a large laboratory scale by means of displacement development on ion exchange resins. The eighth paper [PARTRIDGE and BRIMLEY (131)] provides a systemization of the methods and describes its full scale application to egg albumin and to yeast protein.

In contrast to the methods of MOORE and STEIN in which elution chromatography is used and in which every attempt is made to obtain well separated zones of the individual amino acids, PARTRIDGE has employed the technique of displacement development as originally described by TISELIUS (205) and CLAESSON (21) in which the zones are adjacent to each other. Displacement development enjoys the advantage of greater capacity as is shown by the fact that PARTRIDGE was able to chromatograph the hydrolysate of 280 g. of protein on essentially the same amount of resin that HIRS, MOORE, and STEIN used for the hydrolysate of 2.5 g. of protein, although it should be pointed out that the resins were of different types and that HIRS et al. suggest that the sample could have been doubled.

PARTRIDGE and BRIMLEY used three ion exchange resins in the course of the separations: Zeo-Karb 215, a sulfonated polystyrene resin of their own preparation, and Dowex-2. The columns ranged in size from 61 × 7.6 cm. to 4.8 × 0.8 cm. One great difficulty of displacement development on large columns is the tendency of the front to become distorted. This difficulty can be surmounted by passing the filtrate successively through coupled columns of diminished diameter and length (22, 76, 133). The apparatus is shown in outline in *Fig. 4*. The solution and then the developer pass first through the largest column, successively through two smaller columns, and finally to the fraction collector.

Before the primary fractionation of a hydrolysate was made, tyrosine and phenylalanine which behave irregularly on Zeo-Karb 215 were removed by treatment with charcoal and this treated solution was then passed into four columns of which the first is sulfonated polystyrene and the others are Zeo-Karb 215. After the sample had been placed on the columns and a little water passed through, the polystyrene section was disconnected and the Zeo-Karb 215 sections were developed with 44 liters of 0.15 N-NH₃ at the rate of 20 ml. per min. and 250-ml. fractions were collected. The sulfonated polystyrene section which contains the basic amino acids was set up as part of a new multiple column and developed with 10 liters of 0.075 N-NaOH at the rate of 6 ml. per min. in 250-ml. fractions.

Fig. 5 shows the results which were obtained from such a primary fractionation of the hydrolysate of 280 g. of egg albumin. The progress of the separation was assessed through a paper chromatogram of each fraction by means of which the qualitative nature of the amino acids was determined. On this basis, the various fractions were

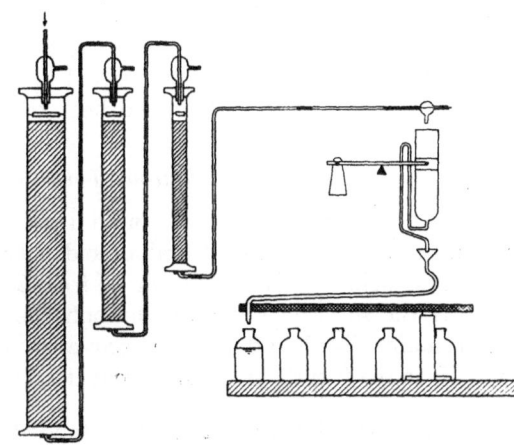

Fig. 4. Apparatus for fractionation of amino acid mixtures by displacement chromatography. [From: Biochemic. J. **51**, 628 (1952).]

combined to give Bands I to VII each of which with the exception of Band I was a mixture. The mixtures were then separated by further

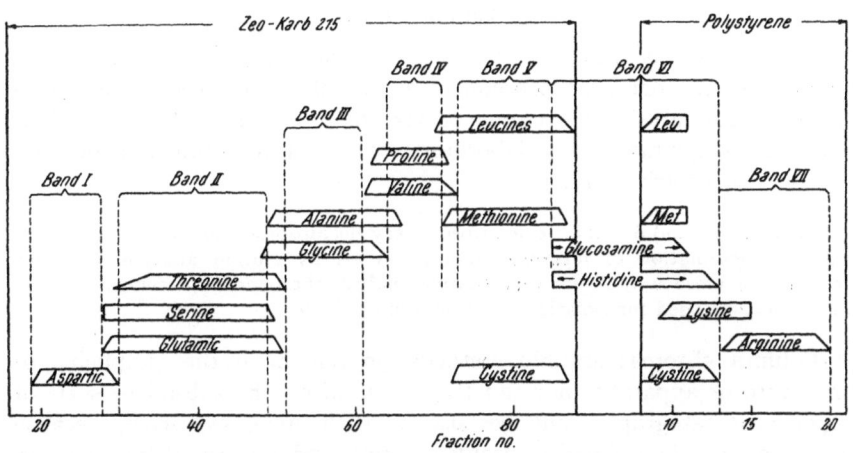

Fig. 5. Primary fractionation of hydrolysis products of 280 g. of commercial egg albumin by displacement chromatography on ion exchange resins. [From: Biochemic. J. **51**, 628 (1952).]

chromatograms under other conditions and finally all were isolated in pure form except the mixture of leucine and isoleucine.

PARTRIDGE and BRIMLEY have applied this systematic procedure on two very different scales: to hydrolysates of 10 g. of yeast protein and of

280 g. of egg albumin. They were able to recover 60 per cent of the dry weight of the yeast protein and 54 per cent of the egg albumin. The percentage of recovery is of the same order as that of HIRS, MOORE, and STEIN.

WESTALL (*213*) has used this method to isolate 3 g. of γ-amino-butyric acid from 18 kg. of fresh trimmed beet roots.

c) Miscellaneous Isolative Methods.

BAKER and SOBER (*3*) have used Dowex-50 or Amberlite XE-64 in the enzymatic resolution of amino acids. Enzymatic action on an acyl-*DL*-amino acid yields a solution of free *L* antipode and acyl-*D*-amino acid. If the solution is passed through the ion exchanger, the *L*-amino acid remains fixed and can be recovered while the acyl-*D*-amino acid passes into the filtrate and can be converted to the free form by hydrolysis.

DOBYNS and BARRY (*42*) have extended the starch chromatographic methods to the separation and isolation of iodinated amino acids. Thyroxine, iodide, monoiodotyrosine, and diiodotyrosine may be separated on starch by developing with $1:2:1$ *n*-butyl alcohol-*n*-propyl alcohol-0.05 N-Na$_2$CO$_3$ followed by $1:2:1$ *n*-butyl alcohol-*n*-propyl alcohol-0.1 N-HCl.

3. Conclusions.

In the methods of MOORE and STEIN and of PARTRIDGE and BRIMLEY, column chromatographic procedures are available for the quantitative amino acid analysis of proteins or for the isolation of the individual amino acids on a rather large laboratory scale. As PARTRIDGE and BRIMLEY themselves remark (*131*):

"... MOORE and STEIN have described a very elegant procedure for the chromatographic fractionation of mixtures of amino-acids by elution analysis ... This procedure is more suitable for quantitative analysis than the displacement method and it may be used for isolation work on a small scale."

Column chromatographic methods do not have the simplicity of procedure or apparatus nor can they be used on the sub-micro scale of paper chromatography but because of their increased scale, because substances can be isolated in usable amount, and because of their greater reproducibility, they permit more direct and definitive conclusions than paper chromatographic methods can. Column chromatographic methods undoubtedly require the expenditure of much time and an appreciable outlay of funds before all necessary equipment has been acquired and the method has been placed into operation, but the repeated use to which ion exchange columns may be put and the automatic mechanisms with which they may be equipped greatly decrease the attention and labor from that point on.

We shall probably see little further simplification in the chromatographic methods for the quantitative analysis of mixtures of amino acids. We may hope that simplification of isolative procedures will be made although even now they are not unduly complex. The usefulness of the methods is only now beginning to be appreciated and the future should demonstrate their application to an increasing number of problems.

II. The Determination of Amino Acid Sequence in Proteins. The Identification of Terminal Residues and the Separation and Identification of Peptides.

Once the qualitative and quantitative amino acid composition of a purified protein has been determined, the interest immediately centers on finer details of structure. If the fundamental accuracy of the FISCHER-HOFMEISTER theory of protein structure is assumed, each open chain of an uncyclized protein will contain a free amino group at one end and a free carboxyl group at the other, exclusive of any ε-amino groups of lysine and of any carboxyl groups of aspartic acid and glutamic acid. The amino acid residue with the free amino group has come to be known as the "N-terminal residue" and that with the free carboxyl group as the "C-terminal residue" [SANGER (*158*)]. Both the N-terminal and the C-terminal group would be absent if the protein were cyclic or either might be absent if the N-terminal group formed an atypical peptide bond with the ω-carboxyl group of aspartic acid or glutamic acid, if the C-terminal group formed an atypical peptide bond with the ε-amino group of lysine, or if various covalent linkages which have been suggested from time to time were present. Certain conditions such as the combination of terminal residues with prosthetic groups as well as the steric inaccessibility of terminal residues may lead to erroneous conclusions about cyclization in proteins.

The desire to determine the nature of N- and C-terminal groups is not new although as recently as 1945, FOX (*57*) reviewed suggested methods of determining such groups but could not give an example of successful application to any protein. Methods for the identification of N-terminal groups have made great progress since that time but those for the identification of C-terminal groups still lag behind.

Even if the N- and C-terminal residues of a protein have been identified, knowledge of its structure is still very meager and attempts to delve further into the molecule and to find the complete sequence of amino acids can follow one of two lines: (1) successive degradation of the protein

from either the N- or C-terminal end coupled with the identification
of the amino acid which is removed at each degradation or (2) the isolation
and identification of peptides from partial hydrolysates in sufficient
variety to permit the sequence of all amino acids to be deduced. The
latter is the more popular method of attack and in the hands of SANGER
and collaborators (159, 160, 161, 162) has led to an almost complete
determination of the structure of insulin. The former method may find
application soon if the promise of the EDMAN method (to be described
below) is fulfilled.

In this Section we shall discuss first the methods for the identification
of N- and C-terminal residues and then methods for the isolation and
identification of peptides from partial hydrolysates of peptides and
proteins.

1. Identification of N-Terminal Amino Acids.

a) SANGER's *Method: The Use of 2,4-Dinitrofluorobenzene.*

Principle. SANGER's method (153) for the identification of N-terminal
residues is the only one which has received extensive successful application.
Briefly, the principle of the method in this: The protein (or peptide)
in aqueous alcoholic solution is reacted with 2,4-dinitrofluorobenzene
("DNFB") in the presence of sodium bicarbonate. The primary reaction
is as follows:

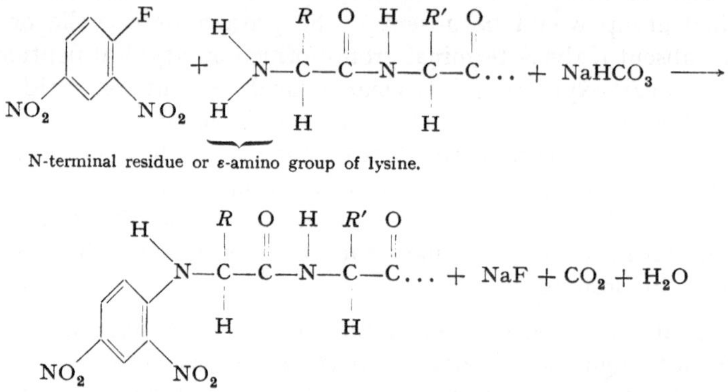

N-terminal residue or ε-amino group of lysine.

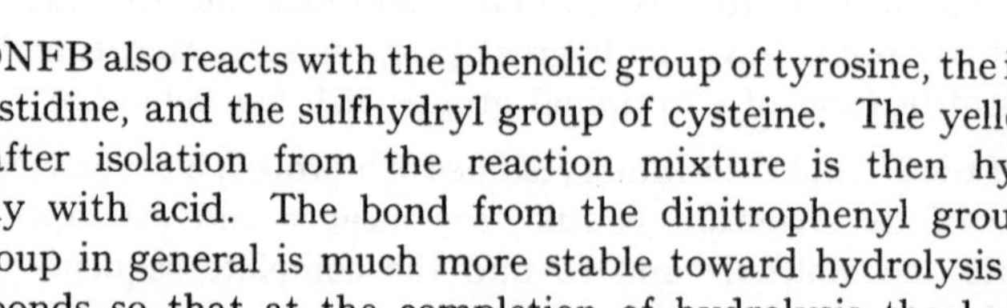

Dinitrophenylprotein ("DNP-protein")

The DNFB also reacts with the phenolic group of tyrosine, the imidazole
ring of histidine, and the sulfhydryl group of cysteine. The yellow DNP-
protein after isolation from the reaction mixture is then hydrolyzed
completely with acid. The bond from the dinitrophenyl group to the
amino group in general is much more stable toward hydrolysis than are
peptide bonds so that at the completion of hydrolysis the hydrolysate

contains the N-terminal DNP-amino acid(s), ε-DNP-lysine, etc., and free amino acids from the remainder of the molecule. The extent of destruction of DNP-amino acids during hydrolysis is dependent upon the nature of the individual amino acid. Because the DNP-amino acids with the exception of ε-DNP-lysine, α-DNP-arginine, and O-DNP-tyrosine are no longer dipolar ions, they may be extracted from the hydrolysate with organic solvents and then identified. It is mainly from this point onward, the identification of the DNP-amino acids, that this review is concerned and we shall not delve into the chemistry of the reaction.

Procedures. After the DNP-amino acids had been extracted from the hydrolysate of the DNP-protein, SANGER turned to partition chromatography on silica gel as a means for separating and identifying the DNP-amino acids and used methods of partition chromatography similar to those of MARTIN and SYNGE (*113*) for the separation of acetylamino acids. SANGER (*153, 155*) and PORTER and SANGER (*148*) describe the details of the fractionations. The scheme involves an initial fractionation into five groups from which single zones of all of the common DNP-amino acids (except DNP-leucine and DNP-isoleucine) may be obtained by rechromatographing at least once and sometimes three times. Unfortunately, as SANGER mentions in these papers, the *R* values of the DNP-amino acids under these chromatographic conditions are greatly influenced by such factors as the particular batch of silica gel, the distance which the zone has travelled down the column, the overloading of the column, and the water content of the silica gel. In addition, the DNP-amino acids are not very soluble in the solvents which are used and the "tailing" of zones sometimes decreases the degree of separation. These difficulties have led several workers to devise extensive modifications of SANGER's original procedure in the hope that more reproducible separations could be obtained. Each of the schemes which has been published will now be discussed briefly.

Both BLACKBURN (*6*) and MIDDLEBROOK (*119, 120*) have modified SANGER's procedure by using buffered columns of silica gel. Buffering apparently does have the advantage that it reduces the differences between various batches of silica gel but BLACKBURN notes that the *R* values are still dependent on the amount of compound and the position on the column. MIDDLEBROOK does not discuss the effects of loading on the *R* values on buffered columns but he provides abundant evidence of the influence of this factor and also of temperature on the *R* values on unbuffered columns.

PERRONE (*135*) has described a method for the separation of the DNP-amino acids on buffered Celite columns. Celite is a purified diatomaceous earth and possibly acts as an inert support only. PERRONE cautions against overloading the column because of the tailing and decreased separation which ensues.

Kieselguhr has been used as a supporting medium by KNESSL, KEIL, MALÝ, and ŠORM (93) and by MILLS (122). The acid-washed and ignited material of the former authors probably differed little from the Celite which PERRONE used. KNESSL et al. do not present a complete scheme of separation. MILLS activated his kieselguhr by shaking it in the presence of solid ammonium carbonate. His procedure is the only one yet devised in which it is possible to separate the entire mixture of common DNP-amino acids by a single passage through one column which is developed with a series of developers. He notes that unexplained failure to achieve certain separations at all times indicates that not all variables have been completely controlled in the the the method. Furthermore, he remarks that "If one band was not seen, it was easy to identify some of the remainder wrongly."

PARTRIDGE and SWAIN (132) do not present a complete scheme of separation but they have chromatographed some DNP-amino acids on a reversed-phase partition chromatogram using chlorinated rubber.

The preceding methods for separating DNP-amino acids all purport to use partition chromatography. The borderline between so-called partition chromatography and adsorption chromatography is indeed a vague one but the method of GREEN and KAY (73) presumably is mainly adsorption chromatography on silicic acid-Celite. The developers are very different from any previously used; most are ternary mixtures in various proportions in which acetic acid or formic acid must be included and which usually contain acetone or ethyl acetate in addition to the main component, namely, ligroin, benzene, or cyclohexane. The present writer admits to prejudice in favor of the method of GREEN and KAY because it was devised in these laboratories and he has used it extensively, but he would like to point out certain features in which it does not have the admitted imperfections of other methods. In common with most other procedures, the scheme of GREEN and KAY will separate 16 common DNP-amino acids with the exception of DNP-leucine and DNP-iso-leucine; this requires nine chromatograms. The method was tested on twelve commercial lots of silicic acid and nine were satisfactory without further treatment although the adsorptive strengths differed somewhat. Recently, because of the depletion of the silicic acid which has been in use for several years in these laboratories, six commercial lots were tested: two were unsatisfactory, three were usable but had certain undesirable features and one was very satisfactory. In this method, the quantity of the DNP-amino acids can be varied over a wide range without appreciable effect on the width or movement of the zone; the zones move uniformly down the column, and the distribution of material in the zones appears to follow the Gaussian pattern found by TRUEBLOOD and MALM-BERG (207) without any tailing.

One failing of the method of GREEN and KAY lies in the lack of quantitative recovery from the column. If known amounts of DNP-amino acids or DNP-peptides are chromatographed and rechromatographed, all compounds show a consistent loss of about 7 per cent per chromatogram (168). The loss is so reproducible than an empirical correction factor can be used with some confidence. Neither SANGER, BLACKBURN, nor MIDDLEBROOK report quantitative experiments of just such a type in which a known quantity of material is placed on the column, removed, and quantitatively determined. PERRONE states that the DNP-amino acids may be recovered quantitatively from the Celite columns, and KROL (94) has used PERRONE's method for the quantitative determination of glycine. KNESSL, KEIL, MALÝ, and ŠORM (93) have obtained recoveries of 94 to 96 per cent. MILLS reports quantitative recovery. From the chlorinated rubber columns, PARTRIDGE and SWAIN recovered DNP-serine quantitatively but failed to do so with di-DNP-lysine and di-DNP-tyrosine and note that DNP-glycine seems to decompose partially.

The above papers contain much detail relative to the application and use of the DNP-method in the study of protein structure and the reader is also referred to the following papers which treat various aspects of the method: DICKMAN and ASPLUND (39); LOWTHER (108); MILLS (121); PORTER (144, 146); SCHROEDER and LE GETTE (176); A. R. THOMPSON (200); and many of the references of Table 2.

Applications. Since SANGER's original description of the use of DNFB in the determination of the N-terminal residues of insulin, the application of the method has become widespread and continues to increase in usefulness. *Table 2* lists the proteins which have been studied and the N-terminal residues which have been detected. A few of these results have been obtained through isolation of the DNP-amino acids by paper chromatography.

The 57 proteins which are listed in Table 2 show a great variety of amino acids as N-terminal residues. In fact, only histidine, hydroxy-proline, and tryptophan of the common amino acids have not been reported as such. In those instances in which the molecular weight is large, one is probably justified in treating the quantitative results with some skepticism. None of the DNP-amino acids is entirely stable to acid hydrolysis and a more or less arbitrary correction factor must be used in calculating quantitative results. When the molecular weight is large so that the actual quantities determined are small, and when the number of calculated residues is of the order of 5, the correction to be applied may well be equivalent to one or more residues per molecule.

Table 2. N-Terminal Residues of Some Proteins as Determined by the Use of 2,4-Dinitrofluorobenzene (DNFB).

Protein	Assumed molecular weight	N-Terminal amino acid(s)	Number per assumed molecular weight	References
Avidin	?	Alanine	3	Fraenkel-Conrat and Porter (60)
Carboxypeptidase	34,000	Asparagine	1	E. O. P. Thompson (203)
α-Casein	100,000	Arginine	10.7	
β-Casein	100,000	Lysine	1.5	Mellon, Korn, and Hoover (117)
		Arginine	5.3	
		Lysine	2.4	
α-Chymotrypsin	21,500	Alanine	1	Desnuelle, Rovery, and Fabre (37)
		Leucine	1	
β-Chymotrypsin	25,000	Alanine	1	
		Leucine	1	
γ-Chymotrypsin	25,000	Alanine	1	Rovery, Fabre, and Desnuelle (152)
		Leucine	1	
Chymotrypsinogen	—	None	—	Desnuelle, Rovery, and Fabre (36)
Clupein	?	Proline	?	Felix, Fischer, Krekels, and Rauen (54)
		Serine	?	
	?	Proline	1?	Šorm and Šormová (186)
		Serine	?	
Collagen	—	None		Bowes and Moss (13, 13a)
		None		Grassmann and Hörmann (72)
Conalbumin	?	Alanine	1	Fraenkel-Conrat and Porter (60)
Corticotropin-A	?	Serine	1	Landmann, Drake, and White (96a)
Edestin	300,000	Glycine	6	Sanger (156)
		Leucine	1	
Fibrin (human)	220,000	Tyrosine	1	
	120,000	Glycine	1	Lorand and Middlebrook (107)
Fibrinogen (human)	220,000	Tyrosine	1	
	300,000	Alanine	1	

Protein	Molecular weight	N-terminal amino acid	Number	Reference
Fibrin (bovine)	450,000	Tyrosine	2	LORAND and MIDDLEBROOK (105)
		Glutamic acid	1	
Fibrinogen (bovine)	450,000	Tyrosine	2	LORAND and MIDDLEBROOK (106)
		Glycine	4	BOWES and MOSS (13, 13a)
Fibrino-peptide	?	Glutamic acid	—	GRASSMANN and HÖRMANN (72)
Gelatin	—	Dependent on starting material	—	
γ-Globulin				
Human	?	Aspartic acid	1	McFADDEN and SMITH (116)
		Glutamic acid	1—2	PUTNAM (149)
Rabbit (active)	160,000	Alanine	1	PORTER (145)
(inactive)	160,000	Alanine	1	SANGER (154)
Gramicidin S	—	δ-Amino group of Ornithine	1	
Hemoglobins				
Horse and donkey	66,000	Valine	6	
Ox, sheep, and goat	66,000	Valine	2	PORTER and SANGER (148)
Human, adult	66,000	Methionine	2	
		Valine	5	
Human, fetal	66,000	Valine	2—3	
Insulin				
Ox, pig, and sheep	12,000	Phenylalanine	2	SANGER (153)
		Glycine	2	
β-Lactoglobulin	40,000	Leucine	3	PORTER (143)
Lysozyme	14,000	Lysine	1	F. C. GREEN and SCHROEDER (74)
			0.5	A. R. THOMPSON (200)
			1	JUTISZ and PÉNASSE (92)
Myeloma proteins	?	Aspartic acid	2	PUTNAM (149)
Myoglobin				
Horse	17,000	Glycine	1	PORTER and SANGER (148)
Whale	17,000	Valine	1	SCHMID (165)
Myosin (rabbit)	—	None		BAILEY (2)
Ovalbumin	—	None		DESNUELLE and CASAL (34)

Protein	Assumed molecular weight	N-Terminal amino acid(s)	Number per assumed molecular weight	References
Ovomucoid	?	Alanine	1	FRAENKEL-CONRAT and PORTER (60)
"Old" yellow enzyme	?	Aspartic acid	?	WEYGAND and JUNK (214)
		Glutamic acid	?	
Pancreatic trypsin inhibitor	9,000	Arginine	1	N. M. GREEN and WORK (75)
Papain	20,000	Isoleucine	1	E. O. P. THOMPSON (204)
Pepsin	35,000	Leucine	1	WILLIAMSON and PASSMANN (216)
Ribonuclease	?	Lysine	1	PORTER (147)
Salmine	—	Proline	?	PORTER and SANGER (148)
Serum albumin				
Bovine, human, and horse	69,000	Aspartic acid	1	DESNUELLE, ROVERY, and FABRE (35) / VAN VUNAKIS (209)
Pork	69,000	Aspartic acid	1	DESNUELLE, ROVERY, and FABRE (35)
Somatotropin	47,000	Alanine	1	LI and ASH (101)
		Phenylalanine	1	
Subtilin	?	Two unidentified sulfur diamino acids ?		CARSON (17) / BAILEY (2)
Tropomyosin (rabbit)	—	None		
α-Trypsin	20,000	Isoleucine	1	ROVERY, FABRE, and DESNUELLE (151)
Trypsinogen	20,000	Valine	1	
Wool keratin	1,600,000	Valine	4	MIDDLEBROOK (120)
		Alanine	2	
		Glycine	8	
		Threonine	8	
		Serine	2	
		Glutamic acid	2	
		Aspartic acid	1	

b) EDMAN's *Method: The Use of Phenylisothiocyanate.*

Principle. EDMAN's method (*50*) is based upon the following series of reactions:

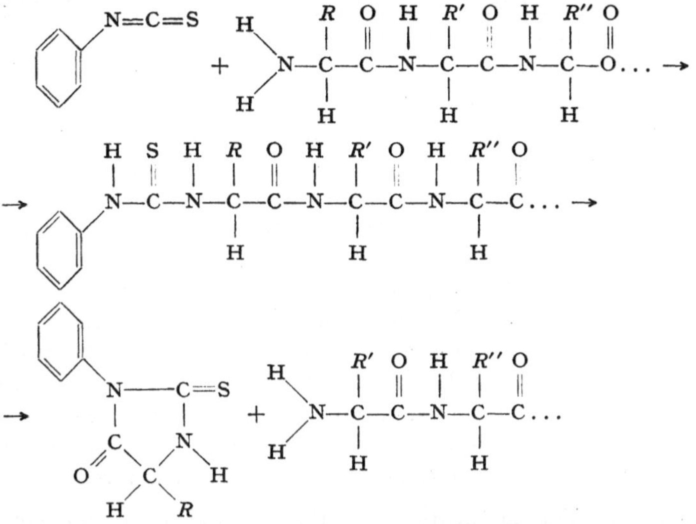

In the first reaction, the coupling of phenylisothiocyanate with the peptide or protein is carried out in pyridine-water 1 : 1 at 40°, and pH 9 is maintained by the addition of small portions of sodium hydroxide. The resulting compound is a phenylthiocarbamyl ("PTC") derivative. The cleavage of the PTC derivative is produced by the action of a saturated solution of hydrochloric acid in nitromethane or glacial acetic acid (*51*). This cleavage results in the formation of the phenylthiohydantoin ("PTH") of the N-terminal amino acid and of a peptide or protein which has been decreased in length by one amino acid residue. The advantage of the EDMAN method lies in this feature that the N-terminal amino acid may be removed under conditions which do not break the other peptide bonds of the molecule. Thus, it is possible to continue to degrade the peptide chain by reacting the remainder with phenylisothiocyanate, again cleaving the molecule and so continuing along the chain.

Procedures. EDMAN has described the preparation of the phenyl-thiohydantoins of the naturally occurring amino acids for use as reference compounds (*49*) and has made model degradations of di-, tri- and tetra-peptides on a scale of 20–30 mg. (*50*). His method for the identification of the phenylthiohydantoin which had been cleaved from the molecule required that it be hydrolyzed to the amino acid and identified by paper chromatography.

Almost simultaneously, SJÖQUIST (*180*) and LANDMANN, DRAKE, and DILLAHA (*96*) have published paper chromatographic methods for the separation and identification of the PTH-amino acids. This improvement should greatly facilitate the use of the method by eliminating the necessity of hydrolyzing the phenylthiohydantoin before identification.

It is gratifying that the methods both of Sjöquist and of Landmann, et al. permit the excellent separation of the PTH-amino acids with one or two unidimensional chromatograms, and yet their procedures are very different. Not only does Sjöquist use starch impregnated paper whereas Landmann et al. use paper buffered with phthalate but the developing solvents and the methods of detecting the colorless PTH-amino acids have little similarity. With two such methods now available, other investigators should find it possible to make rapid application of these methods.

Fox, Hurst, and Itschner (58) have used a microbiological method in conjunction with the Edman technique. The quantitative amino acid composition of the peptide is first determined microbiologically. Another portion of the peptide is then treated, and the N-terminal residue is removed as the phenylthiohydantoin. The amino acid thus removed is identified by determining quantitatively the amino acid composition of the remaining peptide and noting which amino acid is missing.

Applications. The applications of the method have not yet been extensive. Fraenkel-Conrat and Fraenkel-Conrat (59) have obtained results from the study of conalbumin, insulin, β-lactoglobulin, and ovomucoid which are in general agreement with those from the DNP-method (Table 2). Christensen (20) has applied five successive degradations to insulin: the amino acids which were to be expected on the basis of the results of Sanger and Tuppy (161, 162) and Sanger and Thompson (159, 160) indeed were always in major amount but in each successive degradation still others appeared in greater degree. The results of McClure, Schieler, and Dunn (115) on bovine serum albumin are at great variance with those of Van Vunakis (209) and Desnuelle, Rovery, and Fabre (35): these authors found a single N-terminal residue of aspartic acid per mole by the DNP-method whereas McClure et al. find two or three residues of aspartic acid, one residue each of methionine and histidine, and an undetermined number of residues of alanine.

The most successful applications of the method have been made by Ottesen and Wollenberger (129, 129a) and by Landmann, Drake, and Dillaha (96). Ottesen and Wollenberger have used this stepwise degradation to determine the structure of the peptides which are liberated in the transformation of ovalbumin to plakalbumin: in this way they showed that fraction A probably in ala-gly-val-asp-ala-ala, fraction B is ala-gly-val-asp, and fraction C is ala-ala.

Landmann, Drake, and Dillaha have investigated the N-terminal residues of leu-gly, salmine, bovine insulin, B-chain of insulin, β-lactoglobulin, lysozyme, and glutathione and have obtained the expected results in each instance. They have used the procedure to study the sequence of amino acids in the *A* and *B* chains of insulin and in lysozyme. The first five amino acids in each chain of insulin were identified and the sequences agreed with the findings of SANGER and collaborators. The first four residues of lysozyme were lys-val-phe-gly in agreement with SCHROEDER's findings (*168*) and the fifth residue appears to be serine. LANDMANN, DRAKE, and WHITE (*96 a*) have identitied the N-terminal sequence ser-tyr in corticotropin A.

The methodology of EDMAN's procedure is clearly in its formative stages. In all of the references previously cited, many details of method are to be found. In addition, information is presented by DAHLERUP-PETERSEN (*32*) and DAHLERUP-PETERSEN, LINDERSTRÖM-LANG, and OTTESEN (*33*). No thorough study of the quantitative aspects of the method has yet been made although EDMAN (*50*) gives some data on a scale of 20–30 mg. and FRAENKEL-CONRAT and FRAENKELCONRAT (*59*) suggest that losses of 15 to 20 per cent are to be expected. Quantitative determination should be no great problem because the PTH-amino acids show an absorption maximum in the ultraviolet region. One of the greatest problems of the method is to find a suitable solvent for the PTC-derivative during the cleavage step. EDMAN used anhydrous solvents in order to discourage the hydrolysis of the peptide bonds because hydantoin formation does not require water. Nitromethane is not too good a solvent for the longer PTC-peptides and proteins but EDMAN's recent use of glacial acetic acid (*51*) may help to solve this problem. CHRISTENSEN's difficulty (*20*) in applying the method to insulin may in part be due to the use of aqueous solvents whereas LANDMAN *et al.* used anhydrous dioxane apparently without difficulty.

In this reviewer's opinion, the EDMAN method shows great promise and it is to be hoped that it will be more fully exploited in the future.

c) Other Methods.

Although only the methods of SANGER and EDMAN have received practical application to proteins with success, other degradations have recently been proposed and these may well be of use in the future.

We will not discuss them in detail but would call attention to the papers of WESSELY, SCHLÖGL, and KORGER (*212*), SCHLÖGL, SIEGEL, and WESSELY (*164*), HOLLEY and HOLLEY (*87*), FLOWERS and REITH (*56*), and REITH and WALDRON (*150*, *150 a*), EVAN and REITH (*53 a*), and FLETCHER, LOWTHER, and REITH (*55 a*).

d) Conclusions.

The usefulness of SANGER's method for the determination of N-terminal amino acids cannot be questioned. Although only 8 years ago there was

no definite knowledge of N-terminal amino acids in proteins, we now
can list 57 proteins which have been studied more or less satisfactorily
(Table 2). Qualitatively and roughly quantitatively, the SANGER method
is excellent but one can not state that it is an exact quantitative method.
The extent of the destruction of DNP-amino acids during hydrolysis
appears to be greatly dependent upon the nature of the adjacent amino
acid, upon the presence or absence of certain amino acids in the hydrolysate,
upon conditions of hydrolysis, etc., but no very thorough study of these
problems has yet been made. Correction factors to take into account the
extent of the destruction of DNP-amino acids during hydrolysis have been
determined by different authors in various ways and their significance is
difficult to assess. Most authors seem to realize that such correction factors
can only be an approximation. It is this reviewer's opinion that if the molecular weight of the protein is small and the chains are few in number, the
quantitative results probably are significant but if the molecular weight
is high and the chains are numerous, one is justified in regarding the
quantitative results with considerable skepticism.

How useful EDMAN's method will be remains to be seen. His own
experiments have for the most part been on a semi-micro or macro scale
where he found reasonable recoveries and little undesirable cleavage
of other than the N-terminal peptide bond. The greatest usefulness of
the method will come on the micro or submicro level where little information
about the quantitativeness of the reaction is available.

2. Identification of C-Terminal Amino Acids.

At the present time, four methods are being used for the identification
of C-terminal amino acids and two of these are revivals of procedures
which were suggested many years ago. The thiohydantoin method of
SCHLACK and KUMPF (163) was first described in 1926 and carboxypeptidase was used by GRASSMANN, DYCHERHOFF, and EIBELER (71)
in 1930 to show that glycine was the C-terminal residue of gluthathione.
Very recently both FROMAGEOT and CHIBNALL employed metallic hydrides
to reduce the terminal carboxyl group to the alcohol. AKABORI et al.
have devised a hydrazinolytic procedure.

a) The SCHLACK and KUMPF Method.

This method is based on the following series of reactions:

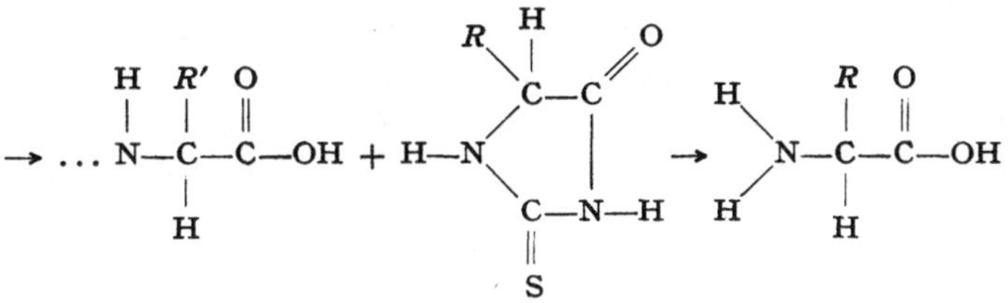

$$\rightarrow \dots \underset{H}{\overset{H}{\underset{|}{N}}}-\underset{|}{\overset{R'}{\underset{|}{C}}}-\overset{O}{\overset{\parallel}{C}}-OH + H-N \underset{\underset{\parallel}{C}-N-H}{\overset{\overset{R}{\underset{|}{\overset{H}{\diagdown}}}\overset{}{C}-\overset{O}{\overset{\diagup\parallel}{C}}}{}} \rightarrow \underset{H}{\overset{H}{N}}\overset{R}{\underset{|}{\overset{|}{C}}}\overset{O}{\overset{\parallel}{C}}-OH$$

The peptide or protein is reacted with ammonium cyanate in acetic anhydride (which blocks the N-terminal position) and the thiohydantoin is then split off, isolated, and usually rehydrolyzed to the free amino

Table 3. C-Terminal Residues of Some Proteins as Determined by the Use of Several Methods.

Protein	C-Terminal Amino Acid(s)	References
	SCHLACK and KUMPF Method	
Bovine plasma albumin	Alanine	EDWARD and NIELSEN (52)
Collagen	None	
Gelatin	Glycine	
	Threonine	GRASSMANN and HÖRMANN (72)
	Alanine	
Insulin	Alanine	WALEY and WATSON (211)
	Alanine	
	Tyrosine?	BAPTIST and BULL (4)
	Phenylalanine?	
Ovalbumin	Alanine	
Ovomucoid	Phenylalanine	TURNER and SCHMERZLER (208)
	Carboxypeptidase Method.	
Carboxypeptidase	None	DAVIE and NEURATH (30)
α-Chymotrypsin	Leucine	
	Tyrosine	GLADNER and NEURATH (66, 67)
	Glycine	
Chymotrypsinogen	None	
Corticotropin A	Phenylalanine	WHITE (215)
Insulin	Alanine	LENS (98)
	Alanine	
	Asparagine	
A chain	Asparagine	HARRIS (78)
B chain	Alanine	
Lysozyme	Leucine	A. R. THOMPSON (201)
	Leucine	HARRIS (78)
Tobacco mosaic virus	Threonine	HARRIS and KNIGHT (79)
Trypsin	None	
Trypsinogen	None	DAVIE and NEURATH (30)
	Reduction Method.	
Insulin	See text	
Ovomucoid	Phenylalanine	PÉNASSE et al. (134)

acid. The exact conditions under which the various reactions are carried out depends upon modifications introduced by several investigators.

Waley and Watson (211), Baptist and Bull (4), and Turner and Schmerzler (208) have all preferred to hydrolyze the thiohydantoin back to the amino acid and identify it by paper chromatography but Edward and Nielsen (52) have devised a method for the separation of the thiohydantoins themselves on paper.

Baptist and Bull (4) have made a rather thorough study of the quantitative aspects of the method. The yields were low and various degradations severely limited the applicability of the procedure.

Very great improvements are required before this method can have even limited usefulness. The applications which have been made are given in *Table 3*, p. 273.

b) The Carboxypeptidase Method.

The carboxypeptidase method enjoys by far the greatest popularity of the methods in use for determining the C-terminal residue. Smith (181) has given an excellent review of the action of the enzyme. Carboxypeptidase releases C-terminal amino acids which contain a free carboxyl group at rates which depend upon the structure of the C-terminal amino acid and to a lesser extent upon the structure of the adjacent amino acid; however, C-terminal proline is released at a very slow rate if at all. In the course of time, the enzyme will liberate amino acids other than C-terminal because the removal of the C-terminal amino acid opens the adjacent amino acid to attack.

In the most general procedure, the protein is acted upon by the enzyme for a period, the protein is removed by trichloracetic acid or ion exchange resin (201), and the amino acid or amino acids in the solution are identified by paper chromatography. Applications of the method are shown in Table 3.

White (215) has been able to identify a C-terminal sequence in corticotropin *A* as -pro-leu-glu-phe by determining the rates at which the various amino acids are released. On the intact corticotropin *A* the reaction ceases after the liberation of leucine as would be expected. White has identified the fourth amino acid as proline by the isolation of the tetrapeptide from partial hydrolysates.

The experience of Steinberg (193, 194) has shown that due caution must be exercised in the use of the method. He originally reported that the C-terminal amino acid of ovalbumin was alanine although Desnuelle and Casal [(34) and Table 2] had detected no N-terminal amino acid by the DNP-method. More recently (194) he has shown that the supposed C-terminal alanine is available to carboxypeptidase only after opening of the molecule by a contaminating enzyme.

c) Reduction Methods.

FROMAGEOT, JUTISZ, MEYER, and PÉNASSE (64) have described a method for the determination of C-terminal amino acids which requires the reduction of free carboxyl groups, the hydrolysis of the reduced protein, and the separation, identification, and estimation of the amino alcohols. The reduction of the free protein is made with lithium aluminum hydride and the separation involves chromatography on silica gel, charcoal, and paper. CHIBNALL and REES (18) use similar reactions except that the esterified instead of free protein is reduced with lithium borohydride. The separation involves columns of the ion exchanger Dowex-2 and paper chromatography. If a neutral or basic amino acid occupies the C-terminal position, it will be reduced to the corresponding amino alcohol but the nature of the product from glutamic or aspartic acid will be determined by its position in the chain or on the C-terminal end as well as by the fact that a carboxyl may be free or in the amide form.

Both groups of workers have applied their methods to a thorough study of insulin. A resume of the results has been given by CHIBNALL and REES (19) and by FROMAGEOT and JUTISZ (63). CHIBNALL and REES state that insulin contains 8 glutamyl, 6 glutaminyl, and 4 asparaginyl residues in the chains in positions other than terminal and 2 asparagine residues and 2 alanine residues in the C-terminal positions. FROMAGEOT and JUTISZ have found 2 C-terminal residues of alanine, glycine, and asparagine. CHIBNALL and REES have also detected aminoethanol and leucinol which are the reduction products of glycine and leucine but the quantities are not equivalent to one residue per mole of insulin and are considered to be derived from the splitting of the chain. JOLLÈS and FROMAGEOT (90) had previously reported the presence of 4 asparaginyl residues in the chains in position other than terminal and 2 C-terminal asparagines.

The fact that both groups have observed artefacts shows that the method must be used with the greatest care.

d) The Hydrazinolytic Method.

Recently the method of AKABORI, OHNO, and NARITA (1) for the determination of C-terminal amino acids has come to the reviewer's attention. Their method depends upon the reaction of anhydrous hydrazine with a peptide or protein to produce the hydrazides of all amino acids except the C-terminal amino acid which is released as the free amino acid. They applied the method to several synthetic peptides and to insulin in which alanine and glycine were found as C-terminal amino acids.

In 1953, OHNO (127 a) published a greatly improved procedure which he applied to lysozyme with at least roughly quantitative results. In his procedure, lysozyme was reacted with anhydrous hydrazine at 100° for periods of 5 to 20 hours. Hydrazine was then removed *in vacuo* over sulfuric acid and the residue was dinitrophenylated to form the DNP-derivative of the C-terminal amino acid and the di-DNP-hydrazides of all other amino acids in the molecule. Because the DNP-amino acid is acidic whereas the di-DNP-hydrazides are not, the DNP-amino acid was readily separated from the mixture by extracting an ethyl acetate solution with aqueous sodium bicarbonate. The DNP-amino acid was then identified chromatographically and estimated spectrophotometrically. Without correction of any kind, OHNO found 0.68 moles of leucine per mole of lysozyme. Control experiments with leucine itself showed that about one-third decomposed. If the appropriate correction factor is applied, the results indicate almost exactly one residue of leucine per mole of lysozyme. By the use of carboxypeptidase, A. R. THOMPSON (*201*) and HARRIS (*78*) showed qualitatively that leucine is the C-terminal amino acid of lysozyme.

The procedure of AKABORI, OHNO, and NARITA differs in principle from the other methods for C-terminal amino acids. It certainly seems worthy of study and development in order to determine its advantages and disadvantages.

e) Conclusions.

The procedures for the identification of C-terminal residues cannot, at present, be considered adequate. It remains to be seen whether any one of them can be developed to a satisfactory state of perfection or whether they will be superseded by an altogether different procedure.

3. The Separation and Identification of Peptides.

Although the ready, systematic separation of the common amino acids has now been achieved by the methods of starch and ion exchange chromatography, it is hardly to be expected that anyone will attempt to devise a systematic procedure for the separation of complex mixtures of peptides despite the admitted desirability of such a procedure. The fractionation of even the approximately 400 dipeptides of the common amino acids would be a task that few would wish to undertake: its magnitude increases rapidly when tripeptides and tetrapeptides only are added. As one would expect, however, methods for the separation of amino acids have also been applied to the separation of peptides. Starch and ion exchange methods have been used for free peptides and the DNP-method for the separation of DNP-peptides. We shall consider three subjects: (*a*) the separation of free peptides, (*b*) the separation of DNP-peptides, and (*c*) the identification of peptides.

a) The Separation of Free Peptides.

STEIN and MOORE (190) point out that peptides higher than approximately tetrapeptides cannot be fractionated efficiently on starch and that still higher peptides tend to emerge with the solvent or to be very tightly fixed.

BORSOOK, DEASY, HAAGEN-SMIT, KEIGHLEY, and LOWY (11) have used starch to separate a peptide fraction from livers of various species of animals, from WITTE's peptone, and from other sources. This peptide "A" when developed on starch with 1 : 2 : 1 n-butyl alcohol-n-propyl alcohol-0.1 N-HCl emerges from the column before any amino acid. It passes from the column with or immediately after the advancing front of developer and for all practical purposes may be considered not to have been adsorbed. This peptide "A" from all sources when hydrolyzed completely was found to contain almost all of the common amino acids. BORSOOK et al. do not claim that peptide "A" is homogeneous nor that all sources yield the same peptide "A" but they allude to certain experimental facts which point in this direction. One is justified in viewing with considerable skepticism the suggestion that peptide "A" is homogeneous or that the peptides from the several sources are identical. The chromatographic properties of peptide "A" are such that it could be composed of a very complex mixture: the efficiency of chromatographic separation is nil when the compounds are not adsorbed. A careful study of the amino acid compositions of peptide "A" from various sources shows that the extremes of amino acid composition may differ as much as 30 to 40 per cent.

When a proteinase from Bacillus subtilis transforms ovalbumin to plakalbumin, peptides are released in the process. These peptides have been separated by OTTESEN and VILLEE (128) through the medium of starch chromatography. Fig. 6 shows the nature of the chromatogram which was obtained on starch by development with 1 : 2 : 1 n-butyl alcohol-n-propyl alcohol-0.1 N-HCl after the enzyme had acted on ovalbumin for three hours. Peptides A and B begin to emerge approximately with the solvent front and continue to emerge through the region of leucine to tyrosine (Fig. 1, p. 244) while peptide C falls between tyrosine and proline. The relative ineffectiveness of starch as a means of separating peptides is clearly evident from these results; peptides A and B are poorly separated and even peptide C has moved through rapidly. It is these peptides which OTTESEN and WOLLENBERGER (129, 129a) have identified by the use of EDMAN's method as mentioned on p. 270: A is ala-gly-val-asp-ala-ala, B is ala-gly-val-asp, and C is ala-ala.

Although the separation of peptides on starch is rather unsatisfactory, Dowex-50 seems to be eminently suited for the task. MOORE and STEIN (126) comment briefly about the chromatography of peptides on Dowex-50 and note that dipeptides such as gly-leu chromatograph well and tetra-peptides such as leu-leu-gly-gly less so on the resin which is used for amino acid separations. This resin is crosslinked with 8 per cent divinyl-benzene. If, however, a resin with 4 per cent divinylbenzene is used, the chromatographic behavior of the tetrapeptide is improved. Dowmont and Fruton (43) have made an excellent study of the separation of peptides on two types of Dowex-50 which are crosslinked with 8 or 4 per cent of divinyl-

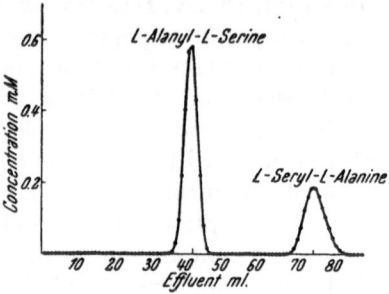

Fig. 7. Chromatographic separation of L-seryl-L-alanine and L-alanyl-L-serine on 8 per cent Dowex-50 at pH 4.05. Column, 30 × 0.9 cm. [From: J. Biol. Chem. **197**, 271 (1952).]

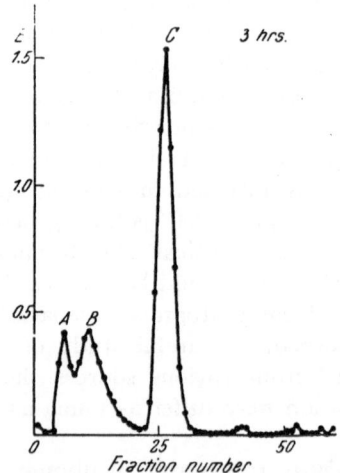

Fig. 6. Isolation by starch chromatography of the peptides released during the conversion of oval-bumin to plakalbumin. Ordinate, extinction coeffi-cient after reaction with ninhydrin; and abscissa, fraction number. [From: C. R. Trav. Lab. Carls-berg, Sér. chim **27**, 421 (1951).]

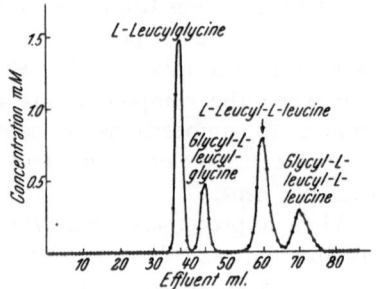

Fig. 8. Chromatographic separation of four peptides on 4 per cent Dowex-50 at pH 4.72. Column, 30 × 0.9 cm. [From: J. Biol. Chem. **197**, 271 (1952).]

benzene. Their methods follow those of Moore and Stein closely in that the same types of resin, buffers, analytical procedure, etc., are used with suitable modifications in size of column, temperature of column, pH of developer, etc. Examples of the excellent separations which they have achieved are presented in *Figs. 7 and 8*. The resolution of peptides which differ only in the sequence of residues is clear evidence of the power of the method.

As Dowmont and Fruton point out, they devised the technique for the analysis of the relatively simple mixtures which are encountered in transamidation reactions. They suggested that more complex mixtures such as those in partial hydrolysates of large peptides and proteins will require long columns and several changes of developer. In this connection,

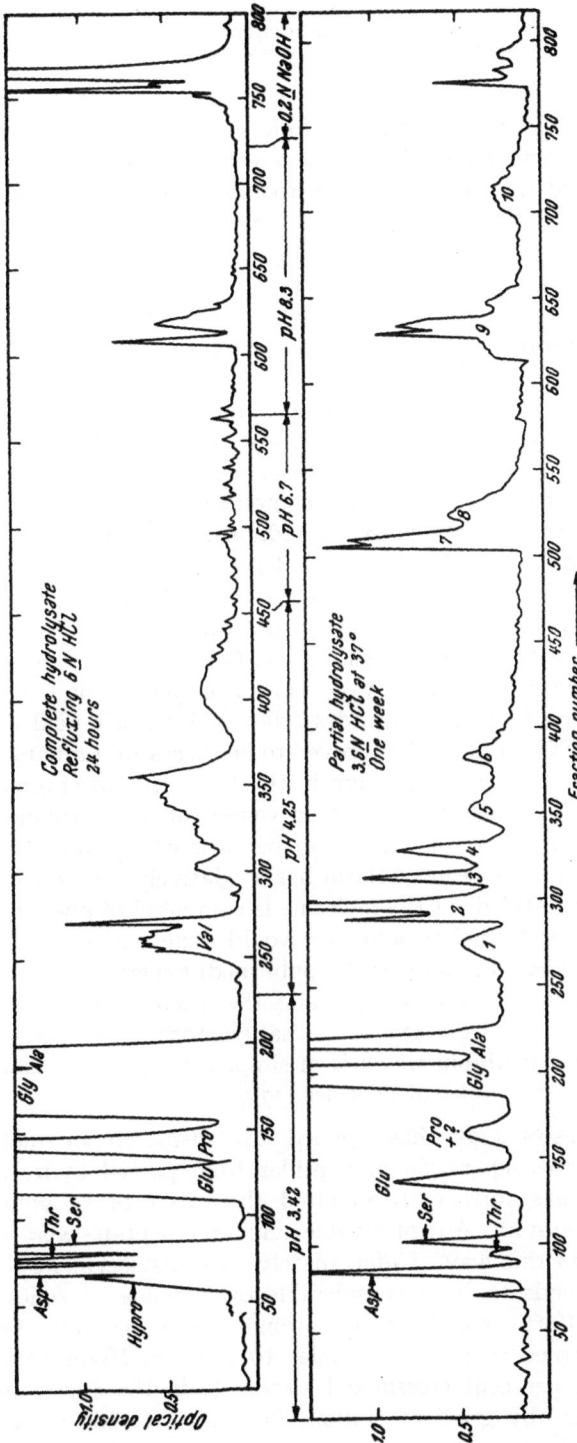

Fig. 9. Comparison of chromatograms of partial and complete hydrolysates of gelatin on Dowex-50. Column, 100 × 0.9 cm. Temperature of chromatogram, 37°.
[From: Proc. Nat. Acad. Sci. (U. S. A.) **39**, 23 (1953).]

SCHROEDER, HONNEN, and GREEN (*172*) have used ion exchange chromatography for the separation of peptides from partial hydrolysates of gelatin. Their work was carried out before the publication of the paper of DOWMONT and FRUTON and they have used the 100-cm. columns and the identical developers of MOORE and STEIN and have altered the procedure only in that the columns were maintained at 37° throughout instead of passing through the cycle of temperatures. The type of separations which were obtained is shown in *Fig. 9* in which identical chromatograms of complete and partial hydrolysates of gelatin are compared. The zones which are numbered 1 through 10 by comparison alone would seem to contain peptides. Further study (*174*) of the zones which may be isolated thus from an acidic as well as from a basic hydrolysate has led to the identification of 34 peptides in partial hydrolysates of gelatin. The investigation of zones which may thus be isolated is complicated by the presence of the buffers which are required when the sodium form of Dowex-50 is used. SCHROEDER, HONNEN, and GREEN (*172*) have circumvented this difficulty by converting the peptides in the zones into the DNP-derivatives in which form they may be extracted from the buffers. These DNP-peptides were then separated further on silicic acid-Celite by an extension of the methods of GREEN and KAY (*73*). This combination of ion exchange chromatography of the free peptides followed by silicic acid chromatography of the DNP-peptides has proved to be successful probably because such very different conditions reign in the two types of chromatograms and enhance the already great effectiveness of the chromatographic method. Whether these methods may be successfully applied to other proteins remains to be seen. Gelatin has a relatively simple composition in that about two-thirds of the molecule is composed of glycine, alanine, proline, and hydroxyproline and one would expect a relatively simple mixture of peptides. Preliminary (unpublished) experiments with partial hydrolysates of lysozyme which contains the usual amino acids in more common proportions have given less satisfactory separations. Partial hydrolysates of silk fibroin which is of simpler composition than gelatin have given very fine zones of peptides (*173*).

A. R. THOMPSON (*202*) has applied the displacement methods of PARTRIDGE to the separation of peptides from partial hydrolysates of lysozyme and has further fractionated the mixtures so obtained by paper chromatography. An appreciable number of peptides were separated and identified in this way. Unfortunately with displacement chromatography there tends to be a considerable overlapping of zones so that duplication of effort occurs in identification of the compounds. THOMPSON has recently turned to methods similar to those of MOORE and STEIN. By means of 4 per cent crosslinked Dowex-50 in the ammonium form and development by gradient elution she has been able to separate a large number of zones which in general are well separated from one another.

Various phases or uses of peptide fractionation have been investigated by LEDERER and KIUN (97), JUTISZ and LEDERER (91), BISERTE and BOULANGER (5), BRENNER and BURCKHARDT (14), and DAVIE and NEURATH (31).

b) The Separation of DNP-Peptides.

SANGER (155) has also made a fine application of his DNP-method to the determination of the N-terminal peptides of insulin. DNP-Peptides were extracted from partial hydrolysates of the DNP-derivatives of the A and B chains and then separated on silica gel. It was simple to determine the N-terminal amino acid of each DNP-peptide by hydrolyzing completely, etc., and likewise to identify the other amino acids in the peptide by chromatographing the extracted hydrolysate on paper. Because the various peptides presumably originated from a single chain, a comparison of their amino acid contents permitted a preliminary conclusion as to the sequence in the longest peptide. This suspected sequence was then substantiated by partial hydrolyses of the individual peptides. In this way, SANGER was able to show that the N-terminal sequence of the A chain is gly-ileu-val-glu-glu-; that the N-terminal sequence of the B chain is phe-val-asp-glu-; and that the B chain also contained the sequence, -thr-pro-lys-ala-.

In a similar way, other authors have studied the terminal peptides of other proteins. PORTER (145) found that both the active and inactive fractions of γ-globulin from rabbit ovalbumin sera have the sequence ala-leu-val-asp-glu- although the identification of glutamic acid is tentative and the aspartic acid may be present as asparagine. ŠORM and ŠORMOVÁ (186) determined that the main N-terminal sequence of clupein is pro-ala-ser- although small amounts of what may have been ser-ala-ser were also detected. SCHROEDER (168) showed that lysozyme has the N-terminal sequence lys-val-phe-gly- and LANDMANN, DRAKE, and DILLAHA (96) have substantiated this by EDMAN's method and have extended the sequence to find that the fifth amino acid is serine. These latter authors also substantiated SANGER's N-terminal sequences for the first five amino acids of the A and B chains of insulin. E. O. P. THOMPSON (203) identified asp-(NH₂)-ser- as the N-terminal sequence of carboxy-peptidase with considerable difficulty because of the lability of the peptide bond between asparagine and serine; the third amino acid may be threonine. He also found (204) that papain has the sequence ileu-pro-glu-. Bovine and human serum albumin both have aspartic acid in the N-terminal position but THOMPSON (204) has shown that they differ in the second amino acid which is threonine in bovine serum albumin and alanine in human serum albumin.

The separation of N-terminal DNP-peptides is greatly simplified by the fact that the DNP-peptides are closely related and differ from each other only in the greater complexity which is produced by the successive addition of residues. The chromatographic properties of compounds so related would be expected to be sufficiently diverse to permit ready separation and such indeed has been the case. It may be much more difficult, however, to fractionate mixtures of unrelated DNP-peptides. As has already been mentioned, SCHROEDER, HONNEN, and GREEN (*172*) have chromatographed DNP-peptides from partial hydrolysates of gelatin after initial separation on Dowex-50. They have used methods similar to those of GREEN and KAY (*73*) for the fractionation of mixtures of DNP-amino acids on silicic acid-Celite and found that certain specific developers [suggested by the work of SCHROEDER and HONNEN (*171*) on the chromatographic properties of synthetic DNP-peptides] were very satisfactory. These studies of synthetic DNP-peptides also show definite correlations between the structure of the DNP-peptide and its chromatographic behavior. These correlations are valuable in aiding the identification of tentatively identified peptides through a comparison of determined and predicted chromatographic behaviors. It has been possible, in the case of certain zones from the ion exchanger, to separate as many as 14 DNP-derivatives, not all of which were necessarily in sufficient amount for characterization (*171*).

KRONER, TABROFF, and McGARR (*95*) have also chromatographed mixtures of unrelated DNP-peptides, in this instance those from a partial hydrolysate of steer hide collagen. After a preliminary separation of the partial hydrolysate into aromatic, neutral, acidic, and basic fractions by means of charcoal and ion exchange resins, the neutral fraction was dinitrophenylated and the DNP-derivatives were separated mainly by means of the buffered Celite columns of PERRONE (*135*). They were able to identify six amino acids and nine peptides from the neutral fraction.

When mixtures of relatively short peptides are to be separated, chromatography of the DNP-derivatives offers a method which may well be exploited with profit.

c) The Identification of Peptides.

When a peptide, particularly a relatively short peptide, has been isolated in pure form, methods which have been discussed above (pp. 262 ff. and 272 ff.) as well as partial hydrolytic procedures should lead to straightforward identification.

d) Conclusions.

Until methods for the degradative determination of sequence have reached a considerably higher state of perfection, we shall probably have to depend upon the study of the partial hydrolytic products of proteins to give us a clue to the sequence of amino acid residues. Therefore, the resolution of mixtures of peptides can be expected to assume greater importance. Column chromatographic methods have made a very promising beginning in the separation of peptides as shown by the work of SCHROEDER, HONNEN, and GREEN (*172*) and of A. R. THOMPSON (*202*) on the initial separation from partial hydrolysates by means of ion exchange methods. No single procedure, however, is likely to be able to resolve completely the peptides in a partial hydrolysate but ion exchange chromatography can produce simpler mixtures which are amenable to further fractionation on paper or in the form of the DNP-derivatives. Although each protein probably will require some modification of techniques, ion exchange chromatography appears to offer an excellent starting point for the initial fractionation of the peptides in partial hydrolysates.

The DNP-method has been useful for elucidating the sequence in the immediate vicinity of the N-terminal residue but it is unlikely that it is capable of much further extension, that is, at best we may expect to determine the sequence of the first four or five residues. Procedural obstacles arise at about the stage of a DNP-tetrapeptide or DNP-pentapeptide in the first place because such higher DNP-peptides extract only with difficulty from partial hydrolysates and in the second place because they become increasingly insoluble in the solvents which are used in further chromatographic work. Another hindrance which may severely limit the applicability of the method is the presence of a labile peptide bond near the N-terminal residue. Such bond may make it impossible to obtain longer peptides. Thus, E. O. P. THOMPSON (*203*) was able to show only with great difficulty that serine was the second amino acid of carboxypeptidase because of the unusual lability of peptide bonds involving the amino group of serine (or threonine).

III. The Separation and Purification of Proteins.

When amino acids or peptides are chromatographed, the desired objective usually is the fractionation of a more or less complex mixture but when proteins are chromatographed, the desired objective may be very different. The isolation of a protein may be a simple or a complex procedure depending upon the starting material but if the protein can be characterized at all, it will in most instances have been obtained in a reasonable state of purity. Although under favorable conditions chromatography may be of great use in the isolation of a protein from the original natural product, yet its greatest usefulness may well rest in the final purification of the substance and in the testing of its homogeneity.

The chromatography of proteins poses problems which are not encountered in the chromatography of amino acids. The tendency of proteins toward denaturation and alteration even under mild conditions may well be the factor which will limit the usefulness of the chromatographic method as applied to proteins themselves. Any successful chromatogram demands reversible adsorption but this requirement may be the most difficult to meet where proteins are concerned. Data of HIRS, MOORE, and STEIN (86) throw some light on this subject. These authors studied the distribution of several proteins between various adsorbents and solvents. The proteins were insulin, β-lactoglobulin, bovine serum albumin, fowl hemoglobin, horse myoglobin, cytochrome c, ribonuclease, and lysozyme; the adsorbents were Celite, potato starch, silica gel, powdered cellulose, tricalcium phosphate, benzoic acid, Dowex-50, Dowex-2, Duolite C-10, and Amberlites IR-4 B, IRA-400, and IRC-50; the solvents were water, aqueous hydrochloric acid, sodium citrate buffers, sodium phosphate buffers, and sodium acetate buffers. In almost all experiments, adsorption of the protein was irreversible but the preliminary tests showed that satisfactory chromatograms might be expected with ribonuclease, lysozyme, and cytochrome c on IRC-50 with buffers of p_H 6 to 7.

Both inorganic adsorbents and ion exchange resins have been used in the chromatography of proteins although the trend seems to favor the resins. Most authors have tried to obtain the protein as a definite zone which moves down the column under proper conditions rather than to adsorb, wash to remove impurities, and then elute. Some frontal analysis has been done.

SHEPARD and TISELIUS (177) studied the adsorption isotherms of several proteins on silica gel and then on application of the data to frontal analysis were able to distinguish serum albumin and immune globulin at p_H 7.0 and 0.1 ionic strength. Likewise, SOBER, KEGELES, and GUTTER (184, 185) have used frontal analysis on Dowex-50. They were able to distinguish ovomucoid, ovalbumin, and conalbumin in a fraction from egg white and also to differentiate between bovine plasma albumin and human carbonmonoxyhemoglobin in an artificial mixture. They detected no irreversible adsorption. Although frontal analysis may be useful in ascertaining the homogeneity of a protein; it can be of limited value in preparative work because only the first compound which emerges from the column can be isolated in pure state.

Tricalcium phosphate of their own preparation was used by SWINGLE and TISELIUS (195) as an adsorbent in an investigation of the chromatographic properties of phycoerythrin. They found that adsorption was promoted by the presence of sodium chloride, and that phosphate solutions of which the p_H was of secondary importance could be used as developers.

Displacement chromatography on tricalcium phosphate and then on silicic acid-Celite was part of the procedure by means of which POLIS and SHMUKLER (*140*) obtained crystalline lactoperoxidase from milk.

MARTIN and PORTER (*112*) and PORTER (*147*) have used Hyflo-Supercel as a support for partition chromatograms of proteins. The solvent systems

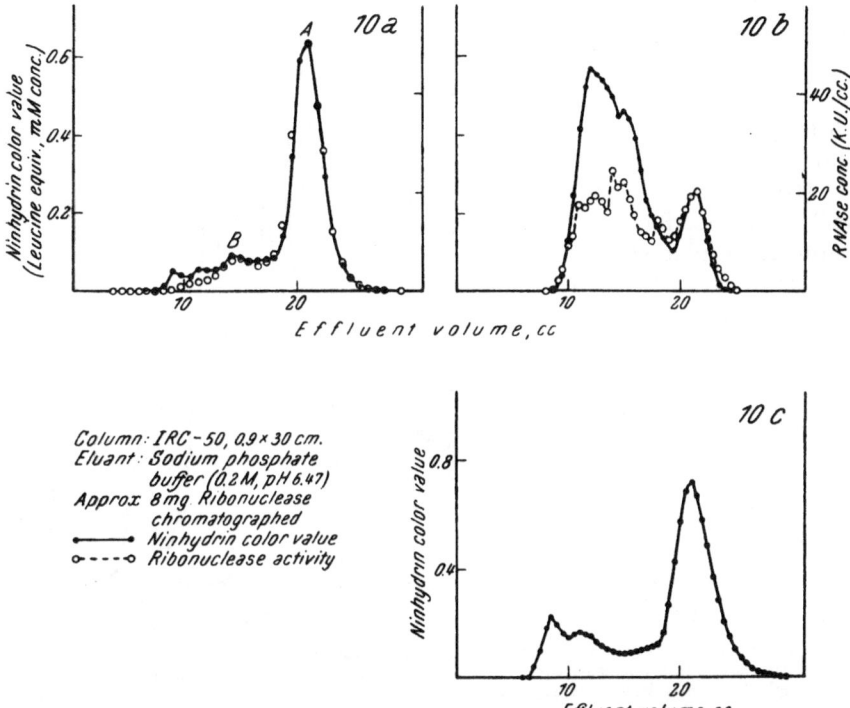

Fig. 10. Chromatographic results from three crystalline preparations of ribonuclease on IRC-50 (XE-64). Column, 30 × 0.9 cm. [From: J. Biol. Chem. **200**, 493 (1953).]

were composed of water, various cellosolves, and salts such as ammonium sulfate, sodium phosphate, or potassium phosphate. MARTIN and PORTER observed that preparations of ribonuclease contained two active components despite efforts to eliminate the lesser component by alteration in the method of preparation; HIRS, MOORE, and STEIN (*86*) have recently substantiated the presence of two constituents with ribonuclease activity. PORTER (*147*) successfully chromatographed insulin, ribonuclease from several species, avidin, chymotrypsinogen, chymotrypsin, trypsin, bovine serum albumin, etc., and showed that ribonuclease and carboxypeptidase form a complex which is capable of acting chromatographically as a single compound.

The ion exchange resin, Amberlite IRC-50, is proving to be very useful in the chromatography of proteins, especially for the final purification to produce a compound which is chromatographically homogeneous. Paleus and Neilands (*130*) were the first to use this adsorbent for proteins and they applied it to the puri-fication of cytochrome *c*. By the use of ammonium hydroxide-ammonium acetate buffers at pH 9, they were able to isolate three fractions from cyto-chrome *c* and to show that they differed in iron content.

Hirs, Moore, and Stein (*86*) have described in detail a chromatographic study of pancreatic ribonuclease on IRC-50. They have used procedures

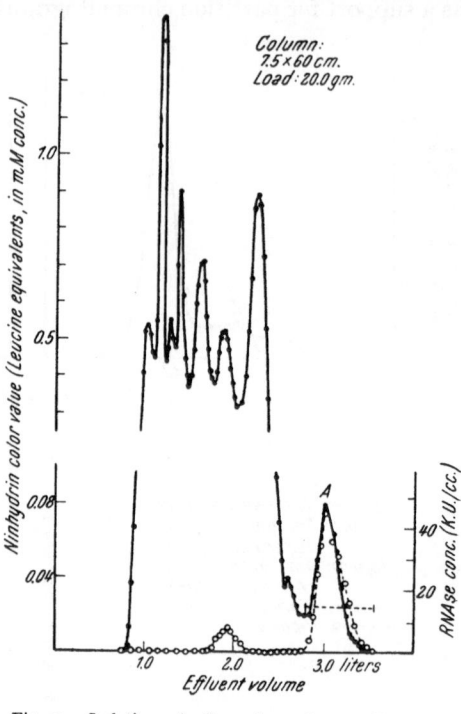

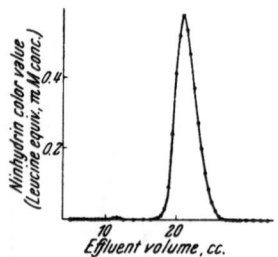

Fig. 11. Chromatographic homogeneity of purified ribonuclease which originally gave the chromatographic results depicted in Fig. 10a. [From: J. Biol. Chem. **200**, 493 (1953).]

Fig. 12. Isolation of ribonuclease from sulfuric acid extracts of beef pancreas on IRC-50 (XE-64) at pH 6.47. The ribonuclease is contained in peak *A*. [From: J. Biol. Chem. **200**, 493 (1953).]

similar to those which have been so effective in the separation of the amino acids. The development of ribonuclease on these columns is effected by means of 0.2 *M* sodium phosphate buffer of pH 6.5. *Fig. 10* shows the nature of the chromatograms which resulted when three crystalline preparations of ribonuclease were passed through individual columns. No evidence of inhomogeneity of these samples had been indicated by electrophoretic or ultracentrifugal studies although solubility investigations had suggested the presence of impurity. The chromatograms, however, clearly revealed the inhomogeneity of the samples. When a larger sample of material which produced the chromatogram in *Fig. 10a* was chromatographed on a preparative scale, the final crystalline product had the homogeneity illustrated in *Fig. 11*. The great power of this method as a means of separating proteins from complex mixtures is depicted in *Fig. 12* which shows the results of a preparative scale chromatogram of sulfuric acid extracts of beef pancreas. Even from this complex mixture in which ribonuclease is a minor component, the definite peak of ribonuclease (*A*) emerges and, in addition, its chromatographic properties have not been influenced by the presence of the other proteins.

TALLAN and STEIN (*199*) by the use of similar methods have found that lysozyme preparations may exhibit varying degrees of homogeneity and that freshly prepared samples crystallized in various ways contain of the order of 95 per cent in one main zone. Isoelectric lysozyme and lysozyme chloride show good stability but lysozyme carbonate at room temperature undergoes progressive transformation into two other active and chromatographically distinct forms.

DIXON, MOORE, STACK-DUNNE, and YOUNG (*41*) have demonstrated that the major protein component of ACTH is biologically inactive and that after chromatography, the active fractions contain less than one per cent of the weight of the starting material but have a potency sometimes 100 times as great the original preparation. LI, TISELIUS, PEDERSEN, HAGDAHL, and CARSTENSEN (*102*) have found that carrier displacement chromatography on charcoal is very effective in purifying adrenocorticotropic peptides. WHITE and FIERCE (*215 a*) have differentiated three active types of material in pituitury adrenocorticotropin by chromatography on Amberlite XE-97 (a finely powdered form of IRC-50).

HIRS (*84*) has extended the methods employing IRC-50 to chymotrypsinogen-α, most preparations of which possessed 95 per cent of the material in a single peak. It could be isolated directly from extracts of bovine pancreas.

BOARDMAN and PARTRIDGE (*9*) point out that the elution of cytochrome c from IRC-50 is dependent not only on the p_H but also the ionic strength of the developer. In addition, they have been able to separate a synthetic mixture of bovine carboxyhemoglobin and sheep fetal hemoglobin.

It is of interest to note that most of the proteins which have been chromatographed are enzymes. Enzymatic properties, of course, are of great aid in such chromatographic studies because they are another and very selective means of assessing the effectiveness of any separation. Most of the papers referred to in this Section have appeared since the publication of the review by ZECHMEISTER and ROHDEWALD (*218*) on enzyme chromatography which appeared in an earlier volume of this Series.

IV. Concluding Remarks.

When a review restricts itself to a specialized procedure which is only one method being applied to a broad field of study, the bias of the reviewer may make it appear that related methods are of much less importance. Nothing could be further from the truth. Column chromatography is only one of the tools which is being applied to the ever expanding study of protein structure. One cannot separate it from other methods because it is used in conjunction with them, and we have referred to them from time to time. However, one cannot but be impressed by the popularity which the chromatographic method (in one form or another) enjoys.

Column chromatography shows evidences of increasing usefulness in the very first stage of the study of protein structure, namely, the isolation of the purified or "pure" protein. After the protein has been isolated, column chromatographic methods are available for the accurate determination of the amino acid composition. And when the study of the finer details of the structure is begun, column chromatography is available for the separation and identification of individual peptides.

References.

1. Akabori, S., K. Ohno and K. Narita: On the Hydrazinolysis of Proteins and Peptides: A Method for the Characterization of Carboxyl-terminal Amino Acids in Proteins. Bull. Chem. Soc. Japan **25**, 214 (1952).

1 a. Åqvist, S. E. G.: The Use of Starch Chromatography and Ion Exchange Resin for Large Scale Separations of N^{15}-Labeled Amino Acids. Acta Chem. Scand. **5**, 1031 (1951).

2. Bailey, K.: End-Group Assay in Some Proteins of the Keratin-Myosin Group. Biochemic. J. **49**, 23 (1951).

3. Baker, C. G. and H. A. Sober: Application of Ion Exchange Chromatography to the Enzymatic Resolution of Amino Acids. J. Amer. Chem. Soc. **75**, 4058 (1953).

4. Baptist, V. H. and H. B. Bull: Determination of the Terminal Carboxyl Residues of Peptides and of Proteins. J. Amer. Chem. Soc. **75**, 1727 (1953).

5. Biserte, G. et P. Boulanger: Fractionnement d'hydrolysats enzymatiques de protéines. C. R. hebd. Séances Acad. Sci. **230**, 583 (1950).

6. Blackburn, S.: The Use of Buffered Columns in the Chromatographic Separation of 2,4-Dinitrophenyl Amino Acids. Biochemic. J. **45**, 579 (1949).

7. Blackburn, S. and A. Robson: A Radiochemical Method for the Microestimation of α-Amino Acids Separated on Paper Partition Chromatograms. Biochemic. J. **54**, 295 (1953).

8. Block, R. J. and D. Bolling: The Amino Acid Composition of Proteins and Foods. 2nd edit. Springfield, Ill.: Charles C. Thomas. 1951.

9. Boardman, N. K. and S. M. Partridge: Separation of Neutral Proteins on Ion Exchange Resins. Nature (London) **171**, 208 (1953).

10. Boggs, L. A., L. S. Cuendet, M. Dubois and F. Smith: Simple Fractionating Device for Chromatographic Analysis. Application to the Study of Carbohydrates. Analyt. Chemistry **24**, 1148 (1952).

11. Borsook, H., C. L. Deasy, A. J. Haagen-Smit, G. Keighley and P. H. Lowy: A Peptide Fraction in Liver. J. Biol. Chem. **179**, 705 (1949).

12. Bowes, J. H.: private communication.

13. Bowes, J. H. and J. A. Moss: Free Amino Groups of Collagen. Nature (London) **168**, 514 (1951).

13 a. — — The Reaction of Fluorodinitrobenzene with the α- and ε-Amino Groups of Collagen. Biochemic. J. **55**, 735 (1953).

14. BRENNER, M. und C. H. BURCKHARDT: Die Adsorption einiger Di- und Tripeptide an synthetischen organischen Ionenaustauschern. Helv. Chim. Acta 34, 1070 (1951).
15. BRENNER, M. und R. FREY: Über die Entsalzung von Lösungen neutraler Aminosäuren mit Hilfe von Ionenaustauschern und ein neues präparatives Verfahren zur Gruppentrennung von Aminosäuren in Eiweißhydrolysaten. Helv. Chim. Acta 34, 1701 (1951).
16. BRIMLEY, R. C. and A. SNOW: Automatic Apparatus for Continuous Collection of Liquid Samples. J. Sci. Instruments 26, 73 (1949).
17. CARSON, J. F.: The Free Amino Groups of Subtilin. J. Amer. Chem. Soc. 74, 1480 (1952).
18. CHIBNALL, A. C. and M. W. REES: The Amide and Free Carboxyl Groups of Insulin. Biochemic. J. 48, xlvii (1951).
19. — — Identification and Estimation of the Amide and C-Terminal Residues in Insulin by Reduction of the Ester with Lithium Borohydride. Ciba Found. Symp. "The Chemical Structure of Proteins" p. 70 (1953).
20. CHRISTENSEN, H. N.: Attempted Successive Applications of the Edman Degradation to Insulin. Acta Chem. Scand. 6, 1555 (1952).
21. CLAESSON, S.: Studies on Adsorption and Adsorption Analysis with Special Reference to Homologous Series. Ark. Kemi, Mineral. Geol. 23 A, No. 1 (1946).
22. — Some Arrangements for Adsorption Analysis with Large Amounts of Substances. Ark. Kemi, Mineral. Geol. 24 A, No. 16 (1947).
22 a. CLOSE, J.. E. L. ADRIAENS, S. MOORE et E. J. BIGWOOD: Composition en acides aminés d'hydrolysats de farine de manioc roui, variété amère. Bull. soc. chim. biol. (Paris) 35, 985 (1953).
23. CONSDEN, R., A. H. GORDON and A. J. P. MARTIN: Qualitative Analysis of Proteins: a Partition Chromatographic Method Using Paper. Biochemic. J. 38, 224 (1944).
24. CORFIELD, M. C. and A. ROBSON: The Amino Acid Composition of Salmine. Biochemic. J. 55, 517 (1953).
25. CRAIG, L. C., J. D. GREGORY and W. HAUSMANN: Versatile Laboratory Concentration Device. Analyt. Chemistry 22, 1462 (1950).
26. CRAIG, L. C., W. HAUSMANN, E. H. AHRENS and E. J. HARFENIST: Automatic Countercurrent Distribution Apparatus. Analyt. Chemistry 23, 1236 (1951).
27. CRAIG, L. C., W. HAUSMANN and J. R. WEISIGER: The Qualitative and Quantitative Amino Acid Content of Bacitracin A. J. Biol. Chem. 199, 865 (1952).
28. CROOK, E. M. and S. P. DATTA: A Liquid Fraction Collector. Chem. and Ind. 1951, 718.
29. CUCKOW, F. W., R. J. C. HARRIS and F. E. SPEED: A Simple Fraction-Collecting Machine for Chromatographic Analysis. J. Soc. Chem. Ind. 68, 208 (1949).
30. DAVIE, E. W. and H. NEURATH: C-Terminal Groups of Trypsinogen, DFP-Trypsin, and Carboxypeptidase. J. Amer. Chem. Soc. 74, 6305 (1952).
31. — — Identification of the Peptide Split from Trypsinogen During Autocatalytic Activation. Biochim. Biophys. Acta 11, 442 (1953).
32. DAHLERUP-PETERSEN, B.: Rate of Ring Closure in the Edman Method. Acta Chem. Scand. 7, 1013 (1953).
33. DAHLERUP-PETERSEN, B., K. LINDERSTRÖM-LANG and M. OTTESEN: Stepwise Degradation of Peptides. Acta Chem. Scand. 6, 1135 (1952).
34. DESNUELLE, P. et A. CASAL: Sur la moindre résistance à l'hydrolyse acide des liaisons peptidiques situées à côté d'une fonction hydroxyle. Biochim. Biophys. Acta 2, 64 (1948).

35. Desnuelle, P., M. Rovery et C. Fabre: Etude des restes N-terminaux dans les serumalbumines de diverses espèces (suivie d'une remarque sur la stabilité des dinitrophénylaminoacides pendant l'hydrolyse). C. R. hebd. Séances Acad. Sci. **233**, 987 (1951).

36. — — — Sur les dérivés dinitrophénylés du chymotrypsinogène et de l'α-chymotrypsine cristallisée. C. R. hebd. Séances Acad. Sci. **233**, 1496 (1951).

37. — — — Extrémités N-terminales de la protéine de l'α-chymotrypsine. Biochim. Biophys. Acta **9**, 109 (1952).

38. Desreux, V.: L'extraction fractionnée systématique des polymères. Rec. trav. chim. Pays-Bas **68**, 789 (1949).

39. Dickman, S. R. and R. O. Asplund: Effect of Xanthylation on the Recovery of DNP-Amino Acids from Protein Hydrolysates. J. Amer. Chem. Soc. **74**, 5208 (1952).

40. Dimler, R. J., J. W. Van Cleve, E. M. Montgomery, L. R. Bair, F. J. Castle and J. A. Whitehead: Fraction Collector with Continuously Rotating Turntable and Improved Receiver Assemblies. Analyt. Chemistry **25**, 1428 (1953).

41. Dixon, H. B. F., S. Moore, M. P. Stack-Dunne and F. G. Young: Chromatography of Adrenotropic Hormone on Ion-Exchange Columns. Nature (London) **168**, 1044 (1951).

42. Dobyns, B. M. and S. R. Barry: The Isolation of Iodinated Amino Acids from Thyroid Tissue by Means of Starch Column Chromatography. J. Biol. Chem. **204**, 517 (1953).

43. Dowmont, Y. P. and J. S. Fruton: Chromatography of Peptides as Applied to Transamidation Reactions. J. Biol. Chem. **197**, 271 (1952).

44. Durso, D. F., E. D. Schall and R. L. Whistler: Automatic Fraction Collector for Chromatographic Separations. Analyt. Chemistry **23**, 425 (1951).

45. Dustin, J. P., C. Czajkowska, S. Moore and E. J. Bigwood: A Study of the Chromatographic Determination of Amino Acids in the Presence of Large Amounts of Carbohydrate. Anal. Chim. Acta **9**, 256 (1953).

46. Du Vigneaud, V., C. Ressler, J. M. Swan, C. W. Roberts, P. G. Katsoyannis and S. Gordon: The Synthesis of an Octapeptide Amide with the Hormonal Activity of Oxytocin. J. Amer. Chem. Soc. **75**, 4879 (1953).

47. Edelman, J. and R. V. Martin: A Simple Automatic Fraction Collector for Preparative Chromatography. Biochemic. J. **50**, xxi (1952).

48. Edman, P.: A Technique for Partition Chromatography on Starch. Acta Chem. Scand. **2**, 592 (1948).

49. — Preparation of Phenyl Thiohydantoins from Some Natural Amino Acids. Acta Chem. Scand. **4**, 277 (1950).

50. — Method for Determination of the Amino Acid Sequence in Peptides. Acta Chem. Scand. **4**, 283 (1950).

51. — Note on the Stepwise Degradation of Peptides *via* Phenyl Thiohydantoins. Acta Chem. Scand. **7**, 700 (1953).

52. Edward, J. T. and S. Nielsen: Chromatography of Thiohydantoins on Paper. Determination of the C-Terminal Amino Acid of Bovine Plasma Albumin. Chem. and Ind. **1953**, 197.

53. Elsden, S. R. and R. L. M. Synge: Starch as a Medium for Partition Chromatography. Biochemic. J. **38**, ix (1944).

53 a. Evans, G. G. and W. S. Reith: Studies on the Determination of the Sequence of Amino Acids in Peptides and Proteins. 3. The Synthesis of 3-(4'-Dimethylamino-3':5'-dinitrophenyl) hydantoin Derivatives of Various Amino Acids, and their Use for the Determination of N-terminal Amino Acids. Biochemic. J. **56**, 111 (1954).

54. FELIX, K., H. FISCHER, A. KREKELS und H. M. RAUEN: Über Clupein. IX. Mitt. Z. physiol. Chem. (Hoppe-Seyler) 286, 67 (1950).
55. FITCH, F. T. and D. S. RUSSELL: Determination of Lanthanum in Rare Earth Mixtures. Analyt. Chemistry 23, 1469 (1951).
55 a. FLETCHER, C. M., A. G. LOWTHER and W. S. REITH: Studies on the Determination of the Sequence of Amino Acids in Peptides and Proteins. 2. The Separation of the Methyl Esters of N-2:4-Dinitrophenyl Derivatives of Amino Acids by Adsorption Chromatography and the Free Amino Groups of Insulin. Biochemic. J. 56, 106 (1954).
56. FLOWERS, H. M. and W. S. REITH: Studies of the Determination of the Sequence of Amino Acids in Peptides and Proteins. I. The Preparation, Properties, and Chromatographic Adsorption of the Azobenzene-p-sulphonyl Derivatives of Various Amino Acids. Biochemic. J. 53, 657 (1953).
57. FOX, S. W.: Terminal Amino Acids in Peptides and Proteins. Adv. Protein Chem. 2, 155 (1945).
58. FOX, S. W., T. L. HURST and K. F. ITSCHNER: A Microbiological Method for the Determination of Sequences of Amino Acid Residues. J. Amer. Chem. Soc. 73, 3573 (1951).
59. FRAENKEL-CONRAT, H. and J. FRAENKEL-CONRAT: A Method for Determination of the Amino Acid Sequence of Proteins. Acta Chem. Scand. 5, 1409 (1951).
60. FRAENKEL-CONRAT, H. and R. R. PORTER: The Terminal Amino Groups of Conalbumin, Ovomucoid, and Avidin. Biochim. Biophys. Acta 9, 557 (1952).
61. FRANTZ, I. D., Jr., H. FEIGELMAN, A. S. WERNER and M. P. SMYTHE: Biosynthesis of Seventeen Amino Acids Labelled with C^{14}. J. Biol. Chem. 195, 423 (1952).
62. FRASER, D. and E. A. JERREL: The Amino Acid Composition of T_3 Bacteriophage. J. Biol. Chem. 205, 291 (1953).
63. FROMAGEOT, C. and M. JUTISZ: Identification of C-End Groups in Proteins by Reduction with Lithium Aluminium Hydride. Ciba Found. Symp. "The Chemical Structure of Proteins" p. 82 (1953).
64. FROMAGEOT, C., M. JUTISZ, D. MEYER et L. PÉNASSE: Méthode pour la caractérisation des groupes carboxyliques terminaux dans les protéines. Application à l'insuline. Biochim. Biophys. Acta 6, 283 (1950).
65. GILSON, A. R.: An Automatic Constant Volume Fraction Collector for Chromatography. Chem. and Ind. 1951, 185.
66. GLADNER, J. A. and H. NEURATH: C-Terminal Groups in Chymotrypsinogen and DFP-α-Chymotrypsin in Relation to the Activation Process. Biochim. Biophys. Acta 9, 335 (1952).
67. — — Carboxyl Terminal Groups of Proteolytic Enzymes. I. The Activation of Chymotrypsinogen to α-Chymotrypsin. J. Biol. Chem. 205, 345 (1953).
68. GLANZMANN, R. und R. SIGNER: Der Aminosäurebestand von Seidenfibroin und von Seidenfibroinfraktionen. Makromolek. Chem. 8, 134 (1952).
69. GORDON, A. H., A. J. P. MARTIN and R. L. M. SYNGE: Partition Chromatography in the Study of Protein Constituents. Biochemic. J. 37, 79 (1943).
70. GRANT, R. A. and S. R. STITCH: A Volume Actuated Fraction Collector for Use in Chromatography. Chem. and Ind. 1951, 230.
71. GRASSMANN, W., H. DYCHERHOFF und H. EIBELER: Über die enzymatische Spaltung des Glutathions. I. Z. physiol. Chem. (Hoppe-Seyler) 189, 112 (1930).
72. GRASSMANN, W. und H. HÖRMANN: Endgruppenbestimmung an Kollagen und Gelatine. Z. physiol. Chem. (Hoppe-Seyler) 292, 24 (1953).
73. GREEN, F. C. and L. M. KAY: Separation of Sixteen Dinitrophenylamino Acids by Adsorption Chromatography on Silicic Acid-Celite. Analyt. Chemistry 24, 726 (1952).

74. GREEN, F. C. and W. A. SCHROEDER: A Terminal Amino Acid Residue of Lysozyme as Determined with 2,4-Dinitrofluorobenzene. J. Amer. Chem. Soc. 73, 1385 (1951).
75. GREEN, N. M. and E. WORK: Pancreatic Trypsin Inhibitor. Biochemic. J. 49, xxxvii (1951).
76. HAGDAHL, L.: Some Technical Improvements in Adsorption Analysis. Acta Chem. Scand. 2, 574 (1948).
76 a. HARFENIST, E. J.: The Amino Acid Composition of Insulins Isolated from Beef, Pork, and Sheep Glands. J. Amer. Chem. Soc. 75, 5528 (1953).
77. HARFENIST, E. J. and L. C. CRAIG: Differences in the Quantitative Amino Acid Composition of Insulins Isolated from Beef, Pork and Sheep Glands. J. Amer. Chem. Soc. 74, 4216 (1952).
78. HARRIS, J. I.: The Use of Carboxypeptidase for the Identification of Terminal Carboxyl Groups in Polypeptides and Proteins. Asparagine as a C-Terminal Residue in Insulin. J. Amer. Chem. Soc. 74, 2944 (1952).
79. HARRIS, J. I. and C. A. KNIGHT: Action of Carboxypeptidase on Tobacco Mosaic Virus. Nature (London) 170, 613 (1952).
80. HARRIS, J. O.: Automatic Chromatogram Fraction Cutter. Chem. and Ind. 1951, 255.
81. HAUSMANN, W.: Amino Acid Composition of Crystalline Inorganic Pyrophosphatase Isolated from Bakers Yeast. J. Amer. Chem. Soc. 74, 3181 (1952).
82. HAUSMANN, W. and L. C. CRAIG: Polypeptin: Purification, Molecular Weight Determination, and Amino Acid Composition. J. Biol. Chem. 198, 405 (1952).
83. HICKSON, J. L. and R. L. WHISTLER: Automatic Fraction Collector for Chromatographic Preparations. Analyt. Chemistry 25, 1425 (1953).
84. HIRS, C. H. W.: A Chromatographic Investigation of Chymotrypsinogen α. J. Biol. Chem. 205, 93 (1953).
85. HIRS, C. H. W., S. MOORE and W. H. STEIN: Isolation of Amino Acids by Chromatography on Ion Exchange Columns; Use of Volatile Buffers. J. Biol. Chem. 195, 669 (1952).
86. — — — A Chromatographic Investigation of Pancreatic Ribonuclease. J. Biol. Chem. 200, 493 (1953).
87. HOLLEY, R. W. and A. D. HOLLEY: A New Stepwise Degradation of Peptides. J. Amer. Chem. Soc. 74, 5445 (1952).
88. HOUGH, L., J. K. N. JONES and W. H. WADMAN: Quantitative Analysis of Mixtures of Sugars by the Method of Partition Chromatography. IV. Separation of the Sugars and their Methylated Derivatives on Columns of Powdered Cellulose. J. Chem. Soc. (London) 1949, 2511.
89. JAMES, A. T., A. J. P. MARTIN and S. S. RANDALL: Automatic Fraction Collectors and a Conductivity Recorder. Biochemic. J. 49, 293 (1951).
90. JOLLÈS, P. et C. FROMAGEOT: Caractérisation du résidu β-aspartique dans l'insuline. Biochim. Biophys. Acta 9, 416 (1952).
91. JUTISZ, M. and E. LEDERER: Quantitative Chromatographic Separation of Synthetic Peptides. Nature (London) 159, 445 (1947).
92. JUTISZ, M. et L. PÉNASSE: Détermination de la lysine N-terminale et totale du lysozyme. Bull. soc. chim. biol. (Paris) 34, 480 (1952).
93. KNESSL, O., B. KEIL, A. MALÝ and F. ŠORM: On Proteins and Amino Acids. IV. Partition Chromatography of DNP-Amino Acids on Kieselguhr and Siliconated Materials. Collect. Czechoslov. Chem. Communs. 15, 918 (1950).
94. KROL, S.: The Quantitative Estimation of Glycine in Small Samples of Proteins. Biochemic. J. 52, 227 (1952).

95. KRONER, T. D., W. TABROFF and J. J. McGARR: Peptides Isolated from a Partial Hydrolysate of Steer Hide Collagen. J. Amer. Chem. Soc. **75**, 4084 (1953).
96. LANDMANN, W. A., M. P. DRAKE and J. DILLAHA: Paper Chromatography of the 3-Phenyl-2-thiohydantoin Derivatives of Amino Acids with Application to End Group and Sequence Studies. J. Amer. Chem. Soc. **75**, 3638 (1953).
96 a. LANDMANN, W. A., M. P. DRAKE, and W. F. WHITE: Studies on Pituitary Adrenocorticotropin. VI. An N-Terminal Sequence of Corticotropin-A. J. Amer. Chem. Soc. **75**, 4370 (1953).
97. LEDERER, E. et T. P. KIUN: Séparations chromatographiques d'acides aminés et de peptides. I. Chromatographie de peptides neutres dans le formol à 10%. Biochim. Biophys. Acta **1**, 35 (1947).
98. LENS, J.: The Terminal Carboxyl Groups of Insulin. Biochim. Biophys. Acta **3**, 367 (1949).
99. LENS, J. and A. EVERTZEN: The Difference Between Insulin from Cattle and from Pigs. Biochim. Biophys. Acta **8**, 332 (1952).
100. LEWIS, J. C., N. S. SNELL, D. J. HIRSCHMANN and H. FRAENKEL-CONRAT: Amino Acid Composition of Egg Proteins. J. Biol. Chem. **186**, 23 (1950).
101. LI, C. H. and L. ASH: The N-Terminal End Groups of Hypophyseal Growth Hormone (Somatotropin). J. Biol. Chem. **203**, 419 (1953).
102. LI, C. H., A. TISELIUS, K. O. PEDERSEN, L. HAGDAHL and H. CARSTENSEN: Chromatography of Adrenocorticotropic Peptides. J. Biol. Chem. **190**, 317 (1951).
103. LIEN, O. G., Jr. and D. M. GREENBERG: Chromatographic Studies on the Interconversion of Amino Acids. J. Biol. Chem. **195**, 637 (1952).
104. LIEN, O. G., Jr., E. A. PETERSEN and D. M. GREENBERG: Automatic Fraction Collector for Column Chromatography. Analyt. Chemistry **24**, 920 (1952).
105. LORAND, L. and W. R. MIDDLEBROOK: The Action of Thrombin on Fibrinogen. Biochemic. J. **52**, 196 (1952).
106. — — Studies on Fibrino-Peptide. Biochim. Biophys. Acta **9**, 581 (1952).
107. — — Species Specificity of Fibrinogen as Revealed by End-Group Studies. Science (New York) **118**, 515 (1953).
108. LOWTHER, A. G.: Identification of N-(2,4-Dinitrophenyl) Amino Acids. Nature (London) **167**, 767 (1951).
109. MADER, C. and G. MADER: Automatic Volume Fraction Collector. Analyt. Chemistry **25**, 1423 (1953).
110. — — Evaporation Error in Volume Fractionation Chromatography. Analyt Chemistry **25**, 1556 (1953).
111. MARSH, D. F. and C. B. SEIBERT: A Simple Electronic Electrolytic Drop Recorder. Science (New York) **108**, 363 (1948).
112. MARTIN, A. J. P. and R. R. PORTER: Chromatographic Fractionation of Ribonuclease. Biochemic. J. **49**, 215 (1951).
113. MARTIN, A. J. P. and R. L. M. SYNGE: A New Form of Chromatogram Employing Two Liquid Phases. 1. A Theory of Chromatography. 2. Application to the Micro-Determination of the Higher Monoamino Acids in Proteins. Biochemic. J. **35**, 1358 (1941).
114. — — Analytical Chemistry of the Proteins. Adv. Protein Chem. **2**, 1 (1945).
115. McCLURE, L. E., L. SCHIELER and M. S. DUNN: The Free Amino Groups of Crystalline Bovine Plasma Albumin. J. Amer. Chem. Soc. **75**, 1980 (1953).
116. McFADDEN, M. L. and E. L. SMITH: The Free Amino Groups of γ-Globulins of Different Species. J. Amer. Chem. Soc. **75**, 2784 (1953).
117. MELLON, E. F., A. H. KORN and S. R. HOOVER: The Terminal Amino Groups of α- and β-Caseins. J. Amer. Chem. Soc. **75**, 1675 (1953).

118. MENDENHALL, R. M.: A Quantitative Amino Acid Analysis of Sheep Adreno-corticotropic (ACTH) Protein. Science (New York) **117**, 713 (1953).

119. MIDDLEBROOK, W. R.: Identification of the End Amino Groups of Wool by Means of their 2,4-Dinitrophenyl Derivatives. Nature (London) **164**, 501 (1949).

120. — The Chain Weight of Wool Keratin. Biochim. Biophys. Acta **7**, 547 (1951).

121. MILLS, G. L.: Identification of Dinitrophenylamino Acids. Nature (London) **165**, 403 (1950).

122. — Observations on the Application of Fluorodinitrobenzene to the Quantitative Analysis of Proteins. Biochemic. J. **50**, 707 (1952).

123. MONIER, R., Y. GENDRON, M. JUTISZ et C. FROMAGEOT: Nouvelle détermination des acides aminés basiques du lysozyme. Biochim. Biophys. Acta **8**, 588 (1952).

124. MOORE, S. and W. H. STEIN: Photometric Ninhydrin Method for Use in the Chromatography of Amino Acids. J. Biol. Chem. **176**, 367 (1948).

125. — — Chromatography of Amino Acids on Starch Columns. Solvent Mixtures for the Fractionation of Protein Hydrolysates. J. Biol. Chem. **178**, 53 (1949).

126. — — Chromatography of Amino Acids on Sulfonated Polystyrene Resins. J. Biol. Chem. **192**, 663 (1951).

127. NYE, W.: Simple Automatic Pipet. Analyt. Chemistry **22**, 848 (1950).

127 a. OHNO, K.: On the Structure of Lysozyme. I. Quantitative Estimation of Carboxyl-terminal Amino Acid by Improved Hydrazinolysis Method. J. Biochem. (Japan) **40**, 621 (1953).

128. OTTESEN, M. and C. VILLEE: The Peptides Released in the Enzymatic Transformation of Ovalbumin to Plakalbumin. C. R. Trav. Lab. Carlsberg, Sér. chim. **27**, 421 (1951).

129. OTTESEN, M. and A. WOLLENBERGER: Stepwise Degradation of the Peptides Liberated in the Transformation of Ovalbumin to Plakalbumin. Nature (London) **170**, 801 (1952).

129 a. — — Stepwise Degradation of the Peptides Liberated in the Transformation of Ovalbumin to Plakalbumin. C. R. Trav. Lab. Carlsberg, Sér chim. **28**, 463 (1953).

130. PALEUS, S. and J. B. NEILANDS: Preparation of Cytochrone *c* with the Aid of Ion Exchange Resin. Acta Chem. Scand. **4**, 1024 (1950).

131. PARTRIDGE, S. M. and R. C. BRIMLEY: Displacement Chromatography on Synthetic Ion-Exchange Resins. 8. A Systematic Method for the Separation of Amino Acids. Biochemic. J. **51**, 628 (1952).

132. PARTRIDGE, S. M. and T. SWAIN: A Reversed-Phase Partition Chromatogram Using Chlorinated Rubber. Nature (London) **166**, 272 (1950).

133. PARTRIDGE, S. M., R. G. WESTALL and J. R. BENDALL: Improvements in or Relating to the Fractionation of the Components of a Mixture in Solution. Brit. Patent 644,382 (1950).

134. PÉNASSE, L., M. JUTISZ, C. FROMAGEOT et H. FRAENKEL-CONRAT: La détermination des groupes carboxyliques des protéines. II. Le groupe carboxylique terminal de l'ovomucoïde. Biochim. Biophys. Acta **9**, 551 (1952).

135. PERRONE, J. C.: Separation of Amino Acids as Dinitrophenyl Derivatives. Nature (London) **167**, 513 (1951).

136. PHILLIPS, D. M. P.: A Simple Automatic Fraction-Cutter for Liquid Columns. Nature (London) **164**, 545 (1949).

137. PIERCE, J. G. and V. DU VIGNEAUD: Preliminary Studies on the Amino Acid Content of a High Potency Preparation of the Oxytocic Hormone of the Posterior Lobe of the Pituitary Gland. J. Biol. Chem. **182**, 359 (1950).

138. — — Studies of High Potency Oxytocic Material from Beef Posterior Pituitary Lobes. J. Biol. Chem. **186**, 77 (1950).

139. Piez, K. A., E. B. Tooper and L. S. Fosdick: Desalting of Amino Acid Solutions by Ion Exchange. J. Biol. Chem. **194**, 669 (1952).

140. Polis, B. D. and H. W. Shmukler: Crystalline Lactoperoxidase. I. Isolation by Displacement Chromatography. II. Physicochemical and Enzymatic Properties. J. Biol. Chem. **201**, 475 (1953).

141. Porath, J.: Purification of Bacitracin and Some Properties of Purified Bacitracin. Acta Chem. Scand. **6**, 1237 (1952).

142. — Structure of Bacitracin A. Nature (London) **172**, 871 (1953).

143. Porter, R. R.: The Unreactive Amino Groups of Proteins. Biochim. Biophys. Acta **2**, 105 (1948).

144. — Reactivity of the Iminazole Ring in Proteins. Biochemic. J. **46**, 304 (1950).

145. — A Chemical Study of Rabbit Antiovalbumin. Biochemic. J. **46**, 473 (1950).

146. — Use of 1,2,4-Fluorodinitrobenzene in Studies of Protein Structure. Methods Med. Res. **3**, 256 (1950).

147. — Partition Chromatography of Insulin and Other Proteins. Biochemic. J. **53**, 320 (1953).

148. Porter, R. R. and F. Sanger: The Free Amino Groups of Haemoglobins. Biochemic. J. **42**, 287 (1948).

149. Putnam, F. W.: N-Terminal Groups of Normal Human γ-Globulin and of Myeloma Proteins. J. Amer. Chem. Soc. **75**, 2785 (1953).

150. Reith, W. S. and N. M. Waldron: On the Determination of the Sequence of Amino Acids in Peptides. Biochemic. J. **53**, xxxv (1953).

150 a. — — Studies on the Determination of the Sequence of Amino Acids in Peptides and Proteins. 4. The Synthesis of 3-(4'-Dimethylamino-3':5'-dinitrophenyl)-2-thiohydantoin Derivatives of Various Amino Acids, and their Use for Amino Acid-Sequence Determinations. Biochemic. J. **56**, 116 (1954).

151. Rovery, M., C. Fabre et P. Desnuelle: Etude des extrémités N-terminales du trypsinogène et de la trypsine de Boeuf. Biochim. Biophys. Acta **9**, 702 (1952).

152. — — — Extrémités N-terminales de la β- et de la γ-chymotrypsines de Boeuf. Biochim. Biophys. Acta **10**, 481 (1953).

153. Sanger, F.: The Free Amino Groups of Insulin. Biochemic. J. **39**, 507 (1945).

154. — The Free Amino Group in Gramicidin S. Biochemic. J. **40**, 261 (1946).

155. — The Terminal Peptides of Insulin. Biochemic. J. **45**, 563 (1949).

156. — Application of Partition Chromatography to the Study of Protein Structure. Biochem. Soc. Symp. **3**, 21 (1949).

157. — Species Differences in Insulins. Nature (London) **164**, 529 (1949).

158. — The Arrangement of Amino Acids in Proteins. Adv. Protein Chem. **7**, 1 (1952).

159. Sanger, F. and E. O. P. Thompson: The Amino-acid Sequence in the Glycyl Chain of Insulin. 1. The Identification of Lower Peptides from Partial Hydrolysates. Biochemic. J. **53**, 353 (1953).

160. — — The Amino-acid Sequence in the Glycyl Chain of Insulin. 2. The Investigation of Peptides from Enzymic Hydrolysates. Biochemic. J. **53**, 366 (1953).

161. Sanger, F. and H. Tuppy: The Amino-acid Sequence in the Phenylalanyl Chain of Insulin. 1. The Identification of Lower Peptides from Partial Hydrolysates. Biochemic. J. **49**, 463 (1951).

162. — — The Amino-acid Sequence in the Phenylalanyl Chain of Insulin. 2. The Investigation of Peptides from Enzymic Hydrolysates. Biochemic. J. **49**, 481 (1951).

163. Schlack, P. und W. Kumpf: Über eine neue Methode zur Ermittlung der Konstitution von Peptiden. Z. physiol. Chem. (Hoppe-Seyler) **154**, 125 (1926).

164. SCHLÖGL, K., A. SIEGEL und F. WESSELY: Konstitutionsermittlung von Peptiden. IV. Papierchromatographische Trennung und Identifizierung der Abbaustufen. Z. physiol. Chem. (Hoppe-Seyler) **291**, 265 (1952).

165. SCHMID, K.: Untersuchungen über das Wal-myoglobin. Helv. Chim. Acta **32**, 105 (1949).

166. SCHRAM, E. and E. J. BIGWOOD: Fraction Collector for Chromatography. Analyt. Chemistry **25**, 1424 (1953).

167. SCHRAM, E., J. P. DUSTIN, S. MOORE et E. J. BIGWOOD: Application de la chromatographie sur échangeur d'ions à l'étude de la composition des aliments en acides aminés. Anal. Chim. Acta **9**, 149 (1953).

168. SCHROEDER, W. A.: Sequence of Four Amino Acids at the Amino End of the Single Polypeptide Chain of Lysozyme. J. Amer. Chem. Soc. **74**, 5118 (1952).

169. — unpublished results.

170. SCHROEDER, W. A. and R. B. COREY: Automatic Weight-driven Time-controlled Fraction Collector. Analyt. Chemistry **23**, 1723 (1951).

171. SCHROEDER, W. A. and L. R. HONNEN: Correlation Between the Structure of Some Dinitrophenylpeptides and their Chromatographic Behavior on Silicic Acid-Celite. J. Amer. Chem. Soc. **75**, 4615 (1953).

172. SCHROEDER, W. A., L. R. HONNEN and F. C. GREEN: Chromatographic Separation and Identification of Some Peptides in Partial Hydrolysates of Gelatin. Proc. Nat. Acad. Sci. (U. S. A.) **39**, 23 (1953).

173. SCHROEDER, W. A. and L. M. KAY: unpublished results.

174. SCHROEDER, W. A., L. M. KAY, J. LeGETTE, L. R. HONNEN and F. C. GREEN: The Constitution of Gelatin. J. Amer. Chem. Soc. (in press).

175. SCHROEDER, W. A., L. M. KAY and I. C. WELLS: Amino Acid Composition of Hemoglobins of Normal Negroes and Sickle-cell Anemics. J. Biol. Chem. **187**, 221 (1950).

176. SCHROEDER, W. A. and J. LeGETTE: A Study of the Quantitative Dinitrophenylation of Amino Acids and Peptides. J. Amer. Chem. Soc. **75**, 4612 (1953).

177. SHEPARD, C. C. and A. TISELIUS: The Chromatography of Proteins. The Effect of Salt Concentration and p_H on the Adsorption of Proteins to Silica Gel. Discuss. Faraday Soc. **7**, 275 (1949).

178. SHULGIN, A. T., O. G. LIEN, Jr., E. M. GAL and D. M. GREENBERG: Synthesis and Chromatographic Separation of Isotopically Labelled *DL*-Threonine and *DL*-Allothreonine. J. Amer. Chem. Soc. **74**, 2427 (1952).

179. SIMMONDS, D. H.: Leucine-Isoleucine Content of Wool. Nature (London) **172**, 677 (1953).

180. SJÖQUIST, J.: Paper Strip Identification of the Phenyl Thiohydantoins. Acta Chem. Scand. **7**, 447 (1953).

181. SMITH, E. L.: Proteolytic Enzymes. In: The Enzymes, Vol. 1, Part 2, p. 802 New York: Academic Press. 1951.

182. SMITH, E. L. and A. STOCKELL: J. Biol. Chem. (in press).

183. SMITH, E. L., A. STOCKELL and J. R. KIMMEL: J. Biol. Chem. (in press).

184. SOBER, H. A., G. KEGELES and F. J. GUTTER: Chromatographic Analysis of Mixture of Proteins from Egg White. Science (New York) **110**, 564 (1949).

185. — — — Chromatography of Proteins. Frontal Analysis on a Cation Exchange Resin. J. Amer. Chem. Soc. **74**, 2734 (1952).

186. ŠORM, F. and Z. ŠORMOVÁ: On Proteins and Amino Acids. VII. On Clupein. Collect. Czechoslov. Chem. Communs. **16**, 207 (1951).

186 a. SOUPART, P., S. MOORE and E. J. BIGWOOD: Amino Acid Composition of Human Milk. J. Biol. Chem. **206**, 699 (1954).

187. STEIN, W. H.: Excretion of Amino Acids in Cystinuria. Proc. Soc. exp. Biol. Med. **78**, 705 (1951).

188. — A Chromatographic Investigation of the Amino Acid Constituents of Normal Urine. J. Biol. Chem. **201**, 45 (1953).

189. STEIN, W. H. and S. MOORE: Chromatography of Amino Acids on Starch Columns. Separation of Phenylalanine, Leucine, Isoleucine, Methionine, Tyrosine, and Valine. J. Biol. Chem. **176**, 337 (1948).

190. — — Chromatographic Determination of the Amino Acid Composition of Proteins. Cold Spring Harbor Sympos. Quant. Biol. **14**, 179 (1949).

191. — — Amino Acid Composition of β-Lactoglobulin and Bovine Serum Albumin. J. Biol. Chem. **178**, 79 (1949).

192. — — Electrolytic Desalting of Amino Acids. Conversion of Arginine to Ornithine. J. Biol. Chem. **190**, 103 (1951).

193. STEINBERG, D.: The Action of Carboxypeptidase on Ovalbumin. J. Amer. Chem. Soc. **74**, 4217 (1952).

194. — The Combined Action of Carboxypeptidase and *B. subtilis* Enzyme on Ovalbumin. J. Amer. Chem. Soc. **75**, 4875 (1953).

195. SWINGLE, S. M. and A. TISELIUS: Tricalcium Phosphate as an Adsorbent in the Chromatography of Proteins. Biochemic. J. **48**, 171 (1951).

196. SYNGE, R. L. M.: Partial Hydrolysis Products Derived from Proteins and their Significance for Protein Structure. Chem. Rev. **32**, 135 (1943).

197. — Analysis of a Partial Hydrolysate of Gramicidin by Partition Chromatography with Starch. Biochemic. J. **38**, 285 (1944).

198. TALLAN, H. H.: Occurrence of a New Amino Acid, 3-Methylhistidine, in Human Urine. Federat. Proc. (Amer. Soc. exp. Biol.) **12**, 278 (1953).

199. TALLAN, H. H. and W. H. STEIN: Chromatographic Studies on Lysozyme. J. Biol. Chem. **200**, 507 (1953).

199 a. TALLAN, H. H., W. H. STEIN and S. MOORE: 3-Methylhistidine, a New Amino Acid from Human Urine. J. Biol. Chem. **206**, 825 (1954).

200. THOMPSON, A. R.: Destruction of DNP-Amino Acids by Tryptophane. Nature (London) **168**, 390 (1951).

201. — The C-Terminal Residue of Lysozyme. Nature (London) **169**, 495 (1952).

202. — private communication.

203. THOMPSON, E. O. P.: The N-Terminal Sequence of Carboxypeptidase. Biochim. Biophys. Acta **10**, 633 (1953).

204. — private communication.

205. TISELIUS, A.: Displacement Development in Adsorption Analysis. Ark. Kemi, Mineral. Geol. **16** A, No. 18 (1943).

206. TRISTRAM, G. R.: Observations Upon the Application of Partition Chromatography to the Determination of the Monoamino Acids in Proteins. Biochemic. J. **40**, 721 (1946).

207. TRUEBLOOD, K. N. and E. MALMBERG: An Experimental Study of Chromatography on Silicic Acid-Celite. The Applicability of the Theory of Chromatography. J. Amer. Chem. Soc. **72**, 4112 (1950).

208. TURNER, R. A. and G. SCHMERZLER: C-Terminal Residues of Ovomucoid and Ovalbumin. Biochim. Biophys. Acta **11**, 586 (1953).

209. VAN VUNAKIS, H.: The Free Amino Groups of Serum Albumins. Univ. Microfilms (Ann Arbor, Mich.) No. 2865 [Chem. Abstr. **46**, 2112 (1952)].

210. VARNER, J. E. and W. A. BULEN: Automatic Constant-Volume Fraction Collector. J. Chem. Education **29**, 625 (1952).

211. WALEY, S. G. and J. WATSON: The Stepwise Degradation of Peptides. J. Chem. Soc. (London) **1951**, 2394.

212. Wessely, F., K. Schlögl and G. Korger: A New Method for the Degradation of Peptides. Nature (London) **169**, 708 (1952).
213. Westall, R. G.: Isolation of γ-Amino Butyric Acid from Beet-root. Nature (London) **165**, 717 (1950).
214. Weygand, F. und R. Junk: Die freien Aminogruppen des ,,alten" gelben Fermentes. Naturwiss. **38**, 433 (1951).
215. White, W. F.: Studies on Pituitary Adrenocroticotropin. VII. A C-Terminal Sequence of Corticotropin-A. J. Amer. Chem. Soc. **75**, 4877 (1953).
215 a. White, W. F. and W. L. Fierce: Studies on Pituitary Adrenocorticotropin. III. Differentiation of Three Active Types on XE-97 Resin. J. Amer. Chem. Soc. **75**, 245 (1953).
216. Williamson, M. B. and J. M. Passmann: The Terminal Group of Pepsin. J. Biol. Chem. **199**, 121 (1952).
217. Wingo, W. J. and I. Browning: Simple and Inexpensive Fraction Collector. Analyt. Chemistry **25**, 1426 (1953).
218. Zechmeister, L. and M. Rohdewald: Some Aspects of Enzyme Chromatography. Fortschr. Chem. organ. Naturstoffe **8**, 341 (1951).

(Received, December 1, 1953.)

Porphyrins in Nature.

By R. LEMBERG, Sydney.

With 2 Figures.

Contents.

I. Introduction.

When the classical book of H. Fischer and H. Orth (*1*) ,,Die Chemie des Pyrrols" was published in 1937, much was known on the chemical structure of porphyrins, but very little on the biochemical and biological significance of the occurrence of free porphyrins in nature. They had been found widespread but generally only in traces, which it was possible to discover because of the extraordinarily characteristic properties of their fluorescence and light absorption. Larger amounts had been found in certain rare diseases of man and mammals, and in isolated instances haphazardly distributed over nature. It was known that these porphyrins were chemically related to, or identical with, the prosthetic groups of haematin compounds or chlorophyll. Practically nothing was known, however, on the biosynthesis of the porphyrins, on their biochemical interrelationship, and on their biochemical relation to the metal complexes of fundamental biological importance. Such hypotheses as were brought forward were entirely speculative, and most of them have by now been shown to be wrong.

Progress came from different directions: Physiological evidence showed that contrary to previous belief, free porphyrins were generally, if not entirely, by-products of the synthesis leading to the metal complexes, particularly of haemopoiesis in vertebrates, and not products of haemoglobin breakdown (Dobriner and Rhoads, Watson, and others). This was supported by chemical and biochemical evidence showing that the pathway of catabolism of haem compounds to bile pigments does not lead to the formation of free porphyrins at any stage (Lemberg). Previous assumption as to the relations between the more highly carboxylated free porphyrins, such as coproporphyrins and uroporphyrins, to protoporphyrin, the prosthetic group of haemoglobin, and many haematin enzymes, were shown to be wrong. It had been postulated that coproporphyrin and uroporphyrin arose from protoporphyrin by successive carboxylations, in a process considered by H. Fischer as detoxication. It is now clear that copro- and uroporphyrins (as such or in the form of their monopyrrolic precursors) are precursors of protoporphyrin, which latter arises from them by decarboxylation [Turner (*239*), Watson (*9, 10*), Lemberg (*6, 7*)]. Finally, isotopic studies [Shemin, Rittenberg (*223*) and Bloch (*191*)] showed that glycine and acetic acid were precursors in the biosynthesis of haemoglobin protoporphyrin. A synthesis of these facts allowed Lemberg and Legge (*6*, there Chapters X–XIII) to give a far clearer picture of the biosynthesis and the interrelationships of the porphyrins. This theory was presented by the writer as a paper read before the Adelaide Congress of the Australian and New Zealand Association of Science in 1946, and was incorporated in the book at that time.

Meanwhile, rapid progress has been made in our knowledge of the biosynthesis of porphyrins which confirmed the hypothesis of LEMBERG and LEGGE that the primary pyrrolic precursor arises by condensation of glycine with a member of the citric acid cycle. It is now evident that the reason for the sporadic and quantitatively mostly scanty occurence of free porphyrins in nature is the great efficiency and specificity of the synthesis of protohaem. This allows in general only 0.1% or less of the precursors to remain as free porphyrins, be it by the synthesis of unsuitable isomerides as by-products of the synthesis, by failure of enzymes which catalyse the decarboxylation of the side-chains, or by failure of enzymes catalysing the incorporation of iron. The study of these enzymes is still a task for the future.

There is good evidence to show that the biosynthesis of chlorophyll follows lines very similar to those of the porphyrin synthesis of haemoglobin; but the synthesis of the prosthetic groups of some of the haematin enzymes—particularly those which contain not protohaem but other still incompletely studied iron porphyrins as prosthetic groups—has so far not been studied adequately.

The literature on porphyrins is immense. To give an exhaustive bibliography in the frame of this article is impossible. Recourse had to be made to the pages of standard works and reviews, on which many more references will be found. The more recent literature (after about 1946) is, however quoted more exhaustively.

II. The Structure of the Porphin Nucleus.

The correct structure of the porphin nucleus (I) was proposed by W. KÜSTER in 1913 on the basis of his studies on the oxidation products (substituted maleinimides) and of studies by NENCKI and ZALESKI, PILOTY, KNORR and WILLSTÄTTER on reduction products (substituted pyrroles) of haemin. This formula was later proved by the classical synthetic work of H. FISCHER. Porphyrins are derivatives of porphin in which all or most of the eight β-hydrogens of the four pyrrole nuclei are replaced by side-chains. Porphin itself was prepared far later by FISCHER and GLEIM and by ROTHEMUND (cf. *1*, there p. 174). Substituents on the methine bridges which combine the four pyrrolic rings to the porphin system, are only found in porphyrins obtained by decomposition from chlorophylls, e. g. phylloporphyrin (XX, p. 306). They are remnants of the isocyclic ring present in the chlorophylls (II) which in turn is biologically probably derived by the oxidative condensation of a propionic acid side-chain in position 6 with the γ methine bridge. Chlorophylls are essentially magnesium complexes derived from dihydro- or tetrahydroporphyrins.

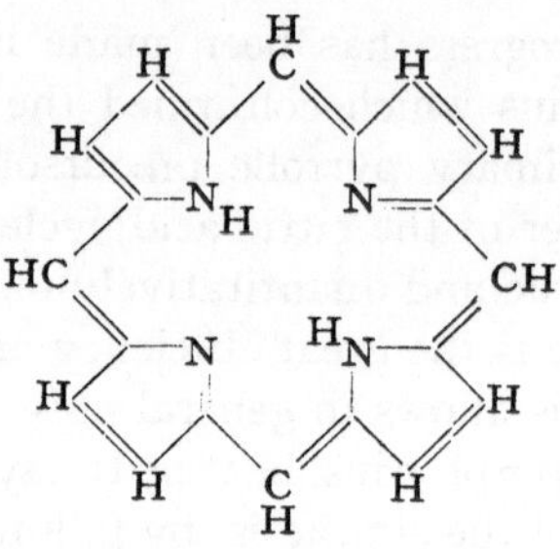

(I.) Porphin nucleus.

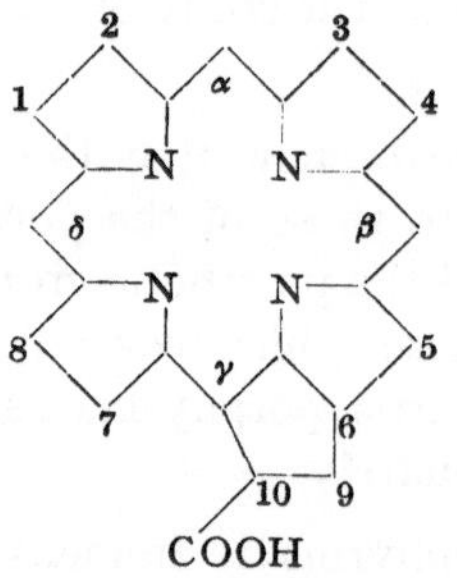

(II.) Porphin system and the isocyclic ring present in chlorophylls.

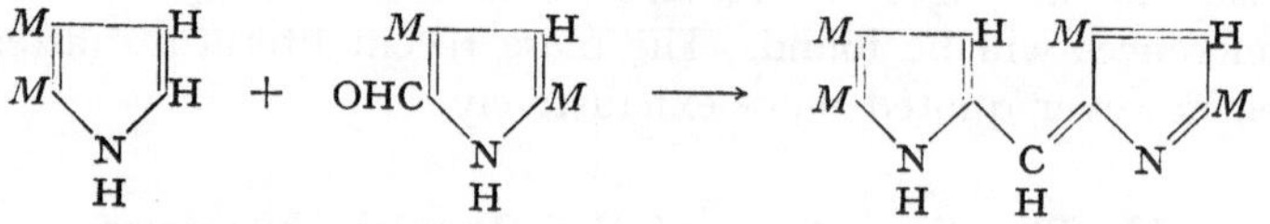

(III.)   (IV.)   (V.) Pyrromethene.

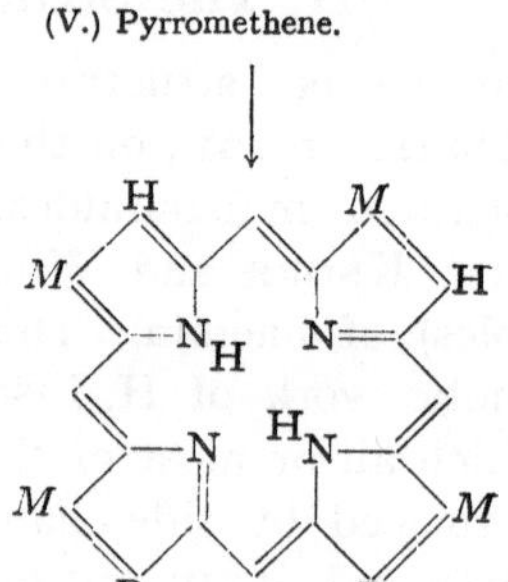

(VI.)

(IX.) Deuteroporphyrin.

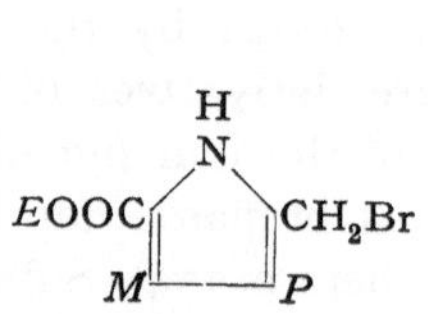

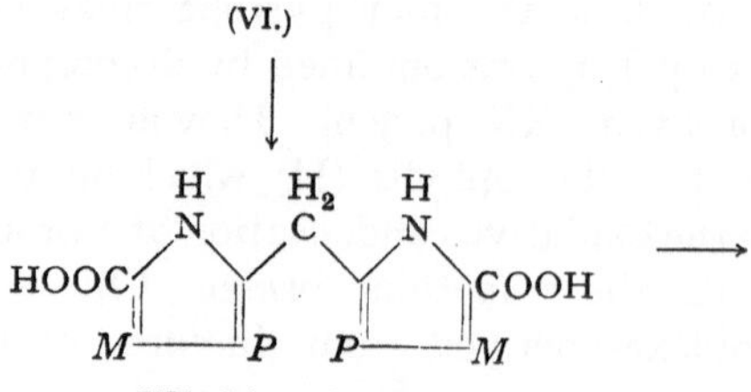

(VII.) Dipyrrylmethane. 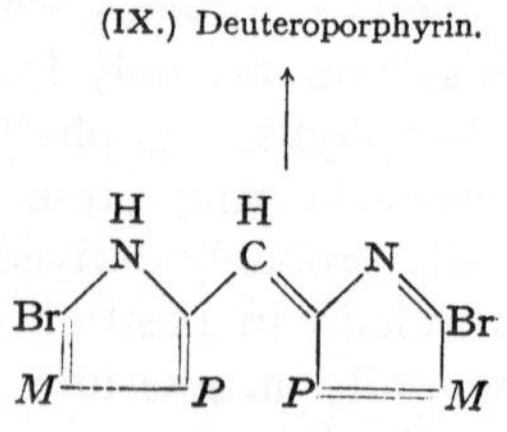   (VIII.) Pyrromethene.

The methods of synthesis of porphyrins have been described by FISCHER and ORTH (*1*, there pp. 1 ff., 160 ff.); see also MACDONALD (*154*). The synthesis of deuteroporphyrin, convertible into protoporphyrin and haemin, may serve as an example ($M = CH_3$; $P = CH_2 \cdot CH_2 \cdot COOH$).

Condensation of an α-methyl-α'-unsubstituted pyrrole (III) with an α-methyl-pyrrole-α'-aldehyde (IV) yields a pyrromethene (V) with two methyl groups in the terminal α-positions, unsymmetrically substituted in the β-positions.* Autocondensation of pyrrole (VI), with bromomethyl and carboxethyl groups in α, yields a dipyrrylmethane (VII),

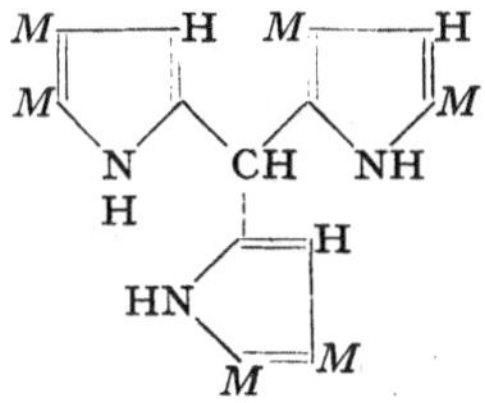

symmetrically substituted in its β-positions. Bromination yields the corresponding pyrromethene (VIII) with two bromine groups in the terminal α positions. Condensation of (V) with (VIII) produces deuteroporphyrin "IX" (IX), with all its side-chains determined in their positions.

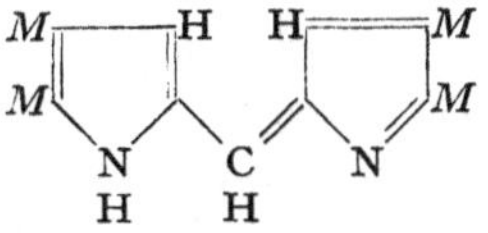

Complications can occur in the synthesis at the stage of condensation of pyrrole (III) with pyrrole α-aldehyde (IV) [CORWIN *et al.* (*45*, *46*, *17*)] by the primary formation of tripyrrylmethane derivatives by condensation, e. g. of one molecule of (IV) with two molecules of (III) to form (X). Decomposition of (X) might conceivably yield (XI) as well as (V). CORWIN and KRIEBLE (*47*) have shown, however, that this does not occur in the synthesis of (IX).

Differences of opinion still exist as to the fine structure of the porphyrins. H. FISCHER and his school assumed not only a definite position of the two central hydrogen atoms on two of the four pyrrolic nuclei, but also definite positions of the double bonds (*1*, there p. 172). In the formula (I) the four pyrrolic nuclei are of three different types, pyrrole (XII), pyrrolene (XIII) and maleinimide (XIV). In the closely related phthalocyanines (tetrabenzo-tetrazaporphin), ROBERTSON and coworkers were, however, unable to discover such a difference between the four

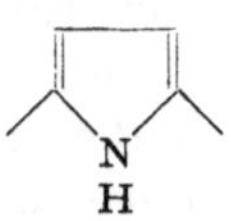

(XII.) Pyrrole type. 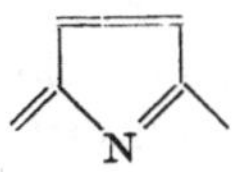   (XIII.) Pyrrolene type. 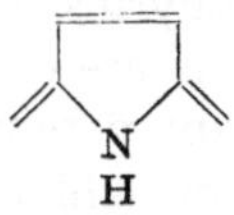   (XIV.) Maleinimide type.

* In these and the following *formulae* the pyrrole rings are no longer drawn in the regular pentagon form which approaches the correct stereochemical arrangement, cf. (I) and (II), but in the more convenient form used by H. FISCHER.

isoindole nuclei connected by $=$N— -linkages, in their Fourier diagram of the molecule obtained by X-ray analysis of metal-free and Ni compound (*205, 206*; cf. *6*, there p. 80). ELEY (*63*) found phthalocyanine crystals to be semi-conductors, and explained this by intermolecular overlap of the π-electron orbitals of the ring. Thus, some workers [e. g. KUHN (*133*)] assume complete "aromaticity" of the porphin ring, with the two central hydrogen atoms equally shared between the four central pyrrole nitrogens. VESTLING and DOWNING (*241*) found some evidence for hydrogen bonds (N—H$\cdots$N) in the infrared spectra of aetioporphyrin in carbon tetrachloride, and FALK and WILLIS (*72*) confirmed this for some other porphyrins in the solid state.

RABINOWITCH (*192*) and SIMPSON (*226*) pointed out that even assuming hydrogen bonds, one might expect the existence of tautomeric forms such as (XV) and (XVI). Earlier evidence for the existence of such

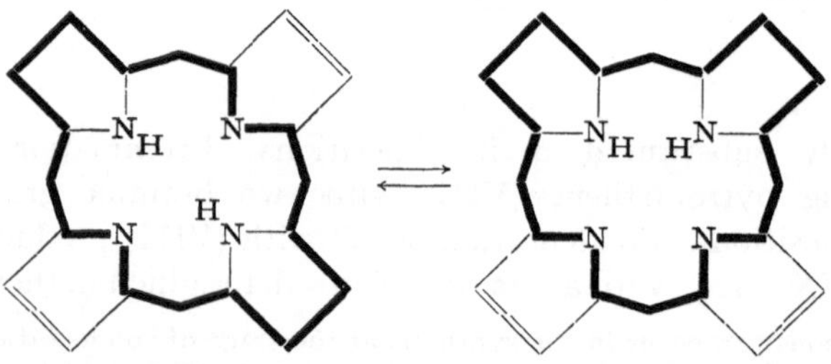

(XV.)   (XVI.)

tautomeric forms has been critically reviewed by LEMBERG and LEGGE (*6*, there p. 83) and found unconvincing. Recently, however, DOROUGH and SHEN (*62*) reported that a study of the absorption and fluorescence spectra of deuterium porphin free base, particularly at liquid air temperature, supported the proposition that porphyrins exist as equilibrium mixtures of tautomeric compounds. The discovery of N-methylporphyrins, with typical absorption spectra of neutral porphyrins by CORWIN *et al.* (*64, 65*, cf. *158*) is also best interpreted by this assumption.

The quantum-mechanical interpretation of the absorption and fluorescence spectra of porphyrins (*113, 115, 133, 226, 149, 190*) has made some progress, but is still far from satisfactory. The spectrum of the porphyrins is typical for a "round-field" molecule in which the π-electrons are confined to move in a closed ring-shaped path in a field of constant potential energy. In contrast, the spectrum of the dihydroporphyrins (chlorins) or tetrahydroporphyrins, in which one or both of the two "isolated double bonds" of porphyrins, [see (XV) or (XVI)], are hydrogenated, is typical for a "longfield" molecule, in which the π-electrons move along a chain. Nevertheless, the findings do not support the assumption of full "aromaticity" of the porphin nucleus. The combustion energies are also not in agreement with it (STERN and KLEBS, see *1*, there p. 597 f.). X-ray diffraction studies on porphyrins (*37*, *125*, *126*) show that the ring is planar, as is that of phthalocyanine. The molecule of coproporphyrin "I" is, however, not strictly centrosymmetrical (*125*); nor has that of phthalocyanine full tetragonal C 4 h symmetry (*205*), which it approaches more closely when the two central hydrogen atoms are replaced by a metal atom (*206*). Porphyrins form di-cations

(XVII.) Di-cation type.   (XVIII.) Di-anion type.

(XVII) and metal complexes, e. g. disodium salts of (XVIII); if PH_2 is the neutral porphyrin, these can be represented by $PH_4{}^{++}$, and P^{--} respectively, leaving out of consideration dissociable groups such as carboxyl groups in the side-chains.

LEMBERG and LEGGE (*6*, there p. 80) drew attention to the fact that porphyrins would be expected to be strong bases, if the neutral porphyrin was far less stabilised by resonance than the di-cation. The measurement of the dissociation constants of porphyrins encounters special difficulties in the insolubility of the neutral porphyrins. Another complicating factor is the presence of dissociable carboxyl groups in most of the porphyrins; since these are, however, generally separated from the ring system by an aliphatic chain (propionic acid groups), this dissociation is not expected to influence the spectra greatly. Recently, NEUBERGER and SCOTT (*175*) found that the pK' of the di-cation formation is considerably higher (about 4.0) than previously found by CONANT (about 2.5). By spectrophotometric measurements, they also confirmed earlier findings of TREIBS, indicating that mono-cations were formed under certain conditions in aqueous buffers. The pK' of the equilibrium between neutral coproporphyrin "I" and its mono-cation was found as high as 7.15. The nature of the observed spectrum as that of a mono-cation requires confirmation, particularly as WALTER (*246*) has been unable to find any evidence for such an intermediate between PH_2 and $PH_4{}^{++}$ with one of the compounds, deuteroporphyrin "IX"-dimethyl ester disulphonic acid, also studied by NEUBERGER and SCOTT. A pK' of 4 is not higher than would be expected for a pyrrolene nitrogen without additional resonance stabilisation in the porphin ring, but this effect may compensated by the electrostatic repulsion of the two central positive charges.

The resonance structure of the molecule is of fundamental biological importance. The absorption of visible light is conditioned by this structure, and the biological action of the chlorophyll depends on it. Moreover, if photosynthesis preceded respiration in evolution, the property of absorption of visible light is of special evolutionary significance. Resonance stabilisation is, however, of importance, in other respects. The function of haematin compounds as respiratory enzymes and oxygen carriers, does not depend on their colour. The problems of integration of the electrons of the chelated metal, the π-electrons of the porphin ring, and electrons from the protein part of the molecule cannot yet be solved, but it is here that finally the explanation of specificity of action must be sought. Resonance stabilisation of monovalent steps between dihydroporphyrin and porphyrin may be important for the action of chlorophyll in photosynthesis [CALVIN and DOROUGH (35); CAHILL and TAUBE (34)].

III. The Naturally Occurring Porphyrins.

Most of the naturally occurring porphyrins differ from porphin and from one another only by the nature and relative positions of their eight side-chains at the β-positions of the four pyrrole nuclei. For these FISCHER's symbol (XIX) is most convenient. LEMBERG and LEGGE (6) have introduced another symbol by which substituents on the methine bridges (the

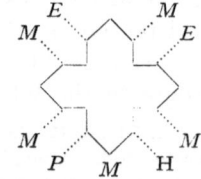

(XIX.) FISCHER's symbol.

(XX.) LEMBERG and LEGGE's symbol: phylloporphyrin.

(XXI.) Phaeoporphyrin a₅.

(XXII.) Numbering system.

four external corners of the central six-membered ring) can be indicated, e. g. phylloporphyrin (XX), and which can be extended to chlorophyll porphyrins with an isocyclic ring, e. g. phaeoporphyrin a₅ (XXI). The usual numbering of rings, β-substituents and methine bridges is given in (XXII).

The following *abbreviations* will be used for the side-chains: $M = $ methyl, $E = $ ethyl, $V = $ vinyl, $HE = $ hydroxyethyl ($-$CHOH $\cdot$ CH$_3$), $F = $ formyl, Ac = acetyl, P = propionic acid (—$CH_2 \cdot CH_2 \cdot COOH$), AC = acetic acid (—$CH_2 \cdot COOH$). In addition to the nature of the side-chains, which characterise a special porphyrin, e. g. coproporphyrin, their relative

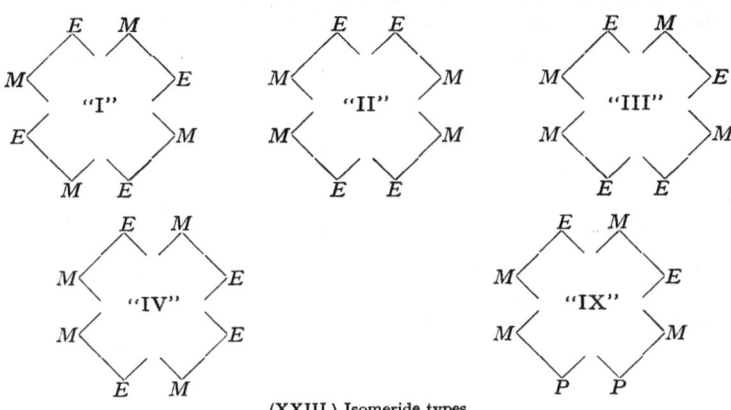

(XXIII.) Isomeride types.

position is of importance. If there are two types of side-chains and each pyrrole ring bears one of each, e. g. M and E in aetioporphyrins, M and P in coproporphyrins, four isomerides, types "I", "II", "III" and "IV" (XXIII) are possible.

All naturally occurring porphyrins, whether as free porphyrins or as prosthetic groups, belong either to the centrosymmetrical type "I" or, more frequently, to the non-symmetrical type "III" ("dualism" of the porphyrins, H. FISCHER). Of the 15 possible isomerides with three different types of side-chains, e. g. mesoporphyrin with M, E and P, only one type, "IX" (XXIII), is found in nature, which can be considered as derived from aetiotype "III" by replacement of $2 E$ by $2 P$.

The number of possible isomerides of chlorophyll porphyrins is still greater, but if the isocyclic ring between the β-position 6 and the γ methine bridge is considered a modified propionic acid side-chain in position 6, they belong, if fact, to the same type as mesoporphyrin "IX". On decomposition, the isocyclic ring is either fully removed leaving an unsubstituted position in 6 as in pyrroporphyrin (XLV), and pyrroaetioporphyrin

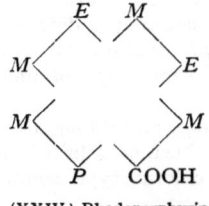

(XXIV.) Rhodoporphyrin.

20*

(which thus has one ethyl group less than aetioporphyrin), or parts of
it remain, e. g. the methyl group on the γ methine bridge in phylloporphyrin, or the carboxyl group in position 6 in rhodoporphyrin (XXIV).

A short summary of the structure of porphyrins without substituents
on methine bridges (uro-, copro-, proto-class) and their occurrence in
nature as prosthetic groups of haemoproteins and metal complexes is

Table 1. Porphyrins without Substituents on Methine Bridges.
Uro-copro-proto Class.

Porphyrin	Formula	Side-chains*	Occurrence as prosthetic group or metal complex
Aetio "III" ...	(XXIII), p. 307	4 M, 4 E	—
Meso "IX" ...	(XXVII), p. 316	4 M, 2 E, 2 P	—
Proto "IX"...	(XXVI), p. 315	4 M, 2 V, 2 P	Fe complex (haem) prosthetic group of haemoglobins, myoglobins, catalases, horse radish peroxidase**, cytochromes b Mg complex in *Chlorella* mutant (*92*)
Deutero "IX"	(XXVIII), p. 316	4 M, 2 H, 2 P	—
Haemato "IX"	(XXIX), p. 317	4 M, 2 HE, 2 P	Fe complex, modified, prosthetic group of cytochrome c***
Chlorocruoro .	(XXX), p. 318	4 M, 1 V, 1 F, 2 P	Fe complex prosthetic group of chlorocruorin, oxygen carrier of sabellid worms (*80*)
Cyto (a)	?	?, 1 F, 2 P****	Fe complex prosthetic group of cytochrome oxidase (cytochrome a_3) and cytochromes a and a_1 (*138*, *195*) †
Copro "I" and "III"	(XXXIII), p. 320 (XXXIV), p. 320	4 M, 4 P	—
Uro "I" and "III"	(XXXV), p. 323 (XXXVI), p. 323	4 AC, 4 P	— Cu complex: turacin in Turaco bird feathers (*198*)

* Abbreviations see on pp. 306–307.

** The prosthetic group of lactoperoxidase probably contains a porphyrin
with an acetyl side-chain [Morell (*167*)], that of myeloperoxidase an oxidised
porphin ring [Lemberg, cf. (*6*), there p. 430].

*** Cytochrome c can be considered a derivative of haematohaem in which
the hydroxyl groups of the α-hydroxyethyl side-chains are replaced by S-linkages
to cysteine of the protein. The same type of linkage probably occurs in cytochrome f (*54*).

**** At least two methyl groups and a long alkyl group with double bond conjugated to the porphin ring have been established also (see p. 319).

† Cytochrome a_2 contains a chlorin type compound [Lemberg and Barrett
(*138 a*)].

Table 2. Porphyrins Derived from Chlorophyll.

Porphyrin	Formula	Side-chains*	Ring C_6—C_γ**		Relation to chlorophylls and occurrence
Vinylphaeo-a_5	(XLIII), p. 328	4 M, 1 E, 1 V, 1 P	—CO—CH(COOM)—		Mg complex of dihydroporphin, esterified with phytol: chlorophyll a.
					Mg complex in *Chlorella* mutant (87), esterified with phytol: protochlorophyll (182, 183).
Vinylphaeo-b_6	—	3 M, 1 F, 1 E, 1 V, 1 P	—CO—CH(COOM)—		Mg complex of dihydroporphin, esterified with phytol: chlorophyll b.
Oxophaeo-a_5	—	4 M, 1 E, 1 Ac, 1 P	—CO—CH(COOM)—		Mg complex of tetrahydroporphin, esterified with phytol: bacteriochlorophyll.
Phylloerythrin	(XLIV), p. 329	4 M, 2 E, 1 P	—CO—CH$_2$—		⎫
Phylloporphyrin ...	(XX,) p. 306	4 M, 2 E, 1 P	—H	CH$_3$—	⎬ Decomposition products of chlorophyll.
Rhodoporphyrin	(XXIV), p. 307	4 M, 2 E, 1 P	—COOH	H—	
Pyrroporphyrin	(XLV), p. 329	4 M, 2 E, 1 P	—H	H—	
Pyrroaetioporphyrin	—	4 M, 3 E	—H	H—	⎭

* Abbreviations see on pp. 306–307.
** See (XXI), (XXII), p. 306.

given in *Table 1*, p. 308. *Table 2* shows the structure of some porphyrins derived from chlorophyll and their relationship to the chlorophylls.

Porphyrins of both classes have been found in bituminous rocks, mineral oils and coals, some as old as Silurian [TREIBS *(237,237a)*; cf. *6*, there p. 648]. The biological role of the tetrapyrroles must have been established rather early in evolution; with the possible exception of obligatory anaerobic microorganisms and of viruses, tetrapyrroles are essential constituents of all living forms, microorganisms, plants and animals.

IV. Methods of Isolation, Separation, Identification and Estimation.

Isolation of Porphyrins from Natural Sources.

With the exception of the highly carboxylated uroporphyrins, porphyrins are moderately soluble in ether at a p_H at which neither their basic groups nor their carboxyl groups are ionised, i. e. at a p_H range of about 2–5, e. g. in acetate buffer, and can be extracted by mixtures of ether and acetic acid, or ethylacetate and acetic acid. The latter solvent also extracts some uroporphyrins. The ether-insoluble uroporphyrins can be adsorbed from urine by aluminium oxide *(1*, there p. 514) or better by talc *(105)* or calcium phosphate in the presence of an excess of Ca^{++} and OH^- ions [GARROD *(83)*, cf. SVEINSSON, RIMINGTON and BARNES *(232)*]; this is best carried out after removal of the ether-soluble porphyrins by extraction with ether. In other instances it is necessary first to esterify uroporphyrins by methanolic hydrochloric acid and to extract the octamethyl ester with ether or chloroform after removal or neutralisation of excess acid. Porphyrins are very readily esterified, either by methanolic hydrochloric acid at room temperature, or by diazomethane. The esters can be re-saponified by allowing them to stand in strong hydrochloric acid at room temperature.

The best general method for the preparation of porphyrins from haemins is the ferrous.acetate-hydrochloric acid method of WARBURG and NEGELEIN *(249)*. Iron is far more readily removed in the ferrous than in the ferric form. Formic acid and iron powder, oxalic acid in acetone, stannous chloride, or sodium amalgam have also been used, in certain instances also hydriodic or hydrobromic acid in acetic acid. PAUL *(188)* recently suggested heating in 5M-pyruvic acid.

Schemes for isolation of various porphyrins have been worked out by SCHUMM, DOBRINER *(61)* and ZEILE and RAU *(262)*, and have been modified by other workers (cf. *6*, there p. 85). It is essential to avoid contamination with heavy metals such as Cu or Zn with which porphyrins readily combine to complex salts; organic solvents, e. g. chloroform, often contain traces of zinc. Ether must be kept peroxide free. Porphyrin solutions must not be exposed to strong light.

Separation.

Solubility. By extraction with ether, ether-soluble and ether-insoluble porphyrins can be separated. Separation of some porphyrins can be achieved by making use of the fact that their sodium salts are not readily soluble in water (e. g. proto-, deutero-, and mesoporphyrin, in contrast to copro- and haematoporphyrin). Chloroform extracts meso-, deutero-, and protoporphyrin, but not coproporphyrin, from the aqueous solution of the hydrochlorides. The methyl esters of the two isomeric copro-porphyrins differ in their solubility in methanol and ether. Esters can often be crystallised from chloroform-methanol or chloroform-ether mixtures.

HCl-Method. The method of WILLSTÄTTER and MIEG (*258*) is of the greatest value (cf. also *262*). It consists in fractional extraction of por-phyrins from their ether solutions by hydrochloric acid of different concentrations. The "HCl-number" is defined as the concentration of hydrochloric acid in g. per 100 ml. which extracts two thirds of the por-phyrin from an equal volume of its ether solution.

The HCl numbers of some porphyrins are as follows: protoporphyrin 2, meso-porphyrin 0.60, deuteroporphyrin 0.36, haematoporphyrin 0.1, coproporphyrin 0.09.

GRANICK and BOGORAD (*89*) used counter current distribution for further separation of the three porphyrins, mesoporphyrin, deutero-porphyrin, and haematoporphyrin, which all have low HCl numbers.

GRANICK found that the dependence of the distribution coefficient on the HCl concentration can be expressed by the formula: $K_a(HCl_a)^2 = K_b(HCl_b)^2$. However, where higher concentrations of HCl are needed for the separation, this formula does not hold, and instead, the distribution coefficient changes in proportion to a much higher power of the HCl concentration than the square (*140*).

Crystallisation. Crystallisation and re-crystallisation, particularly of the methyl ester, until a uniform crystal form and a maximal melting point are reached, is an important way of purification. Nevertheless, neither criterion is a safe guide proving uniformity. Thus, LEMBERG and PARKER (*141*) found that approximately equimolar mixtures of chlorocruoro- and diformyl-deuteroporphyrin esters crystallised as readily and uniformly as the two components alone [cf. also (*119, 181*)].

Chromatography. Adsorption chromatography on alumina, calcium carbonate or magnesium oxide has been used by many workers for the separation of various porphyrin esters (*136, 94, 97, 105, 99, 176, 141, 43*). While this method is often very useful, it fails to separate isomers, and in other instances must be repeated and checked by other methods. Partition chromatography on hydrated silica gel has been used by LUCAS and ORTEN (*151, 134*); proto-, meso-, copro-, and uroporphyrins can be separated by this method, but not isomers.

More important for identification and characterisation are paper-chromatographic methods. The method of NICHOLAS and RIMINGTON (*178, 179*; cf. also *43, 120, 121, 123, 162, 177, 202*) uses lutidine-water systems with or without an atmosphere of ammonia for separation of the free carboxylic acids. The R_f value depends on the number of carboxylic groups and the method thus separates di- from tetra- and octacarboxylic porphyrins. R_f values vary somewhat with conditions; e. g., for octa-, tetra-, di-, and monocarboxylic acids 0.03, 0.62, 0.84 and 0.95 respectively are found. By this method a number of new porphyrins with 3, 5, 6 and 7 carboxyls have also been detected (see below).

While the usefulness of this method has been demonstrated beyond doubt, it may not be out of place to point out its limitations. It fails to separate various porphyrins having the same number of carboxylic acid groups, or isomerides. Porphyrins with a neutral group, very different from those usually found, may not fit into the scheme; thus porphyrin a, which according to the chemical evidence is a dicarboxylic acid, behaves in the paper chromatogram as a monocarboxylic acid (140). Impurities may also simulate the presence of more porphyrins than are actually present; thus uroporphyrin has been observed to give two spots.

An electrophoretic method on agar, which gives a separation similar to that by the NICHOLAS-RIMINGTON method has been described (186).

The method of CHU, GREEN and CHU (39), paper-chromatography of the esters with mixtures of kerosene, chloroform, propanol or dioxane, while more empirical, can separate porphyrins with the same number of carboxyl groups, as well as isomers, particularly in its two-dimensional modifications [KENCH, LANE and VARLEY (122); FALK and BENSON (67)].

Identification of Porphyrins. Melting points and mixed melting points of the methyl esters are an important, though alone not an infallible means of identification.

Absorption Spectra. Absorption spectra are of great importance, though they do not allow differentiation between isomers, nor between closely related porphyrins such as meso- and coproporphyrins.

The absorption spectra of porphyrins in neutral organic solvents, particularly in dioxane, have been studied intensively by A. STERN and coworkers. References to their numerous papers have been given by STERN (230), H. FISCHER (1, there pp. 579f., 587f.) and LEMBERG (6, there p. 75).

Porphyrins having no carbonyl groups attached to the nucleus—and also porphyrins with *two* carbonyl groups attached to vicinal pyrrole rings—have an absorption spectrum with four maxima increasing in strength from the first band in the red to the fourth band in the blue (the "aetiotype spectrum"), occasionally with a weak band between the first two. In addition there is a far stronger absorption maximum in the violet or near ultraviolet named after its discoverer, SORET. The spectra in the more distant ultraviolet require further study. Vinyl groups do not alter this type of spectrum but two vinyls shift the position of the bands about 9 mμ towards the red. Carbonyl groups, e. g. formyl or acetyl, have the same effect ($-CHO > -COCH_3 > -CH=CH_2$) when two occur on vicinal pyrrole rings. One carbonyl (or carboxyl) group alone [cf. rhodoporphyrin (XXIV)] alters the type of the spectrum, making band III stronger than band IV—"rhodotype spectrum". This effect is still more pronounced if two such groups are found on opposite pyrrole rings, cf. oxo-rhodoporphyrin (XXV), which has an acetyl group on the pyrrole ring

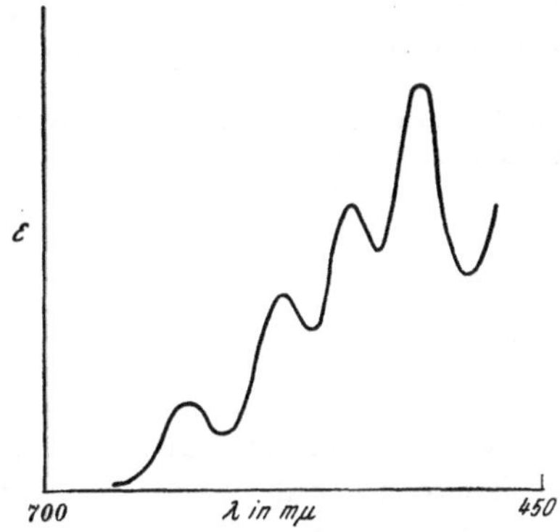

Aetio type.

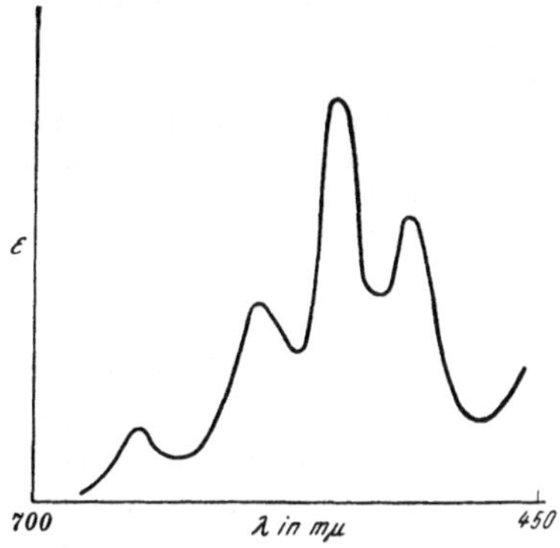

Rhodo type.

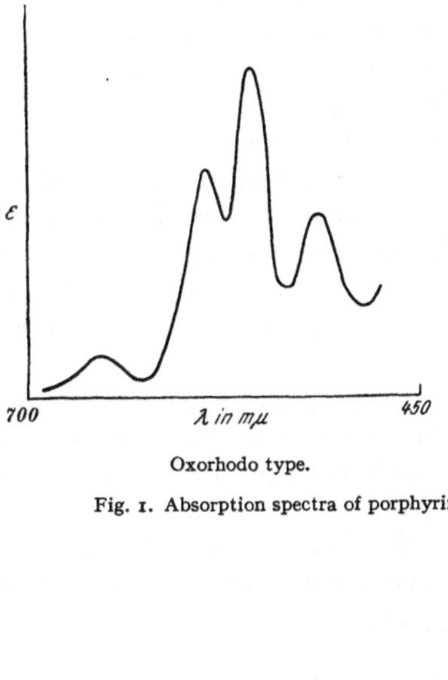

Oxorhodo type.

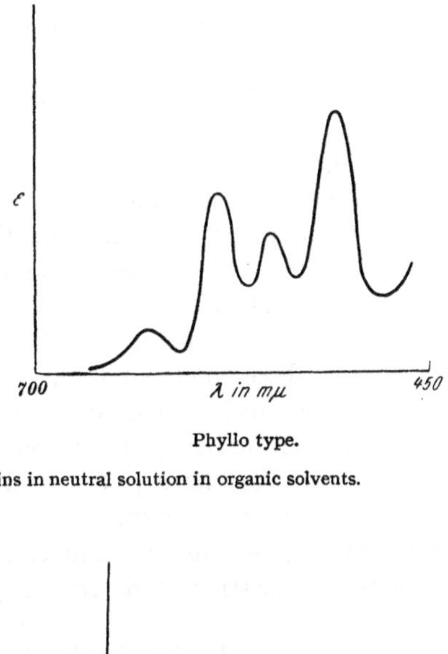

Phyllo type.

Fig. 1. Absorption spectra of porphyrins in neutral solution in organic solvents.

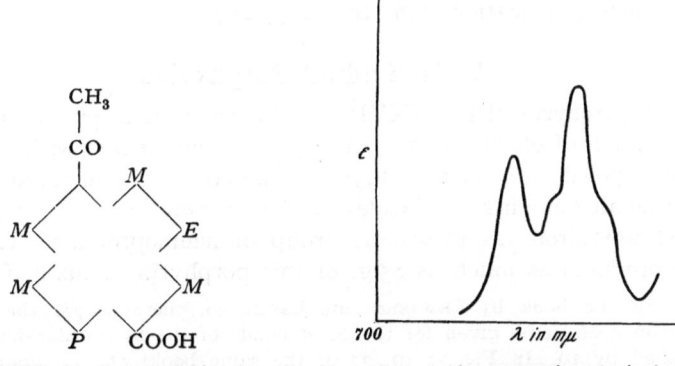

CH₃
|
CO
|

M

M<   >E

M<   >M

P   COOH

(XXV.) Oxo-rhodoporphyrin.

Fig. 2. Acid spectrum of porphyrins in aqueous hydrochloric acid.

opposite to the one bearing carboxyl [FISCHER (76)], when both the second and third bands become stronger than band IV—"oxorhodo-type spectrum" [LEMBERG and FALK (139)].

The spectrophotometric measurement of the ratio of the extinction coefficients at the maxima, in addition to the measurement of the band positions, is of great value for identification and checking the purity of porphyrins of rhodo- and oxo-rhodotype spectrum [LEMBERG (138, 139, 141)]. Substitution on the methine bridges [cf. phylloporphyrin (XX, p. 306)] suppresses band III—"phylloporphyrin type" (Fig. 1, p. 313).

The absorption spectra of porphyrins in aqueous alkaline solutions resemble those of the neutral porphyrins in organic solvents, but the spectra of the di-cations in aqueous hydrochloric acid are rather different (Fig. 2), except for the SORET band. There are only two strong, and two weak bands in the visible, one of the latter being only just demonstrable spectrophotometrically. Unsaturated groups (vinyl as well as carbonyl) shift these bands towards longer wavelengths; their exact position also depends on the concentration of the acid.*

Infrared spectra of porphyrins have been studied by GRAY, NEU-BERGER and SNEATH (96), FALK and WILLIS (72), ALTMAN, MILLER and RICHMOND (14) and CRAVEN, REISMANN and CHINN (49).

Fluorescence spectra have been extensively studied by DHÉRÉ (60) and by A. STERN, and many references are given by LEMBERG and LEGGE (6, there p. 77). Again, the fluorescence spectra of the neutral porphyrins (DHÉRÉ type I) differ from those of the cations (DHÉRÉ type II). The fluorescence is minimal at the isoelectric point; even isomeric porphyrins differ in the shape of their fluorescence-p_H curves (1, there p. 596).

Estimation. Extremely small amounts of porphyrins (0.05–0.5 μg./ml.) can be detected and measured by their strong red fluorescence. This method has been used by numerous workers [(199, 106, 253, 32); cf. also the reviews (6, there p. 88) and (70)].

The photoelectric measurement of the SORET band absorption is, however, almost as sensitive and can avoid losses since it requires less preliminary purification (106, 108, 143, 204).

V. Individual Porphyrins.

Protoporphyrin "IX" (XXVI). Protoporphyrin is prepared from the haemoglobin of blood directly [cf. (99)] or from haemin with formic acid and iron powder (6, there p. 399). Its dimethyl ester melts at 225–230°.

In small amounts it is widespread in nature as free porphyrin and almost ubiquitous as prosthetic group of haemoproteins. The human body produces as much as 85 g. of this porphyrin annually for haemo-

---

* In the book by LEMBERG and LEGGE (6, there p. 76) the millimolar extinction coefficients given for the Soret bands of the hydrochlorides should be multiplied by 10. In Fig. 13 (p. 72 of the same book) the designation of the curves should be reversed.

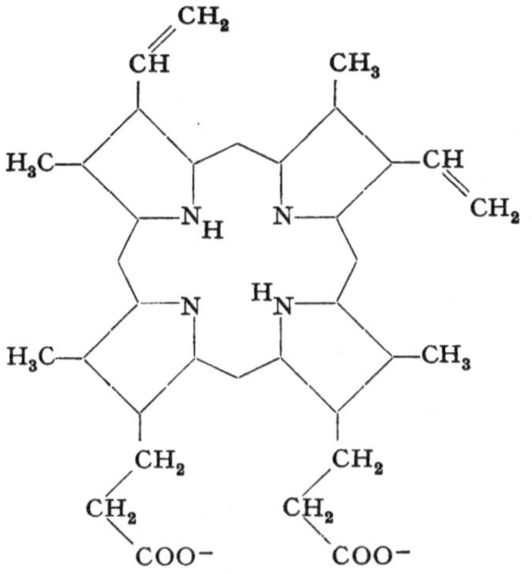

(XXVI.) Protoporphyrin "IX".

globin synthesis alone. SCHUMM recognised the difference of this porphyrin ("haematoporphyroidin") from the haematoporphyrin which HOPPE-SEYLER had obtained by the action of sulphuric acid on haemoglobin. The porphyrin was isolated in pure form by H. FISCHER and renamed protoporphyrin. FISCHER also showed that the porphyrins obtained from putrefying blood (KÄMMERER's), intestine (SNAPPER's), or birds' egg shells (ooporphyrin) were identical with protoporphyrin (cf. *1*, there p. 390; *6*, there p. 60).

The occurrence of the free porphyrin has been summarised by FISCHER (*1*, there p. 398 f.) and LEMBERG and LEGGE (*6*, there pp. 60, 579, 584). It is found in a concentration of about 30 μg. per 100 ml. in normal human erythrocytes (*36*, *251*). It occurs in human faeces, particularly after meat ingestion, and in larger amounts in certain pathological conditions (*159*), and in rats' faeces (*216*). In rodents it occurs in the Harderian glands (*57*). It has been isolated from sheep liver (*197*). As "ooporphyrin" it forms the brown, red and black spots on the egg shells of birds. Together with mesoporphyrin, it has been found in ambergris (*136*). In invertebrates it has been detected in the sea star *Asterias rubens* (*160*, *127*) and in earthworms (*160*, *59*). RAWLINSON and HALE (*195*) have obtained it from *Corynebacterium diphtheriae*, almost certainly from cytochrome *b*.

In protoporphyrin the two hydrophilic carboxyl groups which at physiological p_H are fully dissociated, are found on one edge of the large hydrophobic porphyrin plate (XXVI). This gives protoporphyrin peculiar surface properties (*117*, *11*, *229*) which are probably of biological significance for its adsorption on enzyme surfaces; this may explain the dominance in nature of type "III" porphyrin. In contrast to porphyrins with a different distribution of hydrophilic groups (copro-porphyrins, haematoporphyrin), protoporphyrin forms monolayers of somewhat unstable condensed solid films, with the molecules vertically orientated and closely packed. It is this property which may explain why coproporphyrins and uroporphyrins, which *in vitro* are as easily combined with iron as protoporphyrin, are not found as iron complexes in nature.

Mesoporphyrin "IX" (XXVII). Mesoporphyrin is obtained from protoporphyrin or haemin by hydrogenation of the vinyl side-chains to ethyl side-chains by hydriodic acid, hydrazine, or formic acid plus colloidal palladium (*6*, there p. 59; *86, 169*). The dimethyl ester melts at 216–218°.

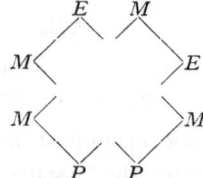

(XXVII.) Mesoporphyrin "IX".

Mesoporphyrin has been found in human faeces (*32, 110, 262*), in pathological urines (*30*), and in ambergris (*136*), probably as product of biological reduction of protoporphyrin.

Deuteroporphyrin "IX" (XXVIII). The synthesis of deuteroporphyrin has been described above; deuteroporphyrin can also be obtained by

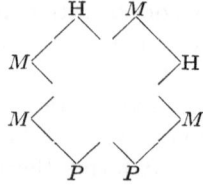

(XXVIII.) Deuteroporphyrin "IX".

removal of iron from deuterohaemin which is formed by heating haemin in molten resorcinol (Schumm, cf. *1*, there p. 414). Its dimethyl ester melts at 224°. Deuterohaemin can be converted into diacetyl-deutero-haemin with acetic anhydride and $SnCl_4$, diacetyl-deuteroporphyrin reduced to haematoporphyrin, and the latter dehydrated to proto-porphyrin. In this way the complete synthesis of protoporphyrin and haemin has been achieved by H. Fischer (*1*, there pp. 371f.). The formyl group can also be introduced into deuterohaemin, and other porphyrins, such as chlorocruoroporphyrin, can be synthesised from it.

Deuteroporphyrin was discovered by Schumm in human faeces after ingestion of blood or after haemorrhage into the gastrointestinal tract. Schumm (217) recognised its difference from coproporphyrin and called it copratoporphyrin; it was renamed deuteroporphyrin by Fischer. It is also formed by prolonged alkaline putrefaction of blood and meat. Thus, bacteria are able either to hydrogenate the vinyl groups of porphyrin, forming mesoporphyrin, or to remove them, forming deuteroporphyrin. Increased amounts have been found in faeces in acute porphyria (162).

Haematoporphyrin "IX" (XXIX). Haematoporphyrin is formed from haemoglobin or protoporphyrin by the action of strong acids,

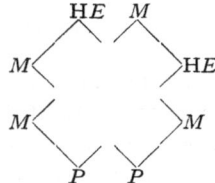

(XXIX.) Haematoporphyrin "IX".

such as sulphuric or hydrobromic, with addition of two molecules of water to the two vinyl groups of protoporphyrin, and can be reconverted into protoporphyrin by heating under reduced pressure. Its dimethyl-ester melts at 212°. Haematoporphyrin does not occur in normal or pathological urine and the old term "haematoporphyrinuria" still found in medical textbooks is a misnomer. Haematoporphyrin has recently been isolated by Granick et al. (24, 91) from a mutant of the alga Chlorella, accompanied by another porphyrin, probably vinylhydroxy-ethyl deuteroporphyrin, i. e. an intermediate between haematoporphyrin with two hydroxyethyl, and protoporphyrin, with two vinyl side-chains. Brugsch (28) has observed formation of such a porphyrin from haemato-porphyrin in vitro. The optical inactivity of the Chlorella haemato-porphyrin makes it appear possible that it is an artifact derived from protoporphyrin. Haematoporphyrin is also of interest on account of its close relationship to cytochrome c, from which it is obtained by the action of hydrobromic acid in acetic acid [Theorell (233)] or by the action of silver salts on the sulphur bridge, followed by removal of iron [Paul (189); cf. also Davenport (53)]. In contrast to haematohaemin obtained from haemoglobin, the haematohaemin c obtained from cyto-chrome c by the action of silver salt is optically active, and differs from haematohaemin in other properties; the corresponding tetramethyl-haematoporphyrins (dimethyl ester dimethyl ethers of haematoporphyrin) also differ.

Chlorocruoroporphyrin (XXX) occurs in nature only as prosthetic group of the green respiratory pigment chlorocruorin [Fox (80)]. In the German literature it is known as "Spirographis porphyrin", after the polychaete worm *Spirographis* in which it occurs. It has been studied by WARBURG and NEGELEIN (*249, 250*) and H. FISCHER and coworkers (cf. *6*, there p. 63). Resorcinol melt of the haemin removes the formyl

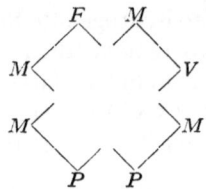

(XXX.) Chlorocruoroporphyrin.

and vinyl groups yielding deuterohaemin. Its structure has been proved by FISCHER and von SEEMANN (*77*), and it has been synthesised from deutero-porphyrin by a complicated introduction of vinyl (via formyl) and formyl groups [FISCHER and WECKER (*78*)]. More conveniently it is accessible by partial oxidation of protoporphyrin ester, best with permanganate in acetone [LEMBERG and PARKER (*141*)].

Porphyrin a (Cytoporphyrin). This porphyrin is the prosthetic group of the cytochromes of type *a* (except a_2); it does not occur as free porphyrin in nature. The problem of its structure is the last great problem of porphyrin chemistry which is still unsolved.

RAWLINSON and HALE (*195*) obtained porphyrin *a* free from proto-porphyrin, but still impure, by a method separating haemin *a* and proto-haemin in heart muscle extracts; they also found evidence for its presence as prosthetic group of cytochrome *a* in *Corynebacterium diphtheriae*. KIESE and DANNENBERG (*52, 128*) obtained it from cytochrome oxidase preparations containing only cytochrome *a* and a_3. LEMBERG (*138*) has recently shown, however, that none of these preparations [cf. also (*71*)] was unaltered or pure porphyrin *a*.

By a modified method (*138*), porphyrin *a* can be obtained in almost quantitative yield (about 15 mg. per kg. of wet defatted ox heart muscle), in amounts which prove that it is the prosthetic group of both cytochromes *a* and a_3. Cryptoporphyrins (see below) were isolated, but their amounts in this preparation are far too small to account for either of the cytochromes. In contrast to impure preparations, the pure porphyrin is stable and has a typical oxorhodotype spectrum [LEMBERG and FALK (*139*); *Fig. 1*, p. 313], with a ratio 2.0—2.1 of the extinction coefficients of the maxima of absorption of bands III and IV in neutral organic solvents. Such a high ratio has so far only been found in porphyrins which have a carbonyl group (formyl, acetyl, ring ketone or carboxyl) on one pyrrole ring, and a second such group, or an unsaturated double bond, attached to the opposite pyrrole ring. One of these has been proved to be a formyl group, the second probably a double bond. The specific extinction could not be raised above a maximum, from which a molecular weight of 930 is calculated, using the molar extinction established by OLIVER and RAWLINSON (*184*) by copper titration, as well as another method (*138*). Similar though somewhat lower molecular weights have been found by GRANICK (*88*) and by WARBURG and GEWITZ (*247*) in iron estimations on the haemin, which WARBURG and GEWITZ obtained crystalline and called cytohaemin. Re-introduction of iron into porphyrin *a* gave this haemin, first obtained by NEGELEIN (*173*), which, as a haemochrome, has a single absorption band in the visible part of the spectrum, at 587 mμ., and its SORET band at 430 mμ.

The low specific extinction and the high molecular weight, combined with the pronounced lipophilic character of haemin and porphyrin *a*, show that the molecule contains a large alkyl group directly attached to the porphin ring which does not contribute to its absorption of visible light. WARBURG and GEWITZ (*248*) have

(XXXI.)   (XXXII.)

recently succeeded in removing this alkyl group by the resorcinol melt, thus obtaining a cytodeuteroporphyrin, different from deuteroporphyrin, whose ester melting point is 189°. They ascribe to it formula (XXXI), assuming that the large alkyl group is connected with the porphin ring by a double bond which causes its removal in the resorcinol melt. The analyses better fit a formula with one or two carbon atoms less than (XXXI), and the corresponding formula (XXXII) for porphyrin *a* (*R* = long alkyl group) would not appear to explain its absorption spectrum.

The present evidence supports the assumption of the following side-chains: at least two methyl groups and two propionic acid side-chains, a formyl group, an unsaturated side-chain, and a large alkyl group, which may contain the unsaturated side-chain vicinal to the ring. The only porphyrins so far found of the same spectral type have the unsaturated group substituting a pyrrole ring opposite to that bearing the carbonyl group [cf. (XLIII), p. 328]. Porphyrin *a* does not contain the isocyclic ring of chlorophyll, nor is it a phytol ester.

The lipophilic character is of significance for the close association of cytochrome oxidase with lipids in the mitochondria. Cytochrome oxidase itself has the character of a lipoprotein. The lipoid character of the prosthetic group is produced by means quite different from those by which chlorophyll acquires wax character.

MORRISON and STOTZ (*168*) have recently claimed in a preliminary note the existence of two haemins *a* differing in their absorption in the infrared, though not in the visible and near ultraviolet. Infrared measurements of porphyrin *a* fractions by Dr. J. B. WILLIS (unpublished) have so far failed to reveal any such differences in our preparations of porphyrin *a*. It is, however, not impossible that porphyrin *a* is a family of substances differing only in the nature of their long alkyl side-chains. We have observed several, spectroscopically identical, fractions which consistently differed in their partition between ether and hydrochloric acid or ether and phosphate buffer.

Cryptoporphyrins. A porphyrin of this type was first obtained by NEGELEIN (*171*) from pigeon breast muscle, but later considered an artifact derived from protohaem (*172*). In contrast to haem *a*, cryptohaem gives a haemochrome with two absorption bands (582 and 533 mμ.) in the visible part of the spectrum. Porphyrins of this type can be, indeed, prepared from haemoglobin [LEMBERG (*138*)] by the action of acetone-HCl and several crystalline methyl esters have been obtained. They differ, however, from the by-product of porphyrin *a* obtained from washed heart muscle. The latter, cryptoporphyrin *a*, obtained in crystalline form as methyl ester, is a formylporphyrin, whereas the cryptoporphyrins *p* from haemoglobin do not appear to have a formyl group. Cryptoporphyrin *a* is probably an alteration product of porphyrin *a* since its yield greatly depends on the method of isolation. The crystalline "verd$_{NO_2}$," dimethyl ester of ALSLEV and KIESE (*12*), prepared in a complicated manner from haemoglobin, is not isochlorocruoroporphyrin, but perhaps a cryptoporphyrin (*138*).

Mono-, Di-, and Tricarboxylic Porphyrins of Unknown Structure.

The "*X*"-porphyrin obtained by GRINSTEIN *et al.* (*107*) from the faeces of lead-intoxicated rabbits is perhaps mesoporphyrin. Its methyl ester melts at 210 to 215° (mesoporphyrin ester, 216–218°).

From faeces of a patient suffering from porphyria cutanea tarda, MACGREGOR *et al.* (*159*) isolated a porphyrin which in the paper chromatogram behaved as a mono-carboxylic acid; its methyl ester melted at 177°. Two new dicarboxylic acids, of ester m. p. 165° and 194–195°, were also present.

Evidence for tricarboxylic porphyrins rests so far entirely on the paper-chromatographic methods of RIMINGTON *et al.* (see above). They have been observed in the urine of lead-intoxicated rabbits and humans (*43*, *179*), and in various forms of porphyria, including the normally porphyrinuric ground squirrel (*Sciurus niger*) (*123*, *134*, *159*, *162*, *179*). MACGREGOR, NICHOLAS and RIMINGTON (*159*) crystallised two different methyl esters of m. p. 214° and 193–194°. Small amounts are found in normal rabbit urine (*43*). BOGORAD and GRANICK (*24*) have also found them in a mutant of the alga *Chlorella*.

Coproporphyrins. Two isomers of coproporphyrin, coproporphyrin "I" (XXXIII) and coproporphyrin "III" (XXXIV) occur. Both have been synthesised by H. FISCHER (*1*, there pp. 475–476). Coproporphyrin was

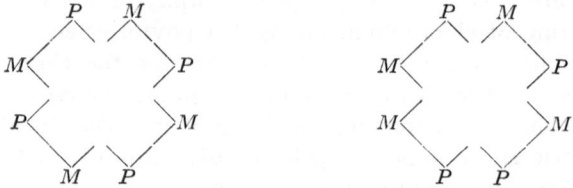

(XXXIII.) Coproporphyrin "I".   (XXXIV.) Coproporphyrin "III".

first found in human faeces, later in urine, where it had previously been wrongly identified with haematoporphyrin [MACMUNN, SAILLET, GÜNTHER (*111*), GARROD (*83*)]. Coproporphyrin "I" is best prepared from faeces of patients with congenital porphyria (see below) [WATSON, SCHWARTZ and HAWKINSON (*255*); LUCAS and ORTEN (*151*)]. Coproporphyrin "III" was first obtained from the urine of a patient with light-sensitive porphyria by VAN DEN BERGH (cf. *1*, there p. 472).

The tetramethyl ester of coproporphyrin "I" melts at 250–258°, that of coproporphyrin "III" has a double melting point, melting first at about 150–160°, resolidifying and melting again at 174–181°. The copper complexes of the esters melt at 284° and 206° respectively (*166, 242*). JOPE and O'BRIEN (*119*) have investigated the melting point curves of mixtures of the "I" and "III" tetramethyl esters*. The melting point does not allow the detection of less than 15% of one isomeride in the other. The "I" isomeride is less soluble in methanol and crystallises from it more readily than the "III" ester.

The two porphyrins can be differentiated by their fluorescence pH curves as free acids (FINK, cf. *1*, there p. 593), or by their DEBYE-SCHERRER X-ray powder diagrams [KENNARD and RIMINGTON (*126*); NICHOLAS and RIMINGTON (*181*)]. While their absorption spectra in the visible and ultraviolet do not differ, differences have been noted in the infrared spectra [FALK and WILLIS (*72*); GRAY, NEUBERGER and SNEATH (*96*), though not by CRAVEN, REISSMANN and CHINN (*49*)]. The two esters can be separated by the paper-chromatographic method of CHU, GREEN and CHU [(*39*); cf. also (*43*)].

Coproporphyrin is almost ubiquitous in man, animals, plants and microorganisms, though usually in very small amounts (*1*, there p. 480). It also occurs occasionally as Zn-complex, and also as Cu-complex (*140*), particularly in acute porphyria. It is excreted in normal human faeces (also in meconium) and in urine, and is found in bile, erythrocytes (*218, 252*) and in traces in the serum (*6*, there p. 580; *55*). The amount present in urine has been underestimated; it has more recently been found that about one half is present in fresh urine as colourless porphyrinogen [WATSON et al. (*219, 253*); ERIKSEN (*66*)]; the average normal coproporphyrin excretion varies between 50 and 250 μg. per day with an average of 175 μg. WEATHERALL's (*256*) findings support the proposition that normal or lead-intoxicated rabbits excrete only colourless precursor in their urine. Earlier claims that coproporphyrin excretion in the urine is increased by ingestion of meat or blood, have not been confirmed by WATSON (*10*); although intestinal bacteria may synthesise coproporphyrin, WATSON found no evidence for its absorption from the intestine. Usually coproporphyrin "I" is found predominant (*10*), while GROTEPASS [(*109*), but cf. (*10*)] found more "III" than "I" in pooled human urine. The coproporphyrin of erythrocytes (reticulocytes) is the "III" isomeride (*252*) and this also predominates in rabbit urine (*43*).

Coproporphyrinuria (increase of urinary coproporphyrin to amounts of a few mg. per day) is found after ingestion of poisons, particularly lead, of aromatic amino

---

* Note that Fig. 10 in LEMBERG and LEGGE (*6*, there p. 65) should have "coproporphyrin tetramethyl ester I" below, and "coproporphyrin tetramethyl ester III" on top of the figure.

compounds e. g. sulphonamides [(*201*), (*27*), but cf. (*137*)], in liver disease, and in a great variety of other diseases. These findings have been reviewed by WATSON and LARSON (*10*), WATSON (*9*), RIMINGTON (*8*) and LEMBERG and LEGGE (*6*, there pp. 590, 628f.). Often coproporphyrin "I" prevails, but in lead intoxication of men and animals, after aromatic amino compounds, in alcoholic liver cirrhosis and in familial erythrocyte elliptocytosis (*29*), coproporphyrin "III" is predominant. Coproporphyrins (both "I" and "III") accompany uroporphyrins in porphyria, particularly that of the chronic congenital type, practically through the whole body (*93*, *97*; *1*, there p. 479, 490).

The central nervous system of warm-blooded animals contains small amounts of coproporphyrin [KLÜVER (*129*); CHU and WATSON (*38*)], particularly in areas poor in cytochrome *c*. BLANSHARD has identified this with coproporphyrin "III" (*23*). The same porphyrin occurs in birds' feathers and hedgehog spines [(DERRIEN and TURCHINI (*58*); see *6*, there p. 50; VÖLKER (*242*, *243*)]. Although the Harderian glands of rats are supposed to contain protoporphyrin, the porphyrin excreted by them in chromodacryorrhoea has been found to be coproporphyrin (*157*).

The coproporphyrin in yeast is predominantly isomer "I" (*1*, there p. 480; *6*, there p. 632), but isomer "III" has been found in starving yeast (*124*). Cell-free yeast extracts can also carry out the coproporphyrin synthesis [MAYER (*164*); RIMINGTON (*200*)]. The results of several investigators (*124*, *200*, *228*, *231*) indicate that coproporphyrin is formed as the product of a deviation from the normal synthesis of the cytochromes, rather than by cytochrome breakdown. Thus not only heat autolysis, but anaerobiosis, cyanide and lack of lactoflavin increase coproporphyrin formation, i. e. factors which cause yeast to adapt itself to anaerobic metabolism.

Bacillus cereus, but not *Escherichia coli*, also produces coproporphyrin in anaerobiosis (*211*), while *Corynebacterium diphtheriae* produces coproporphyrin "III" particularly if iron is lacking in the medium (*48*, *94*, *187*). The same isomeride is synthesised by intestinal bacteria (*127*, *10*) and by mycobacteria (*51*, *236*). Coproporphyrin, probably predominantly type "I", has been isolated from root nodules of leguminous plants [KLÜVER (*130*)].

DOBRINER and RHOADS (*61*) and RIMINGTON (*196*) advanced the theory that coproporphyrins are formed as a rule as by-products of haemoglobin synthesis, and not as breakdown products of haemoglobin. WATSON *et al.* (*210*) were unable to confirm earlier claims of VAN DEN BERGH that protoporphyrin could be converted into coproporphyrin in the liver. Such conversion could in any case yield only coproporphyrin "III", not the more frequent type "I". There is very substantial evidence showing that coproporphyrin "I" is formed as a by-product of proto-haem synthesis, which is not entirely specific. It leads to 0.01–0.1% porphyrins (uro- and copro-) of type "I", whereas the 99.9–99.99% porphyrins of type "III" undergo further transformation into protoporphyrin and haem. Coproporphyrin "III" is also formed by a derangement of the later steps of haem synthesis. This evidence has been summarised by LEMBERG and LEGGE [(6), there Chapters XII and XIII, particularly pp. 582f., 593f., and 628f.].

Porphyrins with Five to Seven Carboxyl Groups.

These porphyrins, intermediates between copro- and uroporphyrins, can be detected by the paper-chromatographic method of NICHOLAS and RIMINGTON (*178*). They have been discovered in porphyrias in urine, faeces and organs by many workers (*123, 134, 159, 162, 179, 181, 202*), though not by KEHL and STICH (*121*). According to NICHOLAS and RIMINGTON (*181*), the ester of m. p. 208° isolated by GRINSTEIN, SCHWARTZ and WATSON (*105*) from "WALDENSTRÖM ester" (see below) is not a hepta-, but a hexacarboxylic ester. Similar porphyrins have also been found in the urine of normal and lead-intoxicated rabbits, in meconium (*43, 179*), in *Corynebacterium diphtheriae* (*178*), and in *Chlorella* mutants (*24*). FISCHER and JORDAN (*75*) had claimed the isolation of a pentacarboxylic porphyrin, concho-porphyrin, from the shell of the pearl mussel, *Pteria*, but this could not be confirmed by NICHOLAS and COMFORT (*177*). Apart from the hexacarboxylic ester of m. p. 208°, mentioned above, the only compounds obtained in pure or approximately pure state, were a heptacarboxylic ester of m. p. 242° (*202*) from congenital porphyria urine, and a pentacarboxylic ester of m. p. 222–27° or 233–35° (*84*), perhaps identical with the ester of similar m. p. obtained by partial decarboxylation *in vitro* of uro-porphyrin ester by GRINSTEIN, SCHWARTZ and WATSON (*105*).

Uroporphyrins. Two isomers of uroporphyrin, uroporphyrin "I" (XXXV) and uroporphyrin "III" (XXXVI) occur. On decarboxylation of the acetic acid side-chains to methyl groups they yield the corresponding coproporphyrins. This reaction is best carried out by heating in aqueous

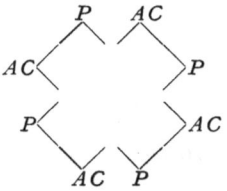

(XXXV.) Uroporphyrin "I".

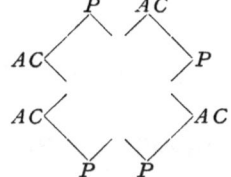

(XXXVI.) Uroporphyrin "III".

hydrochloric acid at 180–190° [FISCHER and ZERWECK (*79*)]. Uro-porphyrin "I" has been isolated as crystalline octamethyl ester of m. p. 293–295° (cf. *181*) by H. FISCHER from the urine of a patient with chronic congenital porphyria (*1*, there p. 505). Uroporphyrin "III" octamethyl ester, of m. p. 264°, has been isolated from acute porphyria urine by WALDENSTRÖM (*244*) and by MERTENS (*165*), but occurs also in chronic porphyria (cf. *185*).

Some doubt has been thrown on the existence of the "WALDENSTRÖM"ester as pure compound by WATSON et al. (*105, 255*; cf. also *191a*), but its existence has recently been confirmed by NICHOLAS and RIMINGTON (*181*). NICHOLAS and RIMING-TON (*180*; cf. also *198*) obtained uroporphyrin "III" from its copper complex turacin, which occurs in the wing feathers of the African *Turaco* bird and which FISCHER had erroneously assumed to be the uroporphyrin "I" copper complex. They found this uroporphyrin identical with the WALDENSTRÖM porphyrin (*181*). Some of the previous confusion in this field is now explained by their observations that the mixed m. p. shows a sharp decline from that of uro ester "I" to a minimum with 75% "I" and 25% "III" esters.

The two uroporphyrins can be differentiated by their fluorescence pH curves (*244*), their infrared spectra (*72*) and DEBYE-SCHERRER diagrams (*126, 181*). A mixture containing 75% ester "III" and 25% ester "I" has a characteristic sharp diffraction pattern indicating mixed crystal formation. The two esters can be separated by paper chromatography [FALK and BENSON (*67*)].

Until recently evidence for the chemical structure of uroporphyrin was based on the fact that decarboxylation yielded coproporphyrin, and that the presence of the extra four carboxyl groups in form of methylmalonic or succinic acid side-chains had been excluded by FISCHER's synthesis of such isouroporphyrins (*1*, there p. 517 f.) and their oxidation products (carboxylated haematinic acid imides). Recently, MACDONALD (*154–156*) has established the correctness of formula (XXXV) for uroporphyrin "I" by its synthesis and careful identification of the natural with the synthetic porphyrin.

The isonitroso compound of ethyl acetone dicarboxylate is condensed in the usual manner of the KNORR synthesis with benzyl acetoacetate, giving the pyrrole (XXXVII) with one acetic ester and one carboxyl ester group in β-position. The carbobenzoxy group is then removed by hydrogenation and decarboxylation, and replaced by the propionic acid side-chain introduced by formylation, condensation

(XXXVII.) (XXXVIII.)

of the formyl group with sodium malonate, and reduction of the acrylic acid side-chain (XXXVIII). Bromination and condensation of the resulting α-bromo-α'-methyl pyrromethene to porphyrin gave uroporphyrin "I" (Formula XXXV).

Porphyrias. Uroporphyrin "I", the predominant uroporphyrin of congenital porphyria, is found in this disease not only in the urine, but also deposited in the bones and present in many organs; it may also accompany uroporphyrin "III" in acute porphyria. Apart from the type of predominant uroporphyrin excreted, chronic congenital and the so-called acute porphyria (actually also hereditary but often symptomless, or with acute attacks between remissions) differ widely in their symptomas; skin photosensitisation characterises chronic porphyria, and gastrointestinal, neuromuscular and psychic symptoms characterise acute porphyria. A third type (porphyria cutanea tarda) has been differentiated [cf. RIMINGTON (*8*)]; this is characterised by slowly developing skin symptoms, excretion of much copro- and protoporphyrin in the faeces during remission and diminution of faecal excretion and increase of urinary excretion in attacks, probably caused by liver damage. It appears that the physiology of the three diseases is also different. Thus porphobilinogen, a colourless monopyrrolic precursor (see below) is only found in acute porphyria. Chronic porphyria appears to affect the erythropoietic system in the bone marrow, acute porphyria perhaps the catalase synthesis in the liver (*214*). Chronic porphyria is also found as congenital disease in cattle, pigs and other animals (cf. *6*, there p. 591).

The American ground squirrel, *Sciurus niger*, has a physiological porphyria, with excretion of uroporphyrin "I" in the urine and deposition of it in the bones [TURNER (*238*)]. FIKENTSCHER (*73*) found uroporphyrin in bones of the normal human embryo; it is still present in postnatal life in the ossicles of the inner ear. It has become increasingly clear that uroporphyrin occurs more widespread in nature than had previously been believed.

Confirming earlier findings of SCHUMM and FISCHER which other workers could not confirm, NICHOLAS and RIMINGTON (*178*) found it in normal human urine; LOCKWOOD (*143*) determined the normal excretion as about 25 µg. per day. Uroporphyrin is present in erythrocytes, both in nucleated erythrocytes of birds (*21, 69*) and in mammalian erythrocytes (*140*). It has been found in ox brain (*23*) and in the bones of certain birds (*58*). In invertebrates it occurs in a number of molluscan shells (*40–42, 235*).

In the plant kingdom, it has been found in the epidermis of *Vicia* shoots (*85*) and in the root nodules of the red kidney bean (*130*). Small amounts of uroporphyrin "I" accompany coproporphyrin "III" in *Corynebacterium diphtheriae* (*94*) and in *Rhodopseudomonas spheroides* (*135*).

As will be seen below, there is good evidence that uroporphyrin, or a monopyrrole precursor of it, is the primary product in porphyrin and haem synthesis.

The existence of at least four different octacarboxylic porphyrins in porphyria urine has been claimed by LARSEN, GLAFKIDES and ORTEN (*134*). RIMINGTON and MILES (*202*) isolated an octamethyl ester of m. p. 244° in congenital porphyria. From turacin, NICHOLAS and RIMINGTON (*100*) isolated an octacarboxylic ester of m. p. 210°, accompanying uroporphyrin "III". Its free acid failed to give coproporphyrin on decarboxylation.

Colourless Precursors of Uroporphyrins.

The presence of a colourless precursor of uroporphyrin has been found as early as 1896 by SAILLET. WALDENSTRÖM [cf. WALDENSTRÖM and VAHLQUIST (*245*)] observed the presence of a colourless chromogen in acute porphyria urine which he called porphobilinogen. It gave a red pigment on condensation with *p*-dimethylaminobenzaldehyde which, in contrast to the similar pigment obtained from urobilinogens, could not be extracted from its aqueous solution by chloroform. This reaction has been introduced as a valuable test for acute porphyria by WATSON and Schwartz (*254*), and for quantitative estimation of porphobilinogen by Vahlquist (*240*).

On heating with hydrochloric acid, porphobilinogen yields uroporphyrin, as well as a bile pigment of urobilinoid character, porphobilin. This reaction has been studied further by Grieg, Askevold and Sveinsson (*98*), Herbert (*116*), and Brockman and Gray (*25, 26*), by the latter with crystalline porphobilinogen. Optimal conversion to porphyrin is obtained on heating in acetate buffer of p_H 5, or on standing or heating in about 0.1 N-HCl. The optimal yield of porphyrin was about 35%, the other products being porphobilins and a violet pigment. Pure porphobilinogen yields uroporphyrin, in contrast to findings of Hawkinson and Watson (*114*); the inability of the latter workers to obtain porphyrin was probably due to the preceding conversion of porphobilinogen by alkali to another chromogen, possibly the dipyrrolic or tetrapyrrolic chromogen of porphobilin.

Porphobilinogen was first obtained in crystalline form by Westall (*257*) and has been intensively studied by Cookson and Rimington (*44*), Kennard (*125*) and Granick and Bogorad (*90*). Waldenström and Vahlquist had concluded from diffusion experiments that porphobilinogen

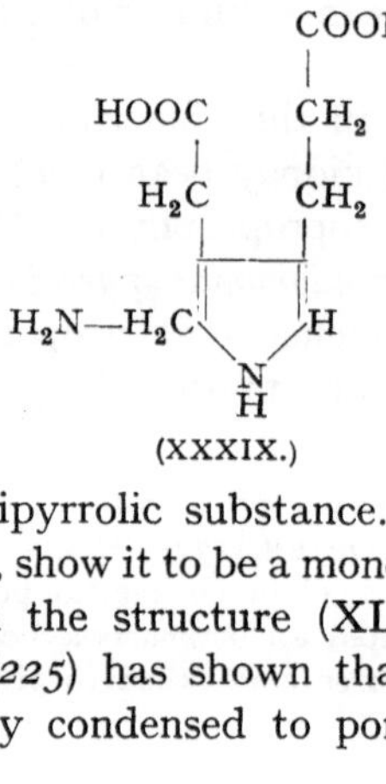

(XXXIX.) 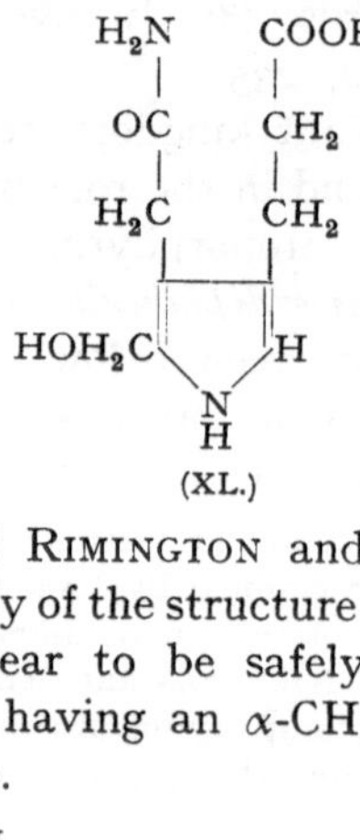　　　　　　(XL.)

was a dipyrrolic substance. The studies of Rimington and Granick, however, show it to be a monopyrrole, probably of the structure (**XXXIX**), although the structure (**XL**) does not appear to be safely excluded. Siedel (*225*) has shown that monopyrroles having an α-CH$_2$OH group are easily condensed to porphyrins *in vitro*.

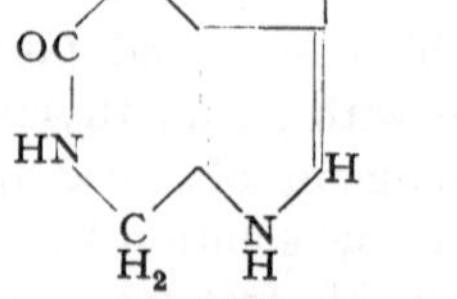

(XLI.)

The free base decomposes at 175–180°. On boiling porphobilinogen with phosphate buffer p_H 8.5, but not at 40°, one molecule of ammonia is developed. Acylation in aqueous pyridine yielded a lactam (XLI), whose methyl ester melts at 248–250°. The X-ray crystallographic studies on the hydrochloride and two forms of hydrobromide indicate that the molecular size is that of a monopyrrole. This has been supported by the isolation of a monoiodo compound with 34% iodine (90).

LEMBERG and LEGGE (6, there p. 67 and p. 587) pointed out that two different dipyrroles were required for the formation of uroporphyrin "III", and that condensation of these two might be expected to yield more than one type uroporphyrin. RIMINGTON and coworkers found no evidence for porphobilinogen being a mixture of isomers. It is still difficult to understand why uroporphyrin "III" and not uroporphyrin "I" is formed on autocondensation of a monopyrrole of the suggested structure, although unexpected reactions have occasionally been found in the porphyrin field. Further study of this problem is required. Uroporphyrin "III" appears to be the main product of the reaction (173, 175), but at least one other isomer was found (26, 44, 84).

Porphobilinogen is a type of compound which was postulated by LEMBERG and LEGGE (6, there p. 642) as primary product of porphyrin and haem synthesis. Further research is required to show whether porphobilinogen itself fulfils this role, or whether it is a product of secondary alteration of this primary precursor; recent results of FALK, DRESEL and RIMINGTON (69) strongly support that porphobilinogen is the direct precursor. Such research will be greatly facilitated by the discovery of SCHMID and SCHWARTZ (214) that an experimental porphyria

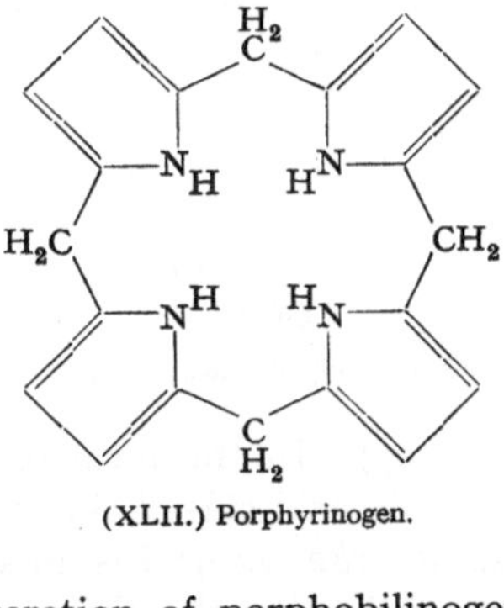

(XLII.) Porphyrinogen.

of acute type, with excretion of porphobilinogen in the urine, can be produced in rabbits by the drug sedormid (cf. 26).

Although it is now certain that porphobilinogen can be transformed into uroporphyrin, the existence of other colourless chromogens as independent precursors, or as intermediates between porphobilinogen and uroporphyrin, has been demonstrated. Porphyrins can be reversibly reduced to porphyrinogens (XLII) which are rather unstable compounds, probably owing to the great deal of distortion of bond angles at the CH_2-groups which this structure requires. HAWKINSON and WATSON (*114*) and HERBERT (*116*) found evidence for the existence of colourless precursors which did not react with *p*-dimethylaminobenzaldehyde. Such precursors would probably be porphyrinogens, as is undoubtedly the chromogen of coproporphyrin (cf. above). Uroporphyrinogen prepared from uroporphyrin by reduction with sodium amalgam gives no colour with *p*-dimethylaminobenzaldehyde. LOCKWOOD (*144*) has recently found evidence that porphobilinogen is converted into another chromogen, probably uroporphyrinogen, on long standing in weakly acid urine under exclusion of air; this second chromogen is oxidised rapidly to uroporphyrin on admission of air, and the yield of uroporphyrin from acute porphyria urine obtained in this manner is considerably greater than by the usual procedures described above. Similarly, FALK, DRESEL and RIMINGTON (*69*) have found anaerobic conversion of porphobilinogen into uro- and copro-porphyrin (probably as chromogens) by chicken blood haemolysates.

Porphyrins derived from Chlorophyll.

Whereas the porphyrins corresponding to chlorophyll *b* or bacterio-chlorophyll have so far not been observed in nature free or as complex salts (cf. Table 2, p. 309), the magnesium complex of the porphyrin corresponding to chlorophyll *a*, vinylphaeoporphyrin a_5 [(XLIII); cf. phaeoporphyrin a_5 (XXI), p. 306] has been found by GRANICK in a mutant

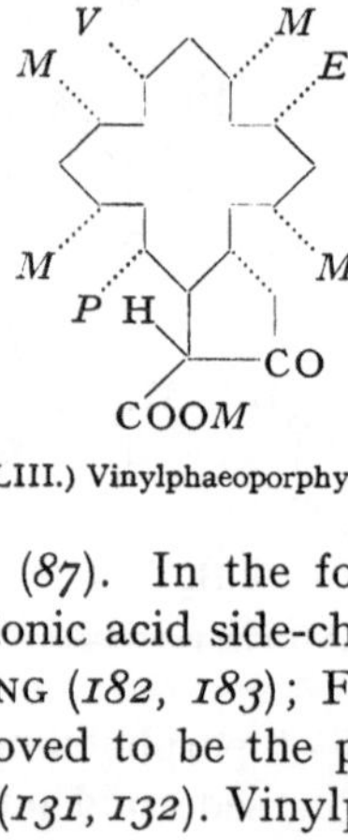

(XLIII.) Vinylphaeoporphyrin a_5.

of the green alga *Chlorella* (*87*). In the form of its phytol ester, with phytol esterifying the propionic acid side-chain, it occurs as protochlorophyll [NOACK and KIESSLING (*182, 183*); FISCHER and STERN (*2*), there p. 321], which has been proved to be the precursor of chlorophyll *a* by KOSKI, FRENCH and SMITH (*131, 132*). Vinylphaeoporphyrin a_5 is obtained from phaeophorbid *a* by boiling in formic acid with iron powder (*2*, there p. 230).

It is generally assumed, and seems to be well supported (cf. e. g. *213*) that chlorins are dihydroporphin derivatives so that the conversion of a chlorin to a porphyrin should be a dehydrogenation. The conversion of phaeophorbid *a* to phaeoporphyrin a_5 is an isomerisation since the hydrogen lost from the nucleus is attached to a vinyl side-chain converting it to ethyl (*2*, there p. 166). Hydriodic acid can occasionally act as an oxidant, owing to the formation of iodine (cf. *76*). It is, however, difficult to explain the formic acid-iron-reaction of FISCHER and NOACK and KIESSLING, which would have to be a straightforward dehydrogenation.

Among the products of biological decomposition of chlorophyll [FISCHER and STERN (*2*), there p. 389ff.] there are several porphyrins. One of these, phylloerythrin (XLIV), still contains the isocyclic nucleus and is a decarboxylated phaeoporphyrin a_5.

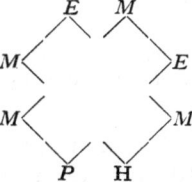

(XLIV.) Phylloerythrin.

This compound was observed as early as 1880 by MACMUNN (*160*) as "chole-haematin", and again by LOEBISCH and FISCHLER (*145*) as "bilipurpurin" in the bile of ruminants, while MARCHLEWSKI (*163*) found it in the excrements of sheep and cattle. He established its derivation from chlorophyll and its porphyrin nature, which was confirmed by FISCHER and HILMER (*74*). Its structure was established by FISCHER (*2*, there p. 189). Chlorophyll is converted to leuco-phylloerythrin by an intestinal bacterial synergism (*19*). The re-absorption of phylloerythrin from the gastrointestinal tract of sheep can cause photosensitising diseases (*203*). Weak-basic chlorophyll porphyrins also occur in human faeces [BRUGSCH (*31*)].

Pyrroporphyrin (XLV), a product of more extensive decomposition which has lost the isocyclic nucleus, has been found in ox bile by ROTHEMUND (*207*).

(XLV.) Pyrroporphyrin.

VI. The Biosynthesis of Porphyrins.

There are very few organisms [LWOFF (*152, 153*); GRANICK (*5*)] which need protoporphyrin as a growth substance. To all other organisms this synthesis, so difficult for the organic chemist, appears to be an easy task; thus, while the organism husbands iron and protein of catabolised haemoglobin, it discards, apparently quantitatively, the prosthetic group in the form of bile pigments [WHIPPLE, cf. (*6*), there p. 509 and 609].

It therefore appeared likely that the precursors must be sought among the most common metabolites. Yeast needs, besides mineral salts, only glucose and ammonia for porphyrin synthesis (*124, 200*), and studies in the rat by THOMAS (*234*) also indicated carbohydrates as a source of porphyrins. Previous theories of porphyrin biosynthesis were entirely speculative, they failed to give account of the typical side-chains of the naturally occurring porphyrins, or dealt only with the later stages of the synthesis [cf. LEMBERG and LEGGE (*6*), there p. 632f.]. The most likely of the assumed precursors, tryptophane, has been shown by isotope experiments not to play this role (*212*).

TURNER (*239*) was the first to assume that the precursor had the acetic acid and propionic acid side-chains which characterise uroporphyrin, but uroporphyrin was then mostly known as a pathological product, and TURNER did not provide evidence for his hypothesis. The discoveries of SHEMIN and RITTENBERG (*223, 220*) and of BLOCH and RITTEN-BERG (cf. *191*) that N^{15} of glycine and deuterium of deuteroacetate (D in the methyl group) were incorporated in the haem of haemoglobin, were the decisive turning point. On this basis, and on the basis of other findings which indicated that the more highly carboxylated porphyrins were not derived from protoporphyrin by carboxylation, LEMBERG and LEGGE (*6*, there p. 637) postulated that a monopyrrolic precursor with acetic and propionic acid side-chains arose by condensation of glycine with two molecules of α-ketoglutarate or a related derivative of the citric acid cycle. The isotope studies of the Columbia University workers, by NEUBERGER, GRAY and MUIR, by ALTMAN *et al.*, GRINSTEIN *et al.* and others have proved this hypothesis to be correct.

The synthesis proceeds not only in the intact animal, but also in surviving nucleated erythrocytes of birds (*222*), in their haemolysates (*68, 146*), in immature erythrocytes, not necessarily reticulocytes, of mammals (*13, 147*), in homogenates of bone marrow and spleen (*15, 16*), in yeast (*261*), and in bacteria (*112*). The synthesis of free erythrocyte porphyrins from glycine in chicken erythrocytes and immature erythrocytes of rabbits has been studied by BÉNARD, GAJDOS and GAJDOS-TÖRÖK (*20—22, 81, 82*).

Not only the nitrogen of the amino group, labelled with N^{15}, but also the α-CH_2 group, labelled with C^{14} is incorporated in the proto-porphyrin of haemoglobin (*13, 102*), but its carboxyl group is not (*103, 193*). The considerable incorporation of deuterium, from deuteroacetate labelled in its methyl group (*191*) indicated that the precursor was an early product of the citric acid cycle, which begins with the condensation of acetate with oxaloacetate to citrate, and then passes via α-keto-glutarate and succinate back to oxaloacetate. The objection was raised against this hypothesis that labelled CO_2 was not incorporated into haem (*33*), but LEMBERG and LEGGE (*7*) pointed out that, according to their theory, the CO_2 which enters by fixation, would be eliminated at a later stage. ALTMAN, MILLER and RICHMOND (*14*) observed that the ε-$C^{14}H_2$ group of lysine, fed to the anaemic and hypoproteinaemic dog, is incorporated in haem, and that the intermediate was not glutamic, but probably α-ketoglutaric acid. SHEMIN and WITTENBERG (*224*) found that succinyl-coenzyme A, rather than α-ketoglutarate was the precursor, but later SHEMIN and KUMIN (*221*) found both substances active, perhaps through another more direct precursor. Using $N^{15}H_2 \cdot C^{14}H_2 \cdot COOH$, MUIR and NEUBERGER (*170*) and RADIN, RITTENBERG and SHEMIN (*193*) demonstrated that twice as many carbon atoms of protoporphyrin are derived from glycine than N atoms. The synthesis of the primary pyrrole (XLVI) (*6*, there p. 642) can therefore be assumed to be brought about by condensation of two molecules of α-ketoglutarate or succinyl-coenzyme A with two molecules of glycine (XLVII), where *R* indicates the carboxyl group, coenzyme A radical or unknown group combined

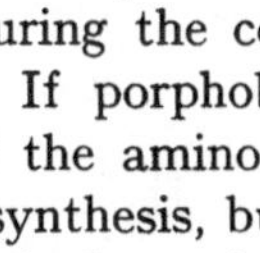

$$\text{(XLVI.)} \qquad \text{(XLVII.)}$$

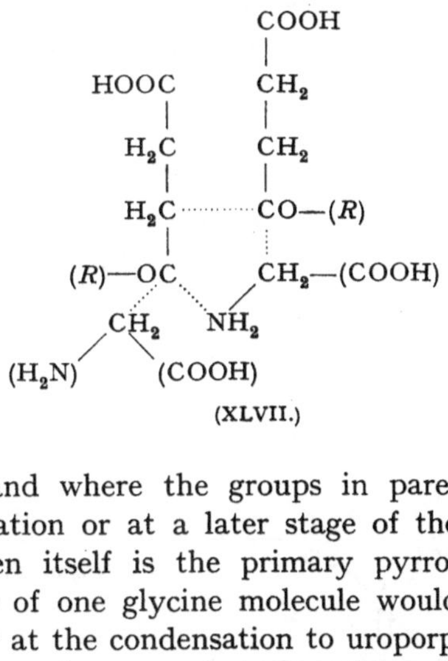

with the succinyl residue, and where the groups in parentheses are removed during the condensation or at a later stage of the porphyrin synthesis*. If porphobilinogen itself is the primary pyrrole (*69*), the removal of the amino group of one glycine molecule would not occur during its synthesis, but only at the condensation to uroporphyrin. The synthesis of the primary pyrrole somewhat resembles that of iso-pelletierine (*18*).

Isotope studies of the Columbia workers (*193, 194, 221, 224, 259, 260*) and MUIR and NEUBERGER (*169, 170, 174*) have further elaborated and confirmed the theory. N^{15} of glycine contributes equally to both basic and acidic pyrrole rings. The same holds for C^{14} of acetate, apart from the carboxyl groups of haem. The methine bridge carbons are derived from glycine, but from neither the methyl nor carboxyl of acetate.

---

* Addition on proof reading. Meanwhile, SHEMIN and RUSSEL (*223a*) have shown that δ-aminolaevulinic acid is an intermediate in this condensation; two molecules ot it condense to form porphobilinogen.

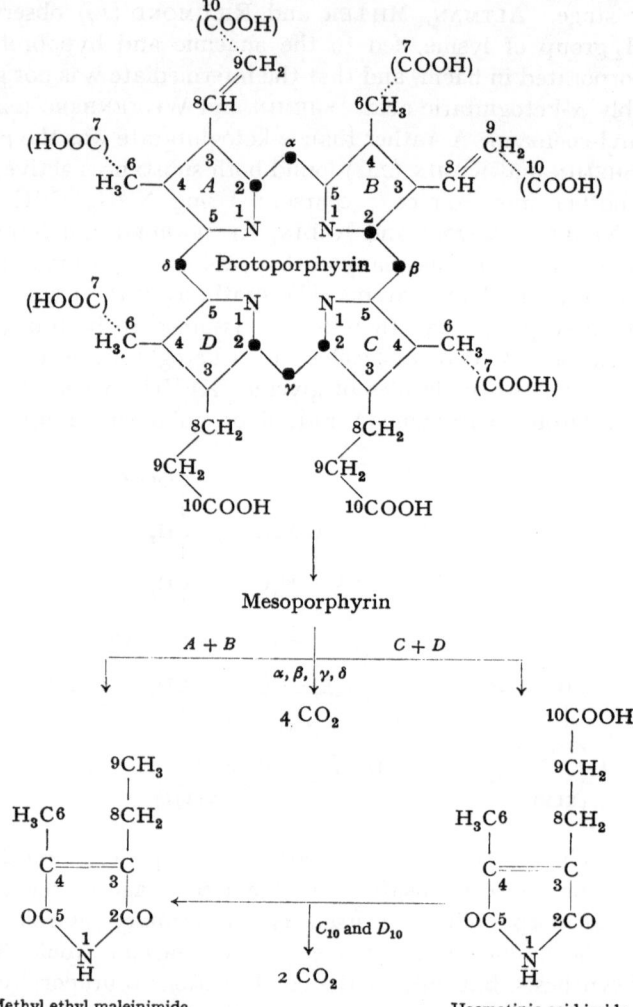

(XLVIII.) Stepwise decomposition of labelled protoporphyrin.

SHEMIN and WITTENBERG (*224, 260*) have carried out a complete stepwise decomposition of the molecule labelled with C^{14} from methyl- or carboxyl-labelled acetate depicted in (XLVIII) and (XLIX). If A_1, B_1, B_2, etc. mean specific isotopic activity of the atoms, thus labelled in (XLVIII), the following equations are found to hold: $A_{(1)} + B_{(1)} = C_{(1)} + D_{(1)}$ and $(A_{(2)} + A_{(3)} \ldots A_{(9)} + B_{(2)} + B_{(3)} \ldots B_{(9)}) = (C_{(2)} + C_{(3)} \ldots C_{(9)} + D_{(2)} + D_{(3)} \ldots D_{(9)})$ which indicates that both basic and acidic pyrrole rings have the same precursor, and that decarboxylation occurs afterwards. Using C^{14} acetate, no C^{14} activity is found in the carbon atoms α,

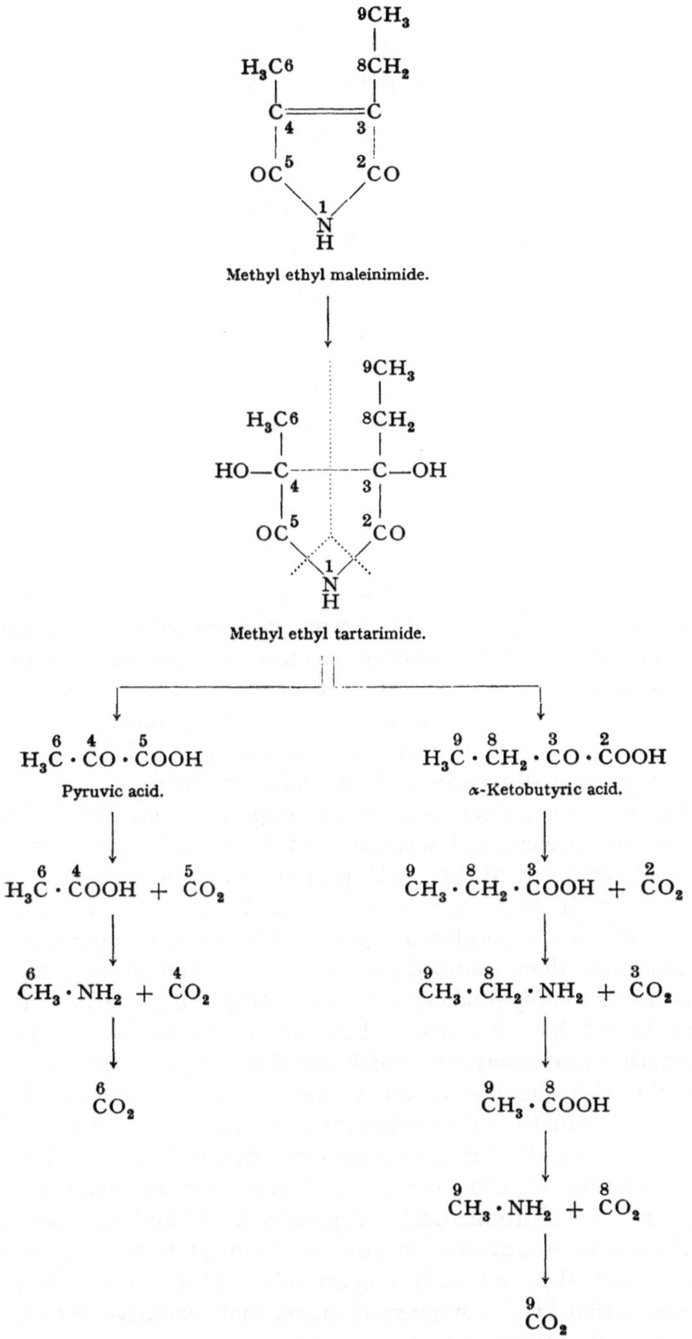

$$9CH_3$$
$$|$$
$$H_3C6 \quad 8CH_2$$
$$C \!=\!=\! C$$
$$|\,4 \qquad 3\,|$$
$$OC^5 \qquad ^2CO$$
$$1$$
$$N$$
$$H$$

Methyl ethyl maleinimide.

$$9CH_3$$
$$|$$
$$H_3C6 \qquad 8CH_2$$
$$|\qquad\qquad |$$
$$HO\!-\!C\!----\!C\!-\!OH$$
$$|\,4 \qquad\quad 3\,|$$
$$OC^5 \qquad\quad ^2CO$$
$$1$$
$$N$$
$$H$$

Methyl ethyl tartarimide.

$$\overset{6}{H_3}\overset{4}{C}\cdot \overset{}{CO}\cdot \overset{5}{COOH} \qquad\qquad \overset{9}{H_3}\overset{8}{C}\cdot \overset{}{CH_2}\cdot \overset{3}{CO}\cdot \overset{2}{COOH}$$

Pyruvic acid. $\qquad\qquad\qquad$ α-Ketobutyric acid.

$$\overset{6}{H_3}\overset{4}{C}\cdot COOH + \overset{5}{C}O_2 \qquad\qquad \overset{9}{C}H_3\cdot \overset{8}{C}H_2\cdot \overset{3}{C}OOH + \overset{2}{C}O_2$$

$$\overset{6}{C}H_3\cdot NH_2 + \overset{4}{C}O_2 \qquad\qquad \overset{9}{C}H_3\cdot \overset{8}{C}H_2\cdot NH_2 + \overset{3}{C}O_2$$

$$\overset{6}{C}O_2 \qquad\qquad\qquad\qquad \overset{9}{C}H_3\cdot \overset{8}{C}OOH$$

$$\overset{9}{C}H_3\cdot NH_2 + \overset{8}{C}O_2$$

$$\overset{9}{C}O_2$$

(XLIX.) Stepwise decomposition of labelled protoporphyrin (contd.).

β, γ, δ of the methine bridges which are lost as CO_2 in the chromic acid oxidation of mesoporphyrin, nor in the carboxyl group of α-keto-butyric acid separated from pyruvic acid by chromatography of the dinitrophenylhydrazones ($A_{(2)} = B_{(2)} = C_{(2)} = D_{(2)} = 0$). These atoms, derived from glycine, are indicated in (XLVIII) by heavy dots. Carbon atoms $A_{(7)}$, $B_{(7)}$, $C_{(7)}$, $D_{(7)}$ and $A_{(10)}$, $B_{(10)}$ are lost in the decarboxylations. Using $CH_3 \cdot C^{14}OOH$, the main activity is found in the two carboxylic acid groups $C_{(10)}$ and $D_{(10)}$; though some is present in the 5 and 3 carbons of all rings. Using $C^{14}H_3 \cdot COOH$, the highest activity is found in the 6 and 9 carbons of all rings, a great deal, however, also in the 4, 8, 5 and 3 carbons and a little in $C_{(10)}$ and $D_{(10)}$. As expected, the specific radio-activity of the total molecule is far higher if methyl-labelled than if carboxyl-labelled acetate is used. There is also evidence that the two halves of the pyrrole which give rise to pyruvic and α-ketoglutaric acid respectively, are derived from the same precursor. Thus:

$$A_{(6)} + B_{(6)} = C_{(6)} + D_{(6)} = A_{(9)} + B_{(9)} = C_{(9)} + D_{(9)}$$

$$A_{(4)} + B_{(4)} = C_{(4)} + D_{(4)} = A_{(8)} + B_{(8)} = C_{(8)} + D_{(8)}$$

$$A_{(5)} + B_{(5)} = C_{(5)} + D_{(5)} = A_{(3)} + B_{(3)} = C_{(3)} + D_{(3)}$$

The fact that $C_{(10)}$ and $D_{(10)}$ from carboxy-labelled acetate equals $C_{(9)} + D_{(9)}$ from methyl-labelled acetate of the same radioactivity indicates that acetic acid enters as a whole, as it does in the citric acid cycle. There is evidence for some, but not very much cycling through the symmetrical succinic acid stage, resulting in labelling of atoms 4, 8, 5 and 3 originally derived from oxalacetic acid.

What is not yet clear, is at which stage the decarboxylations occur, and how the preferential formation of type "III" porphyrin and the absence of symmetrical type "II" porphyrin can be explained. LEMBERG and LEGGE (6, there p. 642) assumed a single precursor (L), the structure now postulated for porphobilinogen (XXXIX), decarboxylated to (LI). The latter was then assumed partly to be condensed by a monocarbon compound, or a compound capable of donating a single carbon atom—from present knowledge this would have to be glycine—to a symmetrical $\alpha\alpha'$-dimethyl pyrromethene which could no longer condense with itself, but could with the loss of an α'-methyl group condense with an un-symmetrical α-methyl-α'-unsubstituted pyrromethene, derived by auto-condensation of (LI), thus giving coproporphyrin "III", or after oxidative decarboxylation affecting only the latter pyrromethene yield proto-porphyrin "IX" (Formula LII). Uroporphyrin "I" and coproporphyrin "I" would be formed by autocondensation of the unsymmetrical pyrromethenes derived from (L) and (LI) respectively. This theory thus assumes decarboxylation at a monopyrrole stage, and oxidative decarboxylation at a dipyrrolic stage.

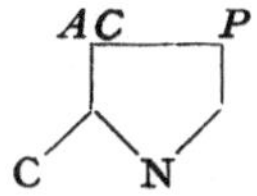

(L.) 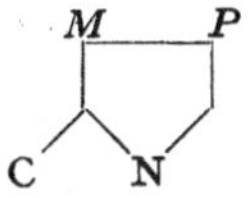   (LI.) 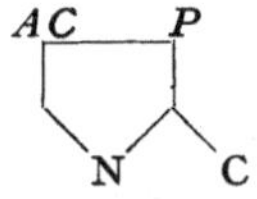   (LIII.)

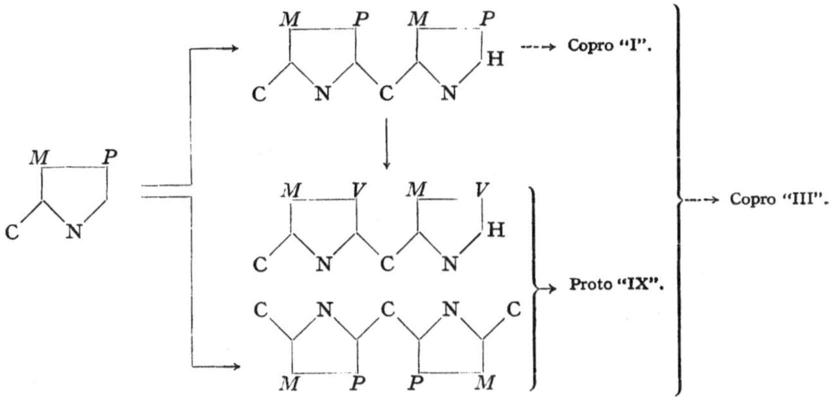

Copro "I".

Copro "III".

Proto "IX".

(LII.)

In the theory of NEUBERGER, MUIR, and GRAY (*174*) the presence of two primary pyrroles, (L) and (LIII), is assumed, and the formation of predominantly type "III" porphyrins is explained by the hypothesis that only in one of these, (L), the methyl group is enzymatically condensed in linear orientation to give a methine bridge. Decarboxylation and oxidative decarboxylation are assumed to occur at the porphyrin level. COOKSON and RIMINGTON (*44*) found, however, no evidence for the existence of isomers of porphobilinogen, which according to FALK, DRESEL and RIMINGTON (*69*) is the primary pyrrole and precursor of uroporphyrin, coproporphyrin and protoporphyrin in the haemolysates of chicken erythrocytes.

While these studies strongly indicate the primary formation of uroporphyrin, or a uroporphyrin precursor, as a precursor of protoporphyrin and haem synthesis, other studies have not given quite as consistent results.

The conversion of protoporphyrin into coproporphyrin and uroporphyrin had become unlikely even before the use of isotopes (cf. *6*, there p. 582). Using N^{15} labelled protoporphyrin in the dog, GRINSTEIN, KAMEN and MOORE (*103*) excluded conversion of proto- into coproporphyrin, and, using $N^{15}H_2 \cdot CH_2 \cdot COOH$, GRINSTEIN, KAMEN, WIKOFF and MOORE (*104*) demonstrated that coproporphyrin "I" in the dog is synthesised from glycine but is not formed in the haemolysis of cells containing labelled protoporphyrin as haemoglobin. Incorporation of glycine N^{15} in the coproporphyrin "III" of lead-poisoned rabbits was demonstrated by GRINSTEIN et al. (*107*). The experiments of HALE et al. (*112*) with *Corynebacterium diphtheriae*, in media varying in iron concentration, however, failed to give decisive results; N^{15} labelling of copro- and uroporphyrin was lower than that of bacterial haem. Isotope experiments in chronic porphyria (*95, 100, 101, 148*) demonstrated that coproporphyrin or uroporphyrin could be considered as precursors of protohaem.

Coproporphyrin, particularly urinary coproporphyrin, reached higher activities and reached maximal activity sooner than uroporphyrin, but this may be due to the uroporphyrin store in this disease being greater than that of coproporphyrin. In a mixed type porphyria, the N^{15} content of uroporphyrin exceeded that of coproporphyrin (*150*).

Uroporphyrin "III" has been shown to be a precursor of haemoglobin protoporphyrin by Salomon, Richmond and Altman (*209*) in rabbit bone marrow, and by Falk, Dresel and Rimington (*69*) in chicken erythrocyte haemolysates.

Watson's experiments (*9, 252*; cf. also *215*) on erythrocyte and urinary coproporphyrin supported the assumption that coproporphyrin "III" was decarboxylated directly to protoporphyrin in contrast to Lemberg and Legge who postulated decarboxylation of a dipyrrolic precursor. Recent experiments have, however, made the role of coproporphyrin as an intermediate in the conversion of uroporphyrin to protoporphyrin doubtful. Falk, Dresel and Rimington (*69*) observed that porphobilinogen was converted in chicken haemolysates to uroporphyrin, coproporphyrin, and protoporphyrin, and that uroporphyrin "III", but not coproporphyrin "III", was converted to protoporphyrin. Similarly, Ycas and Starr (*261*) found conversion of glycine, or of protoporphyrin, but not of coproporphyrin, into haemoproteins, e. g. catalase, in cytochrome-deficient yeast. The process of the conversion of glycine into copro- and uroporphyrin is anaerobic, but aerobiosis is required for the oxidative decarboxylation of two propionic acid side-chains to vinyl in the formation of protoporphyrin and haemin [Bénard, Gajdos, and Gajdos-Török (*22, 81*); Falk, Dresel, and Rimington (*69*)].

The hypothesis of Lemberg and Legge seems still to be the one which fits the facts best. Its unsatisfactory feature which it shares with other theories devised to explain the formation of type "III" porphyrin without recourse to a special "organising enzyme", is that it postulates the formation of the methine bridges from glycine in two different stages, first in the formation of the monopyrrole precursor (e. g. porphobilinogen), and later, in the condensation to the symmetrical dipyrrole, followed by removal of one of these groups in the final ring closure. Experiments so far fail to support any difference in the formation of the four methine bridges.

VII. The Biosynthesis of Chlorophyll.

The studies of Granick on porphyrins in *Chlorella* mutants (see above) have greatly strengthened previous evidence for a close relationship between chlorophyll synthesis and protoporphyrin-haem synthesis (cf. *6*, there p. 650–651). Of particular importance is the finding of the magnesium complex of protoporphyrin, and that of more highly carboxylated porphyrins [Bogorad and Granick (*24*)] in *Chlorella* mutants. These investigations have been reviewed by Granick (*3, 4*). Some recent isotope studies have shown that C^{14} from glutamic acid (LIV) enters chlorophyll (*208*), and, still more important, that glycine N^{15} and acetate C^{14} are precursors of chlorophyll in *Chlorella vulgaris* [Della Rosa, Altman and Salomon (*56*)], with the only difference from haem synthesis that the carboxylic acid group of glycine is also incorporated, probably after conversion into acetate via carbohydrates, rather than through CO_2. Recent studies of LASCELLES (*135*) show that *Rhodopseudomonas spheroides*, an Athiorhodacea, synthesises bacteriochlorophyll as well as porphyrins (coproporphyrin "III" and uroporphyrin "I") from

$$HOOC \cdot CH_2 \cdot CH_2 \cdot CH^{14}(NH_2)C^{14}OOH$$

(LIV.) Glutamic acid.

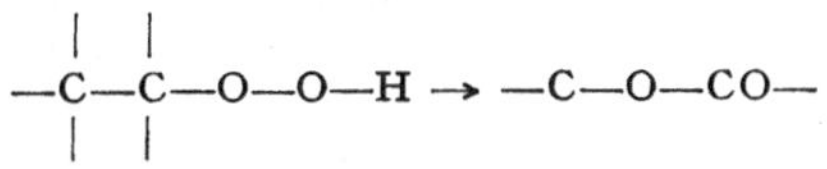

$$-\overset{|}{C}-\overset{|}{C}-O-O-H \rightarrow -\overset{|}{C}-O-CO-$$

(LV.)

glycine and α-ketoglutarate in the presence of Mg^{++}, Mn^{++} and NH_4^+. The organism is able to grow in the dark, but requires light for the synthesis of porphyrins. Iron increases the formation of bacteriochlorophyll, but suppresses porphyrin formation. In view of the possibly early evolutionary stage of this organism and of bacteriochlorophyll, the need of light for porphyrin synthesis supports the view that there was originally a close connection between photosynthesis and porphyrin synthesis.

VIII. Porphyrins are not Intermediates in the Catabolism of Haem Compounds.

Porphyrins may occasionally arise from haematin compounds *in vivo*, but there is so far little evidence that this occurs (cf. *6*, there p. 598).

Porphyrins are not intermediates at any stage during the breakdown of haematin compounds to bile pigments. This reaction begins with the oxidation at a methine bridge of the porphyrin ring in haemoglobin, and the ring is opened with removal of the carbon of this methine bridge, before iron leaves the molecule (cf. *6*, there Chapters X and XI). This is a peroxidative reaction, catalysed by the haem iron of the molecule, leading to the addition of hydrogen peroxide to a double bond at one methine bridge. Similar reactions are known to occur frequently in oxidations *in vivo* of aromatic substances, also in the closely related phthalocyanines [LINSTEAD and WEISS (*142*)]. The subsequent alterations leading to the final elimination of the ring carbon as carbon monoxide [SJÖSTRAND (*227*)] probably begin with a re-arrangement (LV) of peroxides which has been discovered by CRIEGEE (*50*).

References.

Reviews.

1. FISCHER, H. und H. ORTH: Die Chemie des Pyrrols. Bd. II. Pyrrolfarbstoffe. Erste Hälfte. Porphyrine-Hämin-Bilirubin und ihre Abkömmlinge. Leipzig: Akad. Verlagsges. 1937.

2. — — Die Chemie des Pyrrols. Bd. II. Pyrrolfarbstoffe. Zweite Hälfte, von H. FISCHER und A. STERN. Leipzig: Akad. Verlagsges. 1940.

3. GRANICK, S.: The Structural and Functional Relationships Between Haem and Chlorophyll. Harvey Lect. **44**, 220 (1948/49).

4. GRANICK, S.: Biosynthesis of Chlorophyll and Related Pigments. Annu. Rev. Plant Physiol. **2**, 115 (1951).

5. GRANICK, S. and H. GILDER: Distribution, Structure and Properties of the Tetrapyrroles. Adv. Enzymology **7**, 305 (1947).

6. LEMBERG, R. and J. W. LEGGE: Hematin Compounds and Bile Pigments. Their Constitution, Metabolism and Function. New York: Interscience Publ. 1949.

7. — — Pyrrole Pigments. Annu. Rev. Biochem. **19**, 431 (1950).

8. RIMINGTON, C.: Haems and Porphyrins in Health and Disease. I. II. Acta Med. Scand. **143**, 161, 177 (1952).

8a. STOLL, A. und E. WIEDEMANN: Chlorophyll. Fortschr. Chem. organ. Naturstoffe **1**, 159 (1938).

9. WATSON, C. J.: Some Recent Studies on Porphyrin Metabolism and Porphyria. Lancet **1951** I, 539.

10. WATSON, C. J. and E. A. LARSON: Urinary Coproporphyrins in Health and Disease. Physiol. Rev. **27**, 478 (1947).

Original Papers.

11. ALEXANDER, A. E.: Monolayers of Porphyrins and Related Compounds. J. Chem. Soc. (London) **1937**, 1813.

12. ALSLEV, J. und M. KIESE: Darstellung und Eigenschaften von Verdoglobinen. 4. Kristalliner Verd NO_2-Porphyrindimethylester. Arch. exp. Pathol. Pharmakol. **207**, 525 (1949).

13. ALTMAN, K. I., G. W. CASARETT, R. E. MASTERS, T. R. NOONAN, and K. SALOMON: Hemoglobin Synthesis from Glycine Labelled with Radioactive Carbon in its α-Carbon Atom. J. Biol. Chem. **176**, 319 (1948).

14. ALTMAN, K. I., L. L. MILLER, and J. E. RICHMOND: The Role of the Carbon Skeleton of Lysine in the Biosynthesis of Hemoglobin. Arch. Biochem. Biophys. **36**, 399 (1952).

15. ALTMAN, K. I. and K. SALOMON: Hemin Synthesis in Spleen Homogenates. Science (New York) **111**, 117 (1950).

16. ALTMAN, K. I., K. SALOMON, and T. R. NOONAN: Hemin Synthesis in Rabbit Bone Marrow Homogenates. J. Biol. Chem. **177**, 489 (1949).

17. ANDREWS, J. S., A. H. CORWIN, and A. G. SHARP: Porphyrin Studies. VIII. 1. 4, 5, 8-Tetramethyl-2,3,6,7-tetracarbethoxyporphyrin and Some Derivatives. J. Amer. Chem. Soc. **72**, 491 (1950).

18. ANET, E., G. K. HUGHES, and E. RITCHIE: A Synthesis of Isopelletierine and Methylisopelletierine. Nature (London) **164**, 501 (1949).

19. BAUMGÄRTEL, T.: Chlorophyllabbau im menschlichen Darmkanal. Med. Monatsschr. **1**, 401 (1947) [Chem. Zbl. **1948** I, 832].

20. BÉNARD, H., A. GAJDOS et M. GAJDOS-TÖRÖK: Le dosage de la protoporphyrine libre des globules rouges comme moyen d'étude de la synthèse de la protoporphyrine hémoglobinique. C. R. Séances Soc. Biol. **145**, 536 (1951).

21. — — — Biosynthèse expérimentale des uro-, copro- et protoporphyrines par les globules rouges périphériques. C. R. Séances Soc. Biol. **146**, 699 (1952).

22. — — — Action inhibitrice de l'anaérobiose et du fluorure de sodium dans la synthèse des porphyrines par les globules rouges. C. R. Séances Soc. Biol. **146**, 701 (1952).

23. BLANSHARD, T. P.: Isolation from Mammalian Brain of Coproporphyrin III and a uro-type Porphyrin. Proc. Soc. exp. Biol. Med. **82**, 512 (1953).

24. BOGORAD, L. and S. GRANICK: Protoporphyrin Precursors Produced by a Chlorella Mutant. J. Biol. Chem. **202**, 793 (1953).

25. BROCKMAN, P. E. and C. H. GRAY: Studies on Porphobilinogen. Biochemic. J. **54**, 22 (1953).

26. — — Crystalline Porphobilinogen from the Urine of Rabbits Treated with Sedormid. Biochemic. J. (Proc. Biochemic. Soc.) **54**, XXI (1953).

27. BROWNLEE, G.: The Role of the Aromatic Amino Group in Deranged Pigment Metabolism. Biochemic. J. **33**, 697 (1939).

28. BRUGSCH, J.: Fluorometric Method for Detection and Differentiation of Copro- and Haematoporphyrin in Mixtures. Investigation of Properties of Haematoporphyrins. New Derivative: Tetramethylvinyl-α-hydroxyethyl Porphin Dipropionic Acid. Z. ges. inn. Medizin **2**, 645 (1947).

29. — Über das Verhalten des Porphyrinstoffwechsels bei familiärer ErythrocytenElliptocytose (Bemerkungen zur klinisch-wissenschaftlichen Bedeutung der Untersuchung des Porphyrinstoffwechsels bei hämolytischen Anämien). Z. ges. inn. Medizin **6**, 233 (1951) [Chem. Zbl. **1951** II, 3612].

30. — Über Harn-Protoporphyrin- und „Nicht-Koproporphyrin"-Typen des menschlichen Harnes. Z. ges. inn. Medizin **7**, 321 (1952) [Chem. Zbl. **1952** II, 6388].

31. BRUGSCH, J. T.: Untersuchungen des quantitativen Porphyrinstoffwechsels beim gesunden und kranken Menschen; quantitative Untersuchungen der Abbauprodukte des Chlorophylls beim Menschen mittels der Rotfluoreszenz; Untersuchungen beim Stoffwechselgesunden. Z. ges. exp. Medizin **98**, 49 (1936).

32. BRUGSCH, J. T. and A. KEYS: Quantitative Separation and Estimation of Various Porphyrins in Biological Materials. Proc. Soc. exp. Biol. Med. **38**, 557 (1938).

33. BUFTON, A. W. J., R. BENTLEY, and C. RIMINGTON: Non-utilisation of Labelled CO_2 and Formate for the *in vitro* Synthesis of Haem by Surviving Fowl Erythrocytes. Biochemic. J. (Proc. Biochemic. Soc.) **44**, XLIX (1949).

34. CAHILL, A. E. and H. TAUBE: One-Electron Oxidation of Copper Phthalocyanine. J. Amer. Chem. Soc. **73**, 2847 (1951).

35. CALVIN, M. and G. D. DOROUGH: The Possibility of a Triplet State Intermediate in the Photo-Oxidation of a Chlorin. J. Amer. Chem. Soc. **70**, 699 (1948).

36. CARTWRIGHT, G. E. and M. M. WINTROBE: Hematopoiesis. Annu. Rev. Physiology **11**, 335 (1949).

37. CHRIST, C. L. and D. HARKER: X-Ray Crystallographic Studies upon Etioporphyrin I. Amer. Mineralogist **27**, 219 (1942).

38. CHU, E. J. and C. J. WATSON: Experimental Study of the Cerebral Coproporphyrin in Rabbits. Proc. Soc. exp. Biol. Med. **66**, 569 (1947).

39. CHU, T. C., A. A. GREEN, and E. J. CHU: Paper Chromatography of Methyl Esters of Porphyrins. J. Biol. Chem. **190**, 643 (1951).

40. COMFORT, A.: Distribution of Shell Porphyrins in Mollusca. Nature (London) **162**, 851 (1948).

41. — Acid-soluble Pigments of Shells. I. The Distribution of Porphyrin Fluorescence in Molluscan Shells. Biochemic. J. **44**, 111 (1949).

42. — The Pigmentation of Molluscan Shells. Biol. Revs. Cambridge Philos. Soc. **26**, 285 (1951).

43. COMFORT, A. and M. WEATHERALL: Urinary Porphyrins in Lead-treated Rabbits. Biochemic. J. **54**, 247 (1953).

44. COOKSON, G. H. and C. RIMINGTON: Porphobilinogen. Chemical Constitution. Nature (London) **171**, 875 (1953).

45. CORWIN, A. H. and J. S. ANDREWS: Studies in the Pyrrole Series. II. The Mechanism of the Aldehyde Synthesis of Dipyrrylmethenes. J. Amer. Chem. Soc. **58**, 1086 (1936).

46. — — Studies in the Pyrrole Series. III. The Relation of Tripyrrylmethane Cleavage to Methene Synthesis. J. Amer. Chem. Soc. **59**, 1973 (1937).

47. CORWIN, A. H. and R. H. KRIEBLE: A Reinvestigation of the Configuration of Hemin. J. Amer. Chem. Soc. **63**, 1829 (1941).

48. COULTER, C. B. and F. M. STONE: The Occurrence of Porphyrins in Cultures of *C. diphtheriae*. J. Gen. Physiol. **14**, 583 (1930/31).

49. CRAVEN, C. W., K. R. REISSMANN, and H. I. CHINN: Infrared Absorption Spectra of Porphyrins. Analyt. Chemistry **24**, 1214 (1952).

50. CRIEGEE, R.: Die Umlagerung der Dekalin-peroxydester als Folge von kationischem Sauerstoff. Liebigs Ann. Chem. **560**, 127 (1948).

51. CROWE, M. O'L. and A. WALKER: Coproporphyrin III Isolated from the Human Tubercle Bacillus by Chromatographic and Fluorescence Analysis. Fluorescence and Absorption Spectral Data. Brit. J. exp. Pathol. **32**, 1 (1951).

52. DANNENBERG, H. und M. KIESE: Untersuchungen über Cytochrome. I. Die prosthetische Gruppe des sauerstoffübertragenden Ferments (Cytochromoxydase). Biochem. Z. **322**, 395 (1952).

53. DAVENPORT, H. E.: Reduction Cleavage of Cytochrome *c*. Nature (London) **169**, 75 (1952).

54. DAVENPORT, H. E. and R. HILL: The Preparation and Some Properties of Cytochrome *f*. Proc. Roy. Soc. (London) **B 139**, 327 (1952).

55. DE LANGEN, C. D.: Porphyrin in Serum. Acta Med. Scand. **133**, 73 (1949).

56. DELLA ROSA, R. J., K. I. ALTMAN, and K. SALOMON: The Biosynthesis of Chlorophyll as Studied with Labeled Glycine and Acetic Acid. J. Biol. Chem. **202**, 771 (1953).

57. DERRIEN, E.: Note préliminaire sur quelques faits nouveaux pour l'histoire naturelle des porphyrines animales. C. R. Séances Soc. Biol. **91**, 634 (1924).

58. DERRIEN, E. et J. TURCHINI: Sur la biologie des porphyrines naturelles. Bull. soc. chim. biol. (Paris) **8**, 218 (1926).

59. DHÉRÉ, C.: Sur la porphyrine tégumentaire du *Lumbricus terrestris*. C. R. hebd. Séances Acad. Sci. **195**, 1436 (1932).

60. — Progrès récents en spectrochimie de fluorescence des produits biologiques. Fortschr. Chem. organ. Naturstoffe **6**, 311 (1950).

61. DOBRINER, K. and C. P. RHOADS: The Porphyrins in Health and Disease. Physiol. Rev. **20**, 416 (1940).

62. DOROUGH, G. D. and K. T. SHEN: Fundamental Properties of Porphyrin Systems. I. A Spectroscopic Study of N-H Isomerism in Porphyrin free Bases. J. Amer. Chem. Soc. **72**, 3939 (1950).

63. ELEY, D. D.: Phthalocyanines as Semiconductors. Nature (London) **162**, 819 (1948).

64. ELLINGSON, R. C. and A. H. CORWIN: Further Studies on Steric Deformation. J. Amer. Chem. Soc. **68**, 1112 (1946).

65. ERDMAN, J. G. and A. H. CORWIN: The Nature of the N--H Bond in the Porphyrins. J. Amer. Chem. Soc. **68**, 1885 (1946).

66. ERIKSEN, L.: Extraction of Urinary Coproporphyrin Chromogen and its Conversion to Porphyrin. Scand. J. Clin. Labor. Invest. **4**, 55 (1952).

67. FALK, J. E. and A. BENSON: Separation of Uroporphyrin Esters I and III by Paper Chromatography. Biochemic. J. **55**, 101 (1953).

68. FALK, J. E. and E. I. B. DRESEL: Incorporation of Glycine $C^{14}H_2$ into Haem and nett Synthesis of Porphyrin in Chicken Erythrocyte Haemolysates. Resumé Comm. 2nd. Internat. Congress Biochem., Paris, 1952, p. 8.

69. FALK, J. E., E. I. B. DRESEL, and C. RIMINGTON: Porphobilinogen as a Porphyrin Precursor and Interconversion of Porphyrins, in a Tissue System. Nature (London) **172**, 292 (1953).

70. FALK, J. E. and C. RIMINGTON: Porphyrins. Annu. Rep. Chem. Soc. (London) **47**, 271 (1950).

71. — — Attempted Isolation of Haem *a* and Porphyrin *a* from Heart Muscle. Biochemic. J. **51**, 36 (1952).

72. FALK, J. E. and J. B. WILLIS: The Infra-red Spectra of Porphyrins and their Iron Complexes. Austral. J. Sci. Res., Series A **4**, 579 (1951).

73. FIKENTSCHER, R.: Über Porphyrin-Befunde im Serum von Foeten und Neugeborenen. Klin. Wschr. **14**, 569 (1935).

74. FISCHER, H. und H. HILMER: Zur Kenntnis des Phylloerythrins. II. Bemerkungen zur Abhandlung Herrn Dr. KÉMERIS ,,Über einen neuen porphyrinartigen Bestandteil normaler menschlicher Fäces". Z. physiol. Chem. (Hoppe-Seyler) **143**, 1 (1925).

75. FISCHER, H. und K. JORDAN: Zur Kenntnis der natürlichen Porphyrine. XXV. Mitt. Über Konchoporphyrin, sowie Überführung von Protoporphyrin aus Malz in Mesoporphyrin IX. Z. physiol. Chem. (Hoppe-Seyler) **190**, 75 (1930).

76. FISCHER, H., J. RIEDMAIR und J. HASENKAMP: Über Oxo-porphyrine: ein Beitrag zur Kenntnis der Feinstruktur von Chlorophyll *a*. Liebigs Ann. Chem. **508**, 224 (1934).

77. FISCHER, H. und C. v. SEEMANN: Die Konstitution des Spirographishämins. Z. physiol. Chem. (Hoppe-Seyler) **242**, 133 (1936).

78. FISCHER, H. und G. WECKER: Synthese des Spirographisporphyrins. Z. physiol. Chem. (Hoppe-Seyler) **272**, 1 (1941).

79. FISCHER, H. und W. ZERWECK: Zur Kenntnis der natürlichen Porphyrine. 7. Mitt. Über Uroporphyrinogen-heptamethylester und eine neue Überführung von Uro- in Koproporphyrin. Z. physiol. Chem. (Hoppe Seyler) **137**, 242 (1924).

80. Fox, H. M.: Chlorocruorin: a Pigment Allied to Haemoglobin. Proc. Roy. Soc. (London) B **99**, 199 (1926).

81. GAJDOS, A.: Synthesis of Free Protoporphyrin in Peripheral Red Cells. Resumé Commun. 2nd Internat. Congress Biochem., Paris, 1952, p. 9.

82. GAJDOS, A. et M. GAJDOS-TÖRÖK: The Biosynthesis of the Protoporphyrin of Haemoglobin. Exposés ann. biochim. méd. **13**, 201 (1951).

83. GARROD, Q. E.: Inborn Errors of Metabolism, 2nd ed. London: H. Frowde. 1923.

84. GIBSON, Q. H. and D. C. HARRISON: A Note on the Urinary Uroporphyrin in Acute Porphyria. Biochemic. J. **46**, 154 (1950).

85. GOODWIN, R. H., V. M. KOSKI, and O. v. H. OWENS: The Distribution and Properties of a Porphyrin from the Epidermis of Vicia Shoots. Amer. J. Bot. **38**, 629 (1951).

86. GRANICK, S.: Protoporphyrin 9 as a Precursor of Chlorophyll. J. Biol. Chem. **172**, 717 (1948).

87. — Magnesium Vinyl Pheoporphyrin a_5, another Intermediate in the Biological Synthesis of Chlorophyll. J. Biol. Chem. **183**, 713 (1950).

88. — Heme Components of Cytochromes of Horse Heart. Federat. Proc. (Amer. Soc. exp. Biol.) **11**, 221 (1952).

89. GRANICK, S. and L. BOGORAD: Separation of Porphyrins by Countercurrent Distribution. J. Biol. Chem. **202**, 781 (1953).

90. — — Porphobilinogen a Monopyrrole. J. Amer. Chem. Soc. **75**, 3610 (1953).

91. GRANICK, S., L. BOGORAD, and H. JAFFE: Hematoporphyrin IX, a Probable Precursor of Protoporphyrin in the Biosynthetic Chain of Heme and Chlorophyll. J. Biol. Chem. **202**, 801 (1953).

92. GRANICK, S. and R. KETT: Magnesium Protoporphyrin as a Precursor of Chlorophyll in Chlorella. J. Biol. Chem. **175**, 333 (1948).

93. GRAY, C. H.: Report of a Case of Acute Porphyria. Arch. int. Med. **85**, 459 (1950).

94. Gray, C. H. and L. B. Holt: The Isolation of Coproporphyrin III from *Coryne-bacterium diphtheriae* Culture Filtrates. Biochemic. J. **43**, 191 (1948).
95. Gray, C. H. and A. Neuberger: Studies in Congenital Porphyria. 1. Incorporation of ^{15}N into Coproporphyrin, Uroporphyrin and Hippuric Acid. Biochemic. J. **47**, 81 (1950).
96. Gray, C. H., A. Neuberger, and P. H. A. Sneath: Studies in Congenital Porphyria. 2. Incorporation of ^{15}N in the Stercobilin in the Normal and in the Porphyric. Biochemic. J. **47**, 87 (1950).
97. Gray, C. H., C. Rimington and S. Thomson: A Case of Chronic Porphyria Associated with Recurrent Jaundice. Quart. J. Med. **17**, 123 (1948).
98. Grieg, A., R. Askevold, and S. L. Sveinsson: Investigations on Conversion of Porphobilinogen to Porphyrin. Scand. J. Clin. Labor. Invest. **2**, 1 (1950).
99. Grinstein, M.: Studies of Protoporphyrin. VII. A Simple and Improved Method for the Preparation of Pure Protoporphyrin from Hemoglobin. J. Biol. Chem. **167**, 515 (1947).
100. Grinstein, M., R. A. Aldrich, V. Hawkinson, P. Lowry, and C. J. Watson: Photosensitive or Congenital Porphyria with Hemolytic Anemia. Isotopic Studies of Porphyrin and Hemoglobin Metabolism. Blood **6**, 699 (1951).
101. Grinstein, M., R. A. Aldrich, V. Hawkinson, and C. J. Watson: An Isotopic Study of Porphyrin and Hemoglobin Metabolism in a Case of Porphyria. J. Biol. Chem. **179**, 983 (1949).
102. Grinstein, M., M. D. Kamen, and C. V. Moore: The Utilization of Glycine in the Biosynthesis of Hemoglobin. J. Biol. Chem. **179**, 359 (1949).
103. — — — Studies on Globin and Porphyrin Metabolism made with C^{14} and N^{15}. J. Lab. Clin. Med. **33**, 1478 (1948).
104. Grinstein, M., M. D. Kamen, H. M. Wikoff, and C. V. Moore: Isotopic Studies on Porphyrin and Hemoglobin Metabolism. I. Biosynthesis of Coproporphyrin I and its Relationship to Hemoglobin Metabolism. J. Biol. Chem. **182**, 715 (1950).
105. Grinstein, M., S. Schwartz, and C. J. Watson: Studies of the Uroporphyrins. I. The Purification of Uroporphyrin I and the Nature of Waldenström's Uroporphyrin, as Isolated from Porphyria Material. J. Biol. Chem. **157**, 323 (1945).
106. Grinstein, M. and C. J. Watson: Studies of Protoporphyrin. III. Photoelectric and Fluorophotometric Methods for the Quantitative Determination of the Protoporphyrin in Blood. J. Biol. Chem. **147**, 675 (1943).
107. Grinstein, M., H. M. Wikoff, R. Pimenta de Mello, and C. J. Watson: Isotopic Studies on Porphyrin and Hemoglobin Metabolism. II. The Biosynthesis of Coproporphyrin III in Experimental Lead Poisoning. J. Biol. Chem. **182**, 723 (1950).
108. Grinstein, M. and M. M. Wintrobe: Spectroscopic Micromethod for the Quantitative Determination of the Free Erythrocyte Protoporphyrin. J. Biol. Chem. **172**, 459 (1948).
109. Grotepass, W.: Zur Kenntnis der natürlichen Harnporphyrine. Z. physiol. Chem. (Hoppe-Seyler) **253**, 276 (1938).
110. Grotepass, W. und A. Defalque: Über die Porphyrine bei einem Fall von Porphyrie ohne Porphyrinurie. Z. physiol. Chem. (Hoppe-Seyler), **252**, 155 (1938).
111. Günther, H.: Bedeutung des Haematoporphyrins in Physiologie und Pathologie. In: Lubarsch-Ostertag, Ergebnisse der allgemeinen Pathologie und pathologischen Anatomie des Menschen und der Tiere, Bd. **20**, Abt. II, Teil II. München: Bergmann. 1922.
112. Hale, J. H., W. A. Rawlinson, C. H. Gray, L. B. Holt, C. Rimington, and W. Smith: Biosynthesis of Porphyrins and Haems by *Corynebacterium diphtheriae*. Brit. J. exp. Pathol. **31**, 96 (1950).

113. HAUSSER, K. W., R. KUHN und G. SEITZ: Lichtabsorption und Doppelbindung. V. Über die Absorption von Verbindungen mit konjugierten Kohlenstoffdoppelbindungen bei tiefer Temperatur. Z. physik. Chem. B **29**, 391 (1935).

114. HAWKINSON, V. E. and C. J. WATSON: The Separation of Porphobilinogen and an Ehrlich Negative Precursor of Uroporphyrin. Science (New York) **115**, 496 (1952).

115. HELLSTRÖM, H.: Beziehungen zwischen Konstitution und Spektren der Porphyrine. Ark. Kemi, Mineral. Geol. **12 B**, No. 13 (1936).

116. HERBERT, F. K.: Precursors of Porphyrin in the Urine in Idiopathic Porphyria. Biochemic. J. (Proc. Biochemic. Soc.) **52**, XII (1952).

117. HUGHES, A.: Discussion on Surface Phenomena: Films. Proc. Roy. Soc. (London) **A 155**, 710 (1936).

118. JACOB, A.: Über den Abbau von Blutfarbstoff zu Porphyrinen durch Reinkulturen von Bakterien, und über eine neue biologische Synthese von Koproporphyrin III. Klin. Wschr. **18**, 1024 (1939).

119. JOPE, E. M. and J. R. P. O'BRIEN: Spectral Absorption and Fluorescence of Coproporphyrin Isomers I and III and the Melting Points of Their Methyl Esters. Biochemic. J. **39**, 239 (1945).

120. KEHL, R. und W. STICH: Die papierchromatographische Analyse der Porphyrine. Z. physiol. Chem. (Hoppe-Seyler) **289**, 6 (1951).

121. — — Über die papierchromatographische Analyse der Porphyrine und einiger Gallenfarbstoffe. 2. Mitt. Z. physiol. Chem. (Hoppe-Seyler) **290**, 151 (1952).

122. KENCH, J. E., R. E. LANE, and H. VARLEY: Urinary Coproporphyrin I in Lead Poisoning. Biochemic. J. (Proc. Biochemic. Soc.) **51**, IX (1952).

123. KENCH, J. E. and S. C. PAPASTAMATIS: A Study of Congenital Porphyria. Biochemic. J. (Proc. Biochemic. Soc.) **51**, XLI (1952).

124. KENCH, J. E. and J. F. WILKINSON: Porphyrin Formation by Yeast. Nature (London) **155**, 579 (1945).

125. KENNARD, O.: Porphobilinogen. X-Ray Crystallographic Determination of Molecular Weight. Nature (London) **171**, 876 (1953).

126. KENNARD, O. and C. RIMINGTON: Identification and Classification of some Porphyrins on the Basis of their X-Ray Diffraction Patterns. Biochemic. J. **55**, 105 (1953).

127. KENNEDY, G. Y. and H. G. VEVERS: Protoporphyrin in the Integument of *Asterias rubens*. Nature (London) **171**, 81 (1953).

128. KIESE, M.: Über das Porphyrin des sauerstoffübertragenden Fermentes. Naturwiss. **39**, 403 (1952).

129. KLÜVER, H.: On Naturally Occuring Porphyrins in the Central Nervous System. Science (New York) **99**, 482 (1944).

130. — On a Possible Use of the Root Nodules of Leguminous Plants for Research in Neurology and Psychiatry (Preliminary Report on a Free Porphyrin-Hemoglobin System). J. Psychology **25**, 331 (1948).

131. KOSKI, V. M., C. S. FRENCH, and J. H. C. SMITH: Action Spectrum for the Transformation of Protochlorophyll to Chlorophyll *a* in Normal and Albino Corn Seedlings. Arch. Biochem. Biophys. **31**, 1 (1951).

132. KOSKI, V. M. and J. H. C. SMITH: Chlorophyll Formation in a Mutant, White Seedling-3. Arch. Biochem. Biophys. **34**, 189 (1951).

133. KUHN, H.: A Quantum-Mechanical Theory of Light Absorption of Organic Dyes and Similar Compounds. J. Chem. Physics **17**, 1198 (1949).

134. LARSEN, E. G., C. M. GLAFKIDES, and J. M. ORTEN: Porphyrins Present in Porphyria Urine and in Nucleated Erythrocytes. Federat. Proc. (Amer. Soc. exp. Biol.) **12**, 236 (1953).

135. LASCELLES, J.: Porphyrin Synthesis by *Rhodopseudomonas spheroides*. Biochemic. J. (Proc. Biochemic. Soc.) **55**, IV (1953).

136. Lederer, E. et R. Tixier: Sur les porphyrines de l'ambre gris. C. R. hebd. Séances Acad. Sci. **225**, 531 (1947).

137. Legge, J. W.: A Note on the Disturbance of the Haemoglobin Metabolism of the Rat by Sulphanilamide. Biochemic. J. **44**, 105 (1949).

138. Lemberg, R.: Porphyrin *a* and other Porphyrins from Heart Muscle. Nature (London) **172**, 619 (1953).

138 a. Lemberg, R. and J. Barrett: The Prosthetic Group of Cytochrome a_2. Nature (London) **173**, 213 (1954).

139. Lemberg, R. and J. E. Falk: Comparison of Haem *a*, the Dichroic Haem of Heart Muscle and of Porphyrin *a* with Compounds of known Structure. Biochemic. J. **49**, 674 (1951).

140. Lemberg, R. and W. H. Lockwood: Unpublished.

141. Lemberg, R. and J. Parker: Porphyrins with Formyl Groups. II. Preparation of Chlorocruoroporphyrin and Diformyldeuteroporphyrin. Austral. J. exp. Biol. and Med. Sci. **30**, 163 (1952).

142. Linstead, R. P. and F. T. Weiss: The Oxidation of Phthalocyanine, Tetrabenzoporphin and allied Substances. J. Chem. Soc. (London) **1950**, 2981.

143. Lockwood, W. H.: Uroporphyrins. I. Uroporphyrin Content of Normal Urine. Austral. J. exp. Biol. and Med. Sci. **31**, 453 (1953).

144. — Uroporphyrins. II. Precursors of Uroporphyrin in Acute Porphyria. Austral. J. exp. Biol. and Med. Sci. **31**, 457 (1953).

145. Loebisch, W. F. und M. Fischler: Monatsh. Chem. **24**, 353 (1904).

146. London, I. M.: Biosynthesis of Heme. Resumé Commun. 2nd Internat. Congress Biochem., Paris, 1952, p. 13.

147. London, I. M., D. Shemin, and D. Rittenberg: Synthesis of Heme *in vitro* by the Immature Non-nucleated Mammalian Erythrocyte. J. Biol. Chem. **183**, 749 (1950).

148. London, I. M., R. West, D. Shemin, and D. Rittenberg: Porphyrin Formation and Hemoglobin Metabolism in Congenital Porphyria. J. Biol. Chem. **184**, 365 (1950).

149. Longuet-Higgins, H. C., C. W. Rector, and J. R. Platt: Molecular Orbital Calculations on Porphine and Tetrahydroporphine. J. Chem. Physics **18**, 1174 (1950).

150. Lowry, P.: An Isotopic Study of a Case of Porphyria of "Mixed" Type. J. Lab. Clin. Med. **36**, 958 (1950).

151. Lucas, J. and J. M. Orten: Isolation of Porphyrins from Biological Materials by Partition Chromatography. J. Biol. Chem. **191**, 287 (1951).

152. Lwoff, A.: Les hématines considérées comme facteurs de croissance. Les antihématines. Bull. soc. chim. biol. (Paris) **30**, 817 (1948).

153. Lwoff, A. et M. Lwoff: La fonction de l'hémine, facteur de croissance pour *Hemophilus influenzae.* C. R. hebd. Séances Acad. Sci. **204**, 1510 (1937).

154. MacDonald, S. F.: A General Route to $\beta\beta'$-Substituted Pyrrole Intermediates for Porphyrin Synthesis. J. Chem. Soc. (London) **1952**, 4176.

155. — The Syntheses of Pyrroles, a Porphyrin, and the Maleinimide Related to the Uroporphyrins. J. Chem. Soc. (London) **1952**, 4184.

156. MacDonald, S. F. and R. J. Stedman: The Synthesis of Uroporphyrin I. J. Amer. Chem. Soc. **75**, 3040 (1953).

157. McElroy, L. W., K. Salomon, F. H. J. Figge, and G. R. Cowgill: The Porphyrin Nature of the Fluorescent "Blood-caked" Wiskers of Pantothenic Acid-deficient Rats. Science (New York) **94**, 467 (1941).

158. McEwen, W. K.: Steric Deformation. The Synthesis of N—Methyl Etioporphyrin I. J. Amer. Chem. Soc. **68**, 711 (1946).

159. MacGREGOR, A. G., R. E. H. NICHOLAS, and C. RIMINGTON: Porphyria cutanea tarda. Arch. int. Med. **90**, 483 (1952).

160. MAC MUNN, C. A.: Observations on some of the Colouring Matters of Bile and Urine, with especial Reference to their Origin, and of an easy Method of Procuring Haematin from Blood. J. Physiol. **6**, 22 (1880).

161. — On the Presence of Haematoporphyrin in the Integument of Certain Invertebrates. J. Physiol. **7**, 240 (1886).

162. McSWINEY, R. R., R. E. H. NICHOLAS, and F. T. G. PRUNTY: Porphyrins of Acute Porphyria. Detection of hitherto unrecognised Porphyrins. Biochemic. J. **46**, 147 (1950).

163. MARCHLEWSKI, L.: Transformations de la chlorophylle dans l'organisme animal. Bull. soc. chim. biol. (Paris) **6**, 464 (1924).

164. MAYER, R. M.: Über den fermentativen Charakter der Koproporphyrin-Synthese der Hefe. Die zellfreie Koproporphyrinvermehrung. Z. physiol. Chem. (Hoppe-Seyler) **179**, 99 (1928).

165. MERTENS, E.: Über das Uroporphyrin bei akuter Hämatoporphyrie. Z. physiol. Chem. (Hoppe-Seyler) **238**, 1 (1936).

166. — Über die bei akuter Porphyrie auftretenden Porphyrine. Z. physiol. Chem. (Hoppe-Seyler) **250**, 57 (1937).

167. MORELL, D. B.: The Prosthetic Group of Lactoperoxidase. Austral. J. exp. Biol. med. Sci. **31**, 567 (1953).

168. MORRISON, M. and E. STOTZ: Prosthetic Groups of Hemoproteins. Federat. Proc. (Amer. Soc. exp. Biol.) **12**, 249 (1953).

169. MUIR, H. M. and A. NEUBERGER: The Biogenesis of Porphyrins. The Distribution of ^{15}N in the Ring System. Biochemic. J. **45**, 163 (1949).

170. — — The Biogenesis of Porphyrins. II. The Origin of the Methene Carbon Atoms. Biochemic. J. **47**, 97 (1950).

171. NEGELEIN, E.: Kryptohämin. Biochem. Z. **248**, 243 (1932).

172. — Über Kryptohämin. Biochem. Z. **250**, 577 (1932).

173. — Über die Extraktion eines vom Bluthämin verschiedenen Hämins aus dem Herzmuskel. Biochem. Z. **266**, 412 (1933).

174. NEUBERGER, A., H. M. MUIR, and C. H. GRAY: Biosynthesis of Porphyrins and Congenital Porphyria. Nature (London) **165**, 948 (1950).

175. NEUBERGER, A. and J. J. SCOTT: The Basicities of the Nitrogen Atoms in the Porphyrin Nucleus; their Dependence on some Substituents of the Tetrapyrrolic Ring. Proc. Roy. Soc. (London) **A 213**, 307 (1952).

176. NICHOLAS, R. E. H.: Chromatographic Methods for the Separation and Identification of Porphyrins. Biochemic. J. **48**, 309 (1951).

177. NICHOLAS, R. E. H. and A. COMFORT: Acid-soluble Pigments of Molluscan Shells. 4. Identification of Shell Porphyrins with particular reference to Conchoporphyrin. Biochemic. J. **45**, 208 (1949).

178. NICHOLAS, R. E. H. and C. RIMINGTON: Qualitative Analysis of Porphyrins by Partition Chromatography. Scand. J. Clin. Lab. Invest. **1**, 12 (1949).

179. — — Paper Chromatography of Porphyrins: Some Hitherto Unrecognized Porphyrins and Further Notes on the Method. Biochemic. J. **48**, 306 (1951).

180. — — Isolation of Unequivocal Uroporphyrin III. A further Study of Turacin. Biochemic. J. **50**, 194 (1951).

181. — — Studies on the "Waldenström Porphyrin" of Acute Porphyria Urines. Biochemic. J. **55**, 109 (1953).

182. NOACK, K. und W. KIESSLING: Zur Entstehung des Chlorophylls und seiner Beziehung zum Blutfarbstoff. 1. Mitt. Z. physiol. Chem. (Hoppe-Seyler) **182**, 13 (1929).

183. — — Zur Entstehung des Chlorophylls und seiner Beziehung zum Blutfarbstoff. 2. Mitt. Z. physiol. Chem. (Hoppe-Seyler) **193**, 97 (1930).

184. Oliver, I. T. and W. A. Rawlinson: The Microestimation of Porphyrins by Copper Titration. Biochemic. J. **49**, 157 (1951).

185. Ottolenghi-Lodigiani, F. and G. Serchi: Porphyrins and their Metabolism in Congenital Porphyria. II. Porphyrin Compounds in Blood, Faeces, Saliva, and in Gastric and Duodenal Juices. Giorn. ital. dermatol. e sifilol. **93**, 1 (1952) [Chem. Abstr. **47**, 3968 (1953)].

186. Papastamatis, S. C. and J. E. Kench: Ionophoretic Separation of Porphyrin Pigments. Nature (London) **170**, 33 (1952).

187. Pappenheimer, A. M., Jr.: Diphtheria Toxin. III. A Reinvestigation of the Effect of Iron on Toxin and Porphyrin Production. J. Biol. Chem. **167**, 251 (1947).

188. Paul, K. G.: The Conversion of Ferri-porphyrins ("Hemins") to their Corresponding Porphyrins by Means of Pyruvic Acid. Acta Chem. Scand. **4**, 1221 (1950).

189. — The Porphyrin Component of Cytochrome *c* and its Linkage to the Protein. Acta Chem. Scand. **5**, 389 (1951).

190. Platt, J. R.: Molecular Orbital Predictions of Organic Spectra (Porphine and Tetrahydroporphine). J. Chem. Physics **18**, 1168 (1950).

191. Ponticorvo, L., D. Rittenberg, and K. Bloch: The Utilization of Acetate for the Synthesis of Fatty Acids, Cholesterol and Protoporphyrin. J. Biol. Chem. **179**, 839 (1949).

191a. Prunty, F. T. G.: Acute Porphyria. Arch. int. Med. **77**, 623 (1946).

192. Rabinowitch, E.: Spectra of Porphyrins and Chlorophylls. Rev. Med. Physics **16**, 226 (1944).

193. Radin, N. S., D. Rittenberg and D. Shemin: The Role of Glycine in the Biosynthesis of Heme. J. Biol. Chem. **184**, 745 (1950).

194. — — — The Role of Acetic Acid in the Biosynthesis of Heme. J. Biol. Chem. **184**, 755 (1950).

195. Rawlinson, W. A. and J. H. Hale: Prosthetic Groups of the Cytochromes Present in *Corynebacterium diphtheriae* with especial Reference to Cytochrome *a*. Biochemic. J. **45**, 247 (1949).

196. Rimington, C.: Enzyme Theory of Hemopoiesis. C. R. Trav. Lab. Carlsberg, Ser. chim. **22**, 454 (1938).

197. — Identification of the Protoporphyrin in Sheep's Liver. Biochemic. J. **32**, 460 (1938).

198. — A Reinvestigation of Turacin, the Copper Porphyrin Pigment of certain Birds belonging to the Musophagidae. Proc. Roy. Soc. (London) **B 127**, 106 (1939).

199. — A Simple Fluorescence Comparator and its Application to the Determination of Porphyrin. Biochemic. J. **37**, 137 (1943).

200. — Formation of Porphyrin by Autolysing Yeast and by Yeast Press Juice. Nature (London) **151**, 393 (1943).

201. Rimington, C. and A. W. Hemmings: Porphyrinuric Action of Drugs related to Sulphanilamide. Comparison with reported Toxicity, Therapeutic Efficiency and Causation of Methaemoglobinaemia. Definition of the Structure responsible for Porphyrinuric Action. Biochemic. J. **33**, 960 (1939).

202. Rimington, C. and P. A. Miles: A Study of the Porphyrins Excreted in the Urine in a Case of Congenital Porphyria. Biochemic. J. **50**, 202 (1951).

203. Rimington, C. and J. I. Quin: Photosensitising Agent in "Geel-dikkop" Phyllo-erythrin. Nature (London) **132**, 178 (1933).

204. Rimington, C. and S. L. Sveinsson: Spectrophotometric Determination of Uroporphyrin. Scand. J. clin. Lab. Invest. **2**, 209 (1950).

205. Robertson, J. M.: An X-Ray Study of the Phthalocyanines. II. Quantitative Structure Determination of the Metal-free Compound. J. Chem. Soc. (London) **1936**, 1195.

206. ROBERTSON, J. M. and I. WOODWARD: An X-Ray Study of the Phthalocyanines. III. Quantitative Structure Determination of Nickel Phthalocyanine. J. Chem. Soc. (London) **1937**, 219.

207. ROTHEMUND, P.: Occurrence of Decomposition Products of Chlorophyll. III. Isolation of Pyrroporphyrin from Beef Bile. J. Amer. Chem. Soc. **57**, 2179 (1935).

208. ROUX, E. et C. HUSSON: Pigments des chloroplastes et photosynthèse. C. R. hebd. Séances Acad. Sci. **234**, 1573 (1952).

209. SALOMON, K., J. E. RICHMOND, and K. I. ALTMAN: Tetrapyrrole Precursors of Protoporphyrin IX. I. Uroporphyrin III. J. Biol. Chem. **196**, 463 (1952).

210. SALZBURG, P. and C. J. WATSON: A Study of the Supposed Conversion of Protoporphyrin to Coproporphyrin by the Liver. II. The Porphyrin Metabolism of Rabbit Liver. J. Biol. Chem. **139**, 593 (1941).

211. SCHAEFFER, P.: Recherches sur le métabolisme bactérien des cytochromes et des porphyrines. II. Excrétion de porphyrines par culture anaérobie chez certaines bactéries aérobies facultatives. Biochim. Biophys. Acta **9**, 362 (1952).

212. SCHAYER, R. W.: Studies of the Metabolism of Tryptophan labeled with N^{15} in the Indole Ring. J. Biol. Chem. **187**, 777 (1950).

213. SCHLESINGER, W., A. H. CORWIN and L. J. SARGENT: Porphyrin Studies. IX. Synthetic Chlorins and Dihydrochlorins. J. Amer. Chem. Soc. **72**, 2867 (1950).

214. SCHMID, R. and S. SCHWARTZ: Disturbance of Catalase Metabolism in Experimental Porphyria. J. Lab. clin. Med. **40**, 939 (1952).

215. SCHMID, R., S. SCHWARTZ, and C. J. WATSON: Porphyrins in Bone Marrow and Circulating Erythrocytes in Experimental Anemias. Proc. Soc. exp. Biol. Med. **75**, 705 (1950).

216. SCHULTZE, M. O.: The Isolation of Protoporphyrin IX from Feces of Normal and Anemic Rats. J. Biol. Chem. **142**, 89 (1942).

217. SCHUMM, O.: Über ein aus α-Hämatin bei der Darmfäulnis entstehendes Umwandlungsprodukt (,,Kopratin") und das zugehörige Porphyrin. Kopratin und Pyridinblutprobe. Z. physiol. Chem. (Hoppe-Seyler) **149**, 1 (1925).

218. SCHWARTZ, S. and H. M. WIKOFF: The Relation of Erythrocyte Coproporphyrin and Protoporphyrin to Erythropoiesis. J. Biol. Chem. **194**, 563 (1952).

219. SCHWARTZ, S., L. ZIEVE, and C. J. WATSON: An Improved Method for the Determination of Urinary Coproporphyrin and an Evaluation of Factors Influencing the Analysis. J. Lab. clin. Med. **37**, 843 (1951).

220. SHEMIN, D.: The Biosynthesis of Porphyrins. Cold Spring Harbor Sympos. Quant. Biol. **13**, 185 (1948).

221. SHEMIN, D. and S. KUMIN: The Mechanism of Porphyrin Formation. The Formation of a Succinyl Intermediate from Succinate. J. Biol. Chem. **198**, 827 (1952).

222. SHEMIN, D., I. M. LONDON, and D. RITTENBERG: The Synthesis of Protoporphyrin *in vitro* by Red Blood Cells of the Duck. J. Biol. Chem. **183**, 757 (1950).

223. SHEMIN, D. and D. RITTENBERG: The Biological Utilization of Glycine for the Synthesis of the Protoporphyrin of Hemoglobin. J. Biol. Chem. **166**, 621 (1946).

223a. SHEMIN, D. and CH. S. RUSSELL: δ-Aminolevulinic Acid, its Role in the Biosynthesis of Porphyrins and Purines. J. Amer. Chem. Soc. **75**, 4873 (1953).

224. SHEMIN, D. and J. WITTENBERG: The Mechanism of Porphyrin Formation. The Role of the Tricarboxylic Acid Cycle. J. Biol. Chem. **192**, 315 (1951).

225. SIEDEL, W. und F. WINKLER: Oxydation von Pyrrolderivaten mit Bleitetraacetat. Neuartige Porphyrinsynthesen. Liebigs Ann. Chem. **554**, 162 (1943).

226. SIMPSON, W. T.: Theory of the π-Electron System in Porphines. J. Chem. Physics **17**, 1218 (1949).

348   R. LEMBERG:

227. SJÖSTRAND, T.: Formation of Carbon Monoxide in Connexion with Haemo-globin Catabolism. Nature (London) **168**, 1118 (1951).

228. SLONIMSKI, P. P.: Excrétion de porphyrines par la levure en anaérobiose. C. R. hebd. Séances Acad. Sci. **235**, 1064 (1952).

229. STENHAGEN, E. and E. K. RIDEAL: The Interaction Between Porphyrins and Lipoid and Protein Monolayers. Biochemic. J. **33**, 1591 (1939).

230. STERN, A.: Lichtabsorption und Fluoreszenz der Porphyrine. Angew. Chem. **49**, 551 (1936).

231. STICH, W. und H. EISGRUBER: Die Koproporphyrin- und Hämsynthese durch Hefe und ihre Beeinflussung mit B_2-Vitaminen. Z. physiol. Chem. (Hoppe-Seyler) **287**, 19 (1951).

232. SVEINSSON, S. L., C. RIMINGTON, and H. D. BARNES: Complete Porphyrin Analysis of Pathological Urines. Scand. J. clin. Lab. Invest. **1**, 2 (1949).

233. THEORELL, H.: Über die chemische Konstitution des Cytochroms *c*. IV. Mitt. Darstellung von Porphyrin-Cystein-Addukten. Biochem. Z. **301**, 201 (1939).

234. THOMAS, J.: Contribution à l'étude des porphyrines en biologie et pathologie. Lons-Les-Saunier: M. Declume. 1938.

235. TIXIER, R.: Sur les porphyrines de quelques coquilles de Mollusques. Bull. soc. chim. biol. (Paris) **28**, 394 (1946).

236. TODD, C. M.: Occurrence of Cytochrome and Coproporphyrin in Mycobacteria. Biochemic. J. **45**, 386 (1949).

237. TREIBS, A.: Porphyrine in bituminösen Gesteinen, in Erdöl und Kohlen. Zur Entstehung des Erdöls. Angew. Chem. **49**, 551 (1936).

237 a. — Chlorophyll- und Häminderivate in organischen Mineralstoffen. Angew. Chem. **49**, 682 (1936).

238. TURNER, W. J.: Studies on Porphyria. I. Observations on the Fox-Squirrel, *Sciurus niger*. J. Biol. Chem. **118**, 519 (1937).

239. — A Theory of Porphyrinogenesis. J. Lab. clin. Med. **26**, 323 (1940).

240. VAHLQUIST, B.: Die quantitative Bestimmung des Porphobilinogens im Harn von Kranken mit sogenannter akuter Porphyrie. Z. physiol. Chem. (Hoppe-Seyler) **259**, 213 (1939).

241. VESTLING, C. S. and J. R. DOWNING: Infrared Studies of the Porphyrin Molecule. J. Amer. Chem. Soc. **61**, 3511 (1939).

242. VÖLKER, O.: Zur Kenntnis des Porphyrins in Vogelfedern. Z. physiol. Chem. (Hoppe-Seyler) **258**, 1 (1939).

243. — Über die Struktur der bei Vögeln vorkommenden Porphyrine. Z. Natur-forsch. **2 b**, 316 (1947).

244. WALDENSTRÖM, J., H. FINK und W. HOERBURGER: Über ein neues bei der akuten Porphyrie regelmäßig vorkommendes Uroporphyrin. Z. physiol. Chem. (Hoppe-Seyler) **233**, 1 (1935).

245. WALDENSTRÖM, J. und B. VAHLQUIST: Studien über die Entstehung der roten Harnpigmente (Uroporphyrin und Porphobilin) bei der akuten Porphyrie aus ihrer farblosen Vorstufe (Porphobilinogen). Z. physiol. Chem. (Hoppe-Seyler) **260**, 189 (1939).

246. WALTER, R. I.: Simultaneous Dissociation of Two Protons. The Acid-Base Equilibria of Porphyrins. J. Amer. Chem. Soc. **75**, 3860 (1953).

247. WARBURG, O. und H. S. GEWITZ: Cytohämin aus Herzmuskel. Z. physiol. Chem. (Hoppe-Seyler) **288**, 1 (1951).

248. — — Cytodeuteroporphyrin. Z. physiol. Chem. (Hoppe-Seyler) **292**, 174 (1953).

249. WARBURG, O. und E. NEGELEIN: Über das Hämin des sauerstoffübertragenden Ferments der Atmung, über einige Hämoglobine und über Spirographisporphy-rin. Biochem. Z. **244**, 9 (1932).

250. — — Notiz über Spirographishämin. Biochem. Z. **244**, 239 (1932).

251. WARD, E. and H. L. MASON: Free Erythrocyte Protoporphyrin. J. Clin. Investigation **29**, 905 (1950).

252. WATSON, C. J.: The Erythrocyte Coproporphyrin. Arch. int. Med. **86**, 797 (1950).

253. WATSON, C. J., R. PIMENTA DE MELLO, S. SCHWARTZ, V. E. HAWKINSON, and I. BOSSENMAIER: Porphyrin Chromogens or Precursors in Urine, Blood, Bile, and Feces. J. Lab. clin. Med. **37**, 831 (1951).

254. WATSON, C. J. and S. SCHWARTZ: A simple Test for Urinary Porphobilinogen. Proc. Soc. exp. Biol. Med. **47**, 393 (1941).

255. WATSON, C. J., S.SCHWARTZ, and V. HAWKINSON: Studies of the Uroporphyrins. II. Further Studies of the Porphyrins of the Urine, Feces, Bile and Liver in Cases of Porphyria, with Particular Reference to a Waldenström Type Porphyrin Behaving as an Entity on the Tswett Column. J. Biol. Chem. **157**, 345 (1945).

256. WEATHERALL, M.: Fate of Intravenously Administered Coproporphyrin III in Normal and Lead-treated Rabbits. Biochemic. J. **52**, 683 (1952).

257. WESTALL, R. G.: Isolation of Porphobilinogen from the Urine of a Patient with Acute Porphyria. Nature (London) **170**, 614 (1952).

258. WILLSTÄTTER, R. und W. MIEG: Untersuchungen über Chlorophyll. I. Über eine Methode der Trennung und Bestimmung von Chlorophyllderivaten. Liebigs Ann. Chem. **350**, 1 (1906).

259. WITTENBERG, J. and D. SHEMIN: The Utilization of Glycine for the Biosynthesis of both Types of Pyrroles in Protoporphyrin. J. Biol. Chem. **178**, 47 (1949).

260. — — The Location in Protoporphyrin of the Carbon Atoms Derived from the α-Carbon Atom of Glycine. J. Biol. Chem. **185**, 103 (1950).

261. YČAS, M. and T. J. STARR: The Effect of Glycine and Protoporphyrin on a Cytochrome Deficient Yeast. J. Bacteriol. **65**, 83 (1953).

262. ZEILE, K. und B. RAU: Über die Verteilung von Porphyrinen zwischen Äther und Salzsäure und ihre Anwendung zur Trennung von Porphyringemischen. Z. physiol. Chem. (Hoppe-Seyler) **250**, 197 (1937).

(Received, December 30, 1953.)

The Pteridines.

By ADRIEN ALBERT, Canberra.

With 5 Figures.

Contents.

I. Introduction.

It is now nine years since the subject was last reviewed by PURRMANN in this series (*138*). In the meantime the chemistry of the pteridines has undergone dramatic development. The constitutions of more than a dozen naturally-occurring pteridines are now known, as against three in 1945 (xanthopterin, *iso*xanthopterin and leucopterin). Parallel with this development, a more serious biological role has been revealed for the pteridines. From being thought of principally as insect pigments, they are now seen to be important regulators of cell-division, even in the highest forms of life (*80*).

The first pteridines were isolated (from butterfly wings) by HOPKINS in 1891 (*67*), but their constitution remained unknown until PURRMANN showed, in 1940, (*134, 135, 136*) that they were aminohydroxy-derivatives of the nucleus (I), to which WIELAND gave the name "pteridin" in 1941.

(I.) Pteridine.
(Numbering as used in this Review.)

As early as 1895, KÜHLING had synthesized derivatives of (I), but the connexion between these substances and the butterfly-pigments was not demonstrated until 1940. Both before and after this date, a good deal of synthetic work was carried out to explore the chemistry of (I) and its derivatives. In this way, two different systems arose for numbering the pteridine ring: (i) that which KUHN and COOK put forward in 1937, as shown in (I). It is based on the rules customarily applied to all other aza-derivatives of naphthalene*; and (ii) another system (II) designed by WIELAND in 1940 to relate pteridine to purine (III) (*138*). At the present time, the English-speaking countries use the numbering shown in (I), whereas (II) is used elsewhere, but not exclusively. The apparent similarity between pteridines and purines seems to recede

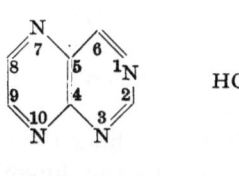

(II.) (Alternative numbering; not recommended.)

(III.) Purine.

(IV.) *Iso*xanthopterin.

---

* „Die Stellung numerieren wir wie beim Naphthalin" (*102*).

each year with increasing knowledge (6), and hence the earlier and more logical numbering (I) is to be preferred.

All naturally-occurring pteridines, as far as is yet known, have an amino-group in the 2-position and a hydroxyl-group in the 4-position. Sometimes, mainly in the biological literature, it is assumed that the intelligent reader will know what is meant if these essential parts of the formula are not described in the name. Those who reason thus describe *iso*xanthopterin (IV), for example, as 7-hydroxypteridine instead of 2-amino-4:7-dihydroxypteridine, although this type of contraction is not used in other branches of chemistry. 7-Hydroxypteridine is a well-known substance with properties very different from those of (IV). Hence, to designate (IV) as 7-hydroxypteridine can only lead to the utter confusion of chemists, biologists, and indexers.

Other irregular practices include (i) the designation (for example) of (IV) as 7-hydroxypterin, as though "pterin" were an abbreviation for 2-amino-4-hydroxypteridine; (ii) the designation (for example) of (IV) as 2-amino-4:7-dihydroxypterin, as though "pterin" was a synonym for pteridine. This does not accord with the original definition of "die Pterine" by SCHÖPF and BECKER: "Als *Pterine* bezeichnen wir bei den Insekten weitverbreitete unter dem Chitin, in Form feiner Körnchen dicht eingelagerte farblose oder farbige Pigmente" (*155*). Thus "pterin", unlike "pteridine" has no precise chemical meaning. The term *pteridoxamine* has been proposed by RAUEN as an abbreviation for 2-amino-4-hydroxypteridine (*142*).

II. Simple Mono- and Di-substituted Pteridines as Models for the Understanding of Naturally-occurring Pteridines.

In 1945, when the last review (*138*) was printed, no mono-substituted pteridines were known, but 32 of them are now available (*7–10, 33a*). These substances, together with additions to the range of di-substituted pteridines, have shed new light on the properties of the natural pteridines.

1. Solubility and Fusibility.

Pteridine is a substance of high solubility in all solvents ranging from light petroleum to water, and it is low-melting (see *Table 1*). The naturally-occurring pteridines, on the other hand, have very low solubility in all solvents, and are unmelted at 350°. A study of mono-substituted pteridines has explained this apparent paradox.

As will be seen from Table 1, the monohydroxypteridines (Nos. 2–5 in Table 1) are far less soluble in water than pteridine, the dihydroxypteridines (Nos. 6–9) are less soluble still, and tri- and tetra-hydroxypteridines (Nos. 10 and 11) are least soluble of all. The hydroxyl-group is a strong hydrogen-bonding group and usually confers increased water-solubility for this reason. However, in the pteridine series, hydrogen bonding between hydroxypteridine molecules can be reinforced by the high dipole moments involved (cf. VII), so that this intermolecular bonding becomes preferred to hydration (9). Hence there is a fall in solubility for every hydrogen-bonding group introduced. When hydrogen-bonding is blocked by O-methylation (Nos. 18 and 19), or by N-methylation (Nos. 20 and 21), solubility in water is greatly increased. As far as has been investigated the same rules apply to primary amino-pteridines (Nos. 12–16), which are poorly soluble in water, and the non hydrogen-bonding tertiary amines (Nos. 22 and 23) which are much more soluble. This paradoxical insolubilizing effect of groups which are usually water-attracting is found in the purine series, and to a much lesser extent among pyrimidines and simpler one-nitrogen heterocyclic substances (9).

That the non-hydrogen bonding pteridines are also more soluble in lipoid solvents follows from first principles, quite independently of the paradoxical effect. Secondary amines (including acetylated primary amines) are intermediate, in aqueous and lipoid solubilities, between the primary and tertiary amines.

Whenever hydroxy-derivatives of nitrogenous hetero-aromatic substances have been submitted to X-ray crystallographic studies, it has been found that the equilibrium exemplified by (V) ⇌ (VI) lies far to the right, although for the corresponding amino-derivatives it favours the left. The evidence (mainly spectrographic) for the hydroxy- and amino-pteridines tends to favour these assumptions, but complete proof is lacking. "Keto" forms such as (VI) contribute to poor solubility only through the dipoles which reside in them, e. g. (VII). When hydrogen-bonding is not possible, as in (VIII) (No. 21 in Table 1), solubility in water can be very high.

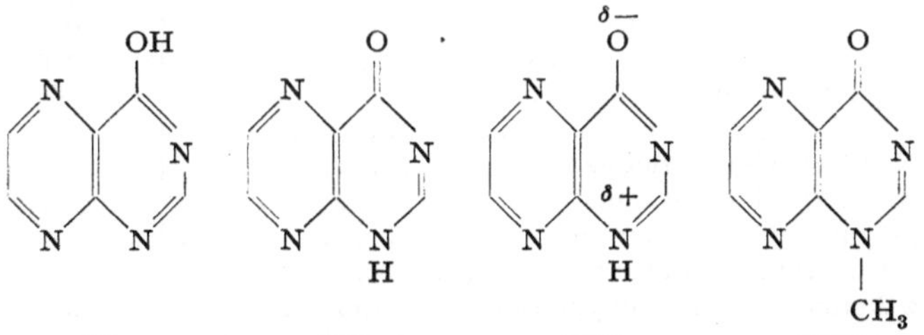

(V.) (VI.) (VII.) (VIII.)

Reference to the last column in Table 1 shows that the non-hydrogen bonding pteridines are low-melting. The pteridines with more than one hydrogen-bonding substituent tend to remain unmelted at 350°. This is consistent with the hypothesis of high lattice-energy for hydrogen-bonded pteridines, as outlined above (*1*). In between the extremes, pteridines with one hydrogen-bonding substituent tend to decompose on heating (it will be seen in the next Section why these substances have low chemical stability).

From the evidence brought forward here, it seems inevitable that substances with the constitution of the naturally-occurring pteridines should be poorly soluble in water (and other solvents), and high-melting. Examples (Nos. 24–27) in Table 1 illustrate this point.

In elucidating the constitution of new pteridines, it may be convenient to make the substances more soluble by acylation, which is specific for amino-groups (hydroxyl-groups only if on side chains), chlorination (specific for hydroxyl-groups), and alkylation (*1*).

Table 1. Solubilities and Melting-points of Pteridines (*9*).

No.	Compound	Solubility in water (20°) 1 part in	M. p. °
1	Pteridine	7	140
2	2-Hydroxypteridine........................	600	dec.
3	4-Hydroxypteridine........................	200	> 350
4	6-Hydroxypteridine........................	3500	dec.
5	7-Hydroxypteridine........................	900	dec.
6	2:4-Dihydroxypteridine	800	dec.
7	4:6-Dihydroxypteridine	5000	dec.
8	4:7-Dihydroxypteridine	4000	> 350
9	6:7-Dihydroxypteridine	3000	> 350
10	4:6:7-Trihydroxypteridine.................	27,000	> 350
11	2:4:6:7-Tetrahydroxypteridine	58,000	> 350
12	2-Aminopteridine..........................	1350	dec.
13	4-Aminopteridine..........................	1400	dec.
14	6-Aminopteridine..........................	1500	dec.
15	7-Aminopteridine..........................	1400	dec.
16	2:4-Diaminopteridine	3000	dec.
17	2-Amino-4-hydroxypteridine	57,000	> 350
18	2-Methoxypteridine........................	80	150
19	4-Methoxypteridine	80	195
20	$N_{(8)}$-Methyl-7-pteridone	50	125
21	$N_{(1)}$-Methyl-4-pteridone	2	221
22	2-Dimethylaminopteridine	3	125
23	4-Dimethylaminopteridine	60	165
24	2-Amino-4:6-dihydroxypteridine (xanthopterin)	40,000	> 350
25	2-Amino-4:7-dihydroxypteridine (*iso*xanthopterin) .	200,000	> 350
26	2-Amino-4:6:7-trihydroxypteridine (leucopterin) ...	750,000	> 350
27	Pteroylglutamic acid	200,000 *	dec. above 250°

* At p_H 3. The addition of alkali to give p_H 5 increases solubility to 1 in 20,000; at p_H 7 it becomes 1 in 70 (*23*).

2. Stability to Acids and Alkalis.

The pteridines are unstable to acid and alkali unless at least two electron-liberating groups are present. The principal effective groups for this purpose are —OH, —NH₂, —SH, and —N(CH₃)₂. The last-named is not hydrogen-bonding, so that the protective effect has a quite different origin from that described in Section 1, above.

Pteridine itself is an extremely unstable substance, as *Table 2* shows. This instability has been found in other hetero-aromatic systems having a high N/C ratio (9) and is attributed to the electron-attracting nature of the nitrogen atoms which tend to deplete the π-layer of its 10 electrons. This π-layer is responsible for the aromatic stabilization of the ring, and does not become properly effective in the pteridine series unless two of the groups specified above are present (9), a point illustrated by Table 2.

Table 2. Decomposition of Pteridines by Acid or Alkali
(1 hour at 110°) (9).

Compound	Percentage decomposed	
	N-H_2SO_4	10 N-NaOH
Pteridine...	74	>57
2-Hydroxypteridine	55	89
4-Hydroxypteridine	60	94
6-Hydroxypteridine	2	100
7-Hydroxypteridine	52	76
2:4-Dihydroxypteridine........................	6	4
6:7-Dihydroxypteridine........................	7	12
4:6:7-Trihydroxypteridine	0	4
2:4:6:7-Tetrahydroxypteridine	0	6

The nature of the decomposition caused by acid or alkali has been investigated mainly in substances containing a hydroxyl- or amino-group in the 4-position. Regardless of whether there is a substituent or not in the 2-position, such substance give 2-aminopyrazine-3-carboxylic acid or a derivative. Thus (V) gives (IX) and, eventually (X) (9); (VIII) gives (XI) (10). This reaction has proved useful in determining the constitution of pteroylglutamic acid (p. 372).

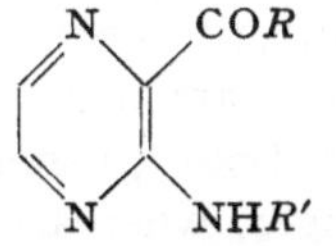

(IX.) $R = NH_2$; $R' = H$.
(X.) $R = OH$; $R' = H$.
(XI.) $R = NH_2$; $R' = CH_3$.

7-Hydroxypteridine, on the other hand, tends to lose the pyrazine ring (*10*).

As would be expected from the foregoing, xanthopterin (2-amino-4:6-dihydroxypteridine) is a highly stable substance because it contains three groups which freely release electrons. It is unaffected by boiling 7 N-hydrochloric acid (*204*) but can be broken down to glycine by this acid at 200° (*156*). It is little affected by boiling with 1.5 N-barium hydroxide for 20 hours (*203*). Leucopterin (7-hydroxyxanthopterin) is almost unaffected by being heated at 150° with concentrated sulphuric acid for 2 hours, but is broken down by 10 N-hydrochloric acid at 165° to glycine, carbon monoxide, carbon dioxide and ammonia in the ratio of 1:1:3:4 (*202*).

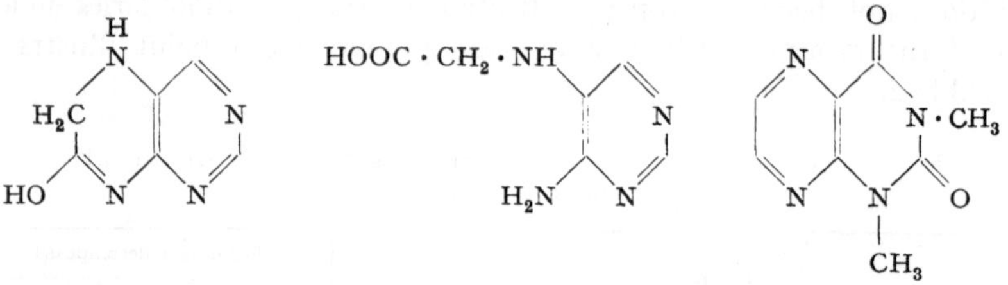

(XII.) 5:6-Dihydro-7-hydroxypteridine. (XIII.) (XIV.) 1:3-Dimethylpterid-2:4-dione.

The elucidation of the structure of new pteridines may be helped by two reactions available for increasing the lability of the more stable pteridines. Hydrogenation facilitates ring-opening (*204*), the most striking example being 5:6-dihydro-7-hydroxypteridine (XII) which gives the pyrimidine (XIII) quantitatively (*8*). Methylation of hydroxy-pteridines by methyl sulphate in water leads to easily hydrolysed N-methyl-derivatives. For example, 2:4-dihydroxypteridine gives 1:3-dimethylpterid-2:4-dione (XIV) which is quantitatively hydrolysed by boiling for one minute with N-sodium hydroxide (*10*), whereas 2:4-di-hydroxypteridine requires 4 N-alkali, for 2 hours at 170°. The products are 2-methylaminopyrazine-3-carboxy-methylamide and (IX) respectively.

3. Ionization; Metal-binding Properties.

Pteridine is a base of pK_a 4.1, that is to say it is a little weaker than quinoline ($pK_a = 4.9$) and distinctly stronger than purine ($pK_a = 2.4$). Unlike purine, pteridine has no acidic properties. The ionization constant of pteridine and a number of substituted pteridines are given in *Table 3*.

The mono-hydroxy-pteridines are acids of widely-spaced strengths. 2-Hydroxypteridine is weaker than phenol ($pK_a = 10.0$), 4-hydroxy-pteridine is considerably stronger, whereas the 6- and 7-isomerides are strongest of all, but not quite so strong as acetic or benzoic acid. Thus in a dihydroxypteridine which contains (on the one hand) a 4-hydroxy-group and (on the other) a 6- or 7-hydroxy-group, the 6- or 7-group will be the first to ionize, with the result that the 4-hydroxy-group will be reduced in acid-strength, by COULOMB's law. This sequence may be seen operating in 4:6- and 4:7-dihydroxypteridine, in xanthopterin and *iso*xanthopterin (Table 3, p. 359). No constants have yet been found for leucopterin because of its complexity and poor solubility, but the example of 6:7-dihydroxypteridine in Table 3 is relevant, for it shows that the coulombic effect operates even in two groups as nearly matched in strength as the 6- and 7-hydroxyl-groups. Presumably, in leucopterin, the pK of the 4-hydroxy-group would be displaced to about 12, but here it may come into conflict with the ionization (as an acid) of the 2-amino-group. No ionization of this group could be detected in 2-amino-pteridine, because of an experimental difficulty (this substance decomposes at p_H 12). As will be gathered from pp. 355–356, this tendency to decomposition would not be expected in hydroxy-derivatives of 2-amino-pteridine, but no anionic species could be detected spectrophotometrically at p_H 13 in 2-amino-4-hydroxypteridine (*108*). Nevertheless an inspection of the spectrum of *iso*xanthopterin (*Table 4*, p. 362) showed that a new species is appearing at p_H 13 that was not present at p_H 12, and this is presumably the 2-amino-group, ionizing as an acid. This topic is thrown into perspective by 4-aminopteridine, which has an easily demonstrable anionic ionization (*7*).

The ionization of the hydroxy-group in 6-hydroxypteridine differs from that of any other known hydroxy-heteroaromatic substance, in that it is not instantaneous, a hysteresis curve being produced (*8*). The same effect is found in 4:6-dihydroxypteridine (*5*), and in xanthopterin where it was first observed (*157*). It does not occur in 6:7-dihydroxy-pteridine, nor in leucopterin. For a discussion see (*3a*).

The ionization of 4-hydroxypteridine is little affected by the presence of a 6-methyl-group, and 4-hydroxy-6-methylpteridine has been included in Table 3 as a model for pteroylglutamic acid. The pK_a of 8.3 in this acid is easily traced to the 4-hydroxyl-group, as it is detected *after* two equivalents of alkali have been added. These two equivalents are used in neutralizing the two —COOH groups of glutamic acid. That the ionization of these two carboxyl-groups has had so little coulombic effect on the pK_a of the 4-hydroxyl-group, shows how distant in space these anionic groups must be, and speaks against writing the formula of pteroylglutamic acid in a condensed form (*30*), or internally hydrogen-bonded.

Turning now to basic strengths, it will be seen that 4-, 6- and 7-hydroxy-pteridines are much weaker bases than pteridine (the exact strengths of 2- and 4-hydroxypteridines, as bases, are not known because they are insufficiently soluble for potentiometric titration in this region, and their cations do not have distinctive spectra). As has been indicated in Section 1, p. 353, the hydroxypteridines presumably resemble other α- and γ-hydroxy-hetero-aromatic substances in that they exist in an equilibrium between truly phenolic forms such as (V, p. 353), and acid amides (or vinylogous acid amides) such as (VI). From what is known of the α- and γ-hydroxy-substances, in general, the equilibrium would favour (VI) at the expense of (V). For example, in 4-hydroxy-quinoline, the ratio is 13,000 to 1 (*187*). The phenolic forms (e. g. V) should, when tautomerizing, transfer the proton from oxygen to the most basic centre, provided that the laws of valence are not contravened. Now, regardless of whether this most basic centre is $N_{(1)}$, $N_{(3)}$, or $N_{(8)}$, 2-, 4- and 7-hydroxy-pteridines could transfer their protons to any of these positions (but not if it were at $N_{(5)}$). The nitrogen that has received this new proton will carry a fractional positive charge, as in (VII), and thus would repel a hydrogen ion, and would tend not to become a cation of the type of (XV). 6-Hydroxypteridine is in quite a different class. It has a hydroxy-group neighbouring on $N_{(5)}$ which is out-of-step with the other nitrogen atoms. It can transfer a proton to $N_{(5)}$, but not to $N_{(1)}$, $N_{(3)}$, or $N_{(8)}$. If the most basic position is among the latter nitrogens, it is almost as free to ionize as a cation regardless of whether a hydroxy-group is present in the 6-position or not. Thus 6-hydroxypteridine is the only mono-hydroxy-pteridine with appreciable basic strength. Following this argument, no evident basic properties would be expected from the poly-hydroxypteridines, and none have been found.

For example, 4:6-dihydroxypteridine will not dissolve in N-mineral acids. Thus the earlier belief (*138*) that *all* 6-hydroxypteridines are basic is incorrect. This belief was founded on the assumption that 5-amino-pyrimidine is more basic than 4-aminopyrimidine. However, the reverse is actually the case (*11*), and this type of comparison is seldom profitable because of the widespread reorganization of π-electrons which can occur when two rings are fused.

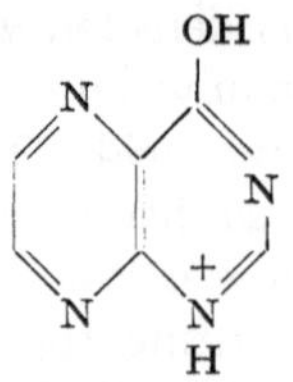

OH

H₂C

CH₃

(XV.) 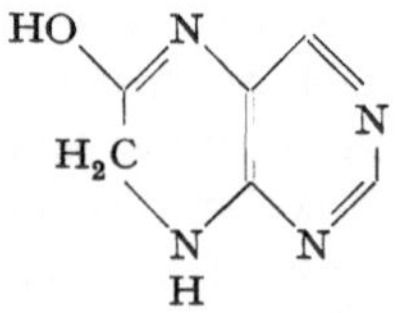   (XVI.) 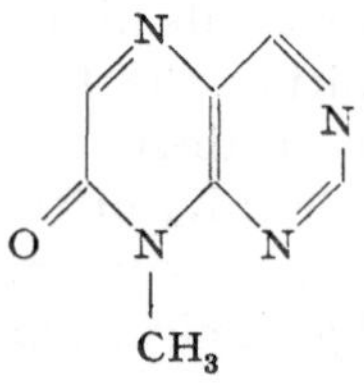   (XVII.)

The mono-aminopteridines are all bases of about the same strength as pteridine. The 2-isomeride, of greatest interest to the worker on natural pteridines, is given as an example in *Table 3*. In 2-amino-4-

hydroxypteridine (an arrangement of substituents found in all known natural pteridines) the amino-group has become one hundred times weaker as a base (see Table 3). Evidently, once again there is conflict between the tautomerizing proton of the hydroxyl-group, and the hydrogen ions of the solution for possession of the most basic position (in hetero-aromatic substances, the proton tends not to go on the —NH$_2$ group).

The foregoing should help to explain why xanthopterin is almost as strong a base as 2-amino-4-hydroxypteridine (because the 6-hydroxyl-group is not appreciably base-weakening), whereas *iso*xanthopterin and leucopterin have basic pKs of < 0. A basic pK$_a$ of about 2 for pteroyl-glutamic acid would be expected, but has not yet been investigated. It is much more soluble in dilute sulphuric acid than in water.

Table 3. Ionization of Pteridines, in Water at 20° (*5, 7, 8, 9, 14*).

Compound	Basic pK$_a$*	Acidic pK$_a$
Pteridine..	4.1	none
2-Hydroxypteridine	< 2	11.1
4-Hydroxypteridine	< 2	7.9
6-Hydroxypteridine	3.7	6.7
7-Hydroxypteridine	1.2	6.4
4:6-Dihydroxypteridine...........................	< 0	6.1; 9.7
4:7-Dihydroxypteridine...........................	< 0	6.1; 9.6
6:7-Dihydroxypteridine...........................	< 0	6.9; 10.0
2-Aminopteridine	4.3	> 12
2-Amino-4-hydroxypteridine.......................	2.3	7.9
2-Amino-4:6-dihydroxypteridine (xanthopterin)	1.6**	6.3; 9.2
2-Amino-4:7-dihydroxypteridine (*iso*xanthopterin) ...	< 0	*ca.* 7; 10.2
4-Hydroxy-6-methylpteridine......................	(?)	8.2
Pteroylglutamic acid	(?)	5.0†; 8.3
6-Hydroxy-7:8-dihydropteridine (XVI)	4.8	10.5
7-Hydroxy-5:6-dihydropteridine (XII)	3.4	9.9
2-Amino-4-hydroxy-7:8-dimethyl-5:6:7:8-tetrahydro-pteridine (*127*)................................	5.6	10.4
5-Formyl-derivative of last named (*127*)...........	< 2	10.0
Leucovorin (5-formyl-5:6:7:8-tetrahydropteroyl-glutamic acid) (*127*)......·...................	< 2	3.1; 4.8; 10.4
N$_{(8)}$-Methyl-7-pteridone (XVII)	1.1	none
N$_{(1)}$-Methyl-4-pteridone (VIII)	1.3	none
N$_{(3)}$-Methyl-4-pteridone	< 1.3	none

* pK$_a$ (the negative logarithm of the ionization constant) corresponds to the p$_H$ at half-neutralization. Thus the stronger the base, the higher the pK$_a$. Likewise the stronger the acid, the lower the pK$_a$.

** c. f. (*157*).

† There is another (carboxylic) group with pK < 5.

The constants for two dihydropteridines are reported in Table 3. The hydroxyl-group is weakened, as an acid, by being placed in a non-aromatic ring (e. g. XVI). The most basic centre of 7-hydroxy-5:6-dihydropteridine (XII, p. 356) may well be $N_{(5)}$ which has been liberated, by hydrogenation, from the base-weakening resonance conferred on nitrogens forming a pyrazine ring (*11*). Some simple analogues of leucovorin, in Table 3, demonstrate that formylation of $N_{(5)}$ abolishes a basic centre in this hydrogenated pteridine.

The last three compounds in Table 3, demonstrate that the N-alkylation of a hydroxypteridine does not raise the basic strength. In fact, comparison of 7-hydroxypteridine and $N_{(8)}$-methyl-7-pteridone (XVII) shows almost no change in pK_a.

No ionization data have been recorded for chrysopterin, ichthyopterin, erythropterin, fluorescyanine, or pterorhodin. Rhizopterin has had preliminary investigation (*209*).

It was first pointed out (*4*) in 1949 that pteridines containing a 4-hydroxyl-group combine with the cations of *heavy metals* to give 1 : 1 complexes of type (XVIII) and, at higher p_H values (about p_H 7), 2 : 1 complexes of type (XIX). All natural pteridines, except the citrovorum factor, combine with metals in this way. So far no metal complexes of pteridines have been isolated from nature, a statement which is equally true for the amino-acids which form metallic complexes of similar strength. In fact, the customary methods of isolation make use of strongly alkaline (or acidic) conditions which would breakup such complexes. Nevertheless the likelihood remains that some of the biological actions of pteridines are effected through these metal complexes. For further data on the complexes, see (*2*).

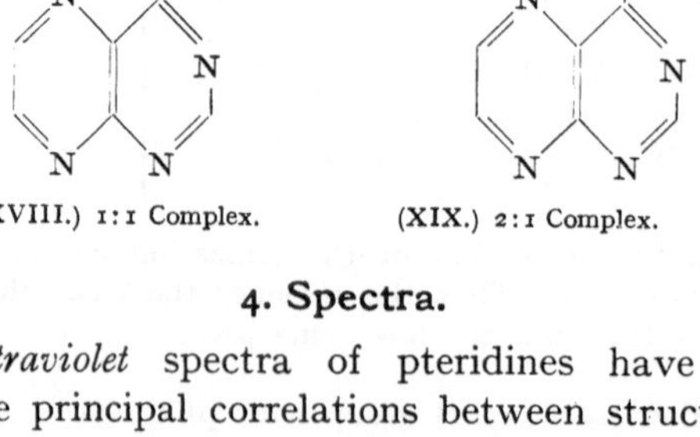

(XVIII.) 1:1 Complex. (XIX.) 2:1 Complex.

4. Spectra.

Only the *ultraviolet* spectra of pteridines have been intensely investigated. The principal correlations between structure and spectra have been reviewed recently (*1*), so that attention can be concentrated here on special aspects relevant to the natural pteridines.

When a substance of high nitrogen-content is isolated from nature, it is likely that it will prove to be either a purine or a pteridine. Sometimes, the wrong diagnosis has been made from the spectrum, e. g. "guanopterin" from butterfly wings turned out to be a purine, *iso*guanine (*134*). This raises a question: to what extent can purines and pteridines be differentiated by their ultraviolet spectra? It may be said at once that pteridines

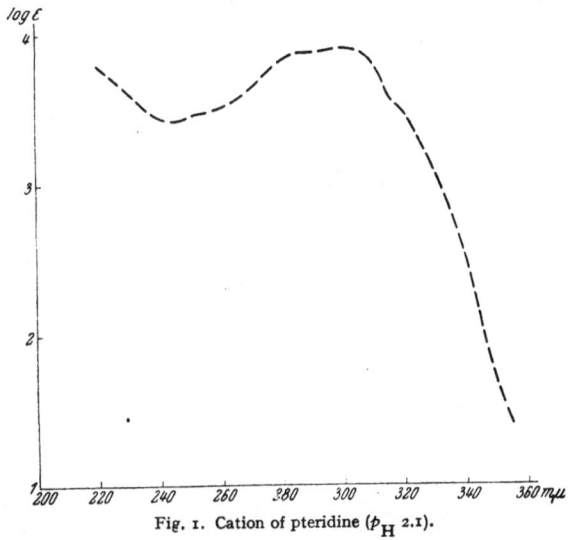

Fig. 1. Cation of pteridine (p_H 2.1).

usually absorb at longer wave-lengths than purines (*134*), but this statement is an over-simplification that requires discussion.

The ultraviolet spectra of pteridines have from 2 to 4 peaks. Sometimes a little fine structure can be seen, as in 7-hydroxypteridine (*Table 4*, p. 362), but this is rare. The cation of pteridine (*Fig. 1*) has one of the simplest spectra, consisting of a peak below 220 mμ. (not further characterized because of technical difficulties) and one at 300 mμ. The short-wave peak may be displaced to longer wave-lengths in other pteridines, and it is not uncommon to find it about 225 mμ., as in 7-hydroxypteridine. The molecule of pteridine resembles that of the cation, but has a shelf of low intensity at the long-wave end. Such shelves are quite common in the natural pteridines and are largely responsible for their visible colour. It will be noted from Table 4 that no pteridines have yet been found to have λ_{max}. higher than 392 mμ. (xanthopterin) unless sulphur is present (4-mercaptopteridine, 408 mμ.). With one exception (6-hydroxypteridine), no pteridine has a longwave λ_{max}. below 300 mμ. The anion of 6-hydroxypteridine

Table 4. Ultraviolet Spectrography of Pteridines in Water.

Compound	λ_{max}, (mμ)	log ε_{max}, (mol.)	pH	Reference
Pteridine	< 220; 298 + 309*; (387)**	> 3.83; 3.87 + 3.83 (1.87)	6.2	(7)
cation	< 220; 300	> 3.84; 3.92	2.1	(7)
4-Hydroxypteridine	230; 265; 310	3.98; 3.54; 3.82	5.6	(7)
anion	242; 333	4.23; 3.79	10.0	(7)
6-Hydroxypteridine (+ 1 H_2O)	< 215; 289	> 4.1; 4.00	5.2	(8)
anion	< 215; 258; 289; 356	> 4.25; 3.90; 3.69; 3.60	8.8	(8)
cation	< 215; 287	> 4.3; 4.09	1.7	(8)
7-Hydroxypteridine	227; 248 + 256*; 303	3.79; 3.44 + 3.45; 4.00	4.0	(8)
anion	226; 260; 326	4.27; 3.76; 4.04	9.0	(8)
6:7-Dihydroxypteridine	< 220; 249; 301	> 4.0; 3.71; 4.18	4.0	(8)
mono-anion	227; 268; 319	4.03; 3.71; 4.29	8.4	(8)
di-anion	< 220; 240; 324 + 338*	> 4.47; 4.13; 4.30 + 4.25	12.0	(8)
2-Aminopteridine	225; 260; 370	4.38; 3.87; 3.82	7.1	(7)
cation	< 210; 302	> 4.0; 3.87	2.1	(7)
2-Amino-4-hydroxypteridine	< 220; 270; 340	> 4.0; 4.05; 3.76	5.3	(9)
cation	< 220; 315	> 4.1; 3.88	0	(9)
anion	< 220; 251; 358	> 4.1; 4.31; 3.83	13	(9)
2-Amino-4:6-dihydroxypteridine (+ 1 H_2O) (xanthopterin)***	< 220; 275; 385	> 4.15; 4.12; 3.39	4.9	(9)
poly-anion	225; 255; 392	> 3.9; 4.27; 3.85	13	(9)
cation	< 220; 245; 355	> 4.0; 4.07; 3.82	(70% H_2SO_4)	(9)
2-Amino-4:7-dihydroxypteridine (isoxanthopterin)	< 220; 286; 340	> 4.3; 3.97; 4.11	1-6	(14)
mono-anion	228; 279; 332	4.50; 3.83; 4.12	9.2	(14)
di-anion	221; 247–252; 339	4.5; 3.94; 3.99	12	(14)
poly-anion	223; 254; 339	4.6; 4.06; 4.15†††	13	(13)
2-Amino-4:6:7-trihydroxypteridine (leucopterin)† ††	240; 285; 340	4.20; 3.84; 4.02	13	(9)

	λ max (mμ)	log ε	pH	Reference
2-Amino-4:6-dihydroxy-7-methylpteridine (chrysopterin)	252; 385	4.20; 3.90	11	(49a)***
2-Amino-4:7-dihydroxypteridine-6-acetic acid (ichthyopterin)	225; 255; 340	4.30; 4.02; 4.11	12.7	(181)
Erythropterin	235; 285; 335	3.89; 3.63; 3.51	12.7	(179, 186)
Fluorescyanine	<220; 255; 335	>4.5; 4.08; 4.17	13	(63)
Pteroic acid	255; 275; 365	4.43; 4.39; 3.92	13	(192)
Pteroylglutamic acid (+ 1 H₂O)	256; 283; 368	4.43; 4.40; 3.96	13	(192, 185)
	<220; 295	>4.36; 4.31	1	(192)
	255; 365	4.50; 3.87	11	(209)
Rhizopterin	245–275; 350	4.27; 3.77	7	(209)
	270; 345	4.32; 3.72	3	(209)
	253; 325	4.32; 3.90	1	(209)
2-Amino-4-hydroxy-6-methylpteridine (for comparison with above) anion	<220; 255; 365	?; 4.36; 3.85	13	(113)
neutral mol. and cation (mixed)	<220; 250; 325	>4.08; 4.02; 3.90	1	(113)
Citrovorum factor	<225; 282	?; 4.46	13	(127)

* Twin-peak.
** Resolved in *cyclohexane*, not in water.
*** A number of incorrect figures are in the literature [see Reference (1)].
† See also (69).
†† Poly-anion.
††† No evidence of decomposition could be found.

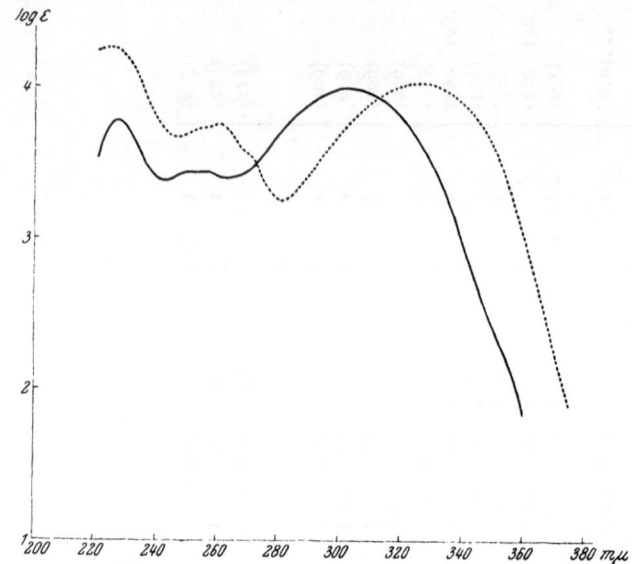

Fig. 2. ——————, Neutral molecule of 7-hydroxypteridine (p_H 4.0) and ......, Anion of 7-hydroxypteridine
(p_H 9.0).

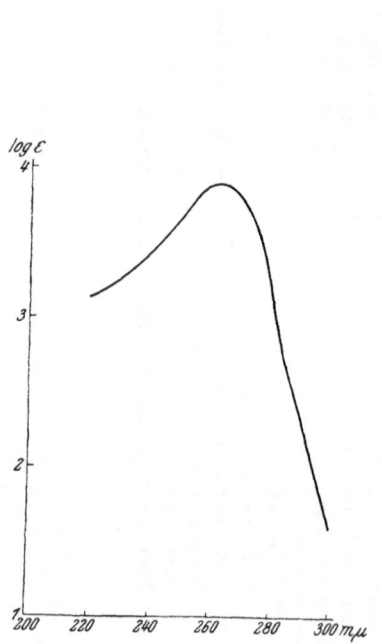

Fig. 3. Neutral molecule of purine (p_H 5.7).

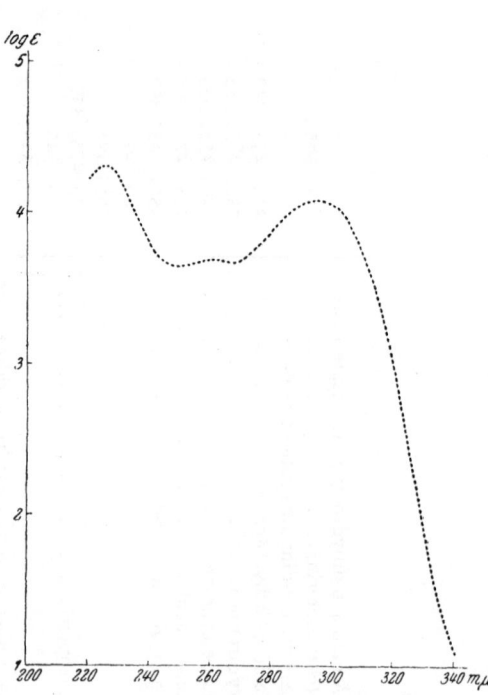

Fig. 4. Anion of 2:6:8-triaminopurine (p_H 13).

is normal, but both the cation and neutral molecule absorb at about 288 mμ. and hence could easily be mistaken for purines (c. f. *iso*guanine, 285 mμ.). Although 6-hydroxypteridine absorbs at shorter wave-lengths than 7-hydroxypteridine, this trend is often reversed in the derivatives, e. g. xanthopterin is bathochromic to *iso*xanthopterin. It is quite common for pteridines to have an extra peak whose λ_{max}. falls somewhere between 245 and 285 mμ., e. g. 7-hydroxy-pteridine (*Fig. 2*).

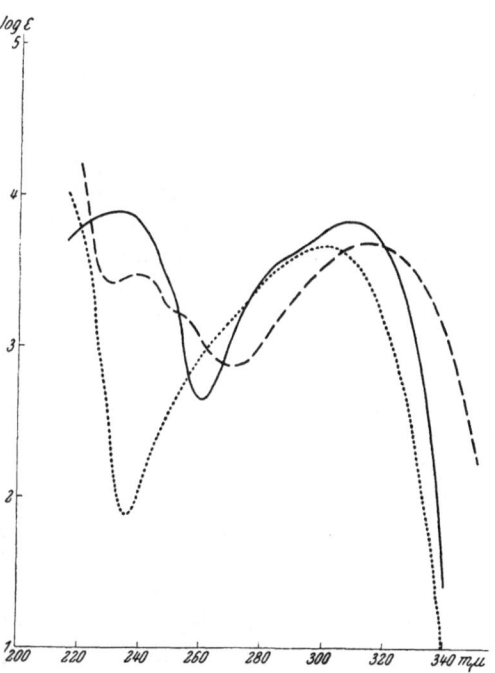

The spectra of purines consist usually of two peaks, e. g. the molecule of purine (*Fig. 3*) (the cation is almost identical). The low-wave peak is rarely displaced to wave-lengths too short to be read reliably on a photo-electric spectrophotometer (cf. 6-hydroxypurine). More often, it is displaced to longer wave-lengths (e. g. 240 mμ. as in xanthine). A few puri-nes have three well-defined peaks (e. g. the anion of 2:6:8-triaminopurine: 226, 261 and 295 mμ.; *Fig. 4*). The longest λ_{max}. values in the purine series occur in the 2-monosubstituted derivati-

Fig. 5. Neutral molecules of: ———, 2-hydroxypteridine (p_H 7.1), ..., 2-hydroxypyrimidine (p_H 6.2); and – – – – –, 2-hydroxypurine (p_H 6.1).

ves, e. g. 2-aminopurine (325 mμ.; 2-mercaptopurine, 347 mμ.). Sub-stances have been assigned a pteridine structure on the (mistaken) grounds that "Purines do not have absorption maxima above 300 mμ." (*165*). Actually, these monosubstituted purines could easily be mistaken for pteridines, not only on account of their λ_{max}. values (which often exceed those of the corresponding pteridines, see *Fig. 5*), but because they far outshine the corresponding pteridines in fluorescence.

These data are emphasized here because of the likelihood that 2-monosubstituted purines will turn up in nature (purine riboside and many 2:6-disubstituted purines have been isolated). It may be relevant to mention here the ready conversion of 2-monosubstituted purines to pteridines, by ring-enlargement under physiological conditions (*3*).

When a further substituent is introduced into these 2-substituted purines, the λ_{max}. recedes to shorter wave-lengths.

In so far as every natural pteridine of known constitution is a 2-amino-4-hydroxypteridine, it is important to note how little the spectra of 2-amino-4-hydroxypteridine owe to 2-aminopteridine and 4-hydroxypteridine (Table 4). Obviously there is a complex electronic interaction between these two groups.

It should be pointed out here that a natural and a synthetic pteridine cannot be declared identical solely on the basis of identical ultraviolet spectra. Often quite large changes can be effected in a pteridine molecule with little or no effect on the spectrum. For example, 4:7-dihydroxy-pteridine has spectra differing little from its 6-methyl-, 6-hydroxy-, 6-carboxy-, 6-carbethoxymethyl- and 6-carbethoxymethyl-5:6-dihydro-derivatives. Similar resemblances among relatives of *iso*xanthopterin have been recorded by ELION and HITCHINGS (*49*).

For spectra of simple dihydro- and tetrahydro-pteridines see (*8, 105*).

5. Chemical Reactions.

Because the syntheses, degradations and transformations of the simpler pteridines were recently reviewed (*1*), mention will be made here of recent developments only.

No new methods of synthesis have come to light. The removal of 6-COOH groups by aluminium amalgam (*114*) offers an agreeable alternative to decarboxylation by heating, which has often caused difficulty (*173, 49*). The following groups can be removed similarly: —CH$_2$NH$_2$, —CH(OH)CH$_3$, and —CH$_2 \cdot$ CO $\cdot$ CH$_3$ (*114*).

It is now realized that mercapto-pteridines alkylate directly on the sulphur atom, but hydroxy-pteridines usually on a nitrogen atom (occasionally diazomethane produces an O-methylpteridine) (*10*). As was pointed out in Section 2 (p. 355), these N-alkylpteridones are more readily degraded than their parent hydroxypteridines.

In a series of brilliant experiments, TAYLOR and his colleagues (*171*) have shown that hydroxy-, amino- and mercapto-groups in the 4-position are interchangeable, the ring usually opening (to well-characterized derivatives of pyrazine) and closing again. Pteridine itself, when warmed with dilute acid is at once decomposed to 2-aminopyrazine-3-aldehyde (*10*).

The further exploration of dihydro- and tetrahydro-pteridines will be discussed under leucovorin (p. 383).

III. The Naturally-occurring Pteridines (excluding the Folic Acid Series).

At the time of the last review in this Series [PURRMANN (*138*)], the constitution of only three natural pteridines was known: xanthopterin, *iso*xanthopterin and leucopterin. Since then a number of other constitutions have been worked out, among them several of substances with relatively low molecular-weight (the more complex folic acid series is dealt with in Chapter IV, p. 372).

1. Glossary of Synonyms.

So many names have been used for the natural pteridines that it is desirable to present a glossary (*Table 5*).

Table 5. Synonyms.

Name	Structure	Reference
Anhydroleucopterin*	*iso*Xanthopterin	(*136*)
Chrysopterin	7-Methylxanthopterin	(*180*)
Desiminoleucopterin*	2:4:6:7-Tetrahydroxypteridine	(*202*)
Desiminoxanthopterin*	2:4:6-Trihydroxypteridine	(*201*)
6-Desoxyleucopterin*	2-Amino-6:7-dihydroxypteridine	(*205*)
8-Desoxyleucopterin*	*iso*Xanthopterin	(*136*)
β-Dihydroxanthopterin	2:4-Diamino-6-hydroxy-*p*-oxazinopyr-imidine	(*49*)
Erythropterin	7-Trihydroxypropenyl-xanthopterin (XXII)	(*179*)
Fluorescyanine	A mixture	(*130*), see also p. 371
Guanopterin*	*iso*Guanine (a purine)	(*134*)
Ichthyopterin	(see p. 370)	(*181*)
Lepidoporphyrin*	Pterorhodin (see below)	
Leucopterin	2-Amino-4:6:7-trihydroxypteridine (XXV)	(*134, 12*)
*iso*Leucopterin*	Formerly used for a fraction of crude leucopterin, later used for 4-amino-2:6:7-trihydroxypteridine, eventually abandoned	(*202, 201*)
Lumazine	2:4-Dihydroxypteridine	(*102*)
Mesopterin*	*iso*Xanthopterin	(*180*)
Pteridine reds	Analogues of pterorhodin	(*86, 88, 89, 91*)
Pteridoxamine	2-Amino-4-hydroxypteridine	(*142*)
Pterorhodin	An artefact (XXVI) synthesized in isolating natural pteridines	(*149*)
Rhodopterin*	Pterorhodin	(*68*)
Uropterin*	Xanthopterin	(*98*)
Urothion	(see p. 372)	(*99, 100*)
Xanthopterin	2-Amino-4:6-dihydroxypteridine (XX)	(*135, 98, 12*)
*iso*Xanthopterin	2-Amino-4:7-dihydroxypteridine	(*136, 14*)

* Name no longer in use.

2. The Xanthopterin Family:
Xanthopterin, Chrysopterin, Erythropterin.

Xanthopterin (XX) has become a more accessible substance since Totter discovered that sodium amalgam reduces leucopterin, an easily synthesized compound, to dihydroxanthopterin, the latter being readily oxidized to xanthopterin (*175*). This process has been developed for production (*12*). The action of diacetyloxyacetic acid on 4-hydroxy-triaminopyrimidine provides a new synthesis of xanthopterin (*97*): it requires specially purified dichloroacetic acid as a starting material.

Any lingering doubts as to the constitution of xanthopterin have been dispelled by BOON and LEIGH's unambiguous synthesis (*27*); at the same time, dihydroxanthopterin was shown to be the 7:8-dihydro-derivative. Catalytic hydrogenation of xanthopterin in acetic acid (or 0.1 N-sodium hydroxide) over palladium or platinum catalysts also gives this substance, but leucopterin and *iso*xanthopterin were found not to take up hydrogen under these conditions (*118*). SCHOU discovered the slow tautomeric change of xanthopterin in acid solution (*157*).

SCHOU's "keto-form" (to which he assigns formula XXI) is non-fluorescent and not attacked by xanthine oxidase, whereas the "enol-form" (XX) is highly fluorescent and oxidized to leucopterin by this enzyme. Heat favours the more rapid attainment of equilibrium, and the enolic form is strongly favoured in hot solutions. This tautomerism, the precise chemical nature of which is still unknown (*3a*), has been demonstrated in 6-hydroxypteridine and is peculiar to the hydroxy-group in this position alone (*8*). Moreover, it does not occur in 6:7-dihydroxypteridines. The study of the action of xanthine oxidase on xanthopterin, initiated by WIELAND and LIEBIG (*201*) has been carried further and used for the assay of xanthopterin in biological fluids (*85*).

A number of physical properties of xanthopterin are given in Tables 1–4 (pp. 354, 355, 362, 367). The two tautomers are seen as separate spots in paper chromatography (*182*).

The „saures Xanthopterin" of SCHÖPF and BECKER (*155*) appears to be identical with *xanthopterin* (*182*).

KALCKAR and his colleagues (*84*, *51*) have shown that xanthopterin is not present as such in fresh human urine, but as a precursor, pro-xanthopterin, which can be released by oxidation (catalysed by light or charcoal). Proxanthopterin is stable to acid. It is extracted from urine with butanol, and can be removed from the latter by aqueous buffers. In paper chromatography (butanol/acetic acid) the spot is much nearer the origin than that of xanthopterin and, unlike the latter, it does not fluoresce. The biological functions of xanthopterin are discussed in Chapter VII, p. 389.

Xanthopterin has been isolated from the skin of the crab *Cancer pagurus* (*132*), in the ascidian *Microcosmus polymorphus* (*87*) and in the Pentatomidae which are plant-devouring hemipterae (*119*). It does not appear to occur in plants (*98*). Examples of the occurrence in nature of substances which are possibly identical with xanthopterin will be found in Chapter V, p. 385.

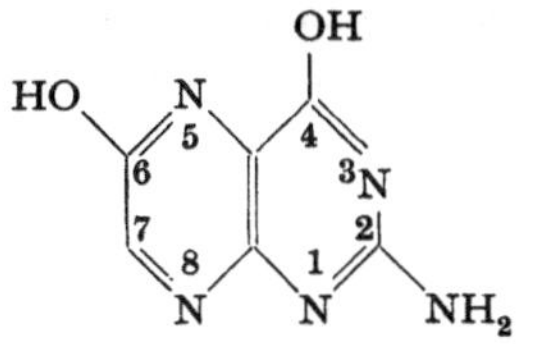

(XX.) Xanthopterin (enol form).

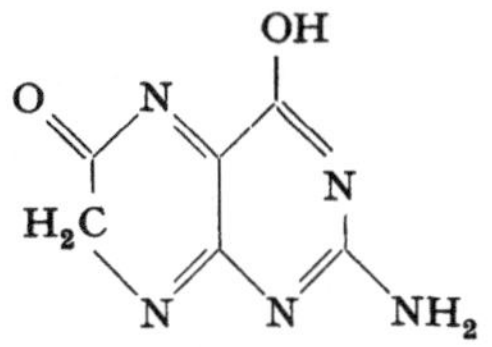

(XXI.) Xanthopterin (SCHOU's keto form).

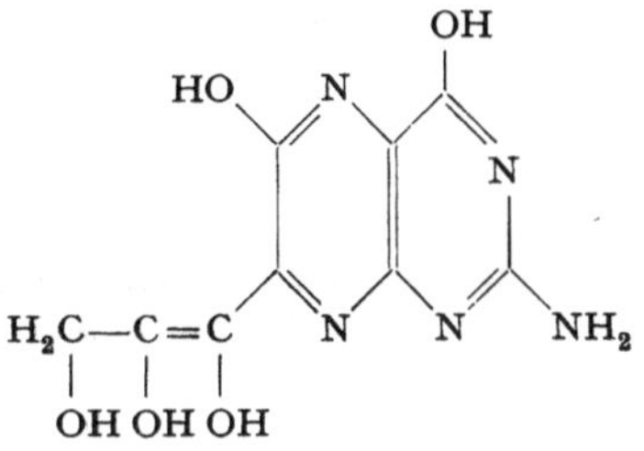

(XXII.) Erythropterin.

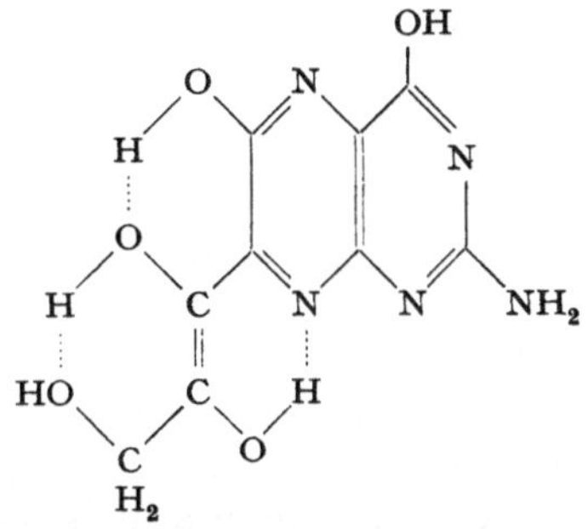

(XXIII.) Erythropterin (hydrogen-bonded formula).

Chrysopterin, a yellow pigment isolated by SCHÖPF and BECKER in 1933 from the citron-yellow butterfly *Gonepteryx rhamni*, has been shown to be 7-methylxanthopterin (*180*). It is a common contaminant of xanthopterin isolated from natural sources, and may be separated from xanthopterin by the fractional crystallization of the barium salts. Paper chromatography in aqueous ammonium chloride (3%) differentiated clearly between the components of the mixture and permitted identification with synthetic specimens. Chrysopterin was further identified by oxidation to 2-amino-4:6-dihydroxypteridine-7-aldehyde. See further under pterorhodin (p. 371).

Erythropterin, the red pigment isolated in 1936 from orange and red pierid butterflies (*155*), was assigned the structure (XXII) by PURRMANN (*139*). This was confirmed through synthesis (*179*): 4-hydroxy-triaminopyrimidine was condensed with ethyl acetoneoxalate, and the resultant 2-amino-4:6-dihydroxy-7-acetonylpteridine was treated in turn with acetic anhydride, bromine and potassium acetate, giving 2-amino-4:6-dihydroxy-7-(α-hydroxyacetonyl)-pteridine which was oxidized (in about 1% yield) with selenium dioxide, or bromosuccinimide, to erythropterin. Because of its high chemical stability, unexpected in an enediol, erythropterin has been assigned the triply hydrogen-bonded formula (XXIII) by KHORANA (cf. *182*).

Erythropterin has been isolated from *Mycobacterium lacticola* and identified through its ultraviolet spectrum, partition coefficient, R_F value, and ability to give pterorhodin when heated with xanthopterin in acid solution (*186*). It is present in the bacteria, apparently, as a glucoside and two or three other substances are present which seem to be pteridines. Erythropterin has also been demonstrated in human tubercle bacilli (*46*).

The examination of butterfly-wings for the occurrence of erythropterin, xanthopterin and leucopterin has been simplified (*55*).

3. The *iso*Xanthopterin Family:
*iso*Xanthopterin, Ichthyopterin, Fluorescyanine, Leucopterin.

*iso*Xanthopterin (IV, p. 351) can now be readily prepared in quantity (*14*). Spectrographically, this synthetic material has been shown to be considerably purer than any natural material whose spectra have been reported (*13, 14*).

The *"mesopterin"* isolated by Schöpf and Becker (*155*) from *Gonepteryx rhamni* is considered by Tschesche and Korte (*180*) to be probably only impure *iso*-xanthopterin.

Ichthyopterin is a blue-fluorescing, white substance isolated by Hüttel and Sprengling in 1943 from the skin of cyprinid fish in Lake Constance (*69*). The cyprinids, of which the carp is a familiar example, are the commonest fresh water fish of Europe, Asia and North America. Tschesche and Korte (*181*) have found that Hüttel's preparation was not a single substance, and that the principal component is 2-amino-4:7-dihydroxypteridyl-6-acetic acid (XXIV). They also prepared (XXIV) synthetically by the condensation of triamino-4-hydroxypyrimidine with ethyl oxaloacetate, but did not obtain it quite free from its decarboxylation product. Ichthyopterin and the slightly impure (XXIV) were compared for R_F, ultraviolet absorption, and the dependence of fluorescence on p_H. They were both decarboxylated in boiling N-sodium hydroxide to 6-methyl*iso*xanthopterin (identified by R_F).

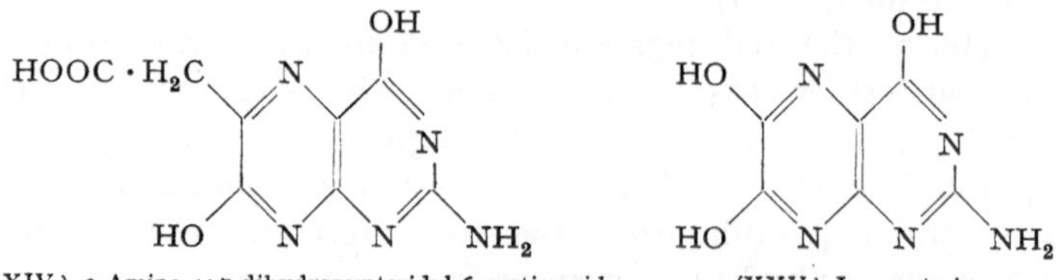

(XXIV.) 2-Amino-4:7-dihydroxypteridyl-6-acetic acid.   (XXV.) Leucopterin.

Fluorescyanine is a blue-fluorescing white substance isolated from melanocytes of the scales of carp *(Cyprinus carpio)* by Polonovski, Busnel, and Pesson in 1943 and from the eggs, wings, and Malpighian tubules of the silkworm *Bombyx mori* by Polonovski and Busnel (*37, 128*) in 1948. This pigment was also found in the larvae of *Bombyx mori* by a group of Japanese workers, who showed that fluorescyanine is not identical with ichthyopterin (*64*) as the French workers had supposed (*37*). In the carp, fluorescyanine is combined with protein (*130*), and in the silkworm with a polypeptide (*128*). It has also been identified in the retina, liver, and kidney of fish, in reptiles, amphibians, crustaceae, and insects (*130*).

Fluorescyanine is soluble in methanol and fluoresces ($\lambda_{max.} = 431$ mμ.) at all pH values (*130*).

Fluorescyanine is not a carboxylic acid but it gives *iso*xanthopterin-6-carboxylic acid on oxidation with potassium permanganate (*64*). It is reduced to a non-fluorescing derivative by sodium dithionite (hydrosulphite) and regenerated by shaking with air (*131*). Professor R. TSCHESCHE and Dr. F. KORTE of Hamburg have kindly informed me that they believe fluorescyanine is a mixture containing a high proportion of *iso*xanthopterin.

Leucopterin (XXV) has so far been found only in insects. Methods for producing it in quantity have been published (*12*, *137*).

4. Substances closely related to the naturally-occurring pteridines: Pterorhodin, Urothion.

Pterorhodin is an artefact formed in the isolation of 7-methylxanthopterin (chrysopterin) under conditions where it can oxidatively condense with xanthopterin present. The constitution of pterorhodin was established as (XXVI) by PURRMANN and MAAS (*140*) in 1944. Later it was synthesized (*149*) from 7-methylxanthopterin and xanthopterin, or from xanthopterin alone in the presence of a "carbon donor" such as acetone, acetaldehyde or acetic anhydride. Similarly xanthopterin when oxidized by sunlight, or by hot 50% sulphuric acid, is decomposed sufficiently to give a "carbon donor", and thence pterorhodin (*182*). Erythropterin and xanthopterin, when heated together in aerated 0.25 N-hydrochloric acid also give pterorhodin (*182*).

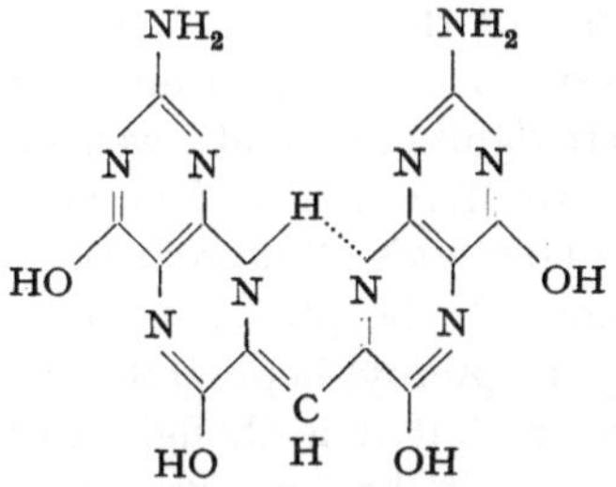

(XXVI.) Pterorhodin.

A number of entirely unnatural analogues of pterorhodin have been prepared from other pairs of derivatives of 2-amino-4-hydroxypteridine, one of each pair containing a methyl-group in either the 6- or 7-position (*149, 86, 88, 89, 91*).

Urothion, $C_{11}H_{13}O_3N_5S_2$, is an optically-active yellow pigment isolated from human urine by Koschara (*99, 100*) in 1943. Koschara observed that it contained a terminal glycol group.

At the Pteridine Conference arranged in Paris 1952 by Professor Polonovski, Tschesche and Korte disclosed that urothion is not strictly a pteridine because the pyrazine ring appears to be fused to a thiazoline ring which carries the glycol group. Professor Tschesche has kindly informed this Reviewer that he hopes shortly to confirm the constitution of this substance by synthesis. Riboflavine is another example of fusion of the pteridine nucleus to another ring.

IV. The Folic Acid Series.

About 1940, a number of related substances were isolated from liver, yeast and green leaves and were found to have anti-anaemic properties and, in many cases, to be growth factors for certain micro-organisms. Thus, Peterson's "norite eluate factor", isolated from liver and yeast, was essential for the growth of *Lactobacillus casei* and of chicks (*163, 71*), and Mitchell's "folic acid", isolated from spinach at this time, promoted the growth of *L. casei, Streptococcus faecalis R* and other bacteria (*111, 112*). Pfiffner's group later described the isolation of "vitamin Bc", a crystalline antianaemic substance from liver (*122*). It was then shown that a substance, obtained from a *Corynebacterium* fermentation residue, was 90% as active as the "norite eluate factor" for *L. casei*, but only 6% as active for *S. faecalis R* (*74, 26, 18*). It had been shown in 1938 that anaemia in monkeys could be cured by "vitamin M" from yeast (*47*). Besides "vitamin Bc", the anaemia of chickens could be cured by a crystalline yeast factor known as "vitamin Bc-conjugate" because it yielded vitamine Bc on enzymic digestion (this conjugate had very little activity for the above bacteria) (*125, 122*).

In 1946, the structure of the "Liver *L. casei* factor" was found by degradation and synthesis to be the pteridine derivative (XXVII), and for this the name "pteroylglutamic acid" was proposed (*18*). This was shown to be identical with Mitchell's "folic acid", Peterson's "norite eluate factor", and Pfiffner's "vitamin Bc". The "Fermentation *L. casei* factor" was shown to be pteroyl-triglutamic acid, whereas the "vitamin Bc-conjugate" is pteroyl-heptaglutamic acid. The nature of "vitamin M" is not known, but it is thought to be a similar conjugate of (XXVII) in so far as it has no growth-promoting activity for *S. faecalis R* until incubated with rat liver (*176*), which contains "Bc-conjugase", an enzyme capable of splitting off γ-glutamic residues in excess of one.

Pteroic acid (XXVIII) has not been found free in nature, but its 10-formyl-derivative was isolated in 1947 and called "rhizopterin". It is a growth-stimulant for *S. faecalis R* but not for *L. casei* (*209*). The existence of such a substance had been suspected in 1943 (*94*).

In 1948, SAUBERLICH and BAUMANN (*153*) found that the bacterium *Leuconostoc citrovorum* would not grow in a synthetic medium unless liver-extract was added. The growth-factor proved to be none of the above pteridines, but the liver could be replaced by a crude substance obtained by hydrogenating 10-formylpteroylglutamic acid over platinum and autoclaving the product (*160*). It was later shown that the auto-claving caused the formyl-group to migrate from the 10- to the 5-position, giving 5-formyl-5:6:7:8-tetrahydropteroylglutamic acid (XLV) (*127, 15*). This "citrovorum factor" has been synthesized. The synthetic material, "leucovorin", is a mixture of two stereoisomers, and has a lower biological activity, until resolved (*44*). The citrovorum factor is a growth-promoter for *S. faecalis* and *L. casei* also, with approximately the same activity as pteroylglutamic acid.

The chemistry of these substances is described below, in Sections 1–5. The term "folic acid" is now used much more widely than was intended by MITCHELL, and many authors use it to cover any or all of the substances described in these five Sections. For example, it is commonly stated that the folic acid of kidney, liver and yeast is a mixture of pteridines in which a substance with the growth-promoting effect of the citrovorum factor predominates (*195*). As it seems rather late in the day to try to confine the use of the term "folic acid" to pteroyl-glutamic acid, it is to be hoped that "folic acid" will be dropped in favour of the more precise chemical names.

1. The Synthesis of Pteroylglutamic Acid ("PGA").

Pteroylglutamic acid (XXVII) is systematically described as *p*-(2-amino-4-hydroxy-6-pteridyl)-methylaminobenzoyl-*L*-glutamic acid. It was first synthesized by WALLER and his colleagues (*192*) by adding 1:2-dibromopropionaldehyde to a solution of triamino-4-hydroxy-pyrimidine and *p*-aminobenzoyl-*L*-glutamic acid in cold pH 4 aqueous buffer. The product was shown by bioassay to contain 20% of PGA. (The crude product was purified by dissolving it in alkali, treating the solution with charcoal, filtering and re-precipitating at pH 3: three such treatments gave material of 92% purity.) Equimolecular quantities of the three reagents were used, but it has been claimed to be advantageous to add a further quantity of the pyrimidine during the reaction (*22*).

Variations in the order of mixing led to failure. Attempts to combine the aldehyde and the acid (with loss of hydrogen bromide) led to the production of a useless anil. An attempt to obtain a 6-bromomethylpteridine, by condensing the aldehyde with the pyrimidine led to substances of no value for the synthesis (*192*).

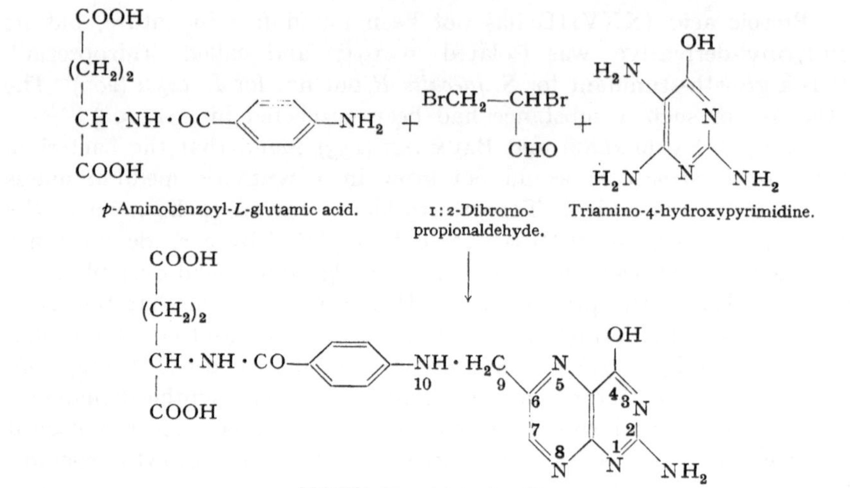

$$\begin{array}{c} COOH \\ | \\ (CH_2)_2 \\ | \\ CH \cdot NH \cdot OC - \langle \rangle - NH_2 \\ | \\ COOH \end{array} + \begin{array}{c} BrCH_2 - CHBr \\ | \\ CHO \end{array} + \begin{array}{c} OH \\ H_2N \\ N \\ H_2N \quad N \quad NH_2 \end{array}$$

p-Aminobenzoyl-L-glutamic acid.　　1:2-Dibromo-　Triamino-4-hydroxypyrimidine.
propionaldehyde.

$$\begin{array}{c} COOH \\ | \\ (CH_2)_2 \\ | \\ CH \cdot NH \cdot CO - \langle \rangle - NH \cdot H_2C \\ | \quad\quad\quad\quad 10 \quad\quad 9 \\ COOH \end{array}$$

(XXVII.) Pteroylglutamic acid.

It will have been noted in the above synthesis that two hydrogen atoms are lost, possibly by auto-oxidation. It was soon realized that the reaction went more smoothly, and in better yield, if 0.5 equivalent of iodine or sodium dichromate were added as an oxidizing agent during the condensation (*16, 158, 148*). Instead of dibromopropionaldehyde in the WALLER synthesis, it is apparently highly satisfactory to use 1:1-dichloro-3-bromoacetone (*96*) or 1:1:3-tribromoacetone (*158, 199*). With these reagents, loss of the elements of hydrogen halide makes it unnecessary to supply an oxidizing agent.

An improved method of purification was devised in 1949: a solution of the crude product in aqueous sodium hydroxide was precipitated first with calcium chloride and then with zinc chloride (precipitates discarded), and the solution brought to pH 3 (*158, 148, 22*).

This WALLER synthesis remained for several years the most favoured method of the pteridine chemists at the American Cyanamid Company, and by means of it they prepared a number of (unnatural) analogues with other amino-acids replacing the glutamic acid, or with altered substituents in the pteridine nucleus, or with substituents in the —NH·CH₂— linkage. Some of these substances have proved to be powerful biological antagonists of PGA. The most notable of them are "Aminopterin", in which the 4-hydroxy-group is replaced by an amino-group (*158*); "Aminoanfol", a similar amino-analogue of pteroyl-aspartic acid (*72*), and "Amethopterin", the 10-methyl-derivative of aminopterin (*158*).

In a *second synthetic method* from the same laboratories, dibromo-propionaldehyde was condensed with pyridine to give N-(2-formyl-2-bromoethyl)-pyridinium bromide (XXVIII), which was combined with triamino-4-hydroxypyrimidine to give a quaternary salt (XXIX). The latter reacted with *p*-aminobenzoyl-*L*-glutamic acid to give PGA (*70*). Here again two hydrogen atoms have been lost during the condensation. This method has been little used.

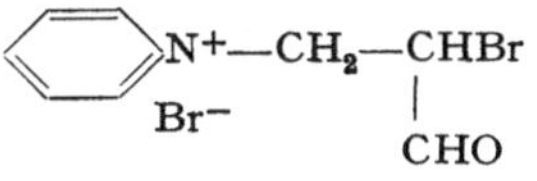

(XXVIII.)
N-(2-Formyl-2-bromoethyl)-pyridinium bromide.

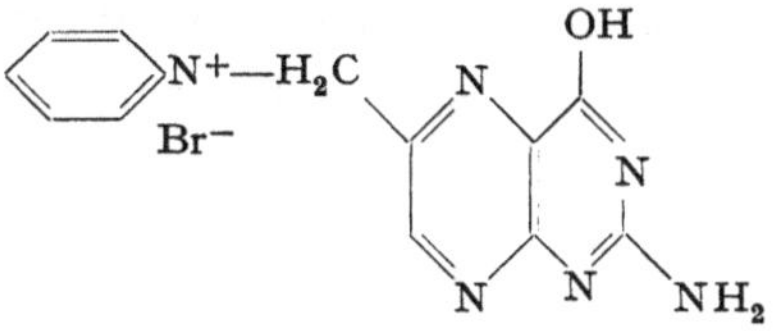

(XXIX.)

A *third synthesis* from the same source, used reductone (XXX) which was condensed with *p*-aminobenzoylglutamic acid to give the anil (XXXI) which reacted with triamino-4-hydroxypyrimidine to give PGA (*19*). In this method, all hydrogen atoms are accounted for. The scarcity of reductone has contributed to the disuse of this method.

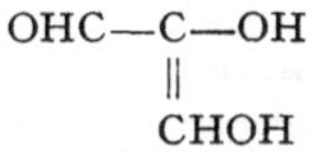

(XXX.) Reductone.

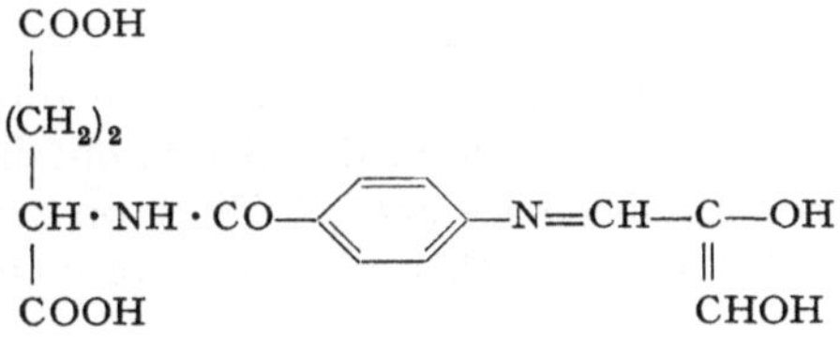

(XXXI.)

In a *fourth synthesis*, the Cyanamid workers discovered a reaction which is probably the most flexible and useful of all for preparing PGA and its analogues. In this reaction, 6-bromomethyl-2-amino-4-hydroxy-pteridine (XXXII) is condensed with *p*-aminobenzoylglutamic acid in glycol at 100° (*29*, *178*, *177*). Although the yields were not good at first, it was found that this, the BOOTHE synthesis, could also be carried out in aqueous alkali (*190*) or in cold formic acid (*146*). The value of this synthesis lies in its flexibility. The majority of syntheses devised for PGA are inapplicable for use with pyrimidines having only three substituents (presumably, because these syntheses operate only on amino-groups having more than a minimal electron-density). However, the BOOTHE synthesis avoids this situation, and hence permits the preparation of PGA analogues having only two substituents in the pteridine ring, e. g. 2-deaminopteroyl-*L*-glutamic acid (*33*). It has also permitted the synthesis of an isomer of PGA in which the side-chain is in the 7- (instead of the 6-) position in the pteridine ring (*190*). The isomer differs greatly from PGA in being very unstable both in the solid state, and in alkaline solution.

The requisite 6-bromomethylpteridines are obtained by brominating the 6-methylpteridines (29, 190, 1, 33). At first the 6-methylpteridines were prepared indirectly, by the reduction of the quaternary pyridinium salts (XXIX) (29). But is was later found possible to prepare them by the action of methylglyoxal (XXXIV) on 4:5-diaminopyrimidines in the presence of sodium bisulphite (158, 9). In the absence of sodium bisulphite (or other aldehyde-binding agent), the 7-methyl-isomer is formed.

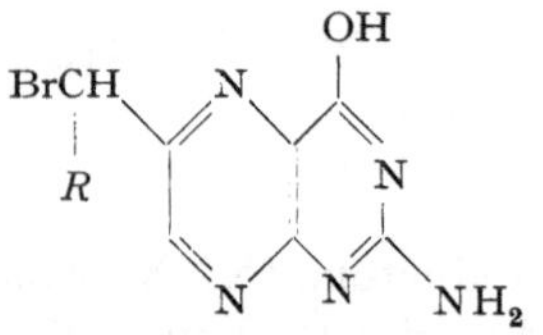

(XXXII.)	(XXXIV.)	(XXXV.)
R = H. 6-Bromomethyl-2-amino-4-hydroxypteridine.	Methylglyoxal.	Methyl 4:4-dimethoxy-acetoacetate.

(XXXIII.)
R = COOH. 2-Amino-4-hydroxypteridyl-6-bromoacetic acid.

A similar synthesis uses 2-amino-4-hydroxypteridyl-6-bromoacetic acid (XXXIII) (185). The latter is readily obtained by brominating the corresponding acetic acid which is obtained by the action of methyl 4:4-dimethoxy-acetoacetate (XXXV) on triamino-4-hydroxypyrimidine (113). PGA, made in this way, is advantageously purified as the silver salt (185).

The only other synthesis claimed to be suitable for trisubstituted pyrimidines is, unfortunately, described only in a patent, wherein α-bromo-acrolein is condensed with p-aminobenzoyl-L-glutamic acid, the product being combined with 2:4:5-triaminopyrimidine (207).

In *another synthesis* of PGA, triamino-4-hydroxypyrimidine is condensed with dihydroxyacetone. In the presence of hydrazine this gives 2-amino-4-hydroxy-6-hydroxymethyl-pyrimidine, which is readily converted to the 6-chloromethyl-analogue which is condensed as (XXXII), above (146). This synthesis with dihydroxyacetone has been highly praised (150). If it is carried out in the absence of hydrazine, there is a strong tendency to split off water instead of hydrogen, resulting in 2-amino-4-hydroxy-6- (and 7-) methylpteridine (90, 200). Replacement of dihydroxyacetone by glyceraldehyde has not given good yields. The 3-carbon keto-alcohol may be replaced by a sugar-derivative, e. g. p-tolyl*iso*glucosamine. The product is oxidized by periodic acid to 2-amino-4-hydroxypteridine-6-aldehyde (XXXVI), which is converted to PGA by hydrogenation in the presence of p-aminobenzoyl-L-glutamic acid (200). This aldehyde can be less wastefully made by the further bromination and hydrolysis of (XXXII) (191).

In a *new synthesis* of PGA, N-toluenesulphonyl-*p*-aminobenzoyl-glutamic acid is alkylated with a substituted propylene oxide molecule such as glycidol acetate which gives (XXXVII). The acetyl group is removed by hydrolysis and the product condensed with triamino-4-hydroxypyrimidine to give the toluenesulphonyl derivative of PGA (*194*).

Pteroylglutamic acid has been produced with ^{14}C in the 2-position by converting radioactive barium carbonate (through guanidine) to tri-amino-4-hydroxypyrimidine. The latter was condensed with $1:1:3$-tribromacetone in the presence of *p*-aminobenzoyl-*L*-glutamic acid (*197*). PGA has also been labelled with ^{14}C in the 9-position by the use of dibromopropionaldehyde [3-^{14}C] in the usual WALLER condensation (*198*).

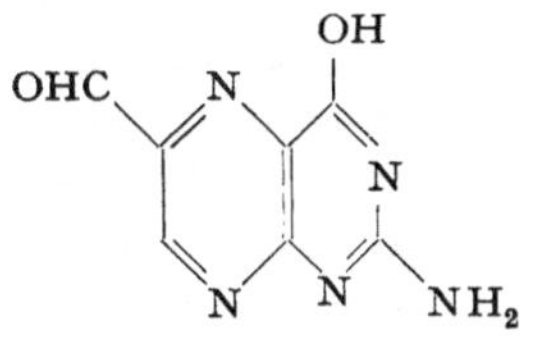

(XXXVI.) 2-Amino-4-hydroxypteridine-6-aldehyde.

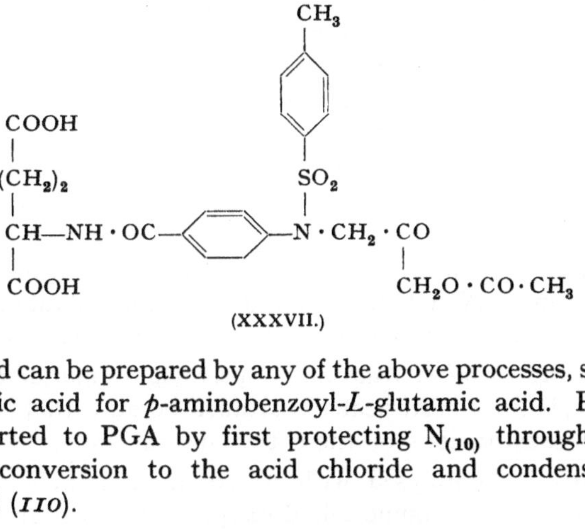

(XXXVII.)

Pteroic acid can be prepared by any of the above processes, substituting *p*-aminobenzoic acid for *p*-aminobenzoyl-*L*-glutamic acid. Pteroic acid can be converted to PGA by first protecting $N_{(10)}$ through acylation, followed by conversion to the acid chloride and condensation with glutamic acid (*110*).

2. The Occurrence and Properties of Pteroylglutamic Acid.

Much of the PGA isolated from natural sources, such as yeast and liver, is thought to be present in the natural state as citrovorum factor (*109*). Pteroylglutamic acid is present (about 5 parts per million) in the dried pollen of various species of plants (*196*).

Pteroylglutamic acid ($C_{19}H_{19}O_6N_7$, anhydrous, M. W. 441.4) is usually seen as an orange-yellow, odourless, microcrystalline powder (well-formed crystals have been obtained) (*166*, *123*). It is soluble in 5000 parts of boiling water (see Table 1, p. 354 for solubilities in cold water) and almost insoluble in alcohol and other organic solvents. It usually contains two molecules of water of crystallization, one of which is loosely bound and the other lost on drying at 75° (below 5 mm. Hg). The specific rotation of a 0.5 per cent. w/v solution in 0.1 N-sodium hydroxide is + 20°. The crystal dimensions have been measured (*123*).

In aqueous solution, it is most stable above p_H 6. The sodium salt is more soluble than 1 in 70 at 20°, and may be sterilized by autoclaving at 15 lb. pressure (such solutions are commonly injected into patients). Below p_H 7, nicotinamide is used to increase solubility (*170*).

PGA is very sensitive towards sunlight, which decomposes it to 2-amino-4-hydroxypteridine-6-aldehyde (XXXVI). This aldehyde then becomes auto-oxidized to the corresponding acid. Finally, photolytic decarboxylation to 2-amino-4-hydroxypteridine occurs (*106, 142*). These three breakdown-products fluoresce, whereas PGA does not (*123, 93*). Fermentative splitting takes a different course: hog-liver homogenate at 38° and p_H 7.3 in the presence of oxygen, gives a blue-fluorescing substance, with at least some of the properties of *iso*xanthopterin (*143*). The decomposition of PGA by light is greatly intensified by riboflavine. As little as 50 μg. of riboflavine per 100 ccm. of solution can accelerate the destruction of 10 mg. of PGA (*154*). Feeding excess PGA to patients produces only a small increase in the bound, and none in the free, urinary xanthopterin (*141, 51*).

The ionization constants have been discussed in Chapter II (p. 356) and in Table 3 (p. 359); the ultraviolet spectra are given in Table 4 (p. 362). For infra-red spectrum see (*192*).

O'DELL and his colleagues (*118*) showed that hydrogenation of PGA, over platinum, in 0.1 N-sodium hydroxide gives the 7:8-dihydro-analogue, whereas in glacial acetic acid 5:6:7:8-tetrahydro-PGA is formed (*15*).

The following *chemical reactions* of pteroylglutamic acid (also given by pteroic acid) have historical importance because they were used to establish the constitutions of these two substances.

When PGA is hydrolysed with N-sodium hydroxide at 100° in a current of oxygen, 2-amino-4-hydroxypteridine-6-carboxylic acid is produced (*192*, cf. *208*). This acid was shown chromatographically to be different from the isomeric 7-carboxylic acid, and both acids were hydrolysed to aminomethylpyrazine carboxylic acids (e. g. XL) of known constitution (*193*). If any doubt remained about the 6- (or 7-) orientation of these substances, it was dispelled when BOON and LEIGH (*27*) unambiguously synthesized 2-amino-4-hydroxy-6-methyl-pteridine [which can easily be oxidized to the acid (*38*)] by the condensation of aminoacetone with 2-amino-6-chloro-4-hydroxy-5-phenyl-azopyrimidine (XXXVIII) to give 2-amino-4-hydroxy-5-phenylazo-6-

pyrimidyl acetone. This was readily reduced and cyclized to 2-amino-4-hydroxy-6-methyl-7:8-dihydropteridine (XXXIX), which was readily dehydrogenated.

N-Benzyl-*p*-aminobenzoic acid, a simple analogue of PGA, is similarly cleaved by aerobic (but not by anaerobic) alkaline hydrolysis (*113*). PGA is oxidized by chlorine to 2-amino-4-hydroxypteridine-6-carboxylic acid (*208*).

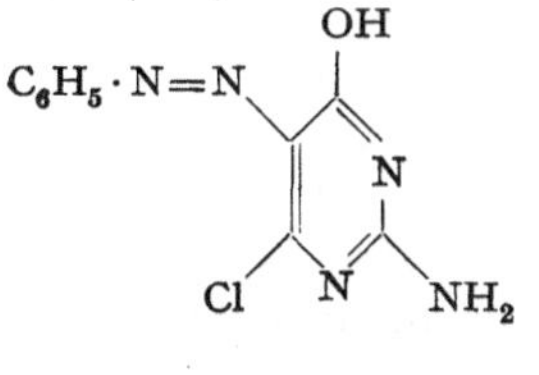

(XXXVIII.)	(XXXIX.)	(XL.)
2-Amino-6-chloro-4-hydroxy-5-phenylazopyrimidine.	2-Amino-4-hydroxy-6-methyl-7:8-dihydropteridine.	

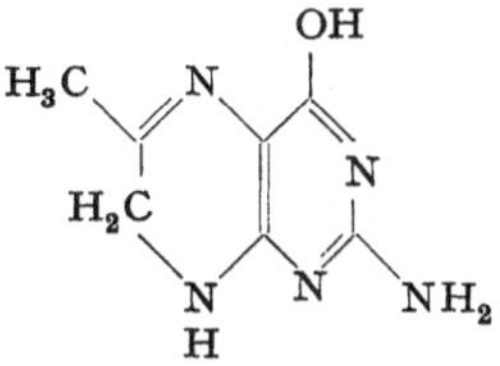

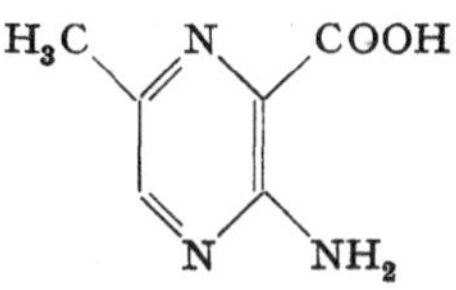

Acid (anaerobic) hydrolysis of pteroyltriglutamic acid by N-sulphuric acid at 100° gives the above-mentioned 2-amino-4-hydroxy-6-methyl-pteridine. Sodium bisulphite splits PGA to a dihydro-derivative of 2-amino-4-hydroxypteridine-6-aldehyde, and this is easily dehydrogenated to (XXXVI) (*191*).

The dimethyl- and diethyl-*esters* of PGA are less soluble in water than PGA, but more soluble in aliphatic alcohols. The molecular weight of the dimethyl-ester was determined by freezing-point depression, in phenol (*123*). The diamide was made from the methyl ester (*123*). These derivatives do not melt on heating. Nitration and halogenation of PGA lead to mono- and di-substitution of the benzene ring (*43*).

Pteroylglutamic acid is commonly assayed by measuring (by diazotization) the amount of aromatic amines liberated by reduction with zinc amalgam and mineral acid. A correction is made for any free *p*-aminobenzoic acid, or aminobenzoylglutamic acid present (*92, 75, 30*). The microbiological assay using *Lactobacillus casei* provides a further and necessary check on the purity of the product (*172*) which may contain pteroic acid, or other substances, liberating *p*-aminobenzoic acid. The need for thorough analyses has recently been demonstrated by NICHOL and his co-workers (*116*) who have used paper chromatography to reveal the presence of pteroic acid in a "highly purified" specimen of PGA. In addition, the purest obtainable specimens of aminopterin and other "folic acid inhibitors" were shown by this technique to contain sufficient PGA or pteroic acid to account for a growth-promoting action on an antagonist-resistant strain of *Streptococcus faecalis*. Clearly, no biological experiment with these antagonists can have quantitative significance until they can be freed from the growth factors which

arise by auto-hydrolysis in the absence of buffers and have been found
to constitute up to 40% of the product.

The biological origin of PGA has been the subject of much speculation.
FORREST and WALKER (52) suppose that reductone (XXX, p. 375)
condenses with *p*-aminobenzoylglutamic acid, to give (XXXI), and
that this then condenses with triamino-4-hydroxypyrimidine (a substance
not yet demonstrated in nature). TSCHESCHE and his colleagues (184)
think it more likely that 2-amino-4:7-dihydroxypteridine-6-aldehyde
(the 7-hydroxy-derivative of XXXVI, p. 377) condenses with *p*-amino-
benzoylglutamic acid, and that the product is reduced to PGA. They
have demonstrated that this aldehyde (which has not yet been found
in nature), as well as the 7-hydroxy-derivative of PGA strongly promote
the growth of *S. faecalis*. The aldehyde is thought to arise from the
N-glucoside of *iso*xanthopterin-6-carboxylic acid, but the evidence is
still only inferential (183). WOOLLEY (212) believes that pterithiamine (XLI), a pteridine analogue of thiamine, is acted on by thiaminase
in the presence of *p*-aminobenzoylglutamic acid, and that (XLII) would
be a by-product of this reaction. WOOLLEY actually synthesized a small
amount of material having PGA activity by this route (211). Pterithiamine had previously been synthesized by ordinary chemical means (57),
but is not known to occur in nature.

(XLI.) Pterithiamine.

(XLII.)

3. Conjugates of Pteroylglutamic Acid.

Teropterin, the "fermentation *L. casei* factor" was isolated from
Corynebacterium-induced fermentations and its properties have been
briefly described (including the solubility, ultraviolet spectrum (similar
to that of PGA) and the preparation of the methyl ester) (73). HUTCHINGS
and his colleagues (165, 76) showed that anaerobic alkaline hydrolysis

gave PGA and two molecules of glutamic acid. Aerobic alkaline hydrolysis took the same course as with PGA, and splitting by sodium sulphite gave the aldehyde (XXXVI) and p-aminobenzoyltriglutamic acid. From this data, the structure was formulated as pteroyl-γ-L-glutamyl-γ-L-glutamyl-L-glutamic acid (XLIII). This structure was confirmed by synthesis, and a suitable method for large-scale preparation was worked

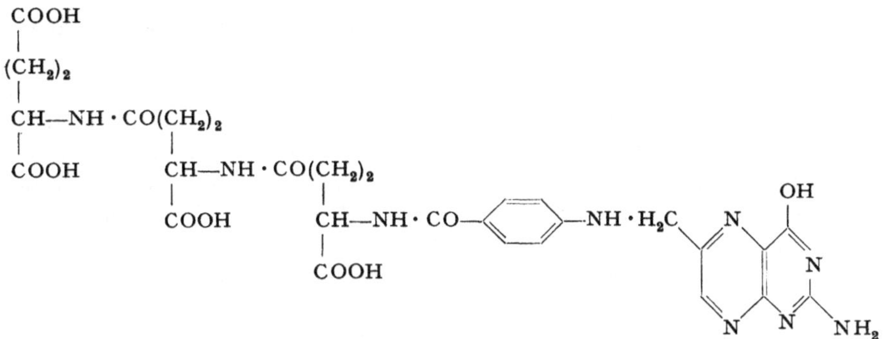

COOH
|
$(CH_2)_2$
|
CH—NH $\cdot$ CO$(CH_2)_2$
| |
COOH   CH—NH $\cdot$ CO$(CH_2)_2$

(XLIII.) Teropterin.

out as follows (28): p-Nitrobenzoylglutamic acid was converted through its γ-hydrazide to the azide. This was mixed with triethyl-γ-glutamylglutamate. The nitro-group in the product was reduced, giving p-aminobenzoyl-γ-glutamyl-γ-glutamylglutamic acid which was condensed with triamino-4-hydroxypyrimidine and 2:3-dibromopropionaldehyde, by the usual WALLER synthesis, to give (XLIII).

The "unnatural" isomers were also prepared, viz.

Pteroyl-α-glutamyl-α-glutamylglutamic acid.
Pteroyl-α-glutamyl-γ-glutamylglutamic acid.
Pteroyl-γ-glutamyl-α-glutamylglutamic acid.

(Natural L-glutamic residues were maintained throughout.) These were shown to be different from the natural product, and they had no growth-promoting effect on L. *casei* or on S. *faecalis* R (159, 20). By similar syntheses, pteroyldiglutamic acids were prepared. They are not known to exist in nature. Diopterin, the γ-isomer, has about 65% of the activity of PGA in promoting growth of both the above organisms (the α-isomer was inactive) (159).

"Vitamin Bc conjugate", obtained from yeast, consists of orange-coloured crystals with a spectrum almost identical with that of PGA. From hydrolysis and electrophoretic behaviour, it has been shown to consist of PGA attached to six more glutamic residues, i. e. it is pteroyl-heptaglutamic acid (124). It is assumed, but not yet established, that only γ-linkages are present.

4. Rhizopterin.

From the fumaric acid fermentation liquors of *Rhizopus nigrans*, a pale yellow substance was obtained by adsorption on charcoal and fuller's earth, followed by chromatography on alumina (*145*). WOLF and his colleagues (*209*) showed that on acid or alkaline hydrolysis it gave formic acid and "*apo*rhizopterin", later identified as pteroic acid. Rhizopterin is easily acylated on the 2-amino-group, which can alternatively be converted to a 2-hydroxy-group with nitrous acid. The acyl-derivatives permitted molecular weight determinations in boiling acetic acid. That rhizopterin is indeed 10-formylpteroic acid (XLIV), was confirmed by synthesis (*209*). During this synthesis, pteroic acid is diformylated, in the 2- and 10-positions, but after-treatment with dilute ammonia removes the formyl-residue from the primary amino-group, and leaves the other formyl-residue on the secondary amine. Removal of the latter formyl-group requires 4 hours refluxing with 2.5 N-sodium hydroxide.

Rhizopterin is best purified by precipitation as red crystals of the double salt which it forms with ethylene-diaminocobaltic chloride. The spectrum of rhizopterin differs from that of pteroic acid (Table 4, p. 362).

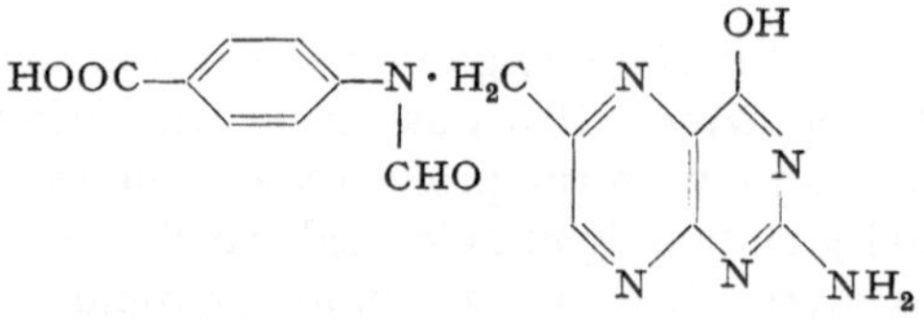

(XLIV.) Rhizopterin.

5. The Citrovorum Factor and Leucovorin.

The citrovorum factor (XLV) is a pteridine, first isolated from liver (*152, 95*), which promotes the growth of the bacterium *Leuconostoc citrovorum*. It is also produced by micro-organisms. A strain of *Streptococcus faecalis* converts PGA to the factor in high yield, particularly in the presence of formic and ascorbic acids (*32*). Several strains of lactobacilli synthesize a bound form of the factor from PGA, in the presence of ascorbic acid (this reaction is inhibited by aminopterin). The factor was liberated from this bound form by an enzyme in chicken pancreas (*62*). *Lactobacillus arabinosus* can convert *p*-aminobenzoic acid to the citrovorum factor, a reaction inhibited by sulphanilamide. An enzyme in liver and kidneys also converts PGA to citrovorum factor, in the presence of ascorbic acid (*115*).

Eight lichens, examined chromatographically on paper (butanol/acetic acid) and developed bio-autographically with *L. citrovorum* and *S. faecalis*, were found to contain rhizopterin, citrovorum factor, formyl-PGA, and pteroyl-triglutamic acid, as well as a substance that remains near the origin and has an intense growth-stimulating effect on both bacteria (PGA was found only in traces) (*162a*).

Three brown, and three red, seaweeds were similarly examined and developed. Citrovorum factor was found in five of them. *Furcellaria fastigiata* contains three other growth promoters for *L. citrovorum*, which remain nearer the origin than the factor. One of these promoters is present in horse liver (*22a*).

Leucovorin (also known as folinic acid SF) is a synthetic substance with the chemical and physical properties of the factor, but, being a racemate (see below), it is only half as active for *Leuconostoc citrovorum* (*161, 95*). ROTH (*147, 42*), FLYNN (*50, 127*), MAY (*109*) and their colleagues have shown how it can be prepared by the catalytic reduction of 10-formyl-PGA or by the hydrogenation of PGA in the presence of formic acid to give 10-formyl-5:6:7:8-tetrahydro-PGA, which is re-arranged to leucovorin, the isomeric 5-formyl-tetrahydro-PGA (XLV), by moist heat at 120°, or by keeping at pH 12 for an hour at 90°. Di- and tetrahydro-pteroylglutamic acids (*118*) also give the same product on reductive formylation (*147*).

Leucovorin forms white needles on precipitation from its salts at pH 3.5. When dried at 30°, a trihydrate is obtained which melts with decomposition at 250°. It becomes anhydrous at 100°, in high vacuum (*147*). It is stable to reduction and to mild oxidative conditions (*42*).

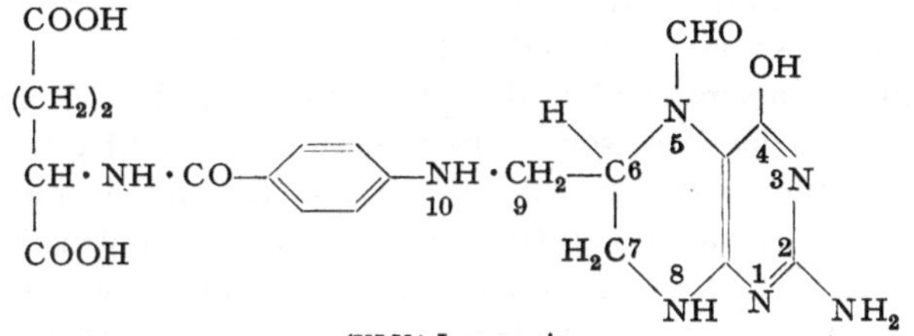

(XLV.) Leucovorin.

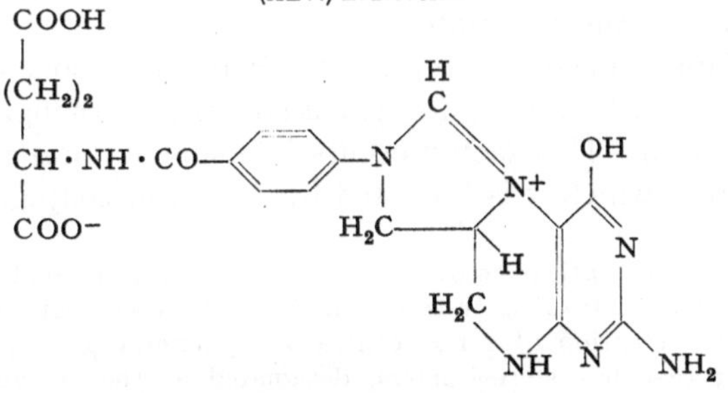

(XLVI.) Anhydro-leucovorin-A.

Leucovorin is only slightly sensitive to alkali, but some destruction occurs on prolonged heating (*127*). It may be concentrated, in a vacuum, at p_H 3, but at p_H 2 it is quickly changed to the imidazolinium compound (XLVI), called anhydro-leucovorin-A (42, 109). (This is zwitterionic but can also form salts, "isoleucovorin chloride", etc. with external anions.) An isomeric anhydroleucovorin-B, of unknown constitution is formed from (XLVI) at p_H 4. None of these derivatives promoted growth of L. citrovorum, but when left at p_H 13 (anaerobically) for an hour, they were all reconverted in good yield to leucovorin. When (XLVI) is allowed to auto-oxidize (at 37° and p_H 2), one molecular equivalent of oxygen is consumed and $N_{(10)}$-formyl-pteroylglutamic acid is produced. On standing in 0.1 N-sodium hydroxide for 24 hours, this is converted to PGA (109).

That leucovorin is hydrogenated in the pyrazine, and not in the pyrimidine ring, has been confirmed polarographically by ALLEN and his colleagues (15). The assignment of the formyl group to the 5-, rather than the 8-position, has been made largely on chemical grounds, with reference to analogues of known constitution. Use has also been made of the fact that leucovorin can be nitrosated in the 10-position, but anhydroleucovorin-A does not react with nitrous acid (42). A number of simply-substituted 7:8-dihydropteridines have more recently been unambiguously synthesized and hydrogenated to tetrahydropteridines which were shown by ultraviolet spectroscopy still to contain an unreduced pyrimidine ring (105).

It has been mentioned above that leucovorin has only half the activity of crystalline citrovorum factor isolated from liver. This difference in potency is attributed to the asymmetry of $C_{(6)}$, the synthetic product being a racemate of both optical isomers, the dL- and the lL-forms*. These two diastereoisomers have been separated by the difference in solubility of their calcium salts, the lL- (or natural) isomer being the less soluble (44). It has a slightly higher growth-promoting action than the natural product on L. citrovorum, although this may not be outside experimental error. The optical rotation $[\alpha]_D$ is — 15.1°, compared with + 14.26° for the racemate.

The acid-labile substance isolated by PFIFFNER and colleagues (123) from horse-liver, with activity for S. faecalis, is now thought to be citrovorum factor (195). This factor occurs in many natural materials in a bound form from which it is liberated by folic acid conjugases (206, 48).

---

* L refers to the optical configuration of the glutamic acid portion of the molecule. This configuration is thus shown to be identical with that of natural glutamic acid. The creation of a new centre of asymmetry at $C_{(6)}$ makes possible the existence of two new stereoisomers, designated as the dL- and the lL-forms respectively (the use of lower-case letters implies that these new optical isomers have been classified by the direction of their rotation and that it has not proved possible to relate them to substances of known configuration, as far as the relationship of the two asymmetric centres is concerned).

Slight autolysis of horse liver gives 10-formyl-pteroylglutamic acid (*162*) and this may be either a degradation product of the factor or an intermediate in its biosynthesis.

6. New members of the Folic Acid Series.

In recent literature, the existence of hitherto undescribed members of the folic acid series is indicated. LASCELLES and WOODS found that cell-suspensions of *Escherichia coli* and *Staphylococcus aureus* could convert *p*-aminobenzoic acid into a substance giving a different response curve with *Lactobacillus casei* from that given by PGA, heavier growth being produced at optimum concentrations. This substance, available only as an aqueous bacterial extract, was not identical with any known member of the folic acid series (*103*).

SARETT (*151*) grew *Lactobacillus arabinosus* in the presence of *p*-aminobenzoic acid and noted the formation of a folic acid-like substance different from PGA, citrovorum factor or 10-formyl-PGA. TOENNIES and colleagues (*174*) suggest that a form of folic acid exists in blood which is broken down to PGA in some analytical techniques and not in others, thus leading to variable results. This conjugate is neither the citrovorum factor nor pteroyltriglutamic acid.

When pyrimethamine ("Daraprim", 5-*p*-chlorophenyl-2:4-diamino-6-ethylpyrimidine) is fed to rats in large doses, the symptoms of folic acid deficiency develop. These can be completely relieved by liver extract out of all proportion to its PGA or citrovorum factor content, and HITCHINGS and his colleagues (*65*) suggest that an unknown form of folic acid is present in the extract. The toxic action of pyrimethamine on bacteria can be completely reversed by the known forms. GREENBERG (*57 a*) demonstrated the presence in pigeon's liver of a formylating co-enzyme which appears to be the ribose phosphate of the citrovorum factor. SWENDSEID and colleagues (*169*) incubated PGA with haemopoietic tissue and obtained "hemafolin", a substance with citrovorum factor activity but differing from this factor in giving a different growth response curve. Moreover, it is more active than the factor in counteracting the toxicity of chloramphenicol for *L. citrovorum*.

V. Substances which are Presumably Pteridines.

From time to time, the presence of a new pteridine in nature is surmised from the demonstration, in an extract, of "a typical pteridine spectrum", or "typical pteridine fluorescence". Doubtless not all of these claims will survive investigation. Some of the better authenticated examples will now be mentioned.

Several pteridine-like substances have been found in eyes, apart from fluorescyanine. From the eyes of dogfish and alligators, PIRIE and Simpson (*126*) isolated fluorescent substances whose chemical, fluorescent and spectroscopic properties were almost identical with those of xanthopterin. Similar fluorescent substances also occur in the eyes of Crustacea and Amphibia. From the eyes of locusts, Goodwin isolated a substance spectroscopically similar to xanthopterin (*56*). Goodwin noted the (at that time) confused state of the literature on the ultraviolet absorption of xanthopterin [this confusion has now been resolved (*1*), in a way that increases the likelihood of Goodwin's pigment being xanthopterin]. The R_F values of presumed pteridines in the retina of frogs have been given (*61*); light decreases the content of these substances.

Lederer (*104*) isolated a red, yellow-fluorescent pigment from the eyes of the wild type of the fruit-fly *(Drosophila melanogaster)*. He suggested that this pigment, drosoptérine, was a pteridine ($\lambda_{max.} = 465$ mμ., in water). It was purified through the silver salt and analysed (C, 42, H, 5.5, and N, 19%). Chan, Heymann, and Clancy (*40*) extracted a red pigment from the red eyed *Drosophila* (mutant "vermillion"). It analysed, approximately, for $C_{10}H_{16}O_5N_4Cl$. They noted that Lederer's figures also are consistent with this formula and suggest that the two pigments are similar or identical. Forrest and Mitchell (*51a*) of the California Institute of Technology isolated a red and an orange pigment from a red-eyed mutant of *Drosophila*. The orange pigment is a dihydro-2-amino-4-hydroxypteridine-6-carboxylic acid with an $N_{(8)}$-lactyl substituent.

A substance "hepatopterin" has been isolated from the hepatopancreas of a crab, *Cancer pagurus*. The spectrum resembles that of xanthopterin, except that $\lambda_{max.}$ occurs at 268 mμ. instead of 276 mμ., in water (*132*). A substance similar, in physical properties, to xanthopterin has been isolated from human tubercle bacilli (*46a*).

Jacobson (*79, 81*) showed that the argentaffine cells of stomach and intestinal epithelium, contain a substance with a xanthopterin-like spectrum and fluorescence. The location of these cells is similar to the distribution of the substance active against pernicious anaemia.

Gowland Hopkins heated uric acid with water (sometimes acidified) under pressure, and obtained at least two yellow, green-fluorescing substances with spectra similar to that of xanthopterin. Upon aeration, in acid solution, they produced a red substance with the properties of pterorhodin (*68*). The most likely course of this reaction would be the degradation of the uric acid to 2:6-dihydroxy-4:5-diaminopyrimidine, and the synthesis of pteridines from this and two- (and three-) carbon atom fragments arising in the decomposition. Such a transformation seems possible in view of the easy conversion of 2-hydroxypurine to 2-hydroxypteridine (*3*).

STREHLER has isolated a cream-coloured substance, luciferesceine, from the head of the firefly *Photinus pyralis* (*167*). It decomposes at 150° and is more soluble in alcohol than in water. It analysed for $C_{11}H_{15}O_4N_5$ (cf. the pigment from *Drosophila*, above). Luciferesceine is not itself chemiluminescent, but it has ultraviolet absorption and fluorescence spectra almost identical with luciferin, which is the source of light in the fire-fly. The fluorescence of luciferesceine is blue, with an emission band of 370–510 mμ. at p$_H$ 8.1 (at p$_H$ 0.2 this becomes 380–600 mμ.). At p$_H$ 2, the absorption spectrum has peaks at 291 and 340 mμ., at p$_H$ 11, they are at 257, 282, and 355 mμ. It contains no sulphur or metals. Potassium hydroxide liberates ammonia. The pK_a is about 8 and this appears to refer to a basic group. STREHLER (*41*) suggests that it may be an imino-ribityl pteridine, and points out the analogy with riboflavine (XLVII), the phosphate of which he claims to be the chemiluminescent substrate of the luminous bacterium *Achromobacter fischeri* and the earthworm *Eisenia submontana*.

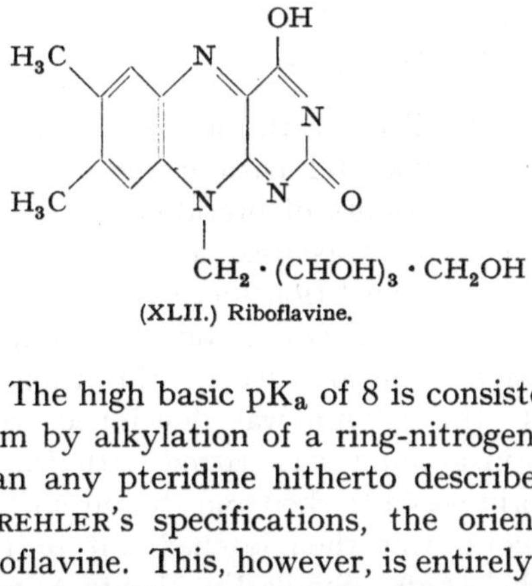

$CH_2 \cdot (CHOH)_3 \cdot CH_2OH$
(XLII.) Riboflavine.

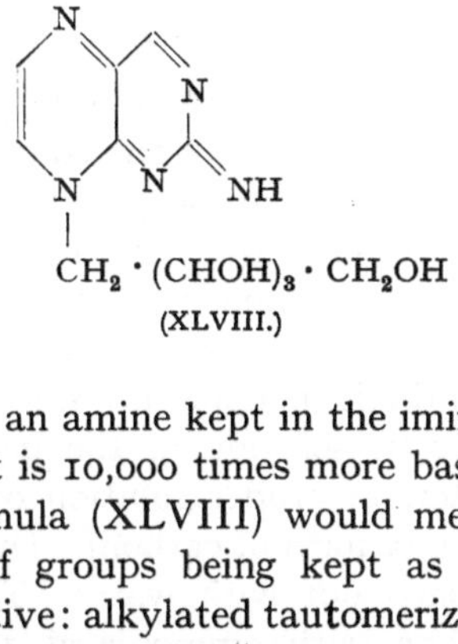

$CH_2 \cdot (CHOH)_3 \cdot CH_2OH$
(XLVIII.)

The high basic pK_a of 8 is consistent with an amine kept in the imino form by alkylation of a ring-nitrogen (*21*). It is 10,000 times more basic than any pteridine hitherto described. Formula (XLVIII) would meet STREHLER's specifications, the orientation of groups being kept as in riboflavine. This, however, is entirely speculative: alkylated tautomerized amino-pteridines are unknown, and we must await the results of degradative studies. Luciferin, the chemiluminescent substrate of fireflies, has been characterized, but not chemically investigated (*107*). Myco-bactin, the growth-factor for *Mycobacterium johnei* (*53*) is another heterocyclic compound with five nitrogen atoms and a somewhat similar spectrum, but it is probably not a pteridine.

The blue fluorescence of human teeth is due to an organic substance with λ_{max}. 270 mμ., and, apparently, a further peak concealed in the continuous absorption of the 300–450 mμ. region of the impure preparation (*60*). The somewhat common embryological origin of fish-scales and teeth might suggest the presence of something related to ichthyopterin.

The fluorescent substance found in the urine of a group of children suffering from leukaemia, and believed to be a pteridine (39) is now believed by these authors to have been an artefact.

The yellow substance, $C_6H_6O_2N_4$ (m. p. 171°), isolated from *Bacterium bongkrek.*, and known as toxoflavin (189), has been classed with the purines. Its $\lambda_{max.}$ of 400 mμ. would seem to be too high for a purine (see p. 360), and the possibility of its being a pteridine has been considered (188).

There is little doubt that pteridines are very widely distributed in nature and that even in the next decade many new examples will be isolated and their constitutions determined. In the light of what has been written above, it is not unreasonable to expect that several of these will deviate from the PGA and xanthopterin types which have tended to dominate the field. In particular, we should expect to find new examples of polyalcoholic side-chains, whether attached to a carbon atom (as in erythropterin, fluorescyanine and urothion), or to a nitrogen atom (as, apparently, in luciferesceine).

· VI. Chromatography and Other Techniques of Isolation and Purification.

SCHÖPF and BECKER (155) introduced fuller's earth for the adsorption of xanthopterin from solution, also alumina and barium sulphate for the separation and recognition of small amounts of pteridines. Hydrochloric acid (0.005 N) in water or methanol was used as the solvent. Since that time, the use of adsorption and partition methods has been greatly extended. ROTH and her colleagues (147) found magnesol (synthetic magnesium silicate) effective to adsorb leucovorin from an aqueous solution of the barium salt; the column was developed with water. A cruder solution of leucovorin at pH 4 was stirred with Darco G-60 (a form of carbon). The leucovorin was adsorbed and the impurities washed out with water. The leucovorin was then leached out of the carbon with ammoniacal 50% alcohol (147). Similarly, Norite A was used to adsorb PGA from a crude solution at pH 3.0.

The carbon was washed with neutral 60% alcohol at 25° to remove impurities and then the PGA was leached out with 0.5 N-ammonia in 60% alcohol at 70°. This solution was concentrated, taken to pH 3.5 and filtered. The filtrate was adsorbed at pH 1.3 on Superfiltrol (an adsorbent of undisclosed composition). After washing with 50% alcohol to remove impurities, the PGA was eluted by percolation with N-ammonia in 60% alcohol. After elution, the column was washed with 0.1 N-hydrochloric acid and used several times. After six such cycles, no decrease in efficiency was observed and almost complete recovery of PGA was obtained, except in the first run. This use of Superfiltrol led to a 17-fold concentration of PGA in the total solids (166, 123).

Superfiltrol was also used to adsorb the methyl ester of PGA from butanol, methanol, acetone or water. Aqueous acetone proved to be the only efficient eluent. Alumina and calcium carbonate proved much less effective (*166*). FLYNN and his colleagues (*50*) used "florisil" (a synthetic adsorbent of undisclosed nature) similarly for leucovorin, at pH 5.5.

Partition chromatography of leucovorin was carried out with butanol/acetic acid/water on a potato starch column (*50*). Amberlite IR-4 was used to remove PGA from an aqueous extract of hog-liver, possibly more by adsorption than by ion exchange. It was eluted with ammonia, and the same batch of resin was used repeatedly (*123*).

The CRAIG countercurrent apparatus (24 tubes) has been used to demonstrate impurities in commercial PGA and to separate the products of the photo-oxidative decomposition of this acid (*142*). Curves were constructed for the partition between butanol and water of 2-amino-4-hydroxypteridine and its 6- and 7- —OH, —CHO and —COOH derivatives (*142*).

Because the majority of pteridines posses no melting points, the use of paper-chromatograms to assess purity and establish identity has become a regular practice among workers in this field, including those engaged in synthesis (*1*). Butanol/acetic acid/water was the first solvent to be recommended for natural pteridines (*55*) and it is still found valuable (*144*). The biggest single advance has been the use of 3% aqueous ammonium chloride (*180, 181*) which permits of faster runs (although the resolution is not so good), gives more sharply defined spots with carboxylic acids, and increases the R_F of substances inclined to stay near the origin when organic solvents are used.

CRAMMER (*45*) adapted paper chromatography for the concentration of butterfly pigments.

VII. The Physiological Action of the Natural Pteridines.
The Simpler Pteridines.

HADDOW (*58*) showed that when various species of mammals are injected with xanthopterin, the kidneys enlarge to many times their normal size. Qualitatively, their function remains unchanged. This enlargement has been traced to an enormous increase in the number of mitoses in the epithelium of the renal tubules. When the injections are stopped, the kidneys slowly revert to normal size. 2-Amino-4-hydroxy- and 2:4-diamino-pteridine act similarly, but further simplification of the molecule produces relatively inert substances (*59*).

In a pharmacological investigation, it was found that both xanthopterin and leucopterin are toxic for mice in a dose of 50 mg./kg. (intravenous). Death does not occur for several days, and damage to kidneys and pancreas is found. After injection of xanthopterin into cats, most of this pteridine is found in the kidneys, but very little appears in the urine (*66*).

In the previous review on pteridines in this Series (*138*), the view was expressed that the butterfly-wing pigments were incompletely metabolised breakdown-products of a growth-hormone active in the transformation of the chrysalis to the imago. Pteroylglutamic acid would admirably fill the role of such a substance (see below) and its oxidation to xanthopterin is an example of the replacement of a carbon-containing side-chain by a hydroxyl-group, such as is well-known in the oxidation of pterorhodin (p. 371). However, erythropterin could hardly be produced in this way. Moreover, as is shown above, xanthopterin is not devoid of metabolic activity. Xanthopterin has also been shown to be active in curing the anaemia of trout fed on deficient diets and that of rats produced by feeding goat's milk (*138*). Some workers have thought that, in these examples, xanthopterin is merely blocking an enzyme which destroys PGA, thus allowing the latter to exert an action out of all proportion to the small amount present (*195*). The xanthopterin and proxanthopterin in human urine increase slightly when a large excess of PGA is taken, but there is no direct relationship (*141*).

Very little biological work has been done with erythropterin, because of its scarcity. Some special biological function may be found for such pteridines with glycol groups (urothion and fluorescyanine are other examples).

The pharmacology of *iso*xanthopterin, now one of the most accessible of pteridines (*14*), calls for investigation. The suggested vitamin-replacing action of fluorescyanine in vitamin B_1 and B_2-deficient rats and pigeons (*36*) awaits independent confirmation because fluorescyanine is no substitute for these vitamins in bacteria (*131*). It is credited with playing a part in melanogenesis (*129*). Actually, various pteridines change the kinetics of the enzyme reactions which convert dihydroxy-phenylalanine to melanine. Fluorescyanine, *iso*xanthopterin, and xanthopterin-7-carboxylic acid accelerate melanogenesis at the expense of intermediary red compounds formed by tyrosinase, but they moderate the initial oxidation of DOPA. In contrast, *iso*xanthopterin-6-carboxylic acid and riboflavine accelerate all the stages (*133*). Xanthopterin, also, interferes with the oxidation of DOPA (*77*), whereas PGA can intensify the acceleration caused by cupric ions (*78*). Chelation is believed to be involved in some of these instances.

An interesting biological antagonism between pteridines has been reported. The rate of proliferation of cells removed from rat embryos is said to be accelerated by 2-amino-4-hydroxy-7-methylpteridine and inhibited by xanthopterin, whereas the reverse is found with cells obtained from newly-born rats (*117*). However this antagonism could not be demonstrated on the bone-marrow cells of rabbits (*24*), and it remains to be shown if it is statistically significant.

The Folic Acid Series.

As was indicated on pp. 382–385, it is still far from clear whether PGA shows any physiological effect other than that of the citrovorum factor formed from it. The analogues of PGA are proving helpful in investigating these problems and a picture is emerging of the conversion of PGA to citrovorum factor in two steps (*83, 120*). The first of these appears to be the reduction of PGA to 5:6:7:8-tetrahydro-PGA, possibly by ascorbic acid; the second step, the introduction of the formyl group, appears to be carried out by formic acid or serine. 9-Methylpteroyl-glutamic acid (and other antagonists which retain the 4-OH group) block the first step, whereas aminopterin (and other 4-amino-analogues of PGA) block not only the first step, but also the utilization of the citrovorum factor (*83*).

At the biochemical level, the principal function of the citrovorum factor seems to be its participation in the transfer of single carbon atoms from formic acid or serine to form (a) the 2- and 8-carbon-atoms in purines, (b) the 5-methyl-group in thymine, (c) the S-methyl-group in methionine (which can also arise from choline), (d) histidine, and (e) serine (reversibly). It has been suggested that citrovorum factor, through its 5-formyl-group, is a coenzyme F which (analogously to coenzyme A, the transporter of acetyl groups) is an acceptor and donor of the formyl group (*34*). The lesser activity of tetrahydro-PGA shows that the situation is more complex than this (*31*). GREENBERG (*57 a*) has shown that boiled pigeon's liver contains a factor that converts the riboside of 4-aminoimidazole-5-carboxamide to inosic acid. This factor, apparently the true coenzyme F, was then synthesized from citrovorum factor and adenosine triphosphate. It was shown that the formyl-group, labelled with C^{14}, is quantitatively transferred to the above riboside even in the presence of excess unlabelled formate ions.

No enzyme requiring a pteridine as coenzyme has been isolated. The clearest indication is in the conversion of serine to glycine by a solution prepared from pigeon's liver: this reaction is catalysed by PGA and an equivalent of formaldehyde is liberated. That this PGA-catalysed reaction is freely reversible was demonstrated by the use of substrates tagged with radioactive carbon (*25*). Detailed studies have failed to confirm the former views that a member of the folic acid series participates in the catabolism of tyrosine (*195*).

The inhibition of xanthine oxidase by 2-amino-4-hydroxypteridine-6-aldehyde (XXXVI, p. 377), which is a breakdown product of PGA, occurs at so great a dilution ($10^{-9} M$) as to suggest that the aldehyde is a natural regulator of this widely-distributed enzyme (*85a, 121*).

At the histological level, a most important discovery has been made by JACOBSON (*80*) who showed that the cells of mammals require the citrovorum factor during mitosis, to convert metaphase into anaphase. Apparently the presence of this factor causes nucleoprotein to be shed into the cytoplasm, thus initiating the transformation (*82*).

At the microbiological level, it has been known since 1940 that sulphonamides injure bacteria by competing with *p*-aminobenzoic acid essential for their nutrition. The only known natural occurrence of *p*-aminobenzoic acid is in the folic acids, and hence it is not surprising that the action of sulphonamides on some bacteria is inhibited non-competitively by PGA, and by the citrovorum factor. On the other hand, this is not true of pathogenic bacteria: possibly these are not permeable to preformed folic acids, or their particular folic acid may have a somewhat different chemical nature (*210*).

At the clinical level, PGA and the citrovorum factor rapidly cure the anaemia of pregnancy and the macrocytic anaemia sometimes arising after operations on the small intestine. In these two conditions, the cobalt-containing vitamin B_{12} is ineffective. On the other hand, these pteridines are not so effective in pernicious anaemia as vitamin B_{12}. Both pteridines have been found helpful, but not curative in the tropical bowel disease known as sprue (*164*). The usual daily dose of PGA (the "Folic Acid" of the pharmacopoeias) is 5–20 mg. (*30*). Citrovorum factor has no clinical superiority over PGA, and, being more expensive, is not used in medicine.

The excretion of PGA in the urine of patients suffering from malignant disease is usually abnormally low (*S. faecalis* assay) (*54*). Acute leukaemia in children is frequently treated with the 4-amino-analogue of PGA (*35*). In this malignant disease, four times more citrovorum factor than normal is present in the leucocytes, suggesting that their metabolic requirement is greater, and explaining the usefulness of the antifolic substances (*168*). The earlier use of pteroyltriglutamic acid (teropterin) in cancer proved fruitless.

References.

1. ALBERT, A.: The Pteridines. Quart. Rev. Chem. Soc. (London) 6, 197 (1952).
2. — Quantitative Studies of the Avidity of Naturally-occurring Substances for Trace Metals. 3. Pteridines, Riboflavin and Purines. Biochemic. J. 54, 646 (1953).
3. — The Conversion of Purines to Pteridines under Mild Conditions. Proc. Biochem. Soc. (in press).
3a. — Some Unsolved Problems. Ciba Colloquium on Pteridines, London, 1954.
4. ALBERT, A. and D. J. BROWN: The Chelation of Pteridines. Proc. 1st Internat. Congr. Biochem., Cambridge 1949, 241.
5. — — Pteridine Studies. IV. 4:6- and 4:7-Dihydroxypteridine. J. Chem. Soc. (London) 1953, 74.
6. — — Purine Studies. I. Stability to Acid and Alkali; Solubility; Ionization. A Comparison with Pteridines. J. Chem. Soc. (London) (in press).

7. ALBERT, A., D. J. BROWN, and G. CHEESEMAN: Pteridine Studies. I. Pteridine, and 2- and 4-Amino- and 2- and 4-Hydroxypteridines. J. Chem. Soc. (London) 1951, 474.

8. — — — Pteridine Studies. II. 6- and 7-Hydroxypteridines and their Derivatives. J. Chem. Soc. (London) 1952, 1620.

9. — — — Pteridine Studies. III. The Solubility and the Stability to Hydrolysis of Pteridines. J. Chem. Soc. (London) 1952, 4219.

10. ALBERT, A., D. J. BROWN, and H. C. S. WOOD: Pteridine Studies. V and VI. J. Chem. Soc. (London) 1954 (in press).

11. ALBERT, A., R. GOLDACRE, and J. PHILLIPS: The Strength of Heterocyclic Bases. J. Chem. Soc. (London) 1948, 2240.

12. ALBERT, A. and H. C. S. WOOD: Pteridine Syntheses. I. Leucopterin and Xanthopterin. J. Appl. Chem. (London) 2, 591 (1952).

13. — — Isoxanthopterin. Nature (London) 172, 118 (1953).

14. — — Pteridine Syntheses. II. Isoxanthopterin. J. Appl. Chem. (London) 3, 521 (1953).

15. ALLEN, W., R. L. PASTERNAK, and W. SEAMAN: Polarographic Determination and Evidence for the Structure of Leucovorin. J. Amer. Chem. Soc. 74, 3264 (1952).

16. AMERICAN CYANAMID COMPANY: Improvements in or relating to the Preparation of Pteridines. Brit. Pat. 646,149 (1947).

17. — Improvements in or relating to the Preparation of Pteridines. Brit. Pat. 676,107 (1949).

18. ANGIER, R. B., J. H. BOOTHE, B. L. HUTCHINGS, J. H. MOWAT, J. SEMB, E. L. R. STOKSTAD, Y. SUBBAROW, C. W. WALLER, D. B. COSULICH, M. J. FAHRENBACH, M. E. HULTQUIST, E. KUH, E. H. NORTHEY, D. R. SEEGER, J. P. SICKELS, and J. M. SMITH, Jr.: The Structure and Synthesis of the Liver L. casei Factor. Science (New York) 103, 667 (1946).

19. ANGIER, R. B., E. L. R. STOKSTAD, J. H. MOWAT, B. L. HUTCHINGS, J. H. BOOTHE, C. W. WALLER, J. SEMB, Y. SUBBAROW, D. B. COSULICH, M. J. FAHRENBACH, M. E. HULTQUIST, E. KUH, E. H. NORTHEY, D. R. SEEGER, J. P. SICKELS, and J. M. SMITH, Jr.: Synthesis of Pteroylglutamic Acid. III. J. Amer. Chem. Soc. 70, 25 (1948).

20. ANGIER, R. B., C. W. WALLER, B. L. HUTCHINGS, J. H. BOOTHE, J. H. MOWAT, J. SEMB, and Y. SUBBAROW: Pteroic Acid Derivatives. VI. Unequivocal Syntheses of Some Isomeric Glutamic Acid Peptides. J. Amer. Chem. Soc. 72, 74 (1950).

21. ANGYAL, S. J. and C. L. ANGYAL: The Tautomerism of N-Heteroaromatic Amines. I. J. Chem. Soc. (London) 1952, 1461.

22. BACKER, H. J. et A. C. HOUTMAN: La purification de l'acide ptéroyl-glutamique (acid folique synthétique). Rec. trav. chim. Pays-Bas 70, 730 (1951).

22a. BÁNHIDI, Z. G. and L.-E. ERICSON: Bioautographic Separation of Vitamin B_{12} and Various Forms of Folinic Acid Occurring in Some Brown and Red Seaweeds. Acta Chem. Scand. 7, 713 (1953).

23. BIAMONTE, A. R. and G. H. SCHNELLER: A Study of Folic Acid Stability in Solutions of the B complex Vitamins. J. Amer. Pharmaceut. Assoc. 40, 313 (1951).

24. BIESELE, J. J. and R. E. BERGER: The Effect of Xanthopterin and Related Agents on the Proliferation of Rabbit Marrow Cells in vitro. Cancer Res. 10, 686 (1950).

25. BLAKLEY, R. L.: Folic Acid as Coenzyme for Interconversion of Serine and Glycine. Proc. Biochem. Soc. (London) (in press).

26. Bloom, E. S., J. M. Vandenbelt, S. B. Binkley, B. L. O'Dell, and J. J. Pfiffner: The Ultraviolet Absorption of Vitamin Bc and Xanthopterin. Science (New York) 100, 295 (1944).

27. Boon, W. R. and T. Leigh: Pteridines. III. Unambiguous Syntheses of Xanthopterin and 2-Amino-4-hydroxy-6-methylpteridine. J. Chem. Soc. (London) 1951, 1497.

28. Boothe, J. H., J. Semb, C. W. Waller, R. B. Angier, J. H. Mowat, B. L. Hutchings, E. L. R. Stokstad, and Y. SubbaRow: Pteroic Acid Derivatives. III. Pteroyl-γ-glutamylglutamic Acid and Pteroyl-γ-glutamyl-γ-glutamyl-glutamic Acid. J. Amer. Chem. Soc. 71, 2304 (1949).

29. Boothe, J. H., C. W. Waller, E. L. R. Stokstad, B. L. Hutchings, J. H. Mowat, R. B. Angier, J. Semb, Y. SubbaRow, D. B. Cosulich, M. J. Fahrenbach, M. E. Hultquist, E. Kuh, E. H. Northey, D. R. Seeger, J. P. Sickels, and J. M. Smith, Jr.: Synthesis of Pteroylglutamic Acid. IV. J. Amer. Chem. Soc. 70, 27 (1948).

30. British Pharmacopoeia, 1943, 236.

31. Broquist, H. P., M. J. Fahrenbach, J. A. Brockman, Jr., E. L. R. Stokstad, and T. H. Jukes: "Citrovorum Factor" Activity of Tetrahydropteroylglutamic Acid. J. Amer. Chem. Soc. 73, 3535 (1951).

32. Broquist, H. P., A. R. Kohler, D. J. Hutchison, and J. H. Burchenal: Studies on the Enzymatic Formation of Citrovorum Factor by Streptococcus Faecalis. J. Biol. Chem. 202, 59 (1953).

33. Brown, D. J.: Some Pteridines related to Folic Acid. I. 2-Deamino-analogues. J. Chem. Soc. (London) 1953, 1644.

33a. — The Monosubstituted Pteridines. Ciba Colloquium on Pteridines, London, 1954.

34. Buchanan, J. M. and D. W. Wilson: Biosynthesis of Purines and Pyrimidines. Federat. Proc. (Amer. Soc. exp. Biol.) 12, 646 (1953).

35. Burchenal, J. H., D. A. Karnofsky, E. M. Kingsley-Pillers, C. M. Southam, W. P. L. Myers, G. C. Escher, L. F. Craver, H. W. Dargeon, and C. P. Rhoads: The Effects of the Folic Acid Antagonists and 2,6-Diaminopurine on Neoplastic Disease. Cancer 4, 549 (1951).

36. Busnel, R.-G.: Actions physiologiques de la fluorescyanine, ptérine des vertébrés inférieurs. Bull. soc. zool. France 70, 30 (1945).

37. — La fluorescyanine et l'acide folique, ptérines de Bombyx mori. Trans. Ninth Int. Congr. Ent. 1, 356 (1952).

38. Cain, C. K., M. F. Mallette, and E. C. Taylor, Jr.: Pyrimido-[4,5-b]-pyr-azines (Pteridines). III. Pteridinemono- and -dicarboxylic Acids. J. Amer. Chem. Soc. 70, 3026 (1948).

39. Carter, C. E. and D. L. Horrigan: Isolation of a Pteridine from Urine in Leukemia. Federat. Proc. (Amer. Soc. exp. Biol.) 12, 187 (1953).

40. Chan, F. L., H. Heymann, and C. W. Clancy: Chemical Composition of the Red Eye Pigment of Drosophila melanogaster. J. Amer. Chem. Soc. 73, 5448 (1951).

41. Cormier, M. J. and B. L. Strehler: The Identification of KCF: Requirement of Long-Chain Aldehydes for Bacterial Extract Luminescence. J. Amer. Chem. Soc. 75, 4864 (1953).

42. Cosulich, D. B., B. Roth, J. M. Smith, Jr., M. E. Hultquist, and R. P. Parker: Chemistry of Leucovorin. J. Amer. Chem. Soc. 74, 3252 (1952).

43. Cosulich, D. B., D. R. Seeger, M. J. Fahrenbach, K. H. Collins, B. Roth, M. E. Hultquist, and J. M. Smith, Jr.: Analogs of Pteroylglutamic Acid. IX. Derivatives with Substituents on the Benzene Ring. J. Amer. Chem. Soc. 75, 4675 (1953).

44. COSULICH, D. B., J. M. SMITH, Jr., and H. P. BROQUIST: Diastereoisomers of Leucovorin. J. Amer. Chem. Soc. **74**, 4215 (1952).

45. CRAMMER, J. L.: Paper Chromatography of Flavine Nucleotides. Nature (London) **161**, 349 (1948).

46. CROWE, M. O'L. and A. WALKER: Pterin-like Pigment Derived from the Tubercle Bacillus. Fluorescence and Absorption Spectral Data for Erythropterin-like Pigment Isolated by Ultrachromatographic Analysis. Science (New York) **110**, 166 (1949).

46 a. — — Fluorescence and Absorption Spectral Data for Pterin isolated from the Tubercle Bacillus by Chromatographic Analysis. Brit. J. exper. Path. **35**, 18 (1954).

47. DAY, P. L., W. C. LANGSTON, and W. J. DARBY: Failure of Nicotinic Acid to Prevent Nutritional Cytopenia in the Monkey. Proc. Soc. exp. Biol. Med. **38**, 860 (1938).

48. DOCTOR, V. M. and J. R. COUCH: Occurrence and Properties of a Conjugated Form of *Leuconostoc Citrovorum* Factor. J. Biol. Chem. **200**, 223 (1953).

49. ELION, G. B. and G. H. HITCHINGS: The Identification of "β-Dihydroxanthopterin" as 2,4-Diamino-6-hydroxy-*p*-oxazino-(2,3-d)-pyrimidine. J. Amer. Chem. Soc. **74**, 3877 (1952).

49 a. ELION, G. B., G. H. HITCHINGS, and P. B. RUSSELL: The Formation of 6-Hydroxy- and 7-Hydroxypteridines from 4,5-Diaminopyrimidines and α-Ketoacids and Esters. J. Amer. Chem. Soc. **72**, 78 (1950).

50. FLYNN, E. H., T. J. BOND, T. J. BARDOS, and W. SHIVE: A Synthetic Compound with Folinic Acid Activity. J. Amer. Chem. Soc. **73**, 1979 (1951).

51. FLÖYSTRUP, T., M. A. SCHOU, and H. M. KALCKAR: Urinary Excretion of a Pterin after Administration of Pteroylglutamic Acid. Acta Chem. Scand. **3**, 985 (1949).

51 a. FORREST, H. S. and H. K. MITCHELL: The Pteridines of *Drosophila melanogaster*. Ciba Colloquium on Pteridines, London, 1954.

52. FORREST, H. S. and J. WALKER: Reductone and the Synthesis of Pteridines. Nature (London) **161**, 721 (1948).

53. FRANCIS, J., H. M. MACTURK, J. MADINAVEITIA, and G. A. SNOW: Mycobactin, a Growth Factor for *Mycobacterium johnei*. Biochemic. J. **55**, 596 (1953).

54. GIRDWOOD, R. H.: Folic Acid Excretion Studies in the Investigation of Malignant Disease. Brit. med. J. **2**, 741 (1953).

55. GOOD, P. M. and A. W. JOHNSON: Paper Chromatography of Pterins. Nature (London) **163**, 31 (1949).

56. GOODWIN, T. W. and S. SRISUKH: Biochemistry of Locusts. 6. The Occurrence of a Flavin in the Eggs and of a Pterin in the Eyes of the African Migratory Locust (*Locusta migratoria migratorioides* R. & F.) and the Desert Locust (*Schistocerca gregaria* FORSK.). Biochemic. J. **49**, 84 (1951).

57. GREEN, A. et R. DELABY: Sur les ptérithiamines. Bull. soc. chim. France V, **18**, 585 (1951).

57 a. GREENBERG, G. R.: Transformylation Co-factor. A Mechanism of Activation of Formate. Federat. Proc. (Amer. Soc. exp. Biol.) **13**, 221 (1954).

58. HADDOW, A.: Mode of Action of Chemical Carcinogens. Brit. med. Bull. **4**, 331 (1947).

59. HADDOW, A. and E. HORNING: Personal communication.

60. HARTLES, R. L. and A. G. LEAVER: The Fluorescence of Teeth under Ultraviolet Irradiation. Biochemic. J. **54**, 632 (1953).

61. HAYANO, S.: A Qualitative Estimation of Pterins in the Retina and Choroid. Igaku to Seibutsugaku (Japan) **17**, 322 (1950) [Chem. Abstr. **46**, 2156 (1952)].

62. HENDLIN, D., L. K. KODITSCHEK, and M. H. SOARS: Investigation on the Biosynthesis of Citrovorum Factor by Lactic Acid Bacteria. J. Bacteriol. 65, 466 (1953).

63. HIRATA, Y. et S. NAWA: Sur la fluorescyanine (ichtyoptérine) obtenue des oeufs de *Bombyx mori* et des écailles de carpe. C. R. Séances Soc. Biol. 145, 661 (1951).

64. HIRATA, Y., S. NAWA, S. MATSUURA et H. KAKIZAWA: Synthèses des ptérines et une remarque sur la constitution d'une ptérine de *Bombyx mori*. Experientia 8, 339 (1952).

65. HITCHINGS, G. H., E. A. FALCO, H. VANDERWERFF, P. B. RUSSELL, and G. B. ELION: Antagonists of Nucleic Acid Derivatives. VII. 2,4-Diaminopyrimidines. J. Biol. Chem. 199, 43 (1952).

66. HÖRLEIN, H.: Pharmakologie des Xanthopterins. Arch. exp. Pathol. Pharmakol. 198, 258 (1941).

67. HOPKINS, F. G.: Pigment in Yellow Butterflies. Nature (London) 45, 197 (1891).

68. — A Contribution to the Chemistry of Pterins. Proc. Roy. Soc. (London), Ser. B 130, 359 (1942).

69. HÜTTEL, R. und G. SPRENGLING: Über Ichthyopterin, einen blaufluorescierenden Stoff aus Fischhaut. Liebigs Ann. Chem. 554, 69 (1943).

70. HULTQUIST, M. E., E. KUH, D. B. COSULICH, M. J. FAHRENBACH, E. H. NORTHEY, D. R. SEEGER, J. P. SICKELS, J. M. SMITH, Jr., R. B. ANGIER, J. H. BOOTHE, B. L. HUTCHINGS, J. H. MOWAT, J. SEMB, E. L. R. STOKSTAD, Y. SUBBAROW, and C. W. WALLER: Synthesis of Pteroylglutamic Acid (Liver *L. casei* Factor) and Pteroic Acid. II. J. Amer. Chem. Soc. 70, 23 (1948).

71. HUTCHINGS, B. L., N. BOHONOS, and W. H. PETERSON: Growth Factors for Bacteria. XIII. Purification and Properties of an Eluate Factor required by certain Lactic Acid Bacteria. J. Biol. Chem. 141, 521 (1941).

72. HUTCHINGS, B. L., J. H. MOWAT, J. J. OLESON, A. L. GAZZOLA, E. M. BOGGIANO, D. R. SEEGER, J. H. BOOTHE, C. W. WALLER, R. B. ANGIER, J. SEMB, and Y. SUBBAROW: Synthesis and Some Biological Properties of 4-Aminopteroylaspartic Acid. J. Biol. Chem. 180, 857 (1949).

73. HUTCHINGS, B. L., E. L. R. STOKSTAD, N. BOHONOS, N. H. SLOANE, and Y. SUBBAROW: The Isolation of the Fermentation *Lactobacillus casei* Factor. J. Amer. Chem. Soc. 70, 1 (1948).

74. HUTCHINGS, B. L., E. L. R. STOKSTAD, N. BOHONOS, and N. H. SLOBODKIN: Isolation of a new *Lactobacillus casei* Factor. Science (New York) 99, 371 (1944).

75. HUTCHINGS, B. L., E. L. R. STOKSTAD, J. H. BOOTHE, J. H. MOWAT, C. W. WALLER, R. B. ANGIER, J. SEMB, and Y. SUBBAROW: A Chemical Method for the Determination of Pteroylglutamic Acid and Related Compounds. J. Biol. Chem. 168, 705 (1947).

76. HUTCHINGS, B. L., E. L. R. STOKSTAD, J. H. MOWAT, J. H. BOOTHE, C. W. WALLER, R. B. ANGIER, J. SEMB, and Y. SUBBAROW: Degradation of the Fermentation *L. casei* Factor. II. J. Amer. Chem. Soc. 70, 10 (1948).

77. ISAKA, S.: Inhibitory Effect of Xanthopterin upon the Formation of Melanin *in vitro*. Nature (London) 169, 74 (1952).

78. ISAKA, S. and S. ISHIDA: Effects of some Chelate Compounds upon the Formation *in vitro* of Melanin. Nature (London) 171, 303 (1953).

79. JACOBSON, W.: The Argentaffine Cells and Pernicious Anaemia. J. Pathol. Bacteriol. 49, 1 (1939).

80. — The Role of the *Leuconostoc Citrovorum* Factor (LCF) in Cell Division and the Mode of Action of Folic-Acid Antagonists on Normal and Leukaemic Cells. J. Pathol. Bacteriol. 64, 245 (1952).

81. JACOBSON, W. and D. M. SIMPSON: The Fluorescence Spectra of Pterins and their possible Use in the Elucidation of the Antipernicious Anaemia Factor. I. Biochemic. J. **40**, 3 (1946).

82. JACOBSON, W. and M. WEBB: The Two Types of Nucleoproteins during Mitosis. Exptl. Cell. Res. **3**, 163 (1952).

83. JUKES, T. H.: Folic Acid and Vitamin B_{12} in the Physiology of Vertebrates. Federat. Proc. (Amer. Soc. exp. Biol.) **12**, 633 (1953).

84. KALCKAR, H. M., T. FLÖYSTRUP, and M. A. SCHOU: Enzymic Fluormetric Analysis of Urinary Pteridine Derivatives. Proc. 1st Internat. Congr. Biochem., Cambridge **1949**, 75.

85. KALCKAR, H. M., N. O. KJELDGAARD, and H. KLENOW: Xanthopterin Oxidase. Biochim. Biophys. Acta **5**, 575 (1950).

85 a. — — — 2-Amino-4-hydroxy-6-formylpteridine, an Inhibitor of Purine and Pterine Oxidases. Biochim. Biophys. Acta **5**, 586 (1950).

86. KARRER, P. und H. FEIGL: Dimethylpteridinrot. Helv. Chim. Acta **34**, 2155 (1951).

87. KARRER, P., C. MANUNTA und R. SCHWYZER: Über ein Vorkommen von Purinen und eines Pterins in einer Ascidienart *(Microcosmus polymorphus)*. Helv. Chim. Acta **31**, 1214 (1948).

88. KARRER, P. und B. J. R. NICOLAUS: Zur Kenntnis des Methyl-pteridinrots, Pteridinrots und ähnlicher Farbstoffe. Helv. Chim. Acta **34**, 1029 (1951).

89. KARRER, P., B. NICOLAUS und R. SCHWYZER: Über die Konstitution des Methyl-pteridinrots. Helv. Chim. Acta **33**, 1233 (1950).

90. KARRER, P. und R. SCHWYZER: Über die Konstitution einiger neuer Pteridine. Eine weitere Folsäuresynthese. Helv. Chim. Acta **31**, 777 (1948).

91. — — Beitrag zur Entstehung und Trennung der 2-Amino-6-oxy-8- und 2-Amino-6-oxy-9-methylpteridine. Methyl-pteridinrot. Helv. Chim. Acta **33**, 39 (1950).

92. KASELIS, R. A., W. LEIBMANN, W. SEAMAN, J. P. SICKELS, E. I. STEARNS, and J. T. WOODS: Modified Colorimetric Assay of Pteroylglutamic Acid. Analyt. Chemistry **23**, 746 (1951).

93. KAVANAGH, F. and R. H. GOODWIN: The Relationship between p_H and Fluorescence of Several Organic Compounds. Arch. Biochemistry **20**, 315 (1949).

94. KERESZTESY, J. C., E. L. RICKES, and J. L. STOKES: A New Growth Factor for *Streptococcus lactis*. Science (New York) **97**, 465 (1943).

95. KERESZTESY, J. C. and M. SILVERMAN: Crystalline Citrovorum Factor from Liver. J. Amer. Chem. Soc. **73**, 5510 (1951).

96. KING, F. E. and P. C. SPENSLEY: The Use of Nitro- and Halogeno-ketones in the Synthesis of Pteridines, including Pteroic Acid, from 2:4:5-Triamino-6-hydroxy-pyrimidine. J. Chem. Soc. (London) **1952**, 2144.

97. KORTE, F. und E. G. FUCHS: Notiz zur Synthese des Xanthopterins. Ber. dtsch. chem. Ges. **86**, 114 (1953).

98. KOSCHARA, W.: Über Xanthopterin. Z. physiol. Chem. (Hoppe-Seyler) **277**, 159 (1943).

99. — Die Isolierung des Urothion. Z. physiol. Chem. (Hoppe-Seyler) **277**, 284 (1943).

100. — Urothion: ein Pteridinderivat mit endständiger Glykolgruppe. Z. physiol. Chem. (Hoppe-Seyler) **279**, 44 (1943).

101. KÜHLING, O.: Über die Oxidation des Tolualloxazins. II. Ber. dtsch. chem. Ges. **28**, 1968 (1895).

102. KUHN, R. und A. H. COOK: Über Lumazine und Alloxazine. Ber. dtsch. chem. Ges. **70**, 761 (1937).

103. LASCELLES, J. and D. D. WOODS: The Synthesis of "Folic Acid" by *Bacterium Coli* and *Staphylococcus aureus* and its Inhibition by Sulphonamides. Brit. J. exp. Pathol. **33**, 288 (1952).

104. LEDERER, E.: Les pigments des invertébrés (à l'exception des pigments respiratoires). Biol. Rev. Cambridge Phil. Soc. **15**, 273 (1946).

105. LISTER, J. H. and G. R. RAMAGE: Hydropteridines. I. The Formation of 5:6:7:8-Tetrahydro-4:6-dimethylpteridines. J. Chem. Soc. (London) **1953**, 2234.

106. LOWRY, O. H., O. A. BESSEY and E. J. CRAWFORD: Photolytic and Enzymatic Transformations of Pteroylglutamic Acid. J. Biol. Chem. **180**, 389 (1949).

107. McELROY, W. D. and J. COULOMBRE: The Immobilization of Adenosine Triphosphate in the Bioluminescent Reaction. J. Cell. Compar. Physiol. **39**, 475 (1952).

108. MASON, S. F.: Personal communication.

109. MAY, M., T. J. BARDOS, F. L. BARGER, M. LANSFORD, J. M. RAVEL, G. L. SUTHERLAND, and W. SHIVE: Synthetic and Degradative Investigations of the Structure of Folinic Acid-SF. J. Amer. Chem. Soc. **73**, 3067 (1951).

110. MERCK & Co. Inc.: Folic Acid. Brit. Pat. 653,068 (1951).

111. MITCHELL, H. K., E. E. SNELL, and R. J. WILLIAMS: The Concentration of "Folic Acid". J. Amer. Chem. Soc. **63**, 2284 (1941).

112. — — — Folic Acid. I. Concentration from Spinach. J. Amer. Chem. Soc. **66**, 267 (1944).

113. MOWAT, J. H., J. H. BOOTHE, B. L. HUTCHINGS, E. L. R. STOKSTAD, C. W. WALLER, R. B. ANGIER, J. SEMB, D. B. COSULICH, and Y. SUBBAROW: The Structure of the Liver *L. casei* Factor. J. Amer. Chem. Soc. **70**, 14 (1948).

114. NAWA, S., S. MATSUURA, and Y. HIRATA: Studies on Pteridines. V. Reductive Cleavage of Pteridyl Side Chains. J. Amer. Chem. Soc. **75**, 4450 (1953).

115. NICHOL, C. A. and A. D. WELCH: Synthesis of Citrovorum Factor from Folic Acid by Liver Slices; Augmentation by Ascorbic Acid. Proc. Soc. exp. Biol. Med. **74**, 52 (1950).

116. NICHOL, C. A., S. F. ZAKRZEWSKI, and A. D. WELCH: Resistance to Folic Acid Analogues in a Strain of *Streptococcus faecalis*. Proc. Soc. exp. Biol. Med. **83**, 272 (1953).

117. NORRIS, E. R. and J. J. MAJNARICH: Cell Proliferation Accelerating and Inhibitory Substances in Blood Serum During Pregnancy. Proc. Soc. exp. Biol. Med. **70**, 663 (1949).

118. O'DELL, B. L., J. M. VANDENBELT, E. S. BLOOM, and J. J. PFIFFNER: Hydrogenation of Vitamin Bc (Pteroylglutamic Acid) and Related Pterines. J. Amer. Chem. Soc. **69**, 250 (1947).

119. OKAY, S.: The Green Pigments of Insects. Rev. faculté sci. Univ. Istanbul B **12**, 89 (1947).

120. PETERING, H. G.: Folic Acid Antagonists. Physiol. Rev. **32**, 197 (1952).

121. PETERING, H. G. and J. A. SCHMITT: Studies in Enzyme Inhibition. I. Action of Some Simple Pterines on Xanthine Oxidase. J. Amer. Chem. Soc. **72**, 2995 (1950).

122. PFIFFNER, J. J., S. B. BINKLEY, E. S. BLOOM, R. A. BROWN, O. D. BIRD, A. D. EMMETT, A. G. HOGAN, and B. L. O'DELL: Isolation of the Antianemia Factor (Vitamin Bc) in Crystalline Form from Liver. Science (New York) **97**, 404 (1943).

123. PFIFFNER, J. J., S. B. BINKLEY, E. S. BLOOM, and B. L. O'DELL: Isolation and Characterization of Vitamin Bc from Liver and Yeast. Occurrence of an Acid-labile Chick Antianemia Factor in Liver. J. Amer. Chem. Soc. **69**, 1476 (1947).

124. PFIFFNER, J. J., D. G. CALKINS, E. S. BLOOM, and B. L. O'DELL: On the Peptide Nature of Vitamin Bc Conjugate from Yeast. J. Amer. Chem. Soc. **68**, 1392 (1946).

125. PFIFFNER, J. J., D. G. CALKINS, B. L. O'DELL, E. S. BLOOM, R. A. BROWN, C. J. CAMPBELL, and O. D. BIRD: Isolation of an Antianemia Factor (Vitamin Bc Conjugate) in Crystalline Form from Yeast. Science (New York) **102**, 228 (1945).

126. PIRIE, A. and D. M. SIMPSON: Preparation of a Fluorescent Substance from the Eye of the Dogfish, *Squalus acanthias*. Biochemic. J. **40**, 14 (1946).

127. POHLAND, A., E. H. FLYNN, R. G. JONES, and W. SHIVE: A Proposed Structure for Folinic Acid-SF, a Growth Factor Derived from Pteroylglutamic Acid. J. Amer. Chem. Soc. **73**, 3247 (1951).

128. POLONOVSKI, M. et R.-G. BUSNEL: Sur un pigment à fluorescence bleue des oeufs de *Bombyx mori*. C. R. hebd. Séances Acad. Sci. **226**, 1047 (1948).

129. POLONOVSKI, M., R.-G. BUSNEL et A. BARIL: Rôle d'un pigment ptérinique, la fluorescyanine, dans la mélanogénèse. C. R. hebd. Séances Acad. Sci. **231**, 1572 (1950).

130. POLONOVSKI, M., R.-G. BUSNEL et M. PESSON: La fluorescyanine, pigment à fluorescence bleue des écailles de Cyprinidés. C. R. hebd. Séances Acad. Sci. **217**, 163 (1943).

131. — — — Propriétés biochimiques des ptérines. Helv. Chim. Acta **29**, 1328 (1946).

132. POLONOVSKI, M. et E. FOURNIER: Sur les ptérines de *Cancer pagurus*. C. R. Séances Soc. Biol. **138**, 357 (1944).

133. POLONOVSKI, M., P. GONNARD et A. BARIL: Action de dérivés ptériniques et de la riboflavine sur la mélanogénèse *in vitro*. Enzymologia **14**, 311 (1951).

134. PURRMANN, R.: Über die Flügelpigmente der Schmetterlinge. VII. Synthese des Leukopterins und Natur des Guanopterins. Liebigs Ann. Chem. **544**, 182 (1940).

135. — Über die Flügelpigmente der Schmetterlinge. X. Die Synthese des Xanthopterins. Liebigs Ann. Chem. **546**, 98 (1940).

136. — Über die Flügelpigmente der Schmetterlinge. XII. Konstitution und Synthese des sogenannten Anhydroleukopterins. Liebigs Ann. Chem. **548**, 284 (1941).

137. — Pyrimidine Condensation Products. U. S. Patent 2,345,215 (1944) [cf. German Patent 721,930 (1942)].

138. — Pterine. Fortschr. Chem. organ. Naturstoffe **4**, 64 (1945).

139. PURRMANN, R. und F. EULITZ: Über die Flügelpigmente der Schmetterlinge. XVI. Zur Kenntnis des Erythropterins. Liebigs Ann. Chem. **559**, 169 (1948).

140. PURRMANN, R. und M. MAAS: Über die Flügelpigmente der Schmetterlinge. XV. Zur Kenntnis des Pterorhodins. Liebigs Ann. Chem. **556**, 186 (1944).

141. RAUEN, H. M. und C. V. HALLER: Über die Ausscheidung von Xanthopterin nach Verabreichung von Pteroylglutaminsäure beim Menschen. Z. physiol. Chem. (Hoppe-Seyler) **286**, 96 (1950).

142. RAUEN, H. M. und H. WALDMANN: Über die Gegenstromverteilung von Pterinen. Z. physiol. Chem. (Hoppe-Seyler) **286**, 180 (1950).

143. RAUEN, H. M., H. WALDMANN und M. BUCHKA: Über den fermentativen Abbau der Pteroylglutaminsäure. Z. physiol. Chem. (Hoppe-Seyler) **288**, 10 (1951).

144. RENFREW, A. G. and P. C. PIATT: Paper Chromatography of Some Synthetic Pteridines. J. Amer. Pharmaceut. Assoc. **39**, 657 (1950).

145. RICKES, E. L., L. CHAIET, and J. C. KERESZTESY: Isolation of Rhizopterin, A New Growth Factor for *Streptococcus Lactis* R. J. Amer. Chem. Soc. **69**, 2749 (1947).

146. ROCHE PRODUCTS LTD.: Folic Acid. Brit. Pat. 624,394 (1949).

147. ROTH, B., M. E. HULTQUIST, M. J. FAHRENBACH, D. B. COSULICH, H. P. BROQUIST, J. A. BROCKMAN, Jr., J. M. SMITH, Jr., R. P. PARKER, E. L. R. STOKSTAD, and T. H. JUKES: Synthesis of Leucovorin. J. Amer. Chem. Soc. **74**, 3247 (1952).

148. ROTH, B., J. M. SMITH, Jr., and M. E. HULTQUIST: Analogs of Pteroylglutamic Acid. V. 4-Alkylamino Derivatives. J. Amer. Chem. Soc. **72**, 1914 (1950).

149. RUSSELL, P. B., R. PURRMANN, W. SCHMITT, and G. H. HITCHINGS: The Synthesis of Pterorhodin (Rhodopterin). J. Amer. Chem. Soc. **71**, 3412 (1949).

150. RYDON, H. N.: Pterins. Annu. Rep. Chem. Soc. (London) **47**, 150 (1950).

151. SARETT, H. P.: The Synthesis of Folic Acid-like Compounds by *Lactobacillus arabinosus*. Arch. Biochem. Biophys. **34**, 378 (1951).

152. SAUBERLICH, H. E.: Comparative Studies with the Natural and Synthetic Citrovorum Factor. J. Biol. Chem. **195**, 337 (1952).

153. SAUBERLICH, H. E. and C. A. BAUMANN: A Factor required for the Growth of *Leuconostoc citrovorum*. J. Biol. Chem. **176**, 165 (1948).

154. SCHEINDLIN, S., A. LEE, and I. GRIFFITH: The Action of Riboflavin on Folic Acid. J. Amer. Pharmaceut. Assoc. **41**, 420 (1952).

155. SCHÖPF, C. und E. BECKER: Über neue Pterine. Liebigs Ann. Chem. **524**, 49 (1936).

156. SCHÖPF, C., E. BECKER und R. REICHERT: Die Hydrolyse der Pterine mit Säuren. Liebigs Ann. Chem. **539**, 156 (1939).

157. SCHOU, M. A.: Tautomeric Conversion of Xanthopterin. Arch. Biochemistry **28**, 10 (1950).

158. SEEGER, D. R., D. B. COSULICH, J. M. SMITH, Jr., and M. E. HULTQUIST: Analogs of Pteroylglutamic Acid. III. 4-Amino Derivatives. J. Amer. Chem. Soc. **71**, 1753 (1949).

159. SEMB, J., J. H. BOOTHE, R. B. ANGIER, C. W. WALLER, J. H. MOWAT, B. L. HUTCHINGS, and Y. SUBBAROW: Pteroic Acid Derivatives. V. Pteroyl-α-glutamyl-α-glutamylglutamic Acid, Pteroyl-γ-glutamyl-α-glutamylglutamic Acid, Pteroyl-α-glutamyl-γ-glutamylglutamic Acid. J. Amer. Chem. Soc. **71**, 2310 (1949).

160. SHIVE, W., T. J. BARDOS, T. J. BOND, and L. L. ROGERS: Synthetic Members of the Folinic Acid Group. J. Amer. Chem. Soc. **72**, 2817 (1950).

161. SILVERMAN, M. and J. C. KERESZTESY: Comparison of Citrovorum Factor and a Synthetic Compound with *Leuconostoc Citrovorum* Growth Activity. J. Amer. Chem. Soc. **73**, 1897 (1951).

162. — — Citrovorum Factor Intermediate in Liver Autolysates. Federat. Proc. (Amer. Soc. exp. Biol.) **12**, 268 (1953).

162a. SJÖSTRÖM, A. G. M. and L.-E. ERICSON: The Occurrence in Lichens of the Folic Acid-, Folinic Acid-, and Vitamin B_{12}-Group of Factors. Acta Chem. Scand. **7**, 870 (1953).

163. SNELL, E. E. and W. H. PETERSON: Growth Factors for Bacteria. X. Additional Factors Required by Certain Lactic Acid Bacteria. J. Bacteriol. **39**, 273 (1940).

164. SPIES, T. D., G. GARCIA LOPEZ, F. MILANES, R. LOPES TOCA, A. REBOREDO, and R. E. STONE: The Response of Patients with Pernicious Anemia, with Nutritional Macrocytic Anemia and with Tropical Sprue to Folinic Acid or Citrovorum Factor. South. Med. J. **43**, 1076 (1950).

165. STOKSTAD, E. L. R., B. L. HUTCHINGS, J. H. MOWAT, J. H. BOOTHE, C. W. WALLER, R. B. ANGIER, J. SEMB, and Y. SUBBAROW: The Degradation of the Fermentation *Lactobacillus casei* Factor. I. J. Amer. Chem. Soc. **70**, 5 (1948).

166. STOKSTAD, E. L. R., B. L. HUTCHINGS, and Y. SUBBAROW: The Isolation of the *Lactobacillus casei* Factor from Liver. J. Amer. Chem. Soc. **70**, 3 (1948).

167. STREHLER, B. L.: The Isolation and Properties of Firefly Luciferesceine. Arch. Biochem. Biophys. **32**, 397 (1951).

168. SWENDSEID, M. E., F. H. BETHELL, and O. D. BIRD: The Concentration of Folic Acid in Leukocytes. Observations on Normal Subjects and Persons with Leukemia. Cancer Res. **11**, 864 (1951).

169. SWENDSEID, M. E., P. D. WRIGHT, and F. H. BETHELL: A Growth Factor for *L. citrovorum* synthesized by Hemapoietic Tissue Reversing Aminopterin and Chloromycetin Inhibition. Proc. Soc. exp. Biol. Med. **80**, 689 (1952).

170. TAUB, A. and H. LIEBERMAN: Stability of Vitamin B_{12}-Folic Acid Parenteral Solutions. J. Amer. Pharmaceut. Assoc. **42**, 183 (1953).

171. TAYLOR, E. C., Jr., J. A. CARBON, and D. R. HOFF: Pteridines. X. A New Approach to the Synthesis of Pteridines. J. Amer. Chem. Soc. **75**, 1904 (1953).

172. TEPLY, L. J. and C. A. ELVEHJEM: The Titrimetric Determination of *"Lactobacillus casei* Factor" and "Folic Acid". J. Biol. Chem. **157**, 303 (1945).

173. TIMMIS, G. M.: Personal communication.

174. TOENNIES, G., H. G. FRANK, and D. L. GALLANT: On the Folic Acid Activity of Human Blood. J. Biol. Chem. **200**, 23 (1953).

175. TOTTER, J. R.: A Convenient Method for the Preparation of Synthetic Xanthopterin. J. Biol. Chem. **154**, 105 (1944).

176. TOTTER, J. R., V. MIMS, and P. L. DAY: Further Studies on the Relationship between Xanthopterin, Folic Acid and Vitamin M. Science (New York) **100**, 223 (1944).

177. TSCHESCHE, R., K.-H. KÖHNCKE und F. KORTE: Zur Frage der antibakteriellen Wirkung der Sulfonamide. IV. Z. Naturforsch. **5 b**, 132 (1950).

178. — — — Über Pteridine. II. Die Synthese der 9-Oxy-pteroylglutaminsäure. Ber. dtsch. chem. Ges. **84**, 485 (1951).

179. TSCHESCHE, R. und F. KORTE: Die Synthese des Erythropterins. Ber. dtsch. chem. Ges. **84**, 77 (1951).

180. — — Über Pteridine. IV. Zur Konstitution des Chrysopterins und Mesopterins. Ber. dtsch. chem. Ges. **84**, 641 (1951).

181. — — Über Pteridine. V. Die Konstitution des Ichthyopterins. Ber. dtsch. chem. Ges. **84**, 801 (1951).

182. — — Über Pteridine. VI. Zur Kenntnis der Pterorhodin-Bildung. Ber. dtsch. chem. Ges. **85**, 139 (1952).

183. — — Zum biochemischen Syntheseweg der Pteroylglutaminsäure. Z. Naturforsch. **8 b**, 87 (1953).

184. TSCHESCHE, R., F. KORTE und I. KORTE: Zur Frage der antibakteriellen Wirkung der Sulfonamide. V. Z. Naturforsch. **5 b**, 312 (1950).

185. TSCHESCHE, R., F. KORTE und R. PETERSEN: Über Pteridine. III. Zur Synthese der Pteroyl-glutaminsäure. Ber. dtsch. chem. Ges. **84**, 579 (1951).

186. TSCHESCHE, R. und F. VESTER: Über Pteridine. VIII. Erythropterin aus *Mycobacterium lacticola*. Ber. dtsch. chem. Ges. **86**, 454 (1953).

187. TUCKER, G. F., Jr. and J. L. IRVIN: Apparent Ionization Exponents of 4-Hydroxyquinoline, 4-Methoxyquinoline and N-Methyl-4-quinolone; Evaluation of Lactam-Lactim Tautomerism. J. Amer. Chem. Soc. **73**, 1923 (1951).

188. VEEN, A. G. VAN und J. K. BAARS: Über das Toxoflavin, ein Isomeres von 1-Methyl-xanthin. Rec. trav. chim. Pays-Bas **57**, 248 (1938).

189. VEEN, A. G. VAN und W. K. MERTENS: Die Giftstoffe der sogenannten Bongkrekvergiftungen auf Java. Rec. trav. chim. Pays-Bas **53**, 257 (1934).

190. Waller, C. W., M. J. Fahrenbach, J. H. Boothe, R. B. Angier, B. L. Hutchings, J. H. Mowat, J. F. Poletto, and J. Semb: 7-Isomer of Pteroylglutamic Acid. J. Amer. Chem. Soc. 74, 5405 (1952).
191. Waller, C. W., A. A. Goldman, R. B. Angier, J. H. Boothe, B. L. Hutchings, J. H. Mowat, and J. Semb: 2-Amino-4-hydroxy-6-pteridinecarboxaldehyde. J. Amer. Chem. Soc. 72, 4630 (1950).
192. Waller, C. W., B. L. Hutchings, J. H. Mowat, E. L. R. Stokstad, J. H. Boothe, R. B. Angier, J. Semb, Y. SubbaRow, D. B. Cosulich, M. J. Fahrenbach, M. E. Hultquist, E. Kuh, E. H. Northey, D. R. Seeger, J. P. Sickels, and J. M. Smith, Jr.: Synthesis of Pteroylglutamic Acid (Liver L. casei Factor) and Pteroic Acid. I. J. Amer. Chem. Soc. 70, 19 (1948).
193. Weijlard, J., M. Tishler, and A. E. Erickson: Some New Aminopyrazines and their Sulfanilamide Derivatives. J. Amer. Chem. Soc. 67, 802 (1945).
194. Weisblat, D. I., B. J. Magerlein, A. R. Hanze, D. R. Myers, and S. T. Rolfson: Synthesis of Pteroic and Pteroylglutamic Acids. I. J. Amer. Chem. Soc. 75, 3625 (1953).
195. Welch, A. D. and C. A. Nichol: Water-soluble Vitamins concerned with One- and Two-Carbon Intermediates. Annu. Rev. Biochem. 21, 633 (1952).
196. Weygand, F. und H. Hofmann: Polleninhaltsstoffe. I. Zucker, Folinsäure und Ascorbinsäure. Ber. dtsch. chem. Ges. 83, 405 (1950).
197. Weygand, F., H.-J. Mann und H. Simon: Synthese von Pteroyl-l-glutaminsäure-[2-^{14}C]. Ber. dtsch. chem. Ges. 85, 463 (1952).
198. Weygand, F. und G. Schaefer: Synthese von Pteroyl-l-glutaminsäure-[11-^{14}C] (Folinsäure-[11-^{14}C]). Ber. dtsch. chem. Ges. 85, 307 (1952).
199. Weygand, F. und V. Schmied-Kowarzik: Weitere Folinsäure-Synthesen. Ber. dtsch. chem. Ges. 82, 333 (1949).
200. Weygand, F., A. Wacker und V. Schmied-Kowarzik: Über die Kondensationsprodukte von p-Tolyl-d-isoglucosamin und Zuckern mit 6-Oxy-2.4.5-triaminopyrimidin, eine neue Folinsäure-Synthese. Ber. dtsch. chem. Ges. 82, 25 (1949).
201. Wieland, H. und R. Liebig: Über die Pigmente der Schmetterlingsflügel. XIV. Ergänzende Beiträge zur Kenntnis der Pteridine. Liebigs Ann. Chem. 555, 146 (1944).
202. Wieland, H., H. Metzger, C. Schöpf und M. Bülow: Über Leukopterin, das Flügelpigment der Kohlweißlinge (Pieriden). II. Liebigs Ann. Chem. 507, 226 (1933).
203. Wieland, H. und R. Purrmann: Über die Flügelpigmente der Schmetterlinge. VI. Über Leukopterin und Xanthopterin. Liebigs Ann. Chem. 544, 163 (1940).
204. Wieland, H. und C. Schöpf: Über den gelben Flügelfarbstoff des Citronenfalters (Gonepterix rhamni). Ber. dtsch. chem. Ges. 58, 2178 (1925).
205. Wieland, H., A. Tartter und R. Purrmann: Über die Flügelpigmente der Schmetterlinge. IX. „Anhydroleukopterin" und „Purpuroflavin". Liebigs Ann. Chem. 545, 209 (1940).
206. Wieland, O. P., B. L. Hutchings, and J. H. Williams: Studies on the Natural Occurrence of Folic Acid and the Citrovorum Factor. Arch. Biochem. Biophys. 40, 205 (1952).
207. Wittle, E. L. and G. Moersch (Parke, Davis & Co.): Preparation of Pteridines. U. S. Pat. 2,561,302 (1951).
208. Wittle, E. L., B. L. O'Dell, J. M. Vandenbelt, and J. J. Pfiffner: Oxidative Degradation of Vitamin Bc (Pteroylglutamic Acid). J. Amer. Chem. Soc. 69, 1786 (1947).

209. WOLF, D. E., R. C. ANDERSON, E. A. KACZKA, S. A. HARRIS, G. E. ARTH, P. L. SOUTHWICK, R. MOZINGO, and K. FOLKERS: The Structure of Rhizopterin. J. Amer. Chem. Soc. **69**, 2753 (1947).

210. WOODS, D. D.: Folic Acid and Related Compounds in the Metabolism of Micro-organisms. Brit. Med. Bull. **9**, 122 (1953).

211. WOOLLEY, D. W.: Enzymatic Synthesis of Folic Acid by the Action of Carp Thiaminase. J. Amer. Chem. Soc. **73**, 1898 (1951).

212. — Biosynthesis and Energy Transport by Enzymic Reduction of "Onium" Salts. Nature (London) **171**, 323 (1953).

213. WRIGHT, W. B., Jr., D. B. COSULICH, M. J. FAHRENBACH, C. W. WALLER, J. M. SMITH, Jr., and M. E. HULTQUIST: Analogs of Pteroylglutamic Acid. IV. Replacement of Glutamic Acid by Other Amino Acids. J. Amer. Chem. Soc. **71**, 3014 (1949).

(Received, November 27, 1953.)

Namenverzeichnis. Index of Names. Index des Auteurs.

27*

Sachverzeichnis. Index of Subjects. Index des Matières.

SPRINGER-VERLAG IN WIEN I

Fortschritte der Chemie organischer Naturstoffe

Progress in the Chemistry of Organic Natural Products

Progrès dans la chimie des substances organiques naturelles

Herausgegeben von

L. Zechmeister

California Institute of Technology, Pasadena, U. S. A.

Die seit längerer Zeit vergriffenen Bände I—IV sind jetzt als unveränderte Nachdrucke erhältlich. Preise auf Anfrage. Weitere Bände siehe nächste Seite

Zu beziehen durch jede Buchhandlung

Fortsetzung von vorhergehender Seite

Fünfter Band. Mit 34 Abbildungen im Text. VIII, 417 Seiten. 1948.
DM 47.—, $ 11.20, sfr. 48.80, £4.0.0d.
Ganzleinen DM 50.40, $ 12.—, sfr. 52.20, £4.6.0d.

Inhalt: **Karrer, P.** Carotinoid-Epoxyde und furanoide Oxyde von Carotinoidfarbstoffen. — **Fox, D. L.** Some Biochemical Aspects of Marine Carotenoids. — **Haagen-Smit, A. J.** Azulenes. — **Hilditch, T. P.** Recent Advances in the Study of Component Acids and Component Glycerides of Natural Fats. — **Hassid, W. Z. and M. Doudoroff.** Enzymatically Synthesized Polysaccharides and Disaccharides. — **Pacsu, E.** Recent Developments in the Structural Problem of Cellulose. — **Brauns, F. E.** Lignin. — **Deulofeu, V.** The Chemistry of Constituents of Toad Venoms. — **Geiger, E.** Biochemistry of Fish Proteins. — **Beadle, G. W.** Some Recent Developments in Chemical Genetics. — **Rasmussen, R. S.** Infrared Spectroscopy in Structure Determination and its Application to Penicillin.

Sechster Band. Mit 32 Abbildungen. VIII, 392 Seiten. 1950.
DM 52.50, $ 12.50, sfr. 54.30, £4.10.0d.
Ganzleinen DM 55.80, $ 13.30, sfr. 57.80, £4.15.6d.

Inhalt: **Deuel Jr., H. J. and S. M. Greenberg.** Some Biochemical and Nutritional Aspects in Fat Chemistry. — **Lederer, E.** Odeurs et parfums des animaux. — **Hoffmann-Ostenhof, O.** Vorkommen und biochemisches Verhalten der Chinone. — **Reti, L.** Cactus Alkaloids and Some Related Compounds. — **Bonner, J.** Plant Proteins. — **Dhéré, Ch.** Progrès récents en spectrochimie de fluorescence des produits biologiques.

Siebenter Band. Mit 12 Abbildungen. VII, 330 Seiten. 1950.
DM 50.40, $ 12.—, sfr. 52.—, £4.6.0d.
Ganzleinen DM 53.70, $ 12.80, sfr. 55.50, £4.12.0d.

Inhalt: **Jeger, O.** Über die Konstitution der Triterpene. — **Heusser, H.** Konstitution, Konfiguration und Synthese digitaloider Aglykone und Glykoside. — **Niemann, C.** Thyroxine and Related Compounds. — **Cook, A. H.** Penicillin and its Place in Science. — **Stoll, A. and B. Becker.** Sennosides A and B, the Active Principles of Senna. — **Williams, J. W.** Some Recent Developments in the Chemistry of Antibodies.

Achter Band. Mit 47 Abbildungen. XI, 400 Seiten. 1951.
DM 67.—, $ 16.—, sfr. 68.80, £5.14.6d.
Ganzleinen DM 70.50, $ 16.80, sfr. 72.20, £6.0.6d.

Inhalt: **Frey-Wyssling, A. and K. Mühlethaler.** The Fine Structure of Cellulose. — **Stacey, M. and C. R. Ricketts.** Bacterial Dextrans. — **Leloir, L. F.** Sugar Phosphates. — **Kenner, G. W.** The Chemistry of Nucleotides. — **Schinz, H.** Die Veilchenriechstoffe. — **Asahina, Y.** Neuere Entwicklungen auf dem Gebiete der Flechtenstoffe. — **Galinovsky, F.** Lupinen-Alkaloide und verwandte Verbindungen. — **Paller, M.** Brechwurzel-Alkaloide. — **Corey, R. B.** X-Ray Diffraction Studies of Crystalline Amino Acids and Peptides. — **Zechmeister, L. and M. Rohdewald.** Some Aspects of Enzyme Chromatography.

Neunter Band. Mit 20 Abbildungen. XI, 535 Seiten. 1952.
DM 79.—, $ 18.80, sfr. 81.—, £6.14.6d.
Ganzleinen DM 82.50, $ 19.60, sfr. 84.50, £7.0.0d.

Inhalt: **Inhoffen, H. H. und H. Siemer.** Synthetische Chemie der Carotinoide. — **Baxter, J. G.** Synthesis and Properties of Vitamin A and Some Related Compounds. — **Meunier, P.** Les Antivitamines. — **Stoll, A.** Recent Investigations on Ergot Alkaloids. — **Tomita, M.** Die Alkaloide der Menispermaceae-Pflanzen. — **Dean, F. M.** Naturally Occurring Coumarins. — **Borsook, H.** The Biosynthesis of Proteins and Peptides, including Isotopic Tracer Studies. — **Kalckar, H. M.** The Enzymes of Nucleoside Metabolism. — **McNutt, W. S.** Nucleosides and Nucleotides as Growth Substances for Microorganisms. — **Campbell, D. H. and N. Bulman.** Some Current Concepts of the Chemical Nature of Antigens and Antibodies.

Zehnter Band. Mit 19 Abbildungen. IX, 529 Seiten. 1953.
DM 80.—, $ 19.—, sfr. 81.70, £6.16.0d.
Ganzleinen DM 83.—, $ 19.80, sfr. 85.—, £7.1.6d.

Inhalt: **Alder, K. und Marianne Schumacher.** Anwendungen der Dien-Synthese für die Erforschung von Naturstoffen. — **Mark, H.** Physical Chemistry of Rubbers. — **Asselineau, J. et E. Lederer.** Chimie des lipides bactériens. — **Rosenkranz, G. and F. Sondheimer.** Syntheses of Cortisone. — **Chatterjee, A.** Rauwolfia Alkaloids. — **Feinstein, L. and M. Jacobson.** Insecticides Occurring in Higher Plants.

Zu beziehen durch jede Buchhandlung

Demnächst beginnt zu erscheinen:

Handbuch der mikrochemischen Methoden

Herausgegeben von

Friedrich Hecht und Michael K. Zacherl
Professor an der Universität Wien   Professor an der Tierärztlichen Hochschule in Wien

In fünf Bänden

Jeder selbständig erscheinende Bandteil bzw. Band ist einzeln käuflich.
Bei Verpflichtung zur Abnahme des Gesamtwerkes sowie bei Vorbestellung
der einzelnen Teile ermäßigt sich der Preis um 20%

Zuerst erscheinen:

Band I / Teil 1

Präparative Mikromethoden in der organischen Chemie. Von Dr. **Hans Lieb**, o. Prof., Vorstand des Medizinisch-chemischen Institutes und Pregl-Laboratoriums der Universität Graz, und Dr. **Wolfgang Schöniger**, Dipl.-Ing., Assistent am selben Institut.

Mikroskopische Methoden. Von Prof. Dr. **Ludwig Kofler** † und Dr. **Adelheid Kofler**, Innsbruck.

Mit 277 Textabbildungen. Etwa 250 Seiten. 1954
Bei Vorausbestellung bis zum Erscheinen:
Ganzleinen DM 38.—, $ 9.05, sfr. 38.90, £3.4.6d.
Endgültiger Ladenpreis nach Erscheinen:
Ganzleinen DM 47.50, $ 11.30, sfr. 48.60, £4.1.0d.

Band II / Teil 1

Radiochemische Methoden. Von Priv.-Doz. Dr. **Engelbert Broda**, I. Chemisches Laboratorium der Universität Wien, und Dr. **Thomas Schönfeld**, am selben Institut.

Messung radioaktiver Strahlen in der Mikrochemie. Von Prof. Dr. **Berta Karlik**, Vorstand des Institutes für Radiumforschung der Österreichischen Akademie der Wissenschaften, Wien, Dr. **Traude Bernert**, am selben Institut, und Priv.-Doz. Dr. **Karl Lintner**, Assistent am II. Physikalischen Institut der Universität Wien.

Photographische Methoden in der Radiochemie. Von Dr. **Hanne Lauda**, Institut für Radiumforschung der Österreichischen Akademie der Wissenschaften, Wien.

Mit etwa 90 Textabbildungen. Etwa 330 Seiten. 1954
Bei Vorausbestellung bis zum Erscheinen:
Ganzleinen DM 64.80, $ 15.45, sfr. 66.40, £5.10.6d.
Endgültiger Ladenpreis nach Erscheinen:
Ganzleinen DM 81.—, $ 19.30, sfr. 83.—, £6.18.0d.

Bei Verpflichtung zur Abnahme des gesamten Handbuches gilt der Vor-
bestellpreis weiter als Subskriptionspreis

In der Folge werden erscheinen:

Band I, Teil 2: **Biochemische Methoden einschließlich medizinischer Verfahren.** Von **Th. Leipert**, Wien. **Lebensmittelmikrochemische Methoden.** Von F. **Münchberg**, Wien, und F. **Zaribnicky**, Wien.
Band II, Teil 2: **Polarographie.** Von H. **Hohn**, Wien, und **Otilie Schlager**, Wien.
Band III: **Anorganisch-analytische Methoden.** Von F. **Hecht**, Wien.
Band IV: **Organisch-analytische Methoden.** Von M. K. **Zacherl**, Wien.
Band V: **Mikromethoden zur Bestimmung physikalisch-chemischer Konstanten.** Von **Martha Sobotka**, Graz.